W0051199

High-Pressure Shock Compression of
Condensed Matter

Editor-in-Chief
 Robert A. Graham

Editorial Board
 Roger Chéret, France
 Godfrey Eden, Great Britain
 Jing Fuqian, China
 Vitalii I. Goldanskii, Russia
 James N. Johnson, USA
 Malcolm F. Nicol, USA
 Akira B. Sawaoka, Japan

Roger Chéret

Detonation of Condensed Explosives

With 124 Illustrations

Springer-Verlag

New York Berlin Heidelberg London Paris
Tokyo Hong Kong Barcelona Budapest

Roger Chéret
Directeur de Recherches
Commissariat a l'Energie Atomique
31–33 Rue de la Federation
75752 Paris   Cedex 15
France

Translator

Language Consultants Ltd.
1, Deal's Gateway
Black Heath Road
London SE10 8BX
England

Library of Congress Cataloging-in-Publication Data
Chéret, Roger.
    Detonation of condensed explosives / Roger Chéret.
        p.   cm.
    Includes bibliographical references and index.
    ISBN-13:978-1-4613-9286-6
    1. Explosions. 2. Detonation waves. 3. Explosives. I. Title.
QD516.C544   1992
662'.2—dc20   92-16118

Printed on acid-free paper.

© 1993 Springer-Verlag New York, Inc.
Softcover reprint of the hardcover 1st edition 1993

All rights reserved. This work may not be translated or copied in whole or in part without the
written permission of the publisher (Springer-Verlag New York, Inc., 175 Fifth Avenue, New
York, NY 10010, USA), except for brief excerpts in connection with reviews or scholarly analysis.
Use in connection with any form of information storage and retrieval, electronic adaptation,
computer software, or by similar or dissimilar methodology now known or hereafter developed
is forbidden.
The use of general descriptive names, trade names, trademarks, etc., in this publication, even if
the former are not especially identified, is not to be taken as a sign that such names, as understood by the Trade Marks and Merchandise Marks Act, may accordingly be used freely by
anyone.

Production coordinated by Brian Howe and managed by Christin Ciresi; manufacturing
supervised by Vincent Scelta.
Typeset by Asco Trade Typesetting Ltd., Hong Kong.

9 8 7 6 5 4 3 2 1

ISBN-13:978-1-4613-9286-6   e-ISBN-13:978-1-4613-9284-2
DOI: 10.1007/978-1-4613-9284-2

Foreword

This work marks a stage in the evolution of a scientific and technical field which has been developed by the Commissariat à l'Energie Atomique (CEA) over several decades. Many members of the staff of the CEA have won renown in this field, and their work has brought it to the high degree of excellence for which it is internationally recognized today. These scientists had to consider every aspect of the field, as it concerned:

— modeling, which has recourse to fluid thermodynamics, molecular physics, and chemistry;
— numerical evaluation, which relies on mathematical analysis and data processing; and
— experiments in the firing area, which require specific stress generators and instrumentation.

Whilst this book is a testament to the activity and success of staff of the CEA, it also reviews a number of the advances made in the discipline. However, it is not intended to be an exhaustive account of those advances; it is assumed that the reader can, if desired, consult the standard monographs, and more recent, more specialized works (notably W.C. Davis and W. Fickett, and C.L. Mader).

The history of the discipline is interesting in itself, and also as an illustration of the causes which lead to progress in a coherent body of scientific work. I should like to make some comments on this progress, of which there is a fascinating summary in the introduction, and which will figure largely throughout the work.

As with many disciplines, alongside theory and experiment, a new line of inquiry is developing—Computational Physics—in which models are tested by numerical codes which, as ideas develop, remain as faithful as possible to the models themselves. These codes must be based on a mathematical knowledge of the equations of the model. In Chapter II there is an introduction to the present state of this knowledge. This gives some idea of the difficulties to be overcome when dealing with the problem in three-dimensional real space. By the same token, the importance of Chapter III will be recognized, concerned as it is with a detailed description of detonation waves using the method of asymptotic expansions.

Researchers in this field have often speculated as to whether they should go no further than a continuum model, or whether useful results could come out of a discrete model within the framework of quantum mechanics. This is exactly the question put by Jacques Yvon when he was chairman of the symposium on High Dynamic Pressures held in Paris in 1978. On reading Chapters V and VI, the reader will note that a quantum model of the explosive molecule is well under way.

Among the most pressing factors governing the nature of research is one which has been, and will be, decisive—human safety with regard to accidents resulting from the detonation of an explosive substance at the wrong time. A striking example of this factor is to be found in the chapters which deal with the beginnings of explosive decomposition and the generation of detonation by shock.

Finally, it is impossible to ignore the fact that in this discipline, more than in others, experimentation is a powerful force which dictates the nature of any progress, which is why every chapter, albeit in varying degrees, is imbued with the ideas of experimentation.

It is only right, as I bring this Foreword to an end, that I pay tribute to the authors who have produced a lucid text, of high scientific quality, in spite of the heavy load of their daily work which they have never at any time reduced during the production of this book.

ROBERT DAUTRAY

Introduction

1. The Historical Background

The idea of releasing thermal energy through chemical reactions goes back to the origins of humanity, a fact borne out by the Greek myth of Prometheus, and his avatars in other civilizations. For thousands of years man strove to understand phenomena which today are commonly referred to as *combustion* and *flame*. This was not quite so difficult, because the characteristic rate of this release is of the order of one decimeter per second, and so can be immediately observed.

In comparison, the idea of releasing mechanical energy through chemical reactions is of very recent origin. It was toward 1630 that black powder (a mixture of saltpeter, sulphur, and carbon, whose invention is currently attributed to the Chinese in the early centuries A.D.) was used for the first time for its "shattering" properties in a mining development in Hungary, and shortly afterward, around 1670 in England. But it was not until the middle of the nineteenth century that such a use was to spread, thanks to laboratory preparation of new substances (guncotton by Friedrich Schönbein at Basel in 1845, nitroglycerine by Ascanio Sobrero at Turin in 1847), and to the patents of Alfred Nobel who, from 1864, improved the industrial manufacture of nitroglycerine and had the idea of marketing it, adsorbed on an inert, porous substance—kieselguhr. *Dynamite* was born, and with it there began an era in which the dynamic aspects of chemical reaction were raised to the rank hitherto occupied only by the thermal aspects.

2. Some Important References

It is in the context of current experience and rates of flame of the order of decimeters per second that the experiments of Marcelin Berthelot and Paul Vieille, and the account of these experiments given by the former to the French Academy of Sciences in 1881 (*C.R. Acad. Sci. Paris*, vol. CXIII, p. 18), must be placed: for the first time a reaction rate of the order of some kilometers per second in gaseous mixtures was recorded. The new system of propagation was called *explosive wave*. The general nature of the phenomenon appears when a comparison is made with the results obtained by other scientists (among them F.A. Abel from 1869 to 1874), in both the liquid and solid phases, and leads to the emergence of the notion of *explosive substance* or, more briefly, of *explosive*.

The first stage of the theory is marked by two papers by E. Sarrau, dated 1892 and 1894, respectively. An important step was taken by D.L. Chapman who, in 1899, using the work of W.J. Rankine, put forward a thermodynamic treatment of permanent plane flow, where a chemical reaction is taking place in the flow. Then, from 1901 to 1913, E. Jouguet and J. Crussard published several papers which, inspired by the work of P.H. Hugoniot, proposed a treatment of any flow with a chemical reaction. The state of theoretical interpretation in 1914 is set out in *Mécanique des Explosifs* by E. Jouguet; it is dominated by the existence of two regimes of propagation, commonly distinguished today by the terms *deflagration* (for the slower) and *detonation* (for the faster).

After smouldering under the embers of the First World War, research was revived outside France as witnessed by the works, which have become classics, of N.N. Semenov (1928), B. Lewis and G. Von Elbe (1938), and W. Jost (1939). All three, in contrast to the view of E. Jouguet, which is synthetic and thermodynamic in nature, apply themselves to a review of known phenomena and their kinetic aspects.

From 1940, the needs of nations involved in the Second World War created a rapid and unprecedented expansion in research into solid explosives. The use of such explosives in nuclear weapons necessitated the characterization of their properties and a modeling of their effects, which in turn meant resorting to the most highly developed methods of physics ... and to the most brilliant minds of the time. The papers of Y.B. Zeldovitch (1940), G.I. Taylor (1941), J. Von Neumann (1942), and W. Döring (1943) date from these troubled years. Although the papers were written in an isolation which we may only guess at, they share a common characteristic—they abandon the kinetic aspect envisaged during the interwar years and return to the only existing theory, that of E. Jouguet, to try to prove its obscure points or to draw from it simple conclusions. Many publications from the first years of new-found peace are the result of that systematic exploration. The works of Y.B. Zeldovitch and A.S. Kompaneets (1955), M.A. Cook (1958), and J. Berger and J. Viard (1962) give a good idea of the state of knowledge during the 1950s.

Since then, great progress has caused another rapid expansion in the study of detonation in condensed explosives: the development of the numerical calculation of molecular orbitals in theoretical chemistry, and of equations of state of dense fluids (pure or mixed) in thermodynamics, formalization of the method of asymptotic expansions in fluid mechanics, etc. Moreover, in the course of the last two decades, the physics applied to ultrafast observations has made it possible to increase substantially the precision of measurements of the velocity of propagation, and to arrive at the direct measurement of material velocities with the same precision.

This progress in modeling and in experimental and numerical tools concerns many aspects of the physics of detonation. So it seems opportune to assemble in an autonomous work established knowledge as it now stands. We hope thus to demonstrate the multiplicity of the disciplines and techniques required for the study of these phenomena and, *ipso facto*, to identify the remaining gaps in the context of the evolution appropriate to each of those disciplines and techniques.

3. Plan of this Work

This work is made up of twelve chapters, divided into four parts:

- *Part One* (*Chapters* I–III): the mechanical and thermodynamic aspects of the propagation of detonation waves;
- *Part Two* (*Chapters* IV–VI): the molecular mechanisms of explosive decomposition;
- *Part Three* (*Chapters* VII–IX): the macroscopic mechanisms of the generation of detonation; and
- *Part Four* (*Chapters* X–XII): the dynamic characterization of explosives.

Chapter I comprises reminders (a statement of the fundamental laws for fluids, a molecular model of reactive fluid, and a continuum model of rective fluid) and an introduction to reactive waves. In Chapter II, in the framework of an arbitrary equation of state of the fluid, the general properties of waves (shocks, deflagrations, detonations) in fluids are elucidated, using the properties of the Hugoniot and Crussard curves. Chapter III goes more deeply into the detonation regime, with the help of the method of asymptotic expansions and the concepts of internal and external structures.

The various factors linked to the appearance of the detonation regime taken together form the subject matter of Parts Two and Three. The first aim is to establish the correlation between the sensitivity to "shock" of an explosive substance and the intramolecular mechanisms, concentrating on the atomic groups (Chapter IV) and on the distribution of the electrons (Chapter V). The aim then is to specify the reaction mechanisms which govern the appearance of an explosive decomposition (Chapter VI), and finally to ascertain the unity of the cooperative mechanisms which lead to the propagation of the chemical reaction (Chapter VII). Chapter VIII deals with the coupling of decomposition and motion. These different points of view are brought together in Chapter IX, which is devoted to the simple case of detonation generated by a plane shock.

The description of flow, dominated by the formation of a detonation wave and its propagation, calls on experimental methods and very specific methods of numerical modeling, characterized by high pressures (a few kilobars to several hundred kilobars), high velocities (a few millimetres per microsecond), and short observation times (a few microseconds). The aim of Part Four is to give the basic principles of the methods of measurement (Chapter X) and their use in some typical configurations (Chapter XI), as well as the basic principles of those numerical forecasts that are sanctioned by the codes designed for the computation of the state surface of the detonation products (Chapter XII).

In writing this work, I have received valuable collaboration from:

- Alain Delpuech (Chapters IV, V, and VI),
- Christian Michaud (Chapters IV, V, and VI),
- Noël Camarcat (Chapter VIII), and
- François Olive (Chapter X).

Without their collaboration it would not have been possible to tackle the subject from such different angles.

This work stems from a suggestion by Mr. Robert Dautray, who expressed his confidence in us throughout the undertaking.

ROGER CHÉRET

General Plan

Note for the Reader

— The work is published in one volume, divided into four parts.
— The text is organized into chapters (shown by Chap. followed by Roman numerals) which are subdivided into sections (shown by Sec. followed by one Arabic numeral) and subsections (shown by § followed by two Arabic numerals).
— Equations are numbered within each chapter, as are both tables and figures.
— For practical reasons a list of references may be found at the end of Part One and of Part Two, whilst at the end of each chapter of Parts Three and Four. The authors regret that, in spite of all their efforts, gaps exist in these lists, particularly in the area of publications unavailable in French or English language versions.
— The volume ends with a list of the principal notations and an index.

Summary

Part One, devoted to the mechanical and thermodynamic aspects of the propagation of detonations, uses a fluid modeling of the relevant media. That is why Chapter I is devoted to reminders about fluid, in general, and reactive fluid. Initially, the expression of the basic laws appropriate to fluid behavior is established, taking care to bring out the *jump relations* which result from these laws. Then, so that the reader may gauge the complexity of the reactive fluid with K components, as well as the relevance and significance of an average fluid, there is a standard presentation of the expression of the basic laws for both a molecular and a continuum model of the reactive fluid. Taken together, these laws form an introduction to reactive waves.

Chapter II proceeds to a detailed study of the jump relations for a *perfect fluid with an arbitrary equation of state*. The general properties of shock waves are elucidated through a consideration of the *Hugoniot* curve, and the general properties of deflagration and detonation waves through discussion of the *Crussard* curve. A closer look is then taken at detonations and the so-called *Chapman–Jouguet* condition. Finally, the polytropic fluid model is introduced, and some results are set out, which are used profitably in subsequent chapters.

Chapter III goes more deeply into the detonation regime. The chosen approach considers detonation as a layer of steep pressure gradients in a *dissipative reactive fluid*, and studies this layer using the asymptotic expansions method. A suitable choice of perturbation parameter makes it possible, first, to deal with the phase of propagation when detonation is said to be *built up*. In the three-dimensional, nonstationary configuration, the internal and external zero-order structures are specified, and the laws of propagation are established, through the introduction of the notions of *quasi C–J detonation*, *simple detonation*, and *autonomous detonation*. There follows a more extensive analysis of detonations in one dimension of space. Choosing other perturbation parameters makes it possible to extend the method to other phases of propagation: birth of a simple detonation, extinction/bifurcation of a simple autonomous detonation.

In Part Two, the relationship between the intimate structure and the ability of a substance to decompose explosively is considered.

Chapter IV recounts the failure of attempts to correlate the experimental results, which are traditionally grouped together under the heading *sensitivity to* "*shock*," with the existence and with the properties of certain atomic groups within the molecule in its fundamental state.

Chapter V develops the principal idea which states that, in a chemical reaction propagated in the form of a detonation:

(i) the molecule goes through an excited state;

(ii) there is a special bond, called *explosophore*, whose minimum polarity after excitation determines the sensitivity to "shock."

The way in which this criterion is used is demonstrated: to calculate the experimental sensitivity scale, to interpret the traditional distinction between primary and secondary explosives, and to put the different chemical families of explosives into a single framework.

Between the considerations of Chapter V, which focus on the excited explosive molecule, and those of Chapter VII which are directed towards the first stage of a continuum modeling of chemical reactions, Chapter VI contains results concerning the molecular population after excitation. Here there are assumptions in favor of the existence of two subpopulations of excited molecules; one of these would account for sensitivity, and the other for the conditions in which the cooperative process of explosive decomposition can be created.

In Part Three, devoted to macroscopic detonation mechanisms, we endeavor to bring order and unity to the profusion of terms, experimental results, and models which have accumulated indiscriminately from the legitimate uses and the misuse of explosives. Part Three consists of three chapters in which we range between the local and global features, as well as between the quantitative and qualitative aspects of the subject.

Chapter VII defines and develops the notions of *ignition* phase and *induction* phase. During the former, decomposition of the explosive manifests itself in preferential sites, *hot spots*; possible mechanisms for this dynamic heterogeneity are evoked, and we suggest a return to two principal sources, pore collapse and mesoscopic shear. As far as the second phase is concerned, which leads from ignition to propagation, we show how the variety of situations encountered can depend on orders of magnitude and reaction schemes at the same time.

In Chapter VIII we discuss the link between decomposition and motion from the viewpoint of the "computational" physicist. Applying the general considerations of Chapter I, we deal with a two-component reactive flow and we clarify the alternative which occurs between a unique system for an average fluid and systems for each of the two components. We are then in a position to discuss the simple one-dimensional rectilinear case and to analyze the different modifications with regard to a reference algorithm. We are thus led to the problem of the expression of the *reactivity* of an explosive substance: we recall the current tentative solutions to the problem and we describe the first step on a unified route leading from hot spots to reactivity.

Chapter IX deals with a particular class of geometries, those in which a detonation builds up under the effect of a plane shock, at first from the viewpoint of the "experimental" physicist, before again taking up that of the "computational" physicist. Having defined the field of investigation, we then move on to a general treatment of propagation and we give a precise meaning to the terms *time* and *run to build-up*. We are then in a position to determine, from experimental results of one-dimensional and two-dimensional impact, what are the *sensitivity of an explosive substance*, the *detonability thresholds*, and the *critical pressure*. This last concept and that of *critical diameter* introduced in Chapter III, provide the means for analyzing nonmolecular factors of sensitivity. The deduced experimental laws, combined with the contents of Chapter VIII *in fine*, serve as a basis for a unified formulation of the reactivity of an explosive substance.

In the fourth and last part, we attempt to provide a framework within which to describe the dynamic characterization of explosives.

Chapter X is an exhaustive and detailed review of the means to which the experimental physicist has recourse. We cover the techniques according to the usual classification which depend on the nature of the signal generated at the level of the assembly (optical, electronic, and radiographic measurements), then we deal with stress generators.

Chapter XI provides an opportunity to make use of the concepts introduced and the results obtained throughout the first ten chapters, in order to give a new description of several elementary geometries of simple detonation. Thus, we deal with the traditional "plane" regime, where the detonation is quasi C–J, then an axisymmetric regime with lateral priming where detonation is partly strong and partly quasi C–J. We also deal with spherical explosion regimes where it is quasi C–J and implosion regimes where it is strong. Finally, we deal with the break of a quasi C–J autonomous detonation on the free surface of an explosive structure, while extracting a new theoretical framework for interpreting experimental observations.

Chapter XII is devoted to the numerical prediction of the performances of explosives. Our aim is to determine the sense and value to accord to this prediction, to define an elementary problem and some associated problems, and finally to refine the validity criteria of the prediction. We examine the main options available in the estimation of functions of state which concur with the formation of the prediction, and those which are available for the choice of algorithms and thermochemical data. These considerations are illustrated by several results judged to be significant both in regard to the Chapman–Jouguet state and to the surface of state of the detonation products in its entirety.

Contents

Part Two
Molecular Mechanisms of Explosive Decomposition   . . .   111

Part Three
Macroscopic Mechanisms of Generation of Detonation .   157

CHAPTER VII
Cooperative Mechanisms   159

CHAPTER VIII
Coupling of Decomposition and Motion   170

CHAPTER IX
Generation of Detonation by Plane Shock   189

Part Four
The Dynamic Characterization of Explosives 215

CHAPTER X
Experimental Methods 217

CHAPTER XI
Elementary Configurations of Simple Detonation 307

Part One
The Mechanical and Thermodynamic Aspects of the Propagation of Detonations

Part One

CHAPTER I

Generalities Concerning the Reactive Fluid

1. Expression of Fundamental Laws

1.1. Mathematical Reminders

In a frame $\mathcal{R}$ ($\vec{e}_i$; $i = 1, 2, 3$) of orthogonal rectilinear coordinates, the motion of a system $\mathcal{S}$ may be observed; the velocity vector of the "particle" $M \in \mathcal{S}$ which is at point P (x_i; $i = 1, 2, 3$) at instant t is designated by $\vec{u}$ (u_i; $i = 1, 2, 3$).

Let any connected domain inside $\mathcal{S}$ be D, and let the norm exterior to D be $\vec{n}$ (n_i; $i = 1, 2, 3$). Three conventional results of integral differential calculus are given below. (In order that such statements should not be cumbersome, we will remain deliberately brief insofar as the mathematical expression of hypotheses required for these theorems is concerned.)

Green's Theorem. *A vector $\vec{a}$ (a_i; $i = 1, 2, 3$) is considered, continuous in $\mathcal{S}$(t) except if necessary on a surface Σ, mobile in $\mathcal{R}$, of the norm $\vec{N}$ (N_i; $i = 1, 2, 3$); the intersection of Σ and D [void if necessary] is designated by Σ(D). Thus Green's theorem is designated*

$$\int_{\partial D} a_i n_i \, DS = \int_D a_{i,i} \, DV + \int_{\Sigma(D)} [\![a_i N_i]\!] \, DS, \tag{I.1}$$

where $[\![a_i N_i]\!]$ denotes the difference between the value of a scalar product $\vec{a} \cdot \vec{N}$ on the positive side of $\vec{N}$ and that on the negative side of $\vec{N}$. (N.B. The transition thus defined is independent of the direction of $\vec{N}$.)

It is worth noting that the result (I.1) may be generalized for every tensor $\vec{a}$ which is of order greater than 1.

Particle Derivative of a Volume Integral

A scalar φ is considered continuous in $\mathcal{S}$(t) except, if necessary, on a surface Σ, mobile in $\mathcal{R}$, of norm $\vec{N}$ ($N_i = 1, 2, 3$); the intersection (void if necessary) of

3

Σ and D is denoted by $\Sigma(D)$. The volume integral $\int_D \varphi \, DV$ is a function of time alone, whilst domain D is a *material volume followed in its motion*; its derivative, said to be *particle* because the summation still extends to the same "particles" in the course of time, may be set out as (see [21], for example)

$$\frac{d}{dt}\int_D \varphi \, DV = \int_D \frac{\partial \varphi}{\partial t} DV + \int_{\partial D} \varphi u_i n_i \, DS - \int_{\Sigma(D)} [\![\varphi]\!]\delta_i N_i \, DS, \quad (I.2)$$

where δ_i denotes the rate of propagation of Σ in $\mathscr{R}$ (if $t = f(x_1, x_2, x_3)$ is the equation of Σ, then $\delta_i = f_{,i}|\text{grad } f|^{-2}$); on the right, the first term translates the variation of φ in time, the second expresses the contribution due to the motion of D and is called the *convection* term. The third term is a result of the possible presence of a discontinuity of φ within D and is called *jump*.

Applying Green's theorem to the convection term for vector $\varphi \vec{u}$ gives the expression

$$\frac{d}{dt}\int_D \varphi \, DV = \int_D \left[\frac{\partial \varphi}{\partial t} + (\varphi u_i)_{,i}\right] DV + \int_{\Sigma(D)} [\![\varphi w_i]\!] N_i \, DS \quad (I.3)$$

in which appears the velocity-vector $w_i = u_i - \delta_i$ $(i = 1, 2, 3)$ of a particle of $\mathscr{S}$ relative to Σ. Finally, starting from the identity $d/dt = \partial/\partial t + u_i(\partial/\partial x_i)$, we have

$$\frac{d}{dt}\int_D \varphi \, DV = \int_D \left(\frac{d\varphi}{dt} + \varphi u_{i,i}\right) DV + \int_{\Sigma(D)} [\![\varphi w_i N_i]\!] \, DS. \quad (I.4)$$

Fundamental Lemma of Continuous Media Physics. *Let a defined and continuous function in domain D be $\Psi(M)$, and a dense family in D be $\{d\}$. If for each d the integral of Ψ in d is nil, then the function Ψ is identically nil in D.*

Note 1. By integral must be understood a volume integral if $D \subset R^3$, a surface integral if D belongs to a surface, and a curvilinear integral if D is a curved arc.

Note 2. The lemma remains valid if Ψ is a component of a vector quantity.

1.2. Conservation of Mass

The so-called hypothesis of conservation of mass accepted by conventional physics may be expressed thus:

The mass of an arbitrary domain D of system $\mathscr{S}$ remains invariable when D is followed as it moves.

If the distribution of masses on $\mathscr{S}$ is defined by a volume density $\rho = v^{-1}$, this hypothesis is equivalent to the cancellation of the particle derivative of the integral of ρ over D, and—according to (I.4)—to

$$\int_D \left(\frac{d\rho}{dt} + \rho u_{i,i}\right) DV + \int_{\Sigma(D)} [\![\rho w_i N_i]\!] \, DS = 0. \tag{I.5}$$

Considering successively the cases where $\Sigma(D) = 0$ and $\Sigma(D) \neq 0$, and applying the fundamental lemma to a dense family in D and a dense family in $\Sigma(D)$, leads to an *equation with partial derivatives* which is valid wherever ρ and $\bar{u}$ may be continuously derived

$$\frac{d\rho}{dt} + \rho u_{i,i} = 0 \tag{I.6a}$$

and a *jump equation* which is valid in every point of a surface of discontinuity of ρ and/or $\bar{u}$

$$[\![\rho w_i N_i]\!] = 0. \tag{I.6b}$$

It should be noted that (I.6b) brings in only the normal component $w_i N_i$ of the relative velocity-vector $\vec{W}$ of matter relative to surface Σ, and that (I.6b) can be interpreted by saying that the normal flux of mass through Σ has the same value M on either side of Σ.

1.3. The Law of Dynamics

In its most usual form, the law of dynamics is expressed as follows:

There is at least one referential (a frame of reference in space and time) called Galilean in which—for every physical system and at every moment t—the dynamic torsor is equivalent to the torsor of the exterior actions applied to this system.

In order to apply this law to any system $\mathscr{S}$, an arbitrary domain D within $\mathscr{S}$ must first be considered, then a partition must be made of the effects on D (those effects called *external* which are brought to bear on D by systems external to $\mathscr{S}$, and those called *internal* which are brought to bear on D by those parts of $\mathscr{S}$ which are external to D), and finally a model of these effects must be made. The conclusions that may be drawn from this justify—in the most common physical situations—a representation where:

(i) the external effects are the result of volume distributions, $\langle f_i(P, t); i = 1, 2, 3 \rangle$ of forces and $\langle m_i(P, t); i = 1, 2, 3 \rangle$ of momenta; and
(ii) the internal actions are the result of a surface distribution of forces $\langle T_i = t_{ij}(P, t)n_j; i = 1, 2, 3 \rangle^*$ where $\bar{n}$ conventionally designates the norm external to D.

---

* $\vec{T}$ is called the stress vector in P for the direction $\bar{n}$.

In these conditions, the law of dynamics is wholly expressed on the one hand by the vector equality of the resultants, and on the other of the momenta which result in a point O connected to $\mathscr{R}$

$$\frac{d}{dt}\int_D \rho\vec{u}\,DV = \int_{\partial D} \vec{T}\,DS + \int_D \vec{f}\,DV, \tag{I.7}$$

$$\frac{d}{dt}\int_D \overrightarrow{OM} \wedge \rho\vec{u}\,DV = \int_{\partial D} \overrightarrow{OM} \wedge \vec{T}\,DS + \int_D (\overrightarrow{OM} \wedge \vec{f} + \vec{m})\,DV. \tag{I.8}$$

If the particle derivatives are converted into (I.7) using (I.4), and the surface integrals into (I.7) using (I.1), the result is

$$\int_D \left[\frac{d}{dt}(\rho u_i) + \rho u_i u_{j,j} - t_{ij,j} - f_i\right] DV + \int_{\Sigma(D)} [\![\rho u_i w_j - t_{ij}]\!] N_j\,DS = 0,$$

which, according to (I.6a) and (I.6b), is then written

$$\int_D \left(\rho\frac{du_i}{dt} - t_{ij,j} - f_i\right) DV + \int_{\Sigma(D)} [\![Mw_i - t_{ij}N_j]\!]\,DS = 0.$$

So, a reasoning analogous to that made earlier on the expression (I.5) of the conservation of mass leads to three equations with partial derivatives, called *equations of motion* and three *jump equations*

$$\rho\frac{du_i}{dt} = t_{ij,j} + f_i, \qquad i = 1, 2, 3, \tag{I.9a}$$

$$[\![Mw_i - t_{ij}N_j]\!] = 0, \qquad i = 1, 2, 3. \tag{I.9b}$$

Similar conversions and considerations carried out on (I.8), and taking (I.9) into account, lead to three equations from which partial derivatives and external forces disappear

$$\varepsilon_{ijk}t_{jk} = m_i, \qquad i = 1, 2, 3, \tag{I.10}$$

where ε_{ijk} is the basic antisymmetric tensor in R^3. By introducing the antisymmetric tensor C_{jk} associated with pseudovector m_i by

$$m_i = -\tfrac{1}{2}\varepsilon_{ijk}C_{jk}, \qquad i = 1, 2, 3, \tag{I.11}$$

and the sum $\sigma_{ij} = t_{ij} + \tfrac{1}{2}C_{ij}$, (I.10) can be expressed as

$$\varepsilon_{ijk}\sigma_{jk} = 0, \qquad i = 1, 2, 3, \tag{I.12}$$

which expresses—quite simply—the symmetry of the tensor σ_{jk}, which is called the *Cauchy stress tensor*.

1.4. The First Law of Thermodynamics

Assuming that the concept of mass density of *internal energy e* and the concept of *heat* have been defined elsewhere, and accepted, the first law of thermodynamics is expressed:

For every physical system, at any time, the particle derivative of *total* energy (the sum of *internal* energy and *kinetic* energy) is equal to the power of external effects increased by the rate of heat received).

Here again, as with the law of dynamics, its application leads to an examination of an arbitrary domain D inside $\mathscr{S}$, and the carrying out of a partition of the rate of heat received by D. For the most common phenomena, the conclusions resulting from this justify a representation where the rate of heat received results from:

(i) a surface distribution $\langle q_i(P, t); i = 1, 2, 3 \rangle$*; and
(ii) a volume distribution $\ell(P, T)$.

In these conditions, the first law of thermodynamics is wholly expressed by the scalar equality

$$\frac{d}{dt} \int_D \rho \left(e + \frac{u_i u_i}{2} \right) DV = \int_{\partial D} (u_j t_{ji} - q_i) n_i \, DS + \int_D (u_i f_i + \omega_k m_k + \ell) \, DV, \tag{I.13}$$

where $\vec{\omega}$ is the pseudovector associated with tensor $u_{i,j}$ of the rates of deformation by

$$\omega_i = -\tfrac{1}{2} \varepsilon_{ijk} u_{j,k}. \tag{I.14}$$

It is easy to verify that (I.11) and (I.14) entail

$$\omega_k m_k = \tfrac{1}{2} C_{ij} u_{i,j}. \tag{I.15}$$

Converting the particle derivatives into (I.13) by (I.4), and the surface integrals into (I.13) by (I.1), gives

$$\int_D \left[\frac{d}{dt} \rho \left(e + \frac{u_i u_i}{2} \right) + \rho \left(e + \frac{u_i u_i}{2} \right) u_{i,i} - (u_i f_i + \omega_i m_i + \ell) \right] DV$$

$$- \int_D (u_j t_{ji} - q_i)_{,i} \, DV + \int_{\Sigma(D)} \left[\rho \left(e + \frac{u_i u_i}{2} \right) w_i - (u_j t_{ji} - q_i) \right] N_i \, DS = 0,$$

which, according to (I.6), (I.9), and (I.15), is then expressed

$$\int_D \left[\rho \frac{de}{dt} - u_{i,j}(t_{ij} + \tfrac{1}{2} C_{ij}) + q_{i,i} - \ell \right] DV$$

$$+ \int_{\Sigma(D)} \left[M \left(e + \frac{w_i w_i}{2} \right) - (w_j t_{ji} - q_i) N_i \right] DS = 0. \tag{I.16}$$

A reasoning analogous to that already used on two occasions for the conser-

---

* $\bar{q}$ is called the *heat flux* vector; it has the same dimension as a force per unit of surface.

vation of mass and the law of dynamics leads to an equation with partial derivatives and a jump equation

$$\rho \frac{de}{dt} = u_{i,j}(t_{ij} + \tfrac{1}{2}C_{ij}) - q_{i,i} + \ell, \tag{I.17a}$$

$$M \left[e + \frac{w_i w_i}{2} \right] = [\![w_j t_{ji} - q_i]\!] N_i. \tag{I.17b}$$

1.5. The Second Law of Thermodynamics

Since the first formulations made by Carnot and Clausius, the second law of thermodynamics has taken many forms. Here, to be brief whilst remaining in keeping with preceding presentations, it is assumed that the notions of fields of *absolute temperature* T and mass density s *of entropy*, as defined elsewhere, are accepted. The second law can then be expressed:

For every physical system, at any time, the particle derivative of entropy is superior to or at least equal to the rate of heat received referred locally to the absolute temperature.

In the modeling conditions used above for the rate of heat received, and for an arbitrary domain D inside system $\mathscr{S}$, the second law is wholly expressed by the scalar inequality

$$\frac{d}{dt} \int_D \rho s \, DV \geq \int_D \frac{\ell}{T} DV - \int_{\partial D} \frac{q_i n_i}{T} DS. \tag{I.18}$$

Converting the particle derivative into (I.18) by (I.4), and the surface integral into (I.18) by (I.1), gives

$$\int_D \left[\frac{d}{dt}(\rho s) + \rho s u_{i,i} + \left(\frac{q_i}{T} \right)_{,i} - \frac{\ell}{T} \right] DV + \int_{\Sigma(D)} [\![\rho s w_i + q_i N_i]\!] \, DS \geq 0,$$

which, according to (I.6), is then expressed

$$\int_D \left[\rho \frac{ds}{dt} + \left(\frac{q_i}{T} \right)_{,i} - \frac{\ell}{T} \right] DV + \int_{\Sigma(D)} [\![Ms + q_i N_i]\!] \, DS \geq 0. \tag{I.19}$$

An easily conceivable generalization of the fundamental lemma, mentioned in §I.3, leads to an inequality with partial derivatives and to a jump inequality

$$\rho \frac{ds}{dt} + \left(\frac{q_i}{T} \right)_{,i} - \frac{\ell}{T} \geq 0, \tag{I.20a}$$

$$[\![Ms + q_i N_i]\!] \geq 0. \tag{I.20b}$$

Because of its subsequent interesting consequences, it is worth noting the result of the elimination of ℓ between (I.17a) and (I.20a)

$$\sigma_{ij}u_{i,j} - \rho\left(\frac{de}{dt} - T\frac{ds}{dt}\right) + q_i\left(Log\frac{1}{T}\right)_{,i} \geq 0, \tag{I.21}$$

which is generally known as the Clausius–Duhem inequality.

1.6. The Case of Fluid in the Absence of Moments

Although they result from elementary modelings, the general expressions (I.6), (I.9), (I.12), (I.17), and (I.20) are, nevertheless, of a complexity which suggests that simplifications should be sought, whether in the nature of the stresses $(\vec{f}, \vec{m}, \ell)$ which act on the system, or in the mechanical (σ_{ij}) or thermal (q_i) behavior of the system itself. Below we consider first the consequences of the absence of stresses of type $\vec{m}$, and then of a behavior of perfect fluid or dissipative fluid of the Navier–Fourier type.

The cancellation of field $\vec{m}$ leaves unchanged the equations (I.6) which express the conservation of mass, just as it does those equations (I.20) which express the nondecrease of entropy. On the other hand, this hypothesis markedly simplifies the equations with partial derivatives which stem from the law of dynamics and the first law of thermodynamics, without however influencing the associated jump equations. In fact, (I.6), (I.9), (I.12), and (I.17) taken together become

$$\left.\begin{array}{ll} \dfrac{d\rho}{dt} + \rho u_{i,i} = 0, & [\![\rho w_i N_i]\!] = 0, \\[2ex] \rho\dfrac{du_i}{dt} = \sigma_{ij,j} + f_i, & [\![Mw_i - \sigma_{ij}N_j]\!] = 0, \quad i = 1, 2, 3, \\[2ex] \rho\dfrac{de}{dt} = \sigma_{ij}u_{i,j} - q_{i,i} + \ell, & M\left[\!\!\left[e + \dfrac{w_iw_i}{2}\right]\!\!\right] = [\![w_j\sigma_{ji} - q_i]\!]N_i, \end{array}\right\} \tag{I.22}$$

σ_{ij} symmetrical.

Suppose now that the system behaves like a *fluid*: the mass density e of internal energy is a function of v and s, such that $de - T\,ds$ is proportional to dv. By defining the pressure $p(v, s)$ of the fluid by

$$de - T\,ds = -p\,dv, \tag{I.23}$$

the inequality (I.21) becomes

$$(\sigma_{ij} + p\delta_{ij})u_{i,j} + q_i\,Log\left(\frac{1}{T}\right)_{,i} \geq 0. \tag{I.24}$$

Consider the case where the fluid behaves like a *perfect* medium: each of the left-hand terms in (I.24) is identically nil, whatever the respective gradients of velocity $u_{i,j}$ and of temperature $T_{,i}$. So the tensor of stresses σ_{ij} and the heat flux vector q_i are necessarily in the form

$$\sigma_{ij} = -p\delta_{ij},$$
$$q_i = 0.$$
(I.25)

The expressions (I.25) help to simplify greatly the set of equations (I.22) and inequality (I.20b)

$$\rho \frac{dv}{dt} = u_{i,i},$$

$$\rho \frac{du_i}{dt} = -p_{,i} + f_i, \quad i = 1, 2, 3$$
(I.26a)

$$\rho T \frac{ds}{dt} = \ell,$$

$$[\![\rho w_i N_i]\!] = 0,$$

$$[\![Mw_i + pN_i]\!] = 0, \quad i = 1, 2, 3,$$

$$\left[\!\!\left[M\left(e + \frac{w_i w_i}{2}\right) + pw_i N_i \right]\!\!\right] = 0,$$
(I.26b)

$$M[\![s]\!] > 0.$$

Consider the case where the fluid behaves like a Navier–Fourier *dissipative* medium: each of the left-hand terms in (I.24) is a positive defined quadratic invariant of $u_{i,j}$ and $T_{,i}$, respectively. Then the tensor of stresses σ_{ij} and the heat flux vector q_i are necessarily in the form

$$\sigma_{ij} = -p\delta_{ij} + \pi_{ij},$$
$$\pi_{ij} = \mu' u_{k,k} \delta_{ij} + 2\mu u_{i,j}, \quad \mu, \mu' > 0,$$
(I.27)
$$q_i = -\lambda T_{,i}, \quad \lambda > 0.$$

Equations (I.6a), (I.9a), and (I.17a) become

$$\rho \frac{dv}{dt} = u_{i,i},$$

$$\rho \frac{du_i}{dt} = -p_{,i} + \pi_{ij,j} + f_i,$$
(I.28)

$$\rho T \frac{ds}{dt} = [\lambda T_{,i}]_{,i} + \pi_{ij} u_{i,j} + \ell.$$

In the chapters that follow, we will have only the equations (I.26a, b) and (I.28) with $f_i = 0$ to consider. For (I.26b), in particular, there exists an interesting form which is developed below.

Let the tangential and normal components of $\vec{W}$ over Σ be designated by $\vec{W}_\tau$ and $w\vec{N}$; let the values on either side of Σ be marked by the indices 1 and 2; equations (I.26b) may then be expressed

$$M \equiv \rho_1 w_1 = \rho_2 w_2,$$

$$M(w_2 - w_1) + p_2 - p_1 = 0, \qquad M(\vec{W}_{\tau 1} - \vec{W}_{\tau 2}) = 0, \qquad (I.29)$$

$$M\left(e_2 - e_1 + \frac{w_2^2 - w_1^2}{2} + [\![|W_\tau|^2]\!] \right) + p_2 w_2 - p_1 w_1 = 0.$$

When $M \neq 0$, noting that $w_2 - w_1 = M(v_2 - v_1)$ and eliminating M, we have

$$\vec{W}_{\tau 1} - \vec{W}_{\tau 2} = 0,$$

$$w_1^2 = v_1^2 \frac{p_2 - p_1}{v_1 - v_2},$$

$$\qquad (I.30)$$

$$w_2^2 = v_2^2 \frac{p_2 - p_1}{v_1 - v_2},$$

$$e_2 - e_1 = \tfrac{1}{2}(p_1 + p_2)(v_1 - v_2).$$

It was in the work of W.I. Rankine (*Philosophical Transactions*, vol. CLX, Part I, p. 277, 1870), on the permanent rectilinear flow of a Navier–Fourier dissipative fluid, that the last equation (I.30) appeared for the first time. But its real significance became clear only through the work of Hugoniot [25] and of Crussard [14], which showed that it possessed the outstanding property of linking the thermodynamic states on either side of a surface of discontinuity Σ in a perfect fluid. Depending on the physical phenomena included in Σ, there are various ways in which this equation may be presented, and these are analyzed in Chapter II.

Before going on to this discussion, an examination will be made in greater detail of the physical significance, for *a reactive fluid*, of the quantities σ_{ij} (Cauchy stress tensor) and q_i (heat flux vector), which were introduced earlier to model the mechanical and thermodynamic transfers within the system. In Section 2, this more detailed study is made at the level of molecules, in accordance with a process which is similar to that which leads, classically, to the Boltzmann equation. In Section 3, the study is made on a macroscopic level, in accordance with a process analogous to that which serves as a foundation for the studies of turbulence. These two approaches are elementary and certainly standard in the case of a fluid composed of a single chemical type. They are more complex when, in a fluid, several chemical types meet and react amongst themselves. But this complexity, if not essential, is at least very instructive in the study of the flow of reactive fluids, insofar as it brings to light the relationship and the coexistence of the mechanisms of molecular diffusion and turbulent diffusion, and it also demonstrates how these two mechanisms contribute to transfers of mass, of momentum, and of energy.

2. Molecular Model of the Reactive Fluid

2.1. Introduction

Let us consider a fluid system $\mathscr{S}(t)$ containing molecules distributed in K populations $(k = 1, \ldots, K)$ (N.B. This viewpoint obviously fits the case of a reactive fluid made up of molecules of the same species distributed over K energy levels.) The distribution function f^k is defined such that

$$f^k(\mathbf{x}, \mathbf{u}, t) \, \mathrm{Dx} \, \mathrm{Du}, \qquad k = 1, \ldots, K,$$

is the probable number of (k)-type molecules in the neighborhood Dx Du of point $\mathbf{x}$ $(x_i; i = 1, 2, 3)$ at instant t. (N.B. Throughout §2.3 and §2.4, the index (superscript) k marks a chemical species and is never dummy.)

Let us consider a molecule of type (k) present in x_i with velocity u_i^k at instant t. If there were no intermolecular collisions, monomolecular reactions, radiative jumps, etc., then this molecule would evolve so that, shortly afterwards—at instant $t + dt$—its position would be $(x_i + u_i^k \, dt; i = 1, 2, 3)$ and its velocity $(u_i^k + g_i^k \, dt; i = 1, 2, 3)$ where $(g_i^k(x); i = 1, 2, 3)$ is the mass density of external force applied to the species (k). As a result, the only molecules of type k that would arrive at $(x_i + u_i^k \, dt, u_i^k + g_i^k \, dt)$ at instant $t + dt$ would be those situated at (x_i, u_i^k) at instant t, so that there would be equality in terms of the number of molecules

$$f^k(x_i + u_i^k \, dt; u_i^k + g_i^k \, dt; t + dt) \, \mathrm{Dx} \, \mathrm{Du} = f^k(x_i, u_i^k) \, \mathrm{Dx} \, \mathrm{Du}.$$

But, in fact, the result of collisions, etc., is that molecules present in the neighborhood $(\mathrm{Dx}, \mathrm{Du})$ of (x_i, u_i^k) do not reach the neighborhood $(\mathrm{Dx}, \mathrm{Du})$ of $(x_i + u_i^k \, dt, u_i^k + g_i^k \, dt)$ at instant $t + dt$; let $\Delta f_-^k \, \mathrm{Dx} \, \mathrm{Du}$ be the number of these molecules. Moreover, collisions, etc., result in other molecules of type (k), to the number of $\Delta f_+^k \, \mathrm{Dx} \, \mathrm{Du}$, which are missing in the neighborhood $(\mathrm{Dx}, \mathrm{Du})$ of (x_i, u_i^k) at instant t, occurring in the neighborhood $(\mathrm{Dx}, \mathrm{Du})$ of $(x_i + u_i^k \, dt; u_i^k + g_i^k \, dt)$ at the instant $t + dt$. The above equality is therefore invalid and must be replaced by

$$f^k(x_i + u_i^k \, dt; u_i^k + g_i^k \, dt; t + dt) = f^k(x_i, u_i^k) + (\Delta f_+^k - \Delta f_-^k).$$

In the limit, when $dt \to 0$, the identity is expressed

$$\frac{\partial f^k}{\partial t} + u_j^k f_{,j}^k + g_\alpha^k f_{,\alpha}^k = \frac{\Delta f^k}{\Delta t}, \tag{I.31}$$

where $\Delta f^k / \Delta t$ is the net variation level of f^k due to collisions and mechanisms on the molecular scale. (N.B. The derivative of f^k in relation to the variables x_i is marked by a Roman index, and in relation to the variables u_i is marked by a Greek index.) Equation (I.31) is an extension of the standard Boltzmann equation.

2.2. Kinetic Definition of Thermodynamic Variables

For each type of molecule, the mechanical variables can be defined as integrals over the velocity space of the appropriate quantity weighted by the distribution function.

So the volume density n^k of type (k) molecules present in $(x_i; i = 1, 2, 3)$ at instant t is

$$n^k(x_i, t) = \int f^k(x_i, u_i, t) \, \mathbf{Du}. \qquad (I.32)$$

From n^k and molecular mass m^k, we can define:

$n = \sum_k n^k$ total number of molecules per unit volume of the mixture,

$\rho^k = n^k m^k$ mass of (k) per unit volume of the mixture,

$\rho = \sum_k \rho^k$ density of the mixture, $\qquad (I.33)$

$X^k = \dfrac{n^k}{n}$ molar fraction of (k) in the mixture $(\sum X^k = 1)$,

$Y^k = \dfrac{\rho^k}{\rho}$ mass fraction of (k) in the mixture $(\sum Y^k = 1)$.

Other variables which are secondary but nevertheless useful in simplifying certain expressions can also be introduced using the Avogadro number $\mathcal{N}$

$M^k = \mathcal{N} m^k$ molar mass of (k),

$c^k = \dfrac{n^k}{\mathcal{N}}$ number of (k) moles per unit volume of the mixture.

More generally, the local average of a quantity $G^k(x_i, u_i, t)$ can be defined by

$$\overline{G}^k(x_i, t) = \frac{1}{n^k(x_i, t)} \int G^k(x_i, u_i, t) \cdot f^k(x_i, u_i, t) \, \mathbf{Du}.$$

So, if G^k is the velocity u_i^k of the (k) molecules, we have

$$\overline{u_i^k} = \frac{1}{n^k} \int u_i^k f^k \, \mathbf{Du}$$

from which we can state

$u_i^0 = \sum Y^k \overline{u_i^k}$ average velocity of the mixture,

$\hat{u}_i^k = u_i^k - u_i^0$ proper velocity of a molecule of (k), $\qquad (I.34)$

$\overline{\hat{u}_i^k} = \overline{u_i^k} - u_i^0$ diffusion velocity of (k) in the mixture.

It is essential for what follows to note that the diffusion velocities verify both identities

$$\rho \sum Y^k \overline{\hat{u}_i^k} = \sum \rho^k \overline{\hat{u}_i^k} = \sum n^k m^k \overline{\hat{u}_i^k} = 0. \tag{I.35}$$

The introduction of temperatures T^k and T^0 for the species (k) and the mixture is based on the hypotheses (k = Boltzmann constant)

$$\tfrac{3}{2} kT^k = \tfrac{1}{2} m^k |\overline{\hat{u}^k}|^2,$$

$$\tfrac{3}{2} kT^0 = \tfrac{1}{2} \sum X^k m^k |\overline{\hat{u}^k}|^2,$$

from which the immediate result is

$$T^0 = \sum X^k T^k.$$

N.B. It can be established that, for the Maxwell distribution law,

$$f^k = n^k \left(\frac{m^k}{2\pi kT} \right)^{3/2} \exp \left(-\frac{1}{2} \frac{m^k |\overline{\hat{u}^k}|^2}{kT} \right),$$

temperatures T, T^0, and T^k are equal.

Defining the internal energy of a (k) molecule as the sum

$$\tfrac{1}{2} m^k |\hat{u}^k|^2 + \varepsilon^k,$$

where ε^k is the contribution of the internal degrees of freedom (rotation, vibration, etc.), it is possible to express the average proper energy of a (k) molecule

$$\tfrac{1}{2} m^k |\overline{\hat{u}^k}|^2 + \overline{\varepsilon^k}$$

and then the average internal energy of the $n = \sum n^k$ molecules present in the unit volume of the mixture

$$\sum \tfrac{1}{2} n^k m^k |\overline{\hat{u}^k}|^2 + \sum n^k \overline{\varepsilon^k}.$$

By dividing the preceding expression by $\rho = \sum n^k m^k$, the *specific internal energy* e^0 of the mixture is obtained. In view of $n^k m^k = \rho Y^k$, e^0 takes the form

$$e^0 = \sum Y^k \left(\frac{|\overline{\hat{u}^k}|^2}{2} + \frac{\overline{\varepsilon^k}}{m^k} \right). \tag{I.36}$$

2.3. Equations of Evolution and Invariants by Summation

Multiplying (I.31) by a function $\Psi^k(u)$ and integrating over the entire space of velocities gives

$$\int \Psi^k \frac{\partial f^k}{\partial t} Du + \int \Psi^k u_j^k f_{,j}^k Du + \int \Psi^k g_\alpha^k f_{,\alpha}^k Du = \int \Psi^k \frac{\Delta f^k}{\Delta t} Du.$$

Given the independence of variables x, u, and t and considering only the force fields of the type $g_\alpha^k(\mathbf{x}, t)$, we have

$$\Psi^k u_j^k f_{,j}^k = (\Psi^k u_j^k f^k)_{,j},$$

$$\Psi^k g_\alpha^k f_{,\alpha}^k \, \mathbf{Du} = \psi^k (g_\alpha^k f^k)_{,\alpha} \, \mathbf{Du} = D(\Psi^k g_\alpha^k f^k) - g_\alpha^k f^k \Psi_{,\alpha}^k \, \mathbf{Du}.$$

From which follows

$$\frac{\partial}{\partial t} \int \Psi^k f^k \, \mathbf{Du} + \left(\int \Psi^k u_j^k f^k \, \mathbf{Du} \right)_{,j} - g_\alpha^k \int \Psi_{,\alpha}^k f^k \, \mathbf{Du} = \int \psi^k \frac{\Delta f^k}{\Delta t} \, \mathbf{Du},$$

assuming that $|\mathbf{u}^2| \Psi^k f^k$ tends uniformly towards zero when $|\mathbf{u}| \to \infty$. Then, as a result of the definition of the average value $\bar{G}$ of a function G, we obtain

$$\frac{\partial}{\partial t} n^k \overline{\Psi^k} + (n^k \overline{\Psi^k u_j^k})_{,j} - n^k g_\alpha^k \overline{\Psi_{,\alpha}^k} = \int \Psi^k \frac{\Delta f^k}{\Delta t} \, \mathbf{Du} \qquad (1.37)$$

called the equation of evolution of the function Ψ^k.

For all the processes on the molecular scale (collisions, radiative jumps, etc.) in which $\mathbf{u}$ can change without change in the macroscopic position(s) of the molecules concerned, some Ψ^k quantities can survive *on average* at any instant and at any point, which means in fact that

$$\sum_k \int \Psi^k \frac{\Delta f^k}{\Delta t} \, \mathbf{Du} = 0. \qquad (1.38)$$

Such a Ψ^k quantity is called an invariant by summation. Insofar as we can disregard mass, momentum, and energy removed by radiation in radiative collisions and transitions, then

$$m^k, \quad m^k u_i^k, \quad \tfrac{1}{2} m^k |\mathbf{u}^k|^2 + \varepsilon^k,$$

are invariants by summation. The form taken by the corresponding equations of evolution is examined below.

2.4. Conservation of Mass

By summing the equation of evolution relative to m^k over k, there follows

$$\sum_k \frac{\partial}{\partial t} n^k m^k + \sum_k (n^k m^k \bar{u}_j^k)_{,j} = 0.$$

From the definitions (1.33) of ρ^k and ρ, the definition (1.34) of the average velocity $\bar{u}_j^k$ of k, and the properties (1.35), we have

$$\frac{\partial \rho}{\partial t} + \sum (\rho^k u_j^0)_{,j} = 0,$$

or, alternatively,

$$\frac{\partial \rho}{\partial t} + u_j^0 \rho_{,j} + \rho u_{j,j}^0 = 0.$$

If the operator "particle derivative according to the mixture" is introduced

$$\frac{d^0}{dt} = \frac{\partial}{\partial t} + u_j^0 \frac{\partial}{\partial x_j},$$

the above equation in written

$$\frac{d^0\rho}{dt} + \rho u_{j,j}^0 = 0, \tag{I.39}$$

which is identical to the first equation (I.22) once the particle velocity u_i of the continuum model is identified to the average velocity u_i^0 of the mixture defined in (I.34).

2.5. Conservation of Momentum

By summing the equation of evolution relative to $m^k u_i^k$ over k, we have

$$\sum_k \frac{\partial}{\partial t}(n^k m^k \bar{u}_i^k) + \sum_k (n^k m^k \overline{u_i^k u_j^k})_{,j} - \sum_k n^k m^k g_a^k \delta_{i\alpha} = 0.$$

The average of $u_i^k u_j^k$ can then be written as the sum of four terms

$$u_i^0 u_j^0 + u_i^0 \overline{\hat{u}_j^k} + u_j^0 \overline{\hat{u}_i^k} + \overline{\hat{u}_i^k \hat{u}_j^k}.$$

After a summation weighted by $n^k m^k = \rho\, Y^k$, the second and third terms make a zero contribution according to (I.35). So, by defining σ_{ij}^0 and f_i^0 as

$$\sigma_{ij}^0 = -\sum n^k m^k \overline{\hat{u}_i^k \hat{u}_j^k} \equiv -\sum \rho^k \overline{\hat{u}_i^k \hat{u}_j^k},$$
$$f_i^0 = \sum \rho^k g_i^k, \tag{I.40}$$

we have

$$\frac{\partial}{\partial t}(\rho u_i^0) + (\rho u_i^0 u_j^0)_{,j} - \sigma_{ij,j}^0 - f_i^0 = 0.$$

By a sensible expansion of the first two terms and using the operator d^0/dt, we have

$$\rho \frac{d^0 u_i^0}{dt} + u_i^0 \left[\frac{\partial \rho}{\partial t} + (\rho u_j^0)_{,j} \right] = \sigma_{ij,j}^0 + f_i^0.$$

The second term of the left-hand member is zero in accordance with the equation (I.39) of conservation of mass, with the final result

$$\rho \frac{d^0 u_i^0}{dt} = \sigma_{ij,j}^0 + f_i^0, \tag{I.41}$$

which is identical to the second equation (I.22) when the obvious relations between the sums u_i^0, σ_{ij}^0, f_i^0 defined by (I.34) and (I.40) and the quantities u_i, σ_{ij}, f_i introduced in the continuum model are considered. This relation makes it possible to give the components of the Cauchy stress tensor σ_{ij} a microscopic significance linked to the flow of *proper* momentum: $\vec{N}'$ and $\vec{N}''$ referring to two unit vectors at point P of real space, $\sigma_{ij}(\vec{N}', \vec{N}'')$ results from the summation over k, weighted by the ρ^k, of the proper momentum in the direction $\vec{N}'$ which on average traverses the unit area whose norm is $\vec{N}''$ in the direction $-\vec{N}''$ during a unit of time.

2.6. Conservation of Energy

The quantity Ψ^k considered is $m^k|u^k|^2/2 + \varepsilon^k$ where the last term is assumed to be independent of $(\mathbf{u}, \mathbf{x}, t)$. First, it should be noted that

$$\overline{\Psi^k} = m^k[\tfrac{1}{2}|u^0|^2 + u_i^0\overline{u_i^k} + \tfrac{1}{2}|\hat{u}^k|^2 + \overline{\varepsilon^k}].$$

Considering the expression of e^0 given in (I.36), and also the relations (I.35), summation over k gives

$$\sum n^k\overline{\Psi^k} = \rho(\tfrac{1}{2}|u^0|^2 + e^0).$$

It should then be noted that

$$\Psi^k u_j^k = m^k[\tfrac{1}{2}|u^0|^2 u_j^0 + \tfrac{1}{2}|u^0|^2\hat{u}_j^k + u_i^0\hat{u}_j^k u_j^0 + u_i^0\hat{u}_i^k\hat{u}_j^k]$$
$$+ [\tfrac{1}{2}m^k|\hat{u}^k|^2 + \varepsilon^k](u_j^0 + \hat{u}_j^k),$$

with the result that

$$\sum n^k\overline{\Psi^k u_j^k} = \tfrac{1}{2}\rho|u^0|^2 u_j^0 + 0 + 0 - u_i^0\sigma_{ij}^0 + \rho e^0 u_j^0 + q_j^0,$$

provided that we assume

$$q_j^0 = \sum_k n^k\overline{(\tfrac{1}{2}m^k|\hat{u}|^2 + \varepsilon^k)u_j^k} \equiv \sum_k \rho^k\left[\frac{1}{2}|\hat{u}^k|^2 + \frac{\varepsilon^k}{m^k}\right]\hat{u}_j^k. \qquad (I.42)$$

Finally, it should be noted that

$$\Psi_{,\alpha}^k = m^k[u_\alpha^0 + \hat{u}_\alpha^k]$$

with the result that

$$n^k g_\alpha^k\overline{\Psi_{,\alpha}^k} = m^k g_\alpha^k(u_\alpha^0 + \overline{\hat{u}_\alpha^k}).$$

After all these observations, the equation of evolution of the quantity $\Psi^k = m^k|u^k|^2/2 + \varepsilon^k$, after summation over k, gives

$$\frac{\partial}{\partial t}[\rho(\tfrac{1}{2}|u^0|^2 + e^0)] + [\rho(\tfrac{1}{2}|u^0|^2 + e^0)u_j^0 - u_i^0\sigma_{ij}^0 + q_j^0]_{,j} = u_\alpha^0 f_\alpha^0 + \sum_k \rho^k g_\alpha^k\overline{\hat{u}_\alpha^k}.$$

A sensible expansion of the first two terms of the left-hand member gives

$$\rho\left(\frac{\partial e^0}{\partial t} + u_j^0 e_{,j}^0\right) + (e^0 + \tfrac{1}{2}|u^0|^2)\left[\frac{\partial \rho}{\partial t} + (\rho u_j^0)_{,j}\right]$$
$$+ \rho u_i^0\left(\frac{\partial u_i^0}{\partial t} + u_j^0\frac{\partial u_i^0}{\partial x_j}\right) - (u_i^0\sigma_{ij}^0 - q_j^0)_{,j}.$$

The first term equals $\rho(d^\circ e^\circ/dt)$, the second term is zero according to the equation of conservation of mass; the third term equals $\rho u_i^\circ (d^\circ u_i^\circ/dt) = u_i^\circ (\sigma_{ij,j}^\circ + f_i^\circ)$ according to the equation of conservation of momentum; so the above equation, after simplification, takes the form

$$\rho \frac{d^\circ e^\circ}{dt} = \sigma_{ij}^\circ u_{i,j}^\circ - q_{j,j}^\circ + \sum_k g_i^k \rho^k \overline{\hat{u}_i^k}, \tag{I.43}$$

which is identical to the third equation (I.22) by adding to the relations mentioned above the relation between q_j and the sum q_j° defined in (I.42), and the relation between ℓ and ℓ° defined by

$$\ell^\circ = \sum_k g_i^k \rho^k \overline{\hat{u}_i^k}. \tag{I.44}$$

2.7. Diffusion Equations

In the presence of chemical reactions characterized by a volume rate P^k of production of type (k) mass, the number of type (k) molecules is not a summation invariant. So the equation of evolution of $\Psi^k = 1$ governs the variation in concentration of each species; the corresponding K equations, of which only $K - 1$ are independent as a result of the relation $\sum Y^k = 1$, are called *diffusion equations* in the mixture. From the initial form

$$\frac{\partial n^k}{\partial t} + (n^k \overline{u_j^k})_{,j} = \int \frac{\Delta f^k}{\Delta t} \, \mathbf{Du}$$

it can be deduced, by multiplying by m^k and expanding $\overline{u_j^k}$

$$\frac{\partial}{\partial t} (\rho Y^k) + (\rho Y^k u_j^\circ)_{,j} + (\rho Y^k \overline{\hat{u}_j^k})_{,j} = P^k.$$

As in the above two subsections, a sensible expansion of the first two derivatives leads to

$$\rho \left(\frac{\partial Y^k}{\partial t} + u_j^\circ Y_{,j}^k \right) + Y^k \left[\frac{\partial \rho}{\partial t} + (\rho u_j^\circ)_{,j} \right] + (\rho Y^k \overline{\hat{u}_j^k})_{,j} = P^k.$$

Taking account of the equation of conservation of mass which cancels out the second term, and isolating the flow of mass associated with species k in the mixture

$$\rho Y^k \overline{\hat{u}_j^k} \equiv \rho^k \overline{\hat{u}_j^k},$$

the diffusion equations take the form

$$\rho \frac{d^\circ Y^k}{dt} + (\rho^k \overline{\hat{u}_j^k})_{,j} = P^k. \tag{I.45, k}$$

They show how the variation of Y^k in the mixture is linked to diffusion (term $\rho^k \overline{\tilde{u}_j^k}$) and to chemical reaction (term P^k).

3. Continuum Model of the Reactive Fluid

3.1. Introduction

Let us consider again, as in §2.1, a fluid system $\mathcal{S}(t)$ containing molecules distributed in K populations ($k = 1, \ldots, K$). We shall now abandon the "discontinuous" model associated with the concept of molecules in favor of a continuum model where the mechanical variables appropriate to each component of the fluid describe a continuum and are governed by the laws stated in §§1.2–1.5. In other words, for each (k) component present as a mass quantity ρ^k in a unit volume of the mixture, it is accepted that there exist:

— a velocity u_i^k ($i = 1, 2, 3$);
— a specific internal energy e^k;
— a Cauchy stress tensor σ_{ij}^k ($i, j = 1, 2, 3$); and
— a heat flux vector q_j^k ($j = 1, 2, 3$);

all defined in the same way. Starting from these hypotheses, two notations are essential:

— specific mass (density) of the mixture $\rho = \sum_k \rho^k$; and
— mass fraction of (k) in the mixture $Y^k = \rho^k/\rho$ ($\sum_k Y^k = 1$).

In the expansions which follow, a further hypothesis is made. It states that in every point of $\mathcal{S}$ and at every instant, the velocity u_i^k of each (k) component is the sum of an *average value* $\tilde{u}_i$ (independent of k) and a *variation* (dependent on k):

$$u_i^k = \tilde{u}_i + \tilde{u}_i^k,$$
$$\tilde{u}_i = \sum_k Y^k u_i^k \quad \Leftrightarrow \quad \sum Y_k \tilde{u}_i^k = 0. \tag{I.46}$$

[N.B. Throughout Section 3 the superscript $\sim$ marks an average quantity linked to the decomposition (I.46) of u_i^k.]

Using the hypothesis (I.46) we shall reexamine the expression of the main principles set out in §§1.2–1.4, this time taking into consideration the structure $\mathcal{S}(t)$ with K components and the possibility of chemical reactions by means of the rate of production P^k of the (k) component. In this connection it is essential to note that chemical reactions simply convert matter from one species into another, without creating or destroying, so that

$$\sum_k P^k = 0.$$

Finally, the statement will be simplified by looking only at the case where no discontinuity Σ traverses the reactive fluid.

3.2. Mass Balances: As a Whole and by Species

When following the motion of a material domain D, the mass variation rate of species (k) is due only to chemical reactions, with the result that

$$\frac{d}{dt} \int_D \rho^k \, DV = \int_D P^k \, DV.$$

By the theorem on the particle derivative of a volume integral (see §1.2), the preceding expression (in the absence of discontinuity across $\mathscr{S}$) becomes

$$\int_D \left[\frac{\partial \rho^k}{\partial t} + (\rho^k u_i^k)_{,i} \right] DV = \int_D P^k \, DV,$$

which, according to the fundamental lemma of §1.3, is itself equivalent to

$$\frac{\partial \rho^k}{\partial t} + (\rho^k u_i^k)_{,i} = P^k. \tag{I.47, k}$$

We now take into account the splitting (I.46) of the velocity u_i^k, and introduce the time derivative "following the average flow"

$$\frac{\tilde{d}}{dt} = \frac{\partial}{\partial t} + \tilde{u}_i \frac{\partial}{\partial x_i}. \tag{I.48}$$

Equation (I.47) is then written

$$\frac{\tilde{d}\rho^k}{dt} + \rho^k \tilde{u}_{i,i} + (\rho^k \tilde{u}_i^k)_{,i} = P^k,$$

which reveals a vector $\rho^k \tilde{u}_i^k$, called the diffusion flux and written

$$M_i^k = \rho^k \tilde{u}_i^k. \tag{I.49, k}$$

With this notation, the mass balance for species (k) is written

$$\frac{\tilde{d}\rho^k}{dt} + \rho^k \tilde{u}_{i,i} + M_{i,i}^k = P^k. \tag{I.50, k}$$

By summing (I.49, k) over k, the diffusion and creation terms disappear. This gives the expression

$$\frac{\tilde{d}\rho}{dt} + \rho \tilde{u}_{i,i} = 0, \tag{I.51}$$

which is identical to the first equation (I.22) when u_i and $\tilde{u}_i$ are taken to be one and the same.

Equation (I.50, k) can be converted using (I.51), to reveal the mass fraction Y^k; from which we get

$$\rho \frac{\tilde{d}Y^k}{dt} = P^k - M_{i,i}^k, \tag{I.52, k}$$

whose resemblance to (I.45, k) will be discussed in Section 4.

3.3. Lemma on the Sum of Particle Derivatives

Let us consider for each species the quantity $\rho\varphi^k$ integrated on domain D, when it is moving along. From the reasoning used at the start of §3.2, we have

$$\frac{d^k}{dt}\int_D \rho\varphi^k \, DV = \int_D \left[\frac{\partial}{\partial t}\rho\varphi^k + (\rho\varphi^k u_i^k)_{,i}\right] DV.$$

We turn our attention below to the sum from 1 over K of these particle derivatives

$$\int_D \left[\frac{\partial}{\partial t}\sum \rho\varphi^k + (\sum \rho\varphi^k u_i^k)_{,i}\right] DV$$

and attempt to transform it, taking into account the splitting (I.46) of the velocities. By simply rearranging the derivatives, the above integrand is written first

$$(\sum \varphi^k)\frac{\partial\rho}{\partial t} + \rho\frac{\partial}{\partial t}(\sum \varphi^k) + \rho_{,i}(\sum \varphi^k u_i^k) + \rho\sum \varphi^k u_{i,i}^k + \rho\sum u_i^k\varphi_{,i}^k,$$

then, by using (I.46),

$$(\sum \varphi^k)\frac{d\rho}{dt} + \rho\frac{\partial}{\partial t}(\sum \varphi^k) + (\sum \varphi^k)\tilde{u}_i\rho_{,i} + (\sum \varphi^k\tilde{u}_i^k)\rho_{,i} + (\sum \varphi^k)\rho\tilde{u}_{i,i}$$

$$+ \rho(\sum \varphi^k\tilde{u}_{i,i}^k) + \rho\tilde{u}_i(\sum \varphi^k)_{,i} + \rho(\sum \tilde{u}_i^k\varphi_{,i}^k).$$

The sum of the 1st, 3rd, and 5th terms is zero from the overall mass balance (I.51); the sum of the 2nd and 7th terms is $\rho(\tilde{d}/dt)(\sum \varphi^k)$; the sum of the 4th, 6th, and 8th terms is $(\rho\sum \varphi^k\tilde{u}_i^k)_{,i}$. This gives the identity

$$\sum_k \frac{d^k}{dt}\int_D \rho\varphi^k \, DV = \int_D \left[\rho\frac{\tilde{d}}{dt}(\sum \varphi^k) + (\sum \rho\varphi^k\tilde{u}_i^k)_{,i}\right] DV. \qquad (I.53)$$

3.4. Balance of Momentum

By applying the law of dynamics as stated in §1.3 to a reactive fluid with K species, at the level of the torque resultants, we have

$$\sum_k \left(\frac{d^k}{dt}\int_D \rho^k u_i^k \, DV\right) = \sum_k \int_{\partial D} \sigma_{ij}^k n_j \, DS + \sum_k \int_D \rho^k g_i^k \, DV,$$

where g_i^k denotes the mass density of the external forces acting on species (k).

Using Green's theorem to evaluate the flux which appears in the right-hand member, then using the lemma of §3.3 to evaluate the sum of the particle derivatives in the left-hand member and, finally, using the fundamental lemma of §1.1, the result is

$$\rho\frac{\tilde{d}\tilde{u}_i}{dt} + (\rho\sum Y^k u_i^k\tilde{u}_j^k)_{,j} = \sum (\sigma_{ij,j}^k + \rho^k g_i^k).$$

Then splitting u_i^k according to (I.46), and introducing the sum

$$f_i = \sum_k \rho^k g_i^k \tag{I.54}$$

for the preceding equation, we have the form

$$\rho \frac{d\tilde{u}_i}{dt} = \sum_k (\sigma_{ij}^k - \rho Y^k \tilde{u}_i^k \tilde{u}_j^k)_{,j} + f_i,$$

which—beyond the transposition $u_i \leftrightarrow \tilde{u}_i$ already mentioned in connection with the comparison of (I.22) and (I.51) relating to mass—is identical to the second equation (I.22) on condition that the stress tensor to be taken into account is

$$\tilde{\sigma}_{ij} = \sum_k \sigma_{ij}^k - \rho Y^k \tilde{u}_i^k \tilde{u}_j^k \equiv \sum \sigma_{ij}^k - \tilde{u}_i^k M_j^k. \tag{I.55}$$

For the convenience of the calculations in the following subsection, $\tilde{\tilde{\sigma}}_{ij} = -\sum \rho Y^k \tilde{u}_i^k \tilde{u}_j^k$ should be noted. With the notation (I.55) the balance of momentum takes the form

$$\rho \frac{d\tilde{u}_i}{dt} = \tilde{\sigma}_{ij,j} + f_i. \tag{I.56}$$

3.5. Balance of Energy

Applying the first law of thermodynamics as stated in §1.4 to the reactive fluid with K species gives the scalar equation

$$\sum_k \left(\frac{d^k}{dt} \int_D \rho^k (e^k + \tfrac{1}{2} u_i^k u_i^k) \right) DV = \sum_k \int_{\partial D} (u_i^k \sigma_{ij}^k - q_j^k) n_j \, DS + \sum_k \int_D \rho^k g_i^k u_i^k \, DV,$$

where q_j^k denotes the heat flux through species k.

The left-hand member can be evaluated using the lemma of §3.3 where $\varphi^k = Y^k (e^k + \tfrac{1}{2} u_i^k u_i^k)$ is chosen. By introducing the proper internal energy

$$\tilde{e} = \sum Y^k (e^k + \tfrac{1}{2} \tilde{u}_i^k \tilde{u}_i^k), \tag{I.57}$$

and taking (I.46) into account, the two terms which are under the $\int_D$ sign in (I.53) can be transformed

$$\rho \frac{\tilde{d}}{dt} \left(\sum \varphi^k \right) = \rho \frac{\tilde{d}}{dt} (\tilde{e} + \tfrac{1}{2} \tilde{u}_i \tilde{u}_i),$$

$$\left(\sum \rho \varphi^k \tilde{u}_j^k \right)_{,j} = \left(\sum \rho Y^k e^k \tilde{u}_j^k \right)_{,j} - (\tilde{\tilde{\sigma}}_{ij} \tilde{u}_i)_{,j} + \tfrac{1}{2} \left(\sum \rho Y^k \tilde{u}_i^k \tilde{u}_i^k \tilde{u}_j^k \right)_{,j}.$$

Furthermore, after applying Green's theorem to the right-hand member, the quantity

$$\sum (u_i^k \sigma_{ij}^k - q_j^k)_{,j} + \sum \rho^k g_i^k u_i^k$$

appears under the $\int_D$ sign and which, transformed by (I.46), is written

$$\tilde{u}_i \sum (\rho^k g_i^k + \sigma_{ij,j}^k) + \sum (\sigma_{ij}^k \tilde{u}_{i,j} - q_{j,j}^k) + \sum [\tilde{u}_i^k(\rho^k g_i^k + \sigma_{ij,j}^k) + \sigma_{ij}^k \tilde{u}_{i,j}^k],$$

where, in accordance with (I.54) and (I.56), the first term reduces to

$$\tilde{u}_i\left(\rho\frac{\tilde{d}\tilde{u}_i}{dt} - \tilde{\tilde{\sigma}}_{ij,j}\right).$$

As a consequence of these remarks, the energy balance is written

$$\rho\frac{\tilde{d}\tilde{e}}{dt} = \tilde{\sigma}_{ij}\tilde{u}_{i,j} - \sum [q_j^k + (e^k + \tfrac{1}{2}|\tilde{u}^k|^2)M_j^k]_{,j} + \sum (\sigma_{ij}^k\tilde{u}_i^k)_{,j} + \sum \rho^k g_i^k \tilde{u}_i^k.$$

By comparison with the third equation (I.22), it appears that the heat flux vector $\tilde{q}_j$ to be considered for the mixture is

$$\tilde{q}_j = \sum [q_j^k + (e^k + \tfrac{1}{2}|\tilde{u}^k|^2)M_j^k] - \sum \sigma_{ij}^k\tilde{u}_i^k. \tag{I.58}$$

With this notation, the energy balance equation of the multifluid is

$$\rho\frac{\tilde{d}\tilde{e}}{dt} = \tilde{\sigma}_{ij}\tilde{u}_{i,j} - \tilde{q}_{j,j} + \rho\sum g_i^k Y^k \tilde{u}_i^k, \tag{I.59}$$

which is completely identifiable with the third equation (I.22) when it is accepted that there is a volume rate of heat release

$$\tilde{\ell} = \sum g_i^k \rho^k \tilde{u}_i^k \equiv \sum g_i^k M_i^k. \tag{I.60}$$

4. Introduction to Reactive Waves

Subsections 1.1–1.6 led to the following expression of the equations of motion of a fluid in the absence of an external field of momenta

$$\frac{d\rho}{dt} + \rho u_{i,i} = 0,$$

$$\rho\frac{du_i}{dt} = \sigma_{ij} + f_i, \tag{I.61}$$

$$\rho\frac{de}{dt} = \sigma_{ij}u_{i,j} - q_{j,j} + \ell,$$

which will be referred to as "the standard equations."

Let us now consider a reactive fluid with K components subjected to g_i^k fields independent of k (this hypothesis simplifies the expression given below, in such a way that it will be easy to account for a posteriori without detracting from the conclusions in general).

Analysis of Section 2, and, in particular, comparison of (I.39), (I.41), and (I.43) with the standard equations, makes it possible to identify the kinetic significance of the thermodynamic quantities allocated to the *average fluid* with velocity u_i^0. More precisely, they show how the internal mass energy e^0, the stresses σ_{ij}^0, and the heat flux q_j^0 are linked to the averages over the phase space which introduce the diffusion velocity $\hat{u}_i^k$ of the (k) *type molecule in the average fluid*

$$
\left.
\begin{aligned}
e^0 &= \sum Y^k \left(\frac{|\hat{u}^k|^2}{2} + \frac{\varepsilon^k}{m^k} \right), \\
\sigma_{ij}^0 &= -\sum \rho^k \overline{\hat{u}_i^k \hat{u}_j^k}, \\
q_j^0 &= \sum \rho^k \overline{\left(\frac{|\hat{u}^k|^2}{2} + \frac{\varepsilon^k}{m^k} \right) \hat{u}_j^k}.
\end{aligned}
\right\}
\tag{I.62}
$$

Moreover, analysis of Section 3 and comparison with the standard equations show how to construct thermodynamic quantities allocated to the *average fluid* with the velocity $\tilde{u}_i$, starting both from properties of the components considered in isolation and from the diffusion velocity of the (k) *component in the average fluid*

$$
\left.
\begin{aligned}
\tilde{e} &= \sum Y^k (e^k + \tfrac{1}{2} |\tilde{u}^k|^2), \\
\tilde{\sigma}_{ij} &= \sum \sigma_{ij}^k - \rho^k \tilde{u}_i^k \tilde{u}_j^k, \\
\tilde{q}_j &= \sum q_j^k + \rho^k (e^k + \tfrac{1}{2} |\tilde{u}^k|^2) \tilde{u}_j^k - \sigma_{ij}^k \tilde{u}_i^k.
\end{aligned}
\right\}
\tag{I.63}
$$

It is interesting to correlate the expressions (I.62) and (I.63) using the possible interpretation of e^k, σ_{ij}^k and q^k from the molecular model of each fluid component (k) taken in isolation.

If $u_i'^k$ is the velocity which a type (k) molecule possesses in excess (algebraically) in relation to the average velocity $\bar{u}_i^k$, which is itself calculated using an appropriate distribution function over the phase space, then

$$
u_i^k = \bar{u}_i^k + u_i'^k \quad \Rightarrow \quad \overline{u_i'^k} = 0.
\tag{I.64}
$$

In accordance with (I.62) written for the isolated species (k)

$$
\left.
\begin{aligned}
e^k &= \tfrac{1}{2} \overline{|u'^k|^2} + \frac{\overline{\varepsilon^k}}{m_k}, \\
\sigma_{ij}^k &= -\rho^k \overline{u_i'^k u_j'^k}, \\
q_j^k &= \overline{\left(\tfrac{1}{2} |u'^k|^2 + \frac{\varepsilon^k}{m^k} \right) u_j'^k}.
\end{aligned}
\right\}
\tag{I.65}
$$

If $\tilde{u}_i^k$ is now the average excess velocity of the (k) type relative to the "mixture" which is assumed to move with velocity $\tilde{u}_i = \sum Y^k \bar{u}_i^k$, then

$$
\tilde{u}_i^k = \bar{u}_i^k - \tilde{u}_i.
\tag{I.66}
$$

Then the individual velocity (I.64) of a (k) type molecule is written

$$u_i^k = \tilde{u}_i + \tilde{u}_i^k + u_i'^k,$$

so that, in accordance with (I.62) written this time for the mixture (with $\hat{u}_i^k = \tilde{u}_i^k + u_i'^k$)

$$\left. \begin{array}{l} e^0 = \sum Y^k \overline{\left(\frac{1}{2} |\tilde{\mathbf{u}}^k + \mathbf{u}'^k|^2 + \dfrac{\varepsilon^k}{m^k} \right)}, \\[3mm] \sigma_{ij}^0 = -\sum \rho^k \overline{(\tilde{u}_i^k + u_i'^k)(\tilde{u}_j^k + u_j'^k)}, \\[3mm] q_j^0 = \sum \rho^k \overline{\left(\frac{1}{2} |\tilde{\mathbf{u}}^k + \mathbf{u}'^k|^2 + \dfrac{\varepsilon^k}{m^k} \right)(\tilde{u}_j^k + u_j'^k)}. \end{array} \right\} \tag{I.67}$$

After expanding and taking into account (I.65) and $\overline{u_i'^k} = 0$ according to (I.64), these expressions are exactly those of (I.63) subject to $\tilde{u}_i^k$ having the same meaning, which is the equivalent of assuming that the single velocity u_i^k allotted to the species (k) in Section 3 is the average $\bar{u}_i^k$ in the Section 2 sense of the isolated k species.

This result shows clearly that the difference between the ways in which (I.62) and (I.63) are written is due to the nonidentity of the frames where *diffusion* is estimated. In the first analysis, $\hat{u}_i^k$ is the diffusion velocity of a type (k) molecule in relation to average fluid. In the second analysis, $\tilde{u}_i^k$ represents the diffusion velocity of the average (k) type molecule in average fluid. [In yet other terms, the set $(e^0, \sigma_{ij}^0, q_j^0)$ takes into account the effects of mixture which obviously do not appear in the sets $(e^k, \sigma_{ij}^k, q_j^k)$ relative to isolated k fluids, with the result that the simple summation of these fluids over k could not account for all of the fluxes which occur in the mixture.] It is nevertheless clear that the average value of the first is equal to the second; this explains the identity of the equations of diffusion (I.45, k) and (I.52, k) which are found, respectively, in the two models.

On the subject of these equations, it must be stressed that they make up a set of independent $K - 1$ equations $(\sum Y^k = 1)$ which, with (I.63), do not "close" the system of equations of motion of a mixture even if—and this is a very optimistic hypothesis—on the one hand, the relations of state and behavior of each component taken in isolation can be expressed, and on the other hand, the diffusion velocities $\tilde{u}_i^k$ and the chemical reaction rates P^k as a function of the variables of average fluid and/or their derivatives can be described. In fact, the balance of the equations and unknowns reveals a deficit in $(2K - 1)$ equations which is frequently, but not always justifiably, made good by the addition of:

- $(K - 1)$ equations of thermal equilibrium $T^1 = \cdot = T^k = \cdot = T^K \equiv T$; and
- K equations of ideal state of the mixture (the chemical potential of the k component in the mixture is equal to that which it would have in isolation in the same conditions of p pressure and T temperature).

These indications of the difficulty of the general problem of reactive fluid flows give some measure of the interest which attaches to particular methods adapted to current situations, where the chemical reactivity and/or diffusion is sensitive only in a restricted zone of flow. The chemical reactions which are "propagated by waves" are the subject of the following chapters.

CHAPTER II

Jump Relations in a Perfect Fluid

1. Properties of the Hugoniot Curve (h)

1.1. Introduction

To say that a perfect fluid is the seat of a *shock wave* is to say that it undergoes a discontinuous change in state which does not change its chemical identity (no new type of molecule is created), with the result that two scalar variables are sufficient to describe it throughout the flow. It is in these conditions, and assuming that the index 1 state is fixed, that we shall study the curve (h) called the *Hugoniot* curve (see [25]) defined as

$$e_2 - e_1 = \tfrac{1}{2}(p_1 + p_2)(v_1 - v_2). \tag{II.1}$$

At every point of (h), according to (I.30), the quantities

$$v_1 \left(\frac{p_2 - p_1}{v_1 - v_2} \right)^{1/2}, \tag{II.2}$$

$$v_2 \left(\frac{p_2 - p_1}{v_1 - v_2} \right)^{1/2}, \tag{II.3}$$

are, respectively, the normal velocities w_1 and w_2 *for a suitable orientation of the normal to the discontinuity surface.*

Among states 1 and 2, the one with weaker entropy is called *upstream state*; the other one is called *downstream state*; according to (I.26b) the upstream state is on the negative side of the normal, the downstream state is on the positive side, *for the orientation chosen above.* Throughout this century, two contradictory interpretations of the words upstream and downstream have been in circulation. One, in Jouguet [27], is based on a spatial image (shock, like water in a river, moves from upstream to downstream); the other, in Germain [21], for example, is based on a temporal image (in a given place, upstream and downstream are, respectively, before and after the passage of shock). This is why a definition independent of any reference to motion is preferred here.

The study of (h) for an arbitrary equation of state was carried out by Bethe in 1942 [4]. The main results are set out below.

1.2. General Properties of (h)

(a) Let us consider the transformation which would bring about the change from the state $(v_2, p_2) \in$ (h) to the state $(v_2 + dv_2, p_2 + dp_2) \in$ (h). The laws of thermodynamics and (II.1) and (II.2) allow the description

$$de_2 = \tfrac{1}{2}(v_1 - v_2)\, dp_2 - \tfrac{1}{2}(p_1 + p_2)\, dv_2,$$

$$2\frac{dw_1}{w_1} = \frac{dp_2}{p_2 - p_1} + \frac{dv_2}{v_1 - v_2}, \tag{II.4}$$

$$de_2 = -p_2\, dv_2 + T_2\, ds_2.$$

Eliminating dv_2, dp_2, v_2, and p_2 from (II.2), (II.3), and (II.4) leads to

$$T_2\, ds_2 = (w_2 - w_1)^2 \frac{dw_1}{w_1}, \tag{II.5}$$

which shows that, along (h), s_2 is an increasing monotonic function of w_1.

(b) Let us consider a state 2 adjacent to state 1 and defined by

$$v_2 = v_1 + \Delta v, \qquad s_2 = s_1 + \Delta s.$$

Taking p_2 and e_2 as functions of s_2 and v_2, gives

$$\frac{p_1 + p_2}{2} = p_1 + \frac{1}{2}\left(\frac{\partial p}{\partial v}\right)_1 \Delta v + \frac{1}{4}\left(\frac{\partial^2 p}{\partial v^2}\right)_1 (\Delta v)^2 + \cdots, \tag{II.6}$$

$$e_2 - e_1 = \left(\frac{\partial e}{\partial v}\right)_1 \Delta v + \frac{1}{2}\left(\frac{\partial^2 e}{\partial v^2}\right)_1 (\Delta v)^2$$

$$+ \frac{1}{6}\left(\frac{\partial^3 e}{\partial v^3}\right)_1 (\Delta v)^3 \cdots + \left(\frac{\partial e}{\partial s}\right)_1 \Delta s + \cdots. \tag{II.7}$$

Taking into account identities $p = -\partial e/\partial v(v, s)$ and $T = \partial e/\partial s$ (II.7) can be written

$$\frac{e_2 - e_1}{v_1 - v_2} = p_1 + \frac{1}{2}\left(\frac{\partial p}{\partial v}\right)_1 \Delta v + \frac{1}{6}\left(\frac{\partial^2 p}{\partial v^2}\right)_1 (\Delta v)^2 + \cdots - T_1 \frac{\Delta s}{\Delta v} + \cdots. \tag{II.8}$$

Putting (II.6) and (II.8) together with the Hugoniot relation given in (II.1) gives

$$T_1\, \Delta s = -\frac{1}{12}\left(\frac{\partial^2 p}{\partial v^2}\right)_1 (\Delta v)^3 + 0(\Delta v)^3, \tag{II.9}$$

from which it follows that:

- (h) and the isentrope of state 1 have a contact of order 3 at the point (v_1, s_1); and

- w_1 is equal to the speed of sound a_1 at point (v_1, s_1).

(c) From properties (a) and (b) above, there derives immediately the fundamental law:

> - upstream, the normal velocity is greater than the speed of sound; and
> - downstream, the normal velocity is less than the speed of sound.

In fact, imagine moving along the curve (h) from point (v_1, s_1) in the direction of the increasing entropies. Compared with these states $s > s_1$, the state (v_1, s_1) is necessarily the upstream state, and the velocity w_1 is the normal upstream velocity. This velocity—which takes the value a_1 when $s = s_1 + 0$ according to (b) above—increases when s grows from s_1 according to (a) above. From this, the first part of the conclusion is reached; the second part is reached in a similar way by moving along (h) in the direction of the decreasing entropies.

1.3. Properties if $(\partial^2 p/\partial v^2)(v, s) > 0$ and $v(\partial p/\partial e)(v, e) > -2$

Provided that the fluid satisfies the above thermodynamic inequalities, Bethe demonstrated (see [4]) that (h) is a continuous curve $v(s)$ in the plane (v, s) (see Fig. II.1):

— defined for every value of s greater than or equal to a value $\tilde{s} < s_1$;
— so that $\tilde{v} \equiv v(\tilde{s}) > v_1$;
— so that $v(+\infty) = v_1/4$; and

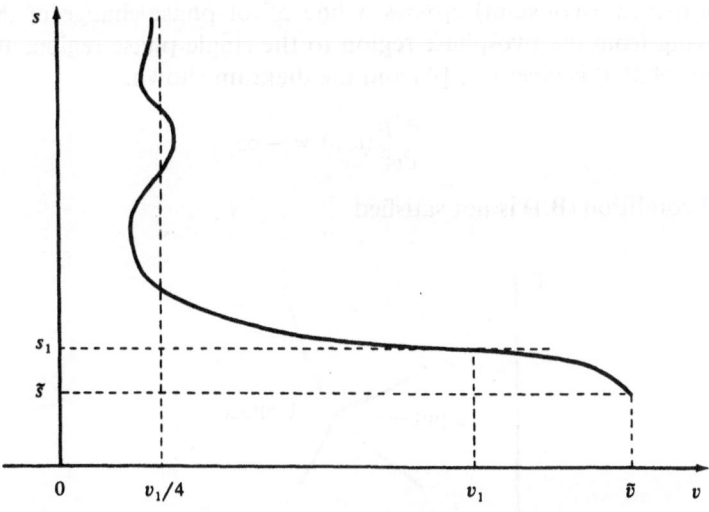

Figure II.1. Projection of the Hugoniot curve (h) on the plane (v, s).

— finally, so that:

a single value of s less than s_1 corresponds to $v \in [v_1, \tilde{v}]$;
$2n + 1$ ($n \geq 0$) values of s greater than s_1 correspond to $v \in [v_{1/4}, v_1]$; and
$2n + 2$ ($n \geq 0$) values of s greater than s_1 correspond to $v < v_1/4$.

In particular, these results show that a state of (h) with an entropy greater than s_1 is necessarily of a lesser mass volume that v_1, which can be expressed

> The downstream state is more dense than the upstream state.

None of these conclusions holds if one or other of the inequalities

$$\frac{\partial^2 p}{\partial v^2}(v, s) > 0 \qquad \text{(B.1)},$$

$$v\frac{\partial p}{\partial e}(v, e) > -2 \quad \text{(B.2)},$$

$$\text{(II.10)}$$

is not satisfied for $s \geq \tilde{s}, v \leq \tilde{v}$. A complete examination of these conditions is difficult. It is, nonetheless, possible to give the following simple results:

(a) Equation (II.10) is satisfied by an ideal gas with constant γ since

$$\frac{\partial^2 p}{\partial v^2}(v, s) = \gamma(\gamma + 1)\frac{p^2}{v} > 0,$$

$$v\frac{\partial p}{\partial e}(v, e) = \gamma - 1 > 0.$$

(b) The speed of sound undergoes a positive jump when an *isentropic dilatation* ($v \nearrow$, s = constant) crosses a line $\mathscr{L}$ of phase-change of the fluid, moving from the two-phase region to the single phase region; at such a point of $\mathscr{L}$ this gives (see [4] and the diagram shown.)

$$\frac{\partial^2 p}{\partial v^2}(v, s) = -\infty$$

and condition (B.1) is not satisfied.

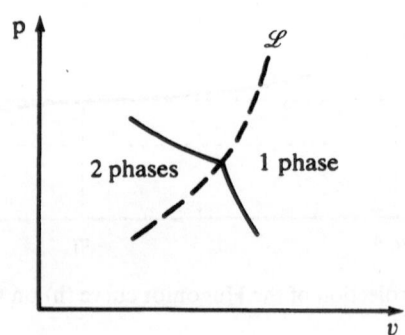

(c) the Grüneisen coefficient $v(\partial p/\partial e)(v, e)$ is of the same sign as the dilatation coefficient at constant pressure because of the thermodynamic identity

$$v\frac{\partial p}{\partial e}(v, e) = \frac{v}{T}\frac{c_p - c_v}{c_v}\left[\frac{\partial v}{\partial T}(p, T)\right]^{-1};\qquad (\text{II.11})$$

reached from the identities (*i*.9), (*i*.3), and (*i*.2) of Appendix A. Thus condition (B.2) will hold except in a region where the fluid contracts when T increases (e.g., H_2O between $0°$ and $4°$ C at normal pressure).

1.4. Properties if, in Addition, $(\partial p/\partial v)(v, e) < 0$

If it is accepted that the fluid responds not only to the inequalities (B.1) and (B.2), but also to

$$\frac{\partial p}{\partial v}(v, e) < 0 \quad (\text{B.3}),\qquad (\text{II.12})$$

it can be shown that (see [4]) along (h):

(a) e is an increasing monotonic function of $s \in [\tilde{s}, +\infty]$;
(b) p is an increasing monotonic function of $s \in [s_1, +\infty]$; and
(c) p remains less than p_1 for $s < s_1$.

Using the above reasoning for the mass volume, it can be seen that

Pressure in the downstream state is greater than in the upstream state.

As with conditions (B.1) and (B.2), a complete examination of condition (B.3) is difficult. Any mention here is limited to pointing out that it is satisfied by an ideal gas with constant γ since

$$\frac{\partial p}{\partial v}(v, e) = -\frac{p}{e} < 0.\qquad (\text{II.13})$$

1.5. Conclusion

Assume that the upstream state is given as well as the normal upstream velocity—which is necessarily greater than the upstream sound velocity. Now consider the downstream entropy; because it is linked to the normal upstream velocity by an increasing monotonic relation owing to the result demonstrated in §1.2(a), it is unambiguously determined by the initial data. Mass volume downstream is also determined according to §1.3; as are pressure and internal mass energy downstream according to §1.4. This gives the law:

If the three conditions (B.1), (B.2), and (B.3) are satisfied, giving the upstream state and the normal upstream velocity determines the downstream state unambiguously.

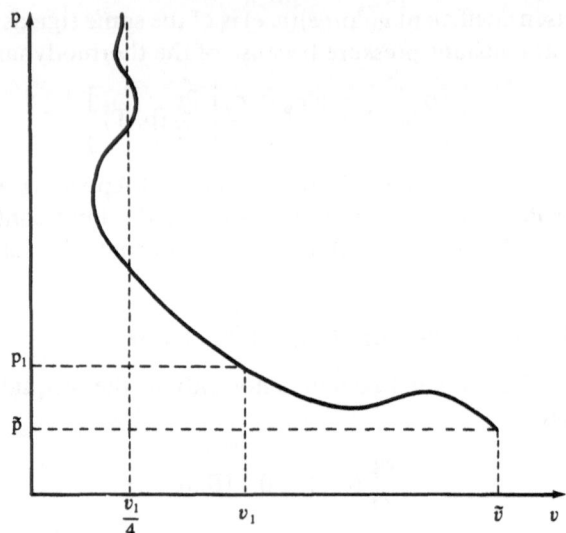

Figure II.2. Projection of the Hugoniot (h) on the plane (v, p).

Taking account of the geometric interpretation which the relation (II.2) makes possible for the normal upstream velocity, this gives the equivalent expression:

The straight line of the plane (v, p) starting from (v_1, p_1) with slope $-(\rho_1 w_1)^2$ which is called the Rayleigh line [33], cuts the projection of (h) on the plane (v, p) at only one point.

Figure II.2 illustrates the behavior of (h) in the plane (v, p) under Bethe's conditions. Since it will be required later in §3.1, we note that the tangent at (v_1, p_1) for the projection of (h) in the plane (v, p), in relation to this projection, will be found on the side opposite the point $(v = v_1, p = +\infty)$.

2. Properties of the Crussard Curve (H)

2.1. Introduction

To say that a perfect fluid is the seat of a *reactive wave* is to say that it undergoes a discontinuous change in its physical state and its chemical identity (new molecules appear), with the result that two scalar variables are no longer sufficient to describe it throughout the flow. *A priori* we might think of introducing, besides the two-dimensional vector **q** which describes the physical state, the Λ-dimensional vector N which represents the concentrations of the Λ chemical species present throughout the flow. However, in the case where it is supposed that the wave separates a medium of composition

N_0 called *upstream* and a medium in chemical equilibrium $\tilde{N}(\mathbf{q})$ called *downstream*, the representation of the fluid can be simplified by generalizing Von Neumann's notion of degree of chemical evolution, or, more precisely, by supposing that there exists a variable of state m such that

$$N = N_0 + m[\tilde{N}(\mathbf{q}) - N_0].$$

In this hypothesis, the reactive wave represents a discontinuity of $(\mathbf{q}, m)$, with m being equal to 0 on the upstream side and 1 on the downstream side, and the domain $m = 0$ decreasing with time.

Let $(\mathscr{D})$ be the surface of the chemical equilibrium states of the *reaction products* of a reactive substance. The *Crussard curve* (H) of this substance is the curve defined on $(\mathscr{D})$ by the equation (see [14])

$$H \equiv e - e_0 - \tfrac{1}{2}(p + p_0)(v_0 - v) = 0, \tag{II.14}$$

where the index 0 refers to the reactive substance itself, whilst the state with no index refers to the reaction products. At every point of (H), the quantities M, w_0, and w are defined by

$$M = \rho_0 w_0 = \rho w = \sqrt{\frac{p - p_0}{v_0 - v}}, \tag{II.15}$$

for a normal $\mathbf{N}$ to $\sum$ oriented from the upstream to the downstream state.

The *detonation* arc of (H) is the arc (H_+) where $p \geq p_0$ and M is real, therefore, where $v \leq v_0$; the *deflagration* arc is the arc (H_-) where $p \leq p_0$ and M is real, therefore, where $v \geq v_0$.

We shall study (H), assuming, as Hayes implicitly assumed (see [23], p. 433), that the reaction products confirm

$$\left.\begin{array}{ll} \dfrac{\partial^2 p}{\partial v^2}(v, s) > 0, & \text{(a)} \\[2mm] \dfrac{\partial p}{\partial e}(v, e) > 0, & \text{(b)} \\[2mm] \dfrac{\partial p}{\partial v}(v, e) < 0. & \text{(c)} \end{array}\right\} \tag{II.16}$$

2.2. Differential Relations on (H)

Let us consider an elementary transformation of the reaction products along (H). In accordance with (I.23), (II.14), and (II.15), we can write

$$de = -p\, dv + T\, ds,$$

$$de = \tfrac{1}{2}(v_0 - v)\, dp - \tfrac{1}{2}(p + p_0)\, dv, \tag{II.17}$$

$$2\frac{dM}{M} = \frac{dp}{p - p_0} + \frac{dv}{v_0 - v}.$$

Through appropriate eliminations, and using the equations (II.15) judiciously, from (II.17) two differential relations can be deduced

$$\frac{ds}{dv} = \frac{(w - w_0)^2}{M}\frac{dM}{dv} = \tfrac{1}{2}(v_0 - v)\left(\frac{dp}{dv} + M^2\right). \tag{II.18}$$

Moreover, by considering the pressure p as a function of v and s, and by using the thermodynamic identities (e.2), (i.11), and (i.14) in Appendix A, it can be shown that the first two equations (II.17) bring about another useful relation in the two forms given below.

$$\left.\begin{array}{c}\dfrac{dp}{dv} = \dfrac{\partial p}{\partial v}(v, s)\dfrac{1 - \dfrac{1}{2}\left(1 - \dfrac{p_0}{p}\right)\left[1 + \dfrac{v^2}{a^2}\dfrac{\partial p}{\partial v}(v, e)\right]}{1 + \dfrac{1}{2}\left(1 - \dfrac{v_0}{v}\right)G}, \quad \text{(a)} \\[3em] \dfrac{\partial p}{\partial v}(v, s) + M^2 = \left(\dfrac{dp}{dv} + M^2\right)\left[1 + \dfrac{1}{2}\left(1 - \dfrac{v_0}{v}\right)G\right]. \quad \text{(b)}\end{array}\right\} \tag{II.19}$$

2.3. Chapman–Jouguet Points on (H)

A Chapman–Jouguet point is the name given to every point such that (see [8], [27])

$$w = a, \tag{II.20}$$

which, according to the definition (II.15) of M^2, is equivalent to

$$M^2 + \frac{\partial p}{\partial v}(v, s) = 0. \tag{II.21}$$

A Chapman–Jouguet (in short C–J) point possesses many properties of which one possible presentation is the following:

(i) According to (II.19b), a C–J point is a point of (H) where

$$\frac{dp}{dv} = \frac{\partial p}{\partial v}(v, s) = -M^2 = \frac{p - p_0}{v - v_0} < 0, \tag{II.22}$$

from which it emerges notably that the C–J points of (H) can be defined in the plane (v, p) as the points of contact of the tangents starting from (v_0, p_0) to the projection of (H).

(ii) According to (II.18), a C–J point is a point of (H) where s and M are both maximum or both minimum.

(iii) At a C–J point, differentiating (II.18) and (II.19b) along (H), taking into account (II.22) and the extremum properties of M and s, yields

$$T\frac{d^2s}{dv^2} = \frac{(w - w_0)^2}{M}\frac{d^2M}{dv^2} = \tfrac{1}{2}(v_0 - v)\frac{d^2p}{dv^2} = \tfrac{1}{2}(v_0 - v)\frac{\dfrac{\partial^2 p}{\partial v^2}(v, s)}{1 + \dfrac{1}{2}\left(1 - \dfrac{v_0}{v}\right)G}. \tag{II.23}$$

(iv) Using the inequality (II.16c) in (II.19a), and then the inequality (II.16a) in (II.23), it can be demonstrated successively that a C–J detonation confirms

$$\left. \begin{array}{c} 1 + \dfrac{1}{2}\left(1 - \dfrac{v_0}{v}\right) G > 0, \quad \text{(a)} \\[3mm] \dfrac{d^2 p}{dv^2} > 0. \quad \text{(b)} \end{array} \right\} \tag{II.24}$$

2.4. Hugoniot Curves over $(\mathscr{D})$

Subsequent expansions show the usefulness of introducing the Hugoniot curve (h^D) of a point $D \in (\mathscr{D})$ defined by

$$h^D \equiv e - e_D - \tfrac{1}{2}(p + p_D)(v_D - v) = 0, \tag{II.25}$$

and the curve R_μ^D defined on $(\mathscr{D})$ by

$$R_\mu^D \equiv p - p_D + \mu^2(v_D - v) = 0, \qquad \mu > 0, \tag{II.26}$$

which is projected on the plane (v, p) according to Rayleigh's line at the point D for the upstream normal relative velocity $w_1 = \mu/\rho_D$.

Let us apply the conclusion of §1.5 to (h^D) and (R_μ^D): (h^D) intersects (R_μ^D), at most, once, at a distinct point D' or coincidental with D according to the modes in Table II.1.

In particular, when $\mu \to \infty$, the point D' tends toward the point at infinity $(v = v_D/4 - 0, p = +\infty)$.

Table II.1. Sections (1)–(4) refer to Figure II.3.

(1) $\mu > (\rho a)_D$	Point D' exists and confirms $(s, \rho, e, p)_{D'} > (s, \rho, e, p)_D; \mu < (\rho a)_{D'}$
(2) $\mu = (\rho a)_D$	Point D' coincides with D
(3) $\tilde{\mu} < \mu < (\rho a)_D$	Point D' exists and confirms $(s, \rho, e, p)_{D'} < (s, \rho, e, p)_D; \mu > (\rho a)_{D'}$
(4) $\mu < \tilde{\mu}$	Point D' does not exist

2.5. Point at Infinity

It is convenient to exclude immediately the case with no physical significance where the reaction products, taken in the conditions (v_0, p_0), would have the same specific internal energy as the initial substance; from now on, then, with A designating the point of $(\mathscr{D})$ defined by $v_A = v_0$ and $p_A = p_0$, it is supposed that

$$e_A - e_0 \neq 0, \tag{II.27}$$

which is the equivalent of saying that A does not belong to (H).

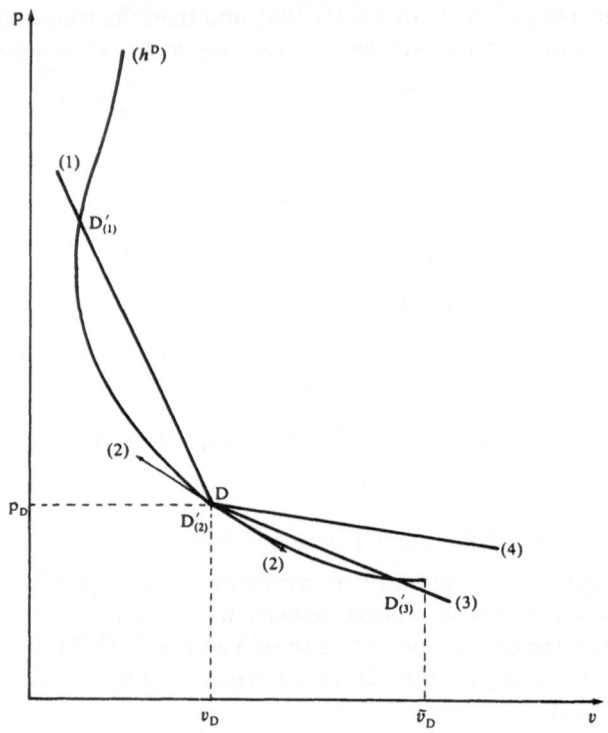

Figure II.3. The four possible configurations for the intersection, on surface $\mathscr{D}$, of the Hugoniot curve (h^D) and the (R_μ^D) curve, projected on the plane (v, p).

Let us consider the current point a of (h^A) and its analytic power $H(a)$ with respect to (H). The equations of (H) and (h^A) lead to the expression

$$H(a) = e_A - e_0. \tag{II.28}$$

Let us consider point B of (H) where $v = v_0$, the current point b of (h^B), and its analytic power $H(b)$ with respect to (H). The equations of (H) and (h^B) lead to the expression

$$H(b) = \tfrac{1}{2}(\mathrm{p}_B - \mathrm{p}_0)(v_0 - v_b). \tag{II.29}$$

Notice that condition (II.16b) implies that point B confirms

$$(\mathrm{p}_B - \mathrm{p}_0)(e_A - e_0) < 0. \tag{II.30}$$

Relations (II.28), (II.29), and (II.30) make it possible to specify the detonation arc (H_+). In fact, for $v < v_0$, $H(b)$ is the same sign as $\mathrm{p}_B - \mathrm{p}_0$, $H(a)$ is the same sign as $e_A - e_0$, and these two quantities are of opposite sign. Therefore (H_+) is included between (h^A) and (h^B). In particular, (H_+) goes through the point $(v = v_0/4 - 0, \mathrm{p} = +\infty)$ (see Fig. II.4).

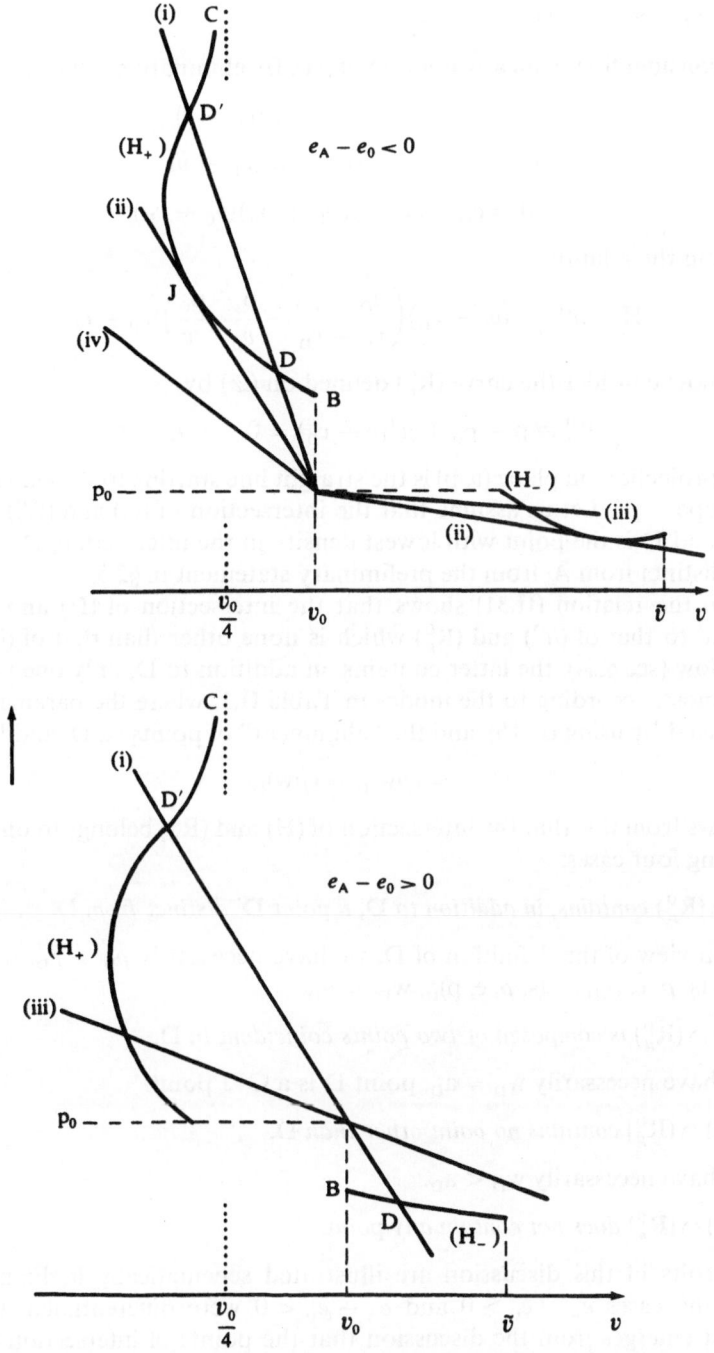

Figure II.4. The possible configurations for the intersection, on surface $\mathscr{D}$, of the Crussard curve ($H = H_+ \cup H_-$) and the "Rayleigh line" (R_μ^0) projected on plane (v, p).

2.6. Intersection of (H) and (R_μ^0)

Let us consider first of all any point D of (H). By eliminating e and e_D between

$$H = e - e_0 - \tfrac{1}{2}(p + p_0)(v_0 - v),$$

$$h^D = e - e_D - \tfrac{1}{2}(p + p_D)(v_D - v),$$

$$0 = e_D - e_0 - \tfrac{1}{2}(p_D + p_0)(v_0 - v_D),$$

we obtain the relation

$$H - h^D = \tfrac{1}{2}(v_0 - v_D)\left(\frac{p_D - p_0}{v_0 - v_D} \cdot \frac{p - p_0}{v_0 - v}\right)(v_0 - v). \qquad (II.31)$$

Let us now consider the curve (R_μ^0) defined on $(\mathscr{D})$ by

$$R_\mu^0 \equiv p - p_0 + \mu^2(v - v_0) = 0, \qquad \mu > 0,$$

whose projection on plane (v, p) is the straight line starting from point (v_0, p_0) with slope $-\mu^2$. Let us assume that the intersection of (H) and (R_μ^0) is nonempty and D is the point with lowest density in the intersection; D is necessarily distinct from A, from the preliminary statement in §2.5.

Then the relation (II.31) shows that the intersection of (H) and (R_μ^0) is identical to that of (h^D) and (R_μ^0) which is none other than that of (h^D) and (R_μ^D). Now (see §2.4), the latter contains, in addition to D, only one point D' at the most, according to the modes in Table II.1, where the parameter μ is interpreted by using (II.15), and the "alignment" of points A, D, and D' by

$$\mu = (\rho w)_D = (\rho w)_{D'}. \qquad (II.32)$$

It follows from this that the intersection of (H) and (R_μ^0) belongs to one of the following four cases:

(i) $H \cap (R_\mu^0)$ *contains, in addition to* D, *a point* D' *distinct from* D.

Thus, in view of the definition of D, we have necessarily $\rho_{D'} > \rho_D$, $w_D > a_D$, so also $(s, \rho, e, p)_{D'} > (s, \rho, e, p)_D$, $w_{D'} < a_{D'}$.

(ii) $(H) \cap (R_\mu^0)$ *is composed of two points coincident in* D.

So we have necessarily $w_D = a_D$; point D is a C–J point.

(iii) $(H) \cap (R_\mu^0)$ *contains no point other than* D.

So we have necessarily $w_D < a_D$.

(iv) $(H) \cap (R_\mu^0)$ *does not contain any point.*

The results of this discussion are illustrated schematically in Figure II.4, where the cases $e_A - e_0 > 0$ and $e_A - e_0 < 0$ were differentiated. In both cases, it emerges from the discussion that the points of intersection (v_0, p_B) and $C(v_0/4 - 0, +\infty)$ of (H) and (R_μ^0) for $\mu = +\infty$ are such that

$$w_B/a_B > 1, \qquad w_C/a_C < 1. \qquad (II.33)$$

3. Considerations Specific to Detonations

The above study of the properties of the Crussard curve has not yet brought out the essential difference between deflagrations and detonations. However, experimental evidence strongly suggests that separate mechanisms govern the two types of propagation and suggests that they be examined separately. That is why we are concerned in what follows to specify the detonation arc (H_+); in particular, we examine the existence and uniqueness of the (C–J) detonation and the role which it may play in the interpretation of observed detonations.

3.1. Properties of the Detonation Arc (H_+)

Intersection Points of (H) and (R_μ^0)

The intersection points of (H) and (R_μ^0) can be defined by the solutions in $v(\mu)$, $p(\mu)$ of the system

$$H \equiv e - e_0 - \tfrac{1}{2}(p + p_0)(v_0 - v) = 0,$$
$$R_\mu^0 \equiv p - p_0 + \mu^2(v - v_0) = 0, \qquad \mu > 0. \tag{II.34}$$

Using the identities $(j.2)$, $(j.3)$, and $(j.5)$ in Appendix A, it is shown that every point of (R_μ^0) confirms

$$\Delta \equiv \frac{D(R_\mu^0, H)}{D(v, p)} = \mu^2 \frac{v}{G}\left(1 - \frac{a^2}{w^2}\right). \tag{II.35}$$

Let us consider the solution $(\bar{v}(\mu), \bar{p}(\mu))$ of (II.34) which tends towards $C(v_0/4 - 0, +\infty)$ when μ decreases from $+\infty$, and let $\bar{N}$ be the corresponding point on (H). When μ decreases from $+\infty$, the point $\bar{N}$, which is situated on (R_μ^0) but cannot be identical to A since $e_A - e_0 \neq 0$, remains in the domain $(v < v_0, p \geq p_0)$ and therefore on (H_+). Moreover, according to (II.35), the arc (H_+) described by $\bar{N}$ can:

— either remain defined for $\mu \in [0, +\infty]$ if $\Delta(\bar{N})$ itself remains strictly negative over this interval;
— or cease to be so defined if $\Delta(\bar{N})$ is nullified for $\mu = \bar{\mu}$ and stops at a C–J point.

In other words, a necessary and sufficient condition of existence of at least one C–J detonation is that $(\bar{H}_+)$ should have no point such that $p = p_0$.

The intersection of (H) and $p = p_0$ is defined by

$$f(v) \equiv e(v, p_0, 1) - e(v_0, p_0, 0) - p_0(v_0 - v) = 0.$$

According to the identity $(j.3)$ of Appendix A, we have

$$f'(v) = \frac{\partial e}{\partial v}(v, p_0, 1) + p_0 = \left[T\frac{c_p c_v}{c_p - c_v}\frac{\partial p}{\partial e}(v, e, 1)\right]_{p=p_0}.$$

According to (II.16b), $f'(v)$ is positive with the result that the maximum of $f(v)$ in the band $v < v_0$ is $f(v_0) = e_A - e_0$. Consequently, in order that *at least one* C–J *detonation* should exist, it is necessary and sufficient that $e_A - e_0 < 0$, that is to say

$$e(v_0, p_0, 1) < e(v_0, p_0, 0). \tag{II.36}$$

From this point on, we operate in the context of the conditions of inequality (II.36).

Uniqueness of the C–J Detonation

According to (II.30) and (II.36) point B belongs to (H_+). Let us consider the solution $(\underline{v}(\mu), \underline{p}(\mu))$ of (II.34) which tends towards (v_0, p_B) when $\mu \to +\infty$; let

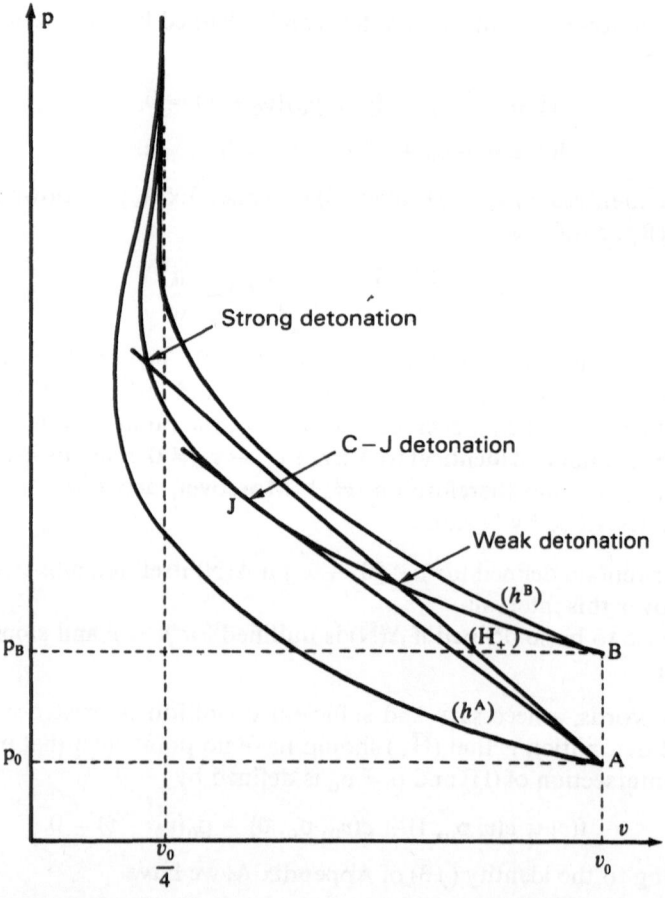

Figure II.5. Positive of the detonation arc (H_+) of the Crussard curve projected on plane (v, p) in the case $e_A < e_0$.

$\underline{N}$ be the corresponding point on (H). When μ decreases from $+\infty$, point $\underline{N}$, which is situated on (R_μ^0) but cannot be identical to A since $e_A - e_0 < 0$, remains in the domain $(v < v_0, p \geq p_0)$ and so on (H_+). The arc $(\underline{H}_+)$ which it describes cannot, according to §3.1(a), end at $p = p_0$. So it ends at a C–J point for a value $\underline{\mu}$ of μ. The two cases $\mu < \underline{\mu}$ and $\mu > \underline{\mu}$ both being incompatible with the results of the discussion in Section 2, we conclude $\mu = \underline{\mu}$. In other words, the arcs $(\overline{H}_+)$ and $(\underline{H}_+)$ come together at a point J of C–J detonation for a value μ_* of μ.

Let us suppose that there exists a point N_α of (H_+) which belongs to neither $(\overline{H}_+)$ nor $(\underline{H}_+)$. According to the discussion in §2.6, the corresponding value μ_α of μ would necessarily be strictly less than μ_* and the solution of (II.34), which goes through N_α, would cease to exist at a C–J point for $\mu = \mu_\beta \in]\mu_\alpha, \mu_*[$. This C–J point would necessarily be such that $d^2p/dv^2 < 0$, which is impossible according to (II.24b).

Consequently, $[H_+]$ is the union of $(\overline{H}_+)$ and $(\underline{H}_+)$ and point J is the *only* C–J detonation point.

From what goes before, it can also be held that $w < a$ on the arc [CJ] (strong detonations), and $w > a$ on the arc [BJ] (weak detonations).

Finally, μ_* being the smallest value of μ for which (R_μ^0) cuts (H_+), and also therefore the smallest value of M on (H_+), it can be concluded that the value D_* of w_0 to which it corresponds by (II.15) is the smallest value which w_0 can have on (H_+).

Sufficient Condition for All Detonations To Be Supersonic

Let us go back to the notations of §2.4 and §2.6

$$h^A(D) = e_D - e_A - \tfrac{1}{2}(p_A + p_D)(v_0 - v_D),$$

where by definition of D

$$0 = e_D - e_0 - \tfrac{1}{2}(p_D + p_0)(v_0 - v_D),$$

from which

$$h^A(D) = e_0 - e_A > 0.$$

Furthermore, let us consider the straight line $(R_\mu^0; \mu = \rho_0 a_A)$ and its current point R. According to the final remark of §1.5, we have

$$h^A(R) \leq 0.$$

By comparing the above inequalities, it can be concluded that (H) and $(R_\mu^0; \mu = \rho_0 a_A)$ share no common point. Therefore $\rho_0 a_A < \mu_*$, or,

$$a(v_0, p_0, 1) < D_*. \tag{II.37}$$

The condition

$$a(v_0, p_0, 1) > a(v_0, p_0, 0) \tag{II.38}$$

is therefore *sufficient* for D_* to be greater than the velocity of sound $a(v_0, p_0, 0)$ in the initial reactive substance, and also for every detonation to be supersonic relative to the upstream state.

The inequality (II.38) is interesting in that it draws attention to the eventuality of a nonsupersonic detonation in an environment having a fairly high velocity $a(v_0, p_0, 0)$. However, we shall ignore this type of propagation, and from now on consider only detonation in *explosives*, defined as substances capable of decomposing in accordance with the inequalities (II.36) and (II.38).

3.2. Jouguet's Conjecture

Apart from some slight differences, the theory which has just been mentioned is that which Jouguet developed—to a much lower level of generality—over sixty years ago. He used it to interpret the experimental results of the time and concluded by considering as *plausible* (see [7], p. 327) that all detonations at constant velocity are represented by the point J of (H_+): this is the so-called "Chapman–Jouguet (C–J) condition."

But as the accuracy of measurements has increased greatly since the beginning of the century, what might have appeared to Jouguet to be a difference due to experimental error today appears to contradict his conjecture. Furthermore, the field of investigation has been extended to detonations of nonconstant velocity, so giving an additional dimension to the problem of interpreting observed detonations.

A survey of current experimental results is given below, followed by some elements which introduce Chapter III.

3.3. Current Experimental Results

Figure II.6 recapitulates certain of the results presented in [5]. For the gaseous mixture $C_3H_8 + 5O_2$, liquid commercial nitromethane, and a solid cyclonite (RDX)-based composition called F.209, respectively, they give the measured value of the detonation velocity D as a function of a geometric parameter whose significance varies according to the propagation geometry. This parameter is:

- the hydraulic diameter d of the tube or cartridge for the propagations called plane ($\Sigma \simeq$ plane surface); and
- the radius X of the wave for propagations called spherical ($\Sigma \simeq$ spherical surface).

Only three observations are made here on these results:

- the velocity D^p of the plane waves can be approximated by a linear relation in $1/d$ of the form

$$D^p = D^p_\infty \left(1 - \frac{a}{d}\right), \qquad a > 0, \tag{II.38'}$$

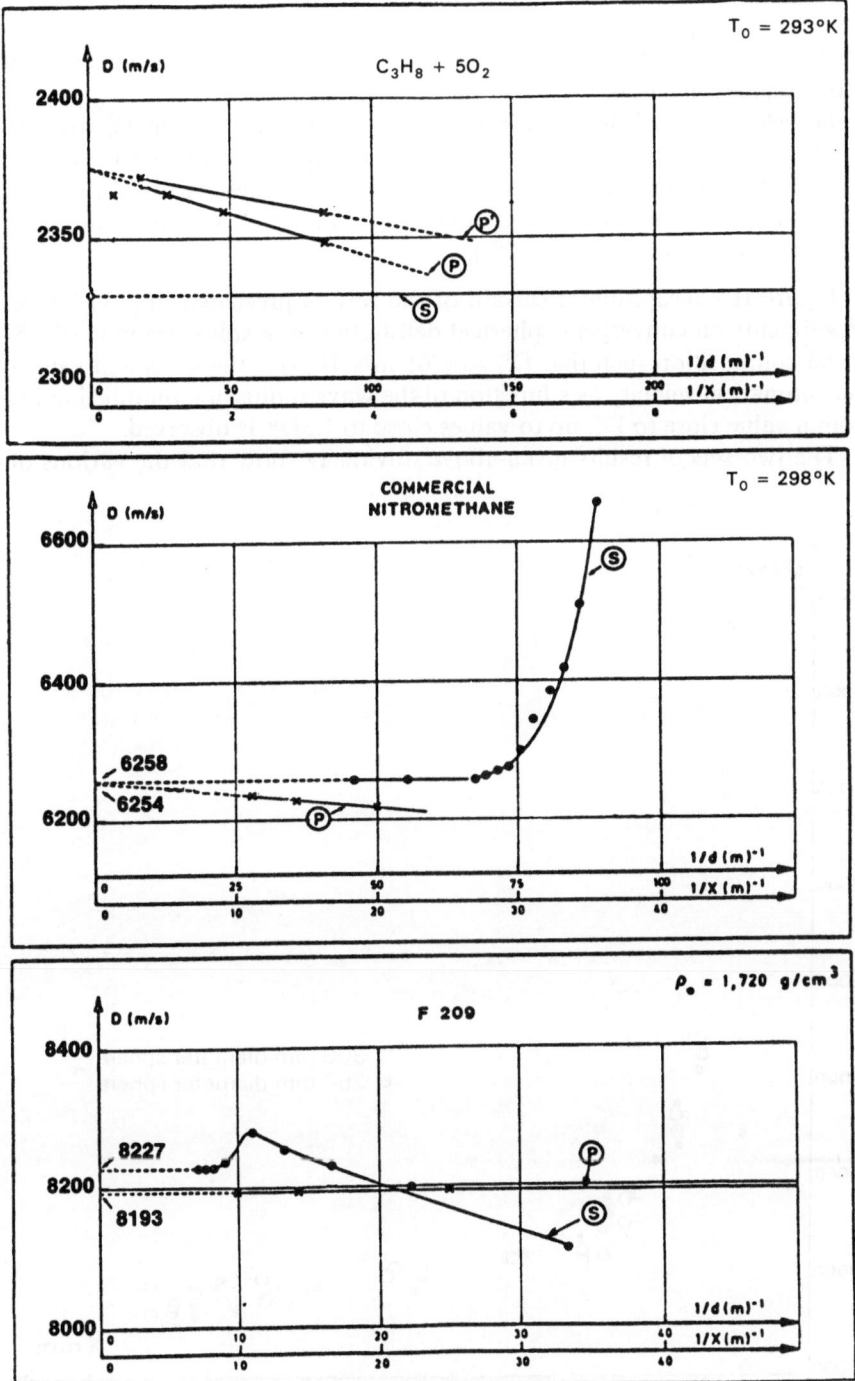

Figure II.6. Plane Ⓟ and spherical Ⓢ detonation velocities D in the stoichiometric propane–oxygen mixture (gas), in commercial nitromethane (liquid) and in the cyclonite (RDX)-based composition F.209 (solid).

where *a* is a length whose value depends on both the reactive medium and the adjacent medium;

- the velocity D^S of the spherical waves tends toward a limit D_∞^S by non-inferior values when the inverse of the radius tends toward zero; and
- for each of the media considered, the measured values are distributed over an interval about ten times greater than that which would result simply from experimental spread (0.2%).

Figure II.7 recapitulates certain of the results presented in [13] for two experiments on convergent spherical detonation in a solid cyclonite (RDX)-based composition such that $D_\infty^p = 8765$ m/s. It gives the measured value of the detonation velocity as a function of the wave radius: a montonic increase from a value close to D_∞^p up to values close to $1.3D_\infty^p$ is observed.

The two sets of results given above obviously show that the various ob-

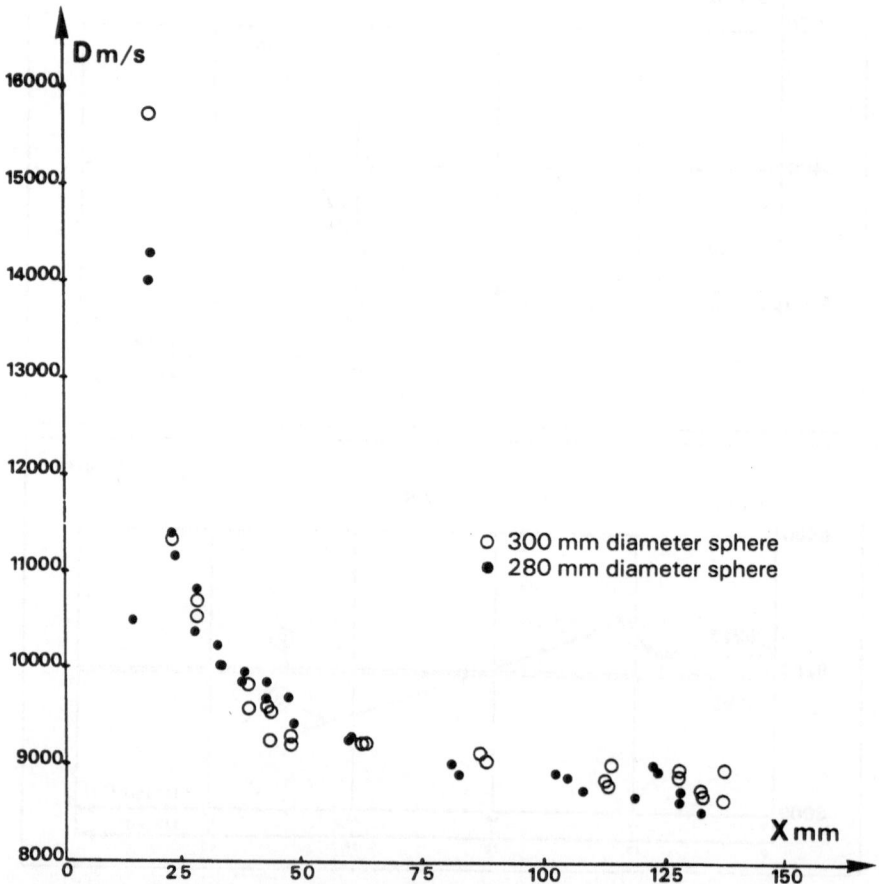

Figure II.7. Velocity D of convergent spherical detonation in a cyclonite (RDX)-based solid explosive ($D_\infty^p = 8765$ m/s) as a function of the radius X of the wave.

served detonations are too diverse to be represented by the single point J on the arc (H_+).

3.4. Interpretation of Observed Detonations

What can we do to be more specific about the description of phenomena if we are not prepared to make do with the Chapman–Jouguet hypothesis? The question is not new. Attempts were made from 1945 to 1950 to account for the variation of D as a function of $1/d$ and the existence of values less than the extrapolated limiting value D_∞^p. Although they were led to a paradox insofar as D_* is the lowest possible velocity on (H_+) (see §2.1), many authors had no hesitation in identifying D_∞^p with the C–J velocity in (II.38'). In doing this, they abandoned—although still not consistently—considering as satisfactory a model where physical and chemical conversions are reduced to a discontinuity.

A decisive advance consists of adapting for detonations the perturbation method applied to shocks by Weyl [37], then by Germain and Guiraud [22]. The account of this method and its results constitute the subject matter of the following chapter which restates and extends the principal idea presented by the author in [19], [20], and [21].

4. Particular Case of a Polytropic Fluid

4.1. Definition and General Properties

The *pressure* of a fluid as the partial derivative of the mass density $e(v, s)$ of internal energy in relation to the mass volume v was introduced above. Physical analysis of the law of state $e(v, s)$ leads, according to the domain of variation of v and s considered, to different models; the *polytropic* model consists of supposing a "factorization" of the form

$$e(v, s) = \frac{pv}{\bar{\gamma} - 1}, \tag{II.39}$$

where $\bar{\gamma}$ is a constant greater than 1, which ensures zero energy in infinite dilutions. In fact, by integration on v, it can be deduced at once that

$$e(v, s) = \frac{A(s)}{\bar{\gamma} - 1} v^{1-\bar{\gamma}}, \tag{II.40}$$

which makes it possible to give p the form

$$p(v, s) = A(s) \cdot v^{-\bar{\gamma}}. \tag{II.41}$$

This definition calls for some important observations:

(i) Taking account of the well-known definitions of the speed of sound, $a^2 =$

$(\partial p/\partial p)(\rho, s)$ and of the mass enthalpy $h = e + pv$, we have

$$\left.\begin{aligned} a^2 &= \tilde{\gamma}pv, \\ h &= \frac{a^2}{\tilde{\gamma} - 1}. \end{aligned}\right\} \tag{II.42}$$

(ii) The entropy s depends only on the product $pv^{\tilde{\gamma}}$, with the result that the isentropes are in the form $pv^{\tilde{\gamma}} = \text{constant}$ and the coefficient $\Gamma \equiv -\partial \log p/\partial \log v(v, s)$ equals $\tilde{\gamma}$.

Certain of the expressions indicated recall the well-known properties of an *ideal gas with constant* γ. In fact, the latter is a particular polytropic fluid where

$$\left.\begin{aligned} &\text{the temperature T is proportional to the product } pv \text{ by} \\ &\text{the factor m/k (k is the Boltzmann constant and m the} \\ &\text{mass of a molecule of the fluid),} \end{aligned}\right\} \tag{II.43}$$

therefore, where

- $\tilde{\gamma} = \gamma$ according to the relation (*i*.8) in Appendix A; and
- $d \log A/ds = (\gamma - 1)/(k/m)$ by virtue of the definition (I.23) of T and the relations (II.40), (II.41), and (II.43).

4.2. Shock: Pressure and Mass Volume Jumps

For a polytropic fluid which would keep the same value of $\tilde{\gamma}$ on either side of a shock, the last equation (I.30) is written

$$\frac{2}{\tilde{\gamma} - 1}(p_2 v_2 - p_1 v_1) = (p_1 + p_2)(v_1 - v_2). \tag{II.44}$$

This equation is a second-degree equation *homogeneous* in p_i and v_i ($i = 1, 2$), and has an obvious solution $p_2 = p_1$ and $v_2 = v_1$. Considering the state 1 as given (so also $\mathfrak{N}_1 = (a_1^2/w_1^2)$), we shall attempt to find a nontrivial solution in p_2 and v_2 of the system formed by the relation (II.44) and the first relation (I.30) which is recalled below

$$w_1^2 = v_1^2 \frac{p_2 - p_1}{v_1 - v_2}. \tag{II.45}$$

To do this, the following relative variations are introduced systematically:

$$\overline{\delta p} = \frac{p_2 - p_1}{p_1}, \qquad \overline{\delta v} = \frac{v_2 - v_1}{v_1},$$

as well as the expression of $\mathfrak{N}_1$ where account is taken of (II.42)

$$\mathfrak{N}_1 = \tilde{\gamma}p_1 v_1/w_1^2.$$

The system considered takes the form

$$\overline{\delta v}\left[\overline{\delta p}\frac{\tilde{\gamma}+1}{\tilde{\gamma}-1}+\frac{2\tilde{\gamma}}{\tilde{\gamma}-1}\right]+\frac{2}{\tilde{\gamma}-1}\overline{\delta p}=0,$$

$$\overline{\delta v}=-\overline{\delta p}\frac{\mathfrak{N}_1}{\tilde{\gamma}}.$$

By eliminating the solution $\overline{\delta p}=0$, we obtain

$$\overline{\delta p}=\frac{2\tilde{\gamma}}{\tilde{\gamma}+1}(\mathfrak{N}_1^{-1}-1),$$

$$\overline{\delta v}=\frac{2}{\tilde{\gamma}+1}(\mathfrak{N}_1-1). \tag{II.46}$$

The expressions (II.46) are simple and convenient; they should however be handled with care, insofar as they have only a local indicative value, like the polytropic approximation itself. Another point is worth stressing: to reach a conclusion on the relative temperature jump $\overline{\delta T}$, a supplementary hypothesis on the function $A(s)$ would have to be added to the base equation (II.40) of the model since, according to (I.23) and (II.40), the temperature equals

$$T=\frac{1}{\tilde{\gamma}-1}\frac{dA}{ds}v^{1-\tilde{\gamma}}.$$

4.3. Shock: Normal Velocity Jump and Deflection

In the form (I.29), jump relations lend themselves to the introduction of the specific enthalpy h, for and on behalf of the internal specific energy e. This gives

$$[\![\rho w]\!]=0,$$

$$\left[\!\left[\frac{pv}{w}+w\right]\!\right]=0,$$

$$\left[\!\left[h+\frac{w^2}{2}\right]\!\right]=0. \tag{II.47}$$

Using the expression (II.42) of h and pv as a function of the square a^2 of the speed of sound, the last two relations are written

$$\left[\!\left[\frac{a^2}{\tilde{\gamma}w}+w\right]\!\right]=0,$$

$$\left[\!\left[\frac{2a^2}{\tilde{\gamma}-1}+w^2\right]\!\right]=0. \tag{II.48}$$

The result of the calculations shows that it is worth putting the value common to $2a^2/(\bar{\gamma} - 1) + w^2$ on either side of the shock in the form

$$\frac{2a^2}{\bar{\gamma} - 1} + w^2 = \frac{\bar{\gamma} + 1}{\bar{\gamma} - 1} a_*^2. \tag{II.49}$$

[N.B. a_* is the common value which a and w would have if they were equal; in the case of permanent irrotational isentropic flow, there exists a simple interpretation of a_* which is without importance here.] So the term inside brackets of the first equation (II.48) is written

$$\frac{a^2}{\bar{\gamma}w} + w = \frac{\bar{\gamma} + 1}{2\bar{\gamma}} \left(\frac{a_*^2}{w} + w\right). \tag{II.50}$$

From this is can be deduced that w_1 and w_2 are linked by

$$\frac{a_*^2}{w_1} + w_1 = \frac{a_*^2}{w_2} + w_2 \, (\equiv \Omega),$$

therefore solutions of the second degree equation

$$w^2 - \Omega w + a_*^2 = 0$$

with the result that their product equals a_*^2. The relation

$$w_1 w_2 = a_*^2 \tag{II.51}$$

is called the Prandtl relation; besides the appeal of simplicity, it has the advantage of allowing a simple geometric discussion of the variation of the deflection $\theta = (\vec{W}_1, \vec{W}_2)$ of the relative velocity vector as a function of

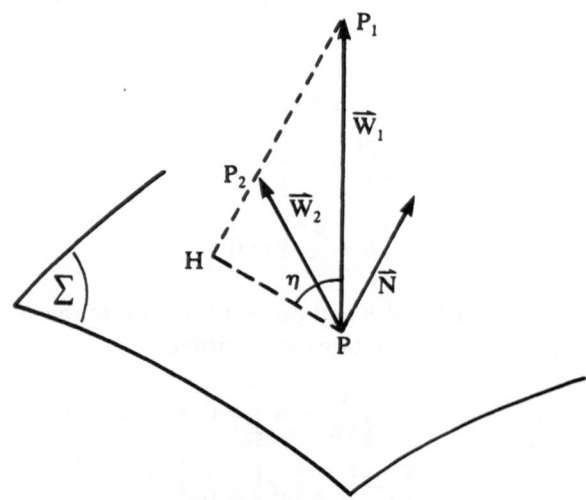

the orientation of the surface Σ with regard to the flow; this orientation is referred to by the angle

$$\eta = \frac{\pi}{2} - (\vec{N}, \vec{W}_1),$$

which is complementary to the angle which $\vec{W}_1$ makes with the normal $\vec{N}$ in P to Σ oriented in such a way that $(\vec{N}, \vec{W}_1)$ is acute, i.e., $w_1 > 0$; it is clear that η is always positive, the value $\pi/2$ corresponding to a nil "incidence."

From P, let us draw the vectors $\overrightarrow{PP_1} = \vec{W}_1$ and $\overrightarrow{PP_2} = \vec{W}_2$. By virtue of the continuity of the tangential component of the relative velocity-vector (see (I.29)), the points P_1 and P_2 have a common projection H in the tangent plane at P to Σ; the point H is therefore on the circle with diameter PP_1 drawn in the plane $PP_1 P_2$. Calculated algebraically

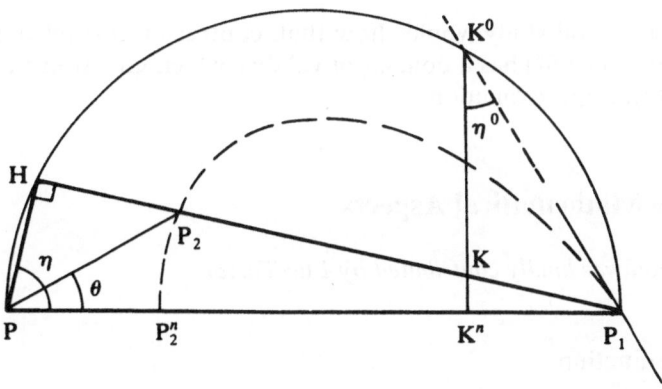

along $\vec{N}$, the distance $\overline{P_1 P_2}$ equals $w_2 - w_1$ which is evaluated using the Prandtl equation (II.51) and the definition of a_* (II.49)

$$\overline{P_1 P_2} = w_2 - w_1 = \frac{a_*^2}{w_1} - w_1 = \frac{2}{\bar{\gamma} + 1}\left(w - \frac{a^2}{w}\right)_1.$$

If the point K is considered homologous to H in the inversion of center P_1 and of strength a_1^2 (which transforms the circle into a straight line), then $\overline{P_1 P_2}$ is written

$$\overline{P_1 P_2} = \frac{2}{\bar{\gamma} + 1}(\overline{HP_1} - \overline{KP_1}) = \frac{2}{\bar{\gamma} + 1}\overline{HK}.$$

When the shock is *normal* (nil incidence, $\eta = \pi/2$), the point H is in P, point P_2 is in P_2^n such that $PP_2^n = a_*^2/|W_1|$, and point K is in K^n so that $P_1 K^n = a_1^2/|W_1|$.

When the shock is at an angle with the flow, the angle η decreases, point H describes the arc PK^0, point K describes the segment of the straight line

$K''K^0$, and point P_2 describes a strophoid arc $P_2''P_1$ whose tangent at P_1 is the straight line P_1K^0.

The result of this geometrical discussion, and the observation $P_1K^0 = a_1$, is the basic physical conclusions:

When the shock bends over the flow:
- the normal velocity jump decreases until it is zero for a limiting angle

$$\eta^0 = \text{Arc sin} \frac{a_1}{|\mathbf{W}_1|}; \tag{II.52}$$

- the deflection increases from zero, reaches a maximum, then decreases to zero when η reaches η^0.

A more general study would show that, contrary to the relations (II.46), the conclusion (II.52) has a domain of validity which greatly exceeds that of the polytropic approximation.

5. Some Mathematical Aspects

This section was kindly contributed by Luc Tartar.

5.1. Introduction

The numerous, and in many ways admirable, properties of the jump relations studied in this chapter depend on a simple physical model: perfect fluid in motion in the absence of a field of moments. For the reader who is not deterred by the language of mathematics, it is of some relevance to give those elements which, in this model, are essential to the birth of jumps (shocks, detonations, etc.) and their propagation. To do this, it is necessary to return to the laws of conservation (I.26a) reproduced below, explaining the particle derivatives

$$\left. \begin{array}{l} \dfrac{\partial v}{\partial t} + u_i v_{,i} - v u_{i,i} = 0, \\[3mm] \dfrac{\partial u_i}{\partial t} + u_j u_{i,j} + p_{,i} = f_i, \qquad i = 1, 2, 3, \\[3mm] \dfrac{\partial s}{\partial t} + u_i s_{,i} = 0. \end{array} \right\}$$

For the mathematician, this system is *quasi linear*, in that:

— the derivatives of the unknowns (v, u_i, s) occur in a linear fashion; and

— the coefficients which modify the derivatives are not constant, but depend on at least one of the unknowns.

This *nonlinear* structure gives the system properties which can be linked, at least qualitatively, to those of the jumps seen above by considering the simplest case, that of the quasi-linear equation

$$\frac{\partial U}{\partial t} + a(U)\frac{\partial U}{\partial x} = 0, \tag{II.53}$$

which concerns the scalar function $U(x, t)$ of a single variable in space x and time t.

Using the characteristic curves (§5.2), it is shown that, even if no discontinuity is involved in the initial condition $U(x, t = 0) = U_0(x)$, after a finite time, a discontinuity arises. Since discontinuities are inevitable, (II.53) should not be considered in the traditional sense, but in the sense of the distributions, so that the search for solutions should itself be extended to discontinuous U functions: thus the necessity for an expression in a conservative form and the notion of a weak solution (§5.3) arise, as well as the nonuniqueness of these solutions and the necessity to have selection criteria (§5.4). Finally, we give some information in the case of systems (§5.5).

5.2. Characteristic Curves

When a first-order linear equation of the form

$$\left.\begin{array}{c} \dfrac{\partial U}{\partial t} + c(x, t)\dfrac{\partial U}{\partial x} = 0, \\[2mm] U(x, t = 0) = U_0(x), \end{array}\right\} \tag{II.54}$$

is considered, where the functions c and U_0 are known, the solution U is obtained by carrying the initial information, given by U_0, along the characteristic curves. These curves are defined as follows: the characteristic curve (for (II.54)) starting at the point $(y, 0)$ is the solution curve of the differential equation

$$\frac{d}{dt}x(t) = c(x(t), t) \quad \text{with} \quad x(t = 0) = y. \tag{II.55}$$

It is still impossible to give an analytic expression of $x(t)$ but it will be shown that if the function c is regular in x this defines a single solution of (II.55), which can be approached by using numerical methods. If the derivative of $U(x(t), t)$ is calculated, we have

$$\frac{d}{dt}U(x(t), t) \equiv \frac{\partial U}{\partial t} + \frac{\partial U}{\partial x}\frac{d}{dt}x(t),$$

which equals 0 according to (II.55).

From this we can deduce that $U(x(t), t) \equiv U_0(y)$ and it can be seen that $c(x, t)$ therefore corresponds to a local information transmission speed (this phenomenon of information propagation with a finite speed is typical of hyperbolic equations).

For equation (II.53) it can be seen that the local information transmission velocity is $c(x, t) \equiv a(U(x, t))$ and therefore is not known beforehand since it depends on the solution which is sought. Nevertheless, the characteristic curve starting at point $(y, 0)$ can be considered

$$\frac{d}{dt}x(t) = a\{U[x(t), t]\} \quad \text{with} \quad x(t = 0) = y, \tag{II.56}$$

from which can be deduced

$$\frac{d}{dt}U(x(t), t) \equiv \frac{\partial U}{\partial t} + \frac{\partial U}{\partial x}\frac{d}{dt}x(t) = 0, \tag{II.57}$$

which gives

$$U(x(t), t) \equiv U_0(y), \tag{II.58}$$

which shows that U is constant along the characteristic; $x(t)$ is therefore a linear function of t and, consequently, the characteristic curves are straight lines. To be more precise about the characteristic starting at point $(y, 0)$ we have

$$\left.\begin{array}{l} x(t) = y + ta(U_0, y), \\ U(x(t), t) = U_0(y). \end{array}\right\} \tag{II.59}$$

It might be thought that, by using this formula, equation (II.53) has been clearly solved; this is not the case, for the characteristics starting at two distinct points $(y, 0)$ and $(z, 0)$ can meet on the plane (x, t).

In the case where y is to the left of z with the speed $a(U_0(y))$ greater than $a(U_0(z))$, the characteristics starting at $(y, 0)$ and $(z, 0)$ meet at the time $t(y, z)$ which equals

$$t(y, z) = \frac{a(U_0(y)) - a(U_0(z))}{z - y} > 0. \tag{II.60}$$

Unless $a(U_0(y))$ is increasing in y it will be impossible to find a consistent solution U for every $t > 0$ using (II.59), for there will of necessity be a conflict between two distinct values of U at the intersection of two characteristics. The maximum time of validity of the method of characteristics can even be deduced from (II.60) and introduces the value β given by

$$\beta = \inf_y \frac{d}{dy}a(U_0(y)). \tag{II.61}$$

Except if $\beta > 0$, in which case the method is applicable for every $t > 0$, the critical time t_c beyond which the method of characteristics cannot be used is

$$t_c = \frac{-1}{\beta}.$$
(II.62)

For every t strictly less than t_c the function g_t of y, defined by $g_t(y) \equiv y + a(U_0(y))t$, is strictly increasing and allows for a reciprocal regular function, but when t tends toward t_c the derivative of this reciprocal function tends toward infinity at certain points. To see this, let us consider the spatial derivative of the local information transmission speed

$$V(x, t) = \frac{\partial}{\partial x} a(U(x, t)).$$
(II.63)

As we have

$$\frac{\partial}{\partial t} a(U) + a(U) \frac{\partial}{\partial x} a(U) \equiv a'(U) \left(\frac{\partial U}{\partial t} + a(U) \frac{\partial U}{\partial x} \right) = 0,$$

by differentiation with respect to x we obtain

$$\frac{\partial V}{\partial t} + a(U) \frac{\partial V}{\partial x} + V^2 = 0,$$
(II.64)

which is an equation of evolution of $V(x(t), t)$ along the characteristics

$$\frac{d}{dt} V(x(t), t) + V^2(x(t), t) = 0,$$
(II.65)

which shows that, on the characteristic starting at y, we have

$$V(x(t), t) = \frac{V(y, t = 0)}{1 + tV(y, t = 0)}.$$
(II.66)

It can be seen that it is the most negative values of $V(y, t = 0)$ which present a danger: if there exists a point z where $V(y, t = 0) \equiv (\partial/\partial y)a(U_0)$ attains its minimum, then on the characteristic starting at z and at the critical time t_c the first derivative of a(U) tends toward $-\infty$, as does the derivative of the reciprocal function of $g_t(y)$ at the point $z + t_c a(U_0(z))$.

Since there is a conflict between the values of $U_0(y)$ and $U_0(z)$ at the meeting point of the characteristics, the obvious idea is to accept discontinuous solutions; as V tends toward $-\infty$ at the critical time, we expect to have to consider only the discontinuities where a(U) decreases when the discontinuity is crossed in the direction of the increasing x's. If the orientation of the axis of the x's is changed, then a must be replaced by $-a$, but the property "the velocity of propagation on the side of the negative x's is greater than the velocity of propagation on the side of the positive x's" remains unchanged; as will be seen later (§5.4), this rule will characterize the "acceptable discontinuities" only in the case where da/dU does not change sign. It should be noted that this property is also indifferent to Galilean transformations, because the same driving velocity is added on both sides; in our example, U can be considered as a scalar quantity whose propagation is studied relative to matter taken to be at rest.

5.3. Weak Solutions

We can speak of discontinuous solutions for equations with partial derivatives provided that we know how to derive discontinuous functions. Physicists talk of "Dirac functions" when they have to derive discontinuous functions whilst mathematicians consider that these "functions" are in fact "measures." The difference between a function and a measure becomes obvious when a distribution of mass, given by a regular density ρ, is compared to a distribution of point masses m_α, at the points z_α. The value of $\rho(x)$ at a given point x has a meaning because ρ is a function and the mass contained in an open set ω can be deduced from it by the integral $\int_\omega \rho(x)\,dx$. For every continuous function Φ we can define the quantity $\int_\rho (x)\Phi(x)\,dx$, which will be written as (ρ, Φ): these quantities make it possible to measure the function using the test functions Φ. For a measure we know how to define only the result on the test functions and not the point value. For the distribution of point masses m_α at points z_α, the corresponding measure will be written $\sum_\alpha m_\alpha \delta_{z_\alpha}$ and its value over Φ will be given by

$$(\mu, \Phi) = \sum_\alpha m_\alpha \Phi(z_\alpha).$$

The measure δ_z is called the Dirac measure or the Dirac mass at point z. To give a precise meaning to the idea of proximity between a point distribution of mass and a continuous distribution, their values must be calculated on the test functions Φ and the results compared. The use of these considerations is implicit in order to move from a particle model to a continuum model; but, in order to establish the conservation equations, it is not enough to describe the system by measures; we must also know how to derive them. To do this, the theory of distributions is used. A distribution on the real straight line R is defined by its values on the test functions Φ which are now taken as indefinitely derivable and nil outside a finite interval. If S is a distribution over R, defined by its values (S, Φ) on the test functions Φ, its derivative dS/dx is the distribution defined by the formula

$$\left(\frac{dS}{dx}, \Phi\right) = -\left(S, \frac{d\Phi}{dx}\right) \quad \text{for any test function } \Phi. \tag{II.67}$$

If the distribution S is linked with a regular function ρ by the formula $(S, \Phi) = \int_R \rho(x)\Phi(x)\,dx$, then integration by parts shows that dS/dx is the distribution linked to the function $d\rho/dx$. Let us apply this definition to the example of a discontinuous function w at the point z

$$w(x) = \begin{cases} w_- & \text{if } x < z, \\ w_+ & \text{if } z < x, \end{cases} \tag{II.68}$$

where w_- and w_+ are constants. To derive this function (in the sense of distributions) we use (II.67) which gives

$$\left(\frac{dw}{dx}, \Phi\right) = -\left(w, \frac{d\Phi}{dx}\right) = -\int_{-\infty}^{z} w_- \frac{\partial\Phi}{\partial x} dx - \int_{z}^{+\infty} w_+ \frac{\partial\Phi}{\partial x} dx,$$

that is, $(w_+ - w_-)\Phi(z)$, which thus reveals a Dirac mass at the point z, written δ_z and defined by

$$(\delta_z, \Phi) = \Phi(z) \quad \text{for any continuous function } \Phi. \tag{II.69}$$

The derivative of w defined by (II.68) is therefore

$$\frac{dw}{dx} = (w_+ - w_-)\delta_z. \tag{II.70}$$

It can therefore be seen that the derivative in the sense of distributions of a discontinuous function exhibits the jump of the function at each point of discontinuity as a coefficient of the Dirac mass at the corresponding point.

But if we can derive distributions indefinitely by repeating the definition (II.67), we cannot multiply them with no restriction. For example, the Dirac measure can be multiplied at point z only by continuous functions in z, with $\psi(z)\delta_z$ being the product of the continuous function ψ by the Dirac mass δ_z. This is the difficulty which we shall now meet for the quasi-linear hyperbolic equation (II.53).

If U is continuous in z, we cannot multiply $\partial U/\partial x$ by a(U) which will generally be discontinuous in z: the expression $a(U)(\partial U/\partial x)$ should not be used for discontinuous U functions. Even if by chance a(U) is continuous in z, we must be wary, as the following "paradox" will show.

The formula

$$\frac{\partial W^3}{\partial x} = 3W^2 \frac{\partial W}{\partial x} \tag{II.71}$$

is valid for regular W functions. Let us now consider the following discontinuous W function

$$W(x) = \begin{cases} -1 & \text{if } x < 0, \\ +1 & \text{if } 0 < x. \end{cases} \tag{II.72}$$

As the function W^2 is constant, the product $3W^2(\partial W/\partial x)$ has a meaning, as has $\partial W^3/\partial x$; but the preceding calculation shows that these two results are different

$$\left. \begin{array}{l} \dfrac{\partial W^3}{\partial x} = 2\delta_0, \\[2mm] 3W^2 \dfrac{\partial W}{\partial x} = 6\delta_0. \end{array} \right\} \tag{II.73}$$

Like all paradoxes, this one has its outcome and shows that the usual rule of calculation (II.71) must not be accepted without hypotheses of regularity on the W function. From this observation we shall accept that, when working with discontinuous functions and their derivatives, formal manipulations are a source of error; this can be eliminated by obeying a simple rule: work only with written equations in a *conservative* form, i.e., the form

$$\frac{\partial U}{\partial t} + \frac{\partial}{\partial x} F(U) = 0. \tag{II.74}$$

We shall therefore take a primitive of a for F and replace (II.53) by (II.74). We are now looking for a solution U which must be a function, we know how to calculate F(U) which is another function, and the two derivatives written in (II.74) are defined in a similar way to (II.67). More precisely we shall say that a function U, possibly discontinuous, is a *weak* solution (i.e., in the sense of distributions) of the problem

$$\left.\begin{array}{l} \dfrac{\partial U}{\partial t} + \dfrac{\partial}{\partial x} F(U) = 0, \\[2mm] U(x, t = 0) = U_0(x), \end{array}\right\} \tag{II.75}$$

if

$$0 = \int_R \int_0^{+\infty} \left(U \frac{\partial \Phi}{\partial t} + F(U) \frac{\partial \Phi}{\partial x} \right) dx\, dt + \int_R U_0(x)\Phi(x, t = 0)\, dx \tag{II.76}$$

for every test function Φ, regular in (x, t) and null outside a finite set.

Once the discontinuous functions in the sense of formula (II.76) are accepted, we shall see which conditions are imposed on the values of the solution on each side of the discontinuity. Let us suppose then that U is a weak solution, regular on each side of a curve of equation x = g(t), and admitting limits on each side of this curve: we can write $U_- \equiv U(g(t)_-, t)$ as the left-hand value and $U_+ \equiv U(g(t)_+, t)$ as the right-hand value.

The plane (x, t) is cut into two open segments, Ω_- on the left and Ω_+ on the right of the curve x = g(t), and the integral (II.76) is calculated on each of these two open segments by integration by parts. Using the fact that U is regular and therefore a solution of (II.75) in a conventional sense on each of the open sets Ω_- and Ω_+, each integral becomes a contour integral by Green's formula. For example, supposing that $\Phi(x, t = 0) = 0$,

$$\iint_{\Omega_-} \left(U \frac{\partial \Phi}{\partial t} + F(U) \frac{\partial \Phi}{\partial x} \right) dx\, dt = \int_{\partial \Omega_-} (U_- \Phi \cdot v_t^- + F(U_-)\Phi \cdot v_x^-)\, d\sigma,$$

where $v^- = (v_x^-, v_t^-)$ is the normal outside Ω_- and $d\sigma$ is the measure of length on the boundary, i.e., on the curve of equation x = g(t). If we note that for Ω_+ we have to take $v^+ = -v^-$ and that $v^-\, d\sigma = (1, -(dg/dt))\, dt$, we can deduce the formula

$$\int \left\{ F(U_-) - F(U_+) - \frac{dg}{dt}(U_- - U_+) \right\} \Phi\, dt = 0, \tag{II.77}$$

for every regular function Φ null for $t = 0$. From this we can deduce the jump formula

$$D(t) \equiv \frac{dg}{dt} = \frac{F(U_+) - F(U_-)}{U_+ - U_-}, \qquad (II.78)$$

which gives $D(t)$, the velocity of discontinuity, according to the left- and right-hand values (a result which is obviously compatible with a change in orientation of the x-axis). If U_- and U_+ are neighboring values we find D next to dF/dU, i.e., to $a(U)$, the data transfer rate in the zones where U is regular.

As we have already pointed out, certain rules of calculation are forbidden when dealing with discontinuous solutions, and we must be careful in the choice of the conservative form of the equation: we can have several equations in conservative form which possess the same regular solutions but different weak solutions.

Let Φ be a regular function, and consider the function Ψ defined by

$$\frac{d\Psi}{dU} \equiv a(U)\frac{d\Phi}{dU}. \qquad (II.79)$$

By multiplying (II.53) by $d\Phi/dU$, and assuming that U is regular, we obtain another conservative form as a result of (II.53)

$$\frac{\partial}{\partial t}\Phi(U) + \frac{\partial}{\partial x}\Psi(U) = 0. \qquad (II.80)$$

If the derivative of Φ is not nullified, $V \equiv \Phi(U)$ can be chosen as a new unknown, and we can substitute

$$\Psi(U) \equiv X(V) \qquad \text{for} \quad V \equiv \Phi(U) \qquad (II.81)$$

to transform (II.80) into

$$\frac{\partial V}{\partial t} + \frac{\partial}{\partial x}X(V) = 0. \qquad (II.82)$$

These equations give the same regular solutions but not the same weak solutions. In fact, the velocity of propagation of the discontinuities for (II.82) is given by

$$\tilde{D}(t) \equiv \frac{X(V_+) - X(V_-)}{V_+ - V_-} \equiv \frac{\Psi(U_+) - \Psi(U_-)}{\Phi(U_+) - \Phi(U_-)}, \qquad (II.83)$$

which, in general, is different from the velocity $D(t)$ given by (II.78) (except in the case where a is independent of U).

To choose between (II.74) and (II.80) a physical law must be used: if the integral of $\Phi(U)$ is conserved over time, (II.80) must be chosen. That is why we have used the laws of conservation of mass, momentum, and total energy, which are valid even in the case of discontinuous solutions, whereas conservation of entropy is valid only for regular solutions.

5.4. Nonuniqueness and Selection Criteria

Armed with this notion of a weak solution which admits discontinuities and demands that they satisfy the jump conditions, we might believe that the difficulties are ended; this is not at all the case, because we have with one stroke broadened the class of solutions too greatly and there may be several which correspond to the same initial datum. To simplify the study of this pheomenon we shall limit ourselves to the Burgers equation

$$\left.\begin{array}{c} \dfrac{\partial U}{\partial t} + \dfrac{\partial}{\partial x}\dfrac{cU^2}{2} = 0, \\[2mm] U(x, t = 0) = U_0(x), \end{array}\right\} \tag{II.84}$$

for which the jump condition (II.78) gives

$$D = \frac{c}{2}(U_- + U_+). \tag{II.85}$$

Let us consider the initial datum

$$U_0(x) = \begin{cases} 0 & \text{for } x < 0, \\ 1 & \text{for } 0 < x, \end{cases} \tag{II.86}$$

and define the functions V and W by

$$V(x, t) = \begin{cases} 0 & \text{for } x \le 0, \\[2mm] \dfrac{x}{ct} & \text{for } 0 \le x \le ct, \\[2mm] 1 & \text{for } ct \le x, \end{cases} \tag{II.87}$$

and

$$W(x, t) = \begin{cases} 0 & \text{for } x < \dfrac{ct}{2}, \\[2mm] 1 & \text{for } \dfrac{ct}{2} < x. \end{cases} \tag{II.88}$$

It can be seen immediately that W is a weak solution, corresponding to the initial datum (II.86), because the jump condition (II.85) is satisfied on the discontinuity $x = ct/2$ whilst V is a solution in the conventional sense, corresponding to the same initial datum. The question is then to discover how to choose the right solution.

If the datum is regular, we have seen that the method of characteristics gives a regular solution, at least for a brief instant. In the case of the initial datum (II.86) it is worth noting that the method of characteristics gave us the formula (II.59) which we can still apply, but which does not define the value of the solution between $x = 0$ and $x = ct$ because the characteristics starting at 0_- and 0_+ diverge. To be sure that the "right solution" is V, we shall use a continuity argument. If we start from the initial datum

$$U_0^\varepsilon(x) = V(x, \varepsilon), \tag{II.89}$$

which is regular, and we use the method of characteristics, we find as a solution

$$U^\varepsilon(x, t) = V(x, t + \varepsilon). \tag{II.90}$$

If we make ε tend towards 0, the initial datum (II.89) converges on (II.86) and the solution (II.90) converges on (II.87); so the continuity argument has "selected" the solution V. If we now consider the Riemann problem

$$U_0(x) = \begin{cases} U_- & \text{for } x < 0, \\ U_+ & \text{for } 0 < x, \end{cases} \tag{II.91}$$

which, in the case $U_- \leq U_+$, possesses a regular solution

$$U(x, t) = \begin{cases} U_- & \text{for } x \leq cU_-t, \\ \dfrac{x}{ct} & \text{for } cU_-t \leq x \leq cU_+t, \\ U_+ & \text{for } cU_+t \leq x, \end{cases} \tag{II.92}$$

the continuity argument would make us prefer this solution to the discontinuous solution (which exists, without conditions on U_- and U_+, with D given by (II.85))

$$U(x, t) = \begin{cases} U_- & \text{for } x < Dt, \\ U_+ & \text{for } 0 < x. \end{cases} \tag{II.93}$$

This continuity argument shows that for the Burgers equation, discontinuities where U_- is less than U_+ should be excluded.

The method of characteristics had suggested a first criterion, saying that the data propagation velocity on the left of the discontinuity should be greater than the data propagation velocity on the right; the discontinuity therefore results from competition between the left and the right which creates a zone of uncertainty where formula (II.59) gives two distinct values; it is natural that the discontinuity should serve as a compromise and be found in this zone. If we now come back to the case of the equation $\partial U/\partial t + (\partial/\partial x) F(U) = 0$ with a general function F, the data propagation velocity will therefore be $a(U) \equiv dF/dU$. The foregoing analysis suggests that we should accept only those discontinuities which result from a compromise in a situation of conflict between left and right: this is the *Lax criterion* (*Comm. Pure Appl. Math.* **33**, (1957), pp. 537–546). To be admissible, a discontinuous solution of (II.74) should necessarily confirm

$$a(U_-) \geq D \geq a(U_+), \tag{II.94}$$

where

$$a(U) \equiv \frac{dF}{dU} \quad \text{and} \quad D = \frac{F(U_+) - F(U_-)}{U_+ - U_-}.$$

But this criterion does not characterize the "right discontinuities" unless a(U) is monotonic. In the general case where a(U) is not monotonic, the criterion we arrive at compares the respective positions of the graph of the function F, locus of points (U, F(U)), and of the chord joining the point (U_-, F(U_-)) to the point (U_+, F(U_+)): this is the *Oleinik criterion* (*Usp. Mat. Nauk* **14**, (1959), pp. 165–170). To be admissible, a discontinuous solution of (II.74) must necessarily confirm

$$\left. \begin{array}{l} \text{the chord is above the graph if } U_- \geq U_+, \\ \text{the chord is below the graph if } U_- \leq U_+. \end{array} \right\} \tag{II.95}$$

Note that in the case where a is monotonic these two criteria coincide with each other and with our first criterion, a(U) decreases on crossing the discontinuity in the direction of the increasing x. In the case where a(U) is not monotonic, the Lax and Oleinik criteria are distinct. Note also that, for the problem (II.75), we can obtain a theorem of existence and of uniqueness in a well-behaved class of functions and that the discontinuities of the solutions obtained are characterized by the Oleinik criterion. It would take too long to go into the details of these mathematical results, but we can state simply the reason why certain discontinuities are retained by the Lax criterion but are finally rejected by the Oleinik criterion; they are unstable and the solution of the Riemann problem, corresponding to the initial datum (II.91), is not obtained using a single discontinuity from U_- to U_+, propagated with the velocity D given by (II.78), but using several smaller discontinuities separated by zones where the solution is continuous.

5.5. The Case of Systems

The general study of the case of hyperbolic (quasi-linear) systems of the conservation laws would take us too far. The mathematical understanding of them is, moreover, not as complete as that of the scalar case, the essential difficulty being to know which discontinuities are admissible.

We know no explicit formula like (II.62) for the time of appearance of a discontinuity, but in order to define discontinuous solutions we must also work with equations written in conservative form

$$\frac{\partial}{\partial t} U_j + \frac{\partial}{\partial x} F_j(U_1, \ldots, U_p) \quad \text{for} \quad j = 1, \ldots, p. \tag{II.96}$$

But there are important differences in relation to the scalar case. The first difference is that the systems of p equations like (II.96) are not all hyperbolic; to exhibit the phenomena of data propagation with a finite velocity it must be possible for the derived matrix, defined by

$$a_{j,k}(U_1, \ldots, U_p) \equiv \frac{\partial}{\partial U_j} F_k(U_1, \ldots, U_p) \qquad \text{for} \quad j, k = 1, \ldots, p, \qquad \text{(II.97)}$$

to be drawn diagonally with real characteristic values $\lambda_j(U_1, \ldots, U_p)$ (these will generally by distinct). These characteristic values λ_j are propagation velocities of certain specific modes and there are therefore p distinct velocities.

The second difference is that we obtain p relations for the velocity D of a discontinuity and so there will be more restrictive jump conditions linking the components U_j^- and U_j^+, the limits of U on the left and right of the discontinuity; we find

$$F_j(U_1^+, \ldots, U_p^+) - F_j(U_1^-, \ldots, U_p^-) = D(U_j^- - U_j^+) \qquad j = 1, \ldots, p. \qquad \text{(II.98)}$$

By way of introduction to the studies which must be made in order to understand better the nature of the solutions of (II.52), let us consider the system

$$\left. \begin{array}{l} \dfrac{\partial}{\partial t} U_1 - \dfrac{\partial}{\partial x} U_2 = 0, \\[3mm] \dfrac{\partial}{\partial t} U_2 - \dfrac{\partial}{\partial x} F(U_1) = 0, \end{array} \right\} \qquad \text{(II.99)}$$

with F increasing. In this case the characteristic values to consider are

$$\lambda_1 = -(F'(U_1))^{1/2} \qquad \text{and} \qquad \lambda_2 = +(F'(U_1))^{1/2}.$$

The jump conditions are

$$U_2^+ - U_2^- = D(U_1^+ - U_1^-) \qquad \text{and} \qquad F(U_1^+ - F(U_1^-) = D(U_2^+ - U_2^-),$$

which leads to

$$D = \left(\frac{F(U_1^+) - F(U_1^-)}{U_1^+ - U_1^-} \right)^{1/2}. \qquad \text{(II.100)}$$

It must be stressed that (II.100) is different from (II.78), but does have a formal connection to those of the relations (I.30) which give the normal relative velocity of a discontinuity in a perfect fluid.

CHAPTER III

The Detonation Layer

1. General Features of the Model

1.1. Introduction

The perturbation methods developed in fluid mechanics over the past three decades have made it possible to resolve many problems of regular or singular perturbation (see, e.g., Van Dyke [36]). Thus Germain and Guiraud [22], in particular, studied shock waves by considering the dissipative effects as a singular perturbation of the Hugoniot model [25] in a perfect fluid. In this chapter we adopt—on the subject of detonations—an analogous approach: detonation is seen as a layer of steep gradients in a Navier–Fourier dissipative fluid, this layer being localized in the neighborhood of the surface $\Sigma(t)$ to which it is reduced by the Crussard model [14] in a perfect fluid. Analysis of flow in such a remarkable zone calls for the use of curvilinear coordinates z, ξ', ξ'' which are associated, respectively, with the normal direction N and the principal directions N′, N″ of $\Sigma(t)$. (N.B. Throughout this chapter the normal N is expressly oriented from upstream to downstream, as in Section II.2.) That is why (§1.2) the corresponding form of the equations (I.28) is established first of all: the expressions are those of Germain and Guiraud, made more manageable by appropriate notations. We then specify the hypotheses adopted to represent the law of chemical evolution (see §1.3). Finally, we go on to the necessary definitions and transformations (§1.4 and §1.5) for development of the perturbation method itself, ending up (§1.6) at zero-order structures and the notion of *quasi C–J detonation*.

1.2. Conservation Equations in Local Variables

In the equations (I.28) of a Navier–Fourier dissipative fluid, let us develop particle derivatives so as to separate the derivations in relation to the variables of space (operator ∇) and the derivation in relation to the time t (operator $\partial/\partial t$) and adopt a vectorial expression. For $f_i = \ell = 0$, we have

$$\left.\begin{array}{c} \dfrac{\partial \rho}{\partial t} + \mathbf{V}\cdot(\rho\mathbf{U}) = 0, \\[12pt] \dfrac{\partial \rho\mathbf{U}}{\partial t} + \mathbf{V}\cdot(\rho\mathbf{U}\otimes\mathbf{U} + p\mathbf{1} + \mathbf{\Pi}) = 0, \\[12pt] \dfrac{\partial s}{\partial t} + \mathbf{U}\cdot\mathbf{V}s + \dfrac{v}{T}[\mathbf{V}\cdot\mathbf{q} + \mathrm{Tr}(\mathbf{\Pi}\cdot\mathbf{V}\mathbf{U})] = 0. \end{array}\right\}$$
(III.1)

In the above equations, transformation of the variables (x_1, x_2, x_3, t) to the variables (z, ξ', ξ'', t) calls for the geometric preliminaries which are clearly established in [22]. We limit ourselves here to mentioning, without proof, the results that are essential for understanding subsequent expansions.

At each point (ξ', ξ'') of $\Sigma(t)$, besides the normal velocity $\delta(\xi', \xi'', t)$ at the surface $\Sigma(t)$, we define the metric elements $H'(\xi', \xi'', t)$ and $H''(\xi', \xi'', t)$ by

$$d\mathbf{P}^2 = (H'\,d\xi')^2 + (H''\,d\xi'')^2 + dz^2.$$
(III.2)

Denoting the principal radii curvature of $\Sigma(t)$ counted algebraically along $\mathbf{N}$ by $r'(\xi', \xi'', t)$ and $r''(\xi', \xi'', t)$, at each point (z, ξ', ξ'') of space, we define the functions h', h'', k of (z, ξ', ξ'', t) by

$$h' = H'\left(1 - \frac{z}{r'}\right), \qquad h'' = H''\left(1 - \frac{z}{r''}\right), \qquad k = \frac{1}{z-r'} + \frac{1}{z-r''}, \quad \text{(III.3)}$$

and the differential vectorial operator

$$\mathbf{J} = \frac{\mathbf{N}'}{h'}\frac{\partial}{\partial\xi'} + \frac{\mathbf{N}''}{h''}\frac{\partial}{\partial\xi''}.$$
(III.4)

With these definitions, the following correspondences can be shown:

(a) For a scalar function $f(z, \xi', \xi'', t)$, the partial derivative with respect to time at a point x_i $(i = 1, 2, 3)$ of space equals

$$\frac{\partial f}{\partial t} - \delta\frac{\partial f}{\partial z} - z\frac{\partial \mathbf{N}}{\partial t}\cdot\mathbf{J}f.$$

(b) For a scalar function $f(z, \xi', \xi'', t)$, the Eulerian gradient with respect to the variables x_i $(i = 1, 2, 3)$ at a given instant equals

$$\mathbf{N}\frac{\partial f}{\partial z} + \mathbf{J}f.$$

(c) For a vector $\mathscr{F} = f\mathbf{N} + f'\mathbf{N}' + f''\mathbf{N}''$, the divergence with respect to the variables x_i $(i = 1, 2, 3)$ at a given instant equals

$$\frac{\partial f}{\partial z} + kf + \mathscr{F}_{\tau,\tau},$$

where the last term is a condensed expression suggestive for

$$\frac{1}{h'h''}\left[\frac{\partial f'h''}{\partial \xi'}+\frac{\partial f''h'}{\partial \xi''}\right].$$

(d) For a second-order symmetric tensor

$$\mathscr{F} = f\mathbf{N}\otimes\mathbf{N} + f'\mathbf{N}'\otimes\mathbf{N}' + f''\mathbf{N}''\otimes\mathbf{N}''$$
$$+ g(\mathbf{N}'\otimes\mathbf{N}'' + \mathbf{N}''\otimes\mathbf{N}')$$
$$+ g'(\mathbf{N}\otimes\mathbf{N}'' + \mathbf{N}''\otimes\mathbf{N}) + g''(\mathbf{N}\otimes\mathbf{N}' + \mathbf{N}'\otimes\mathbf{N}),$$

the divergence on the right with respect to the variables x_i equals

$$\frac{\partial \mathbf{F}}{\partial z} + k\mathbf{F} + \mathscr{F}_{\tau,\tau},$$

which accords with the following expressions:

$$\mathbf{F} = f\mathbf{N} + g'\mathbf{N}' + g''\mathbf{N}'',$$
$$\mathbf{F}' = g'\mathbf{N} + f'\mathbf{N}' + g\mathbf{N}'',$$
$$\mathbf{F}'' = g''\mathbf{N} + g\mathbf{N}' + f''\mathbf{N}'',$$
$$\mathscr{F}_{\tau,\tau} = \frac{1}{h'h''}\left[\frac{\partial \mathbf{F}'h''}{\partial \xi'}+\frac{\partial \mathbf{F}''h'}{\partial \xi''}\right].$$

For greater clarity, let us first use the results (a)–(d) above to transform the nondissipative terms in equations (III.1). This leads to isolating the divergence $\boldsymbol{\varphi} = \nabla\cdot\boldsymbol{\Pi}$ of the dissipative stresses and the rate $\psi = \nabla\cdot(\lambda\nabla T) + \mathrm{Tr}(\boldsymbol{\Pi}\cdot\nabla\mathbf{U})$ of volume energy dissipated thermomechanically, where ∇ denotes here the operator $\mathbf{N}\partial/\partial z + \mathbf{J}$ on the variables (z, ξ', ξ''). Moreover, this leads to the introduction of the components u and $w = u - \delta$ of U and W along N as well as their common component $\mathbf{W}_\tau$ along Σ. Finally, equations (III.1) take the *local form*

$$\left(\frac{\partial}{\partial t} + w\frac{\partial}{\partial z}\right)\mathrm{Log}\, v - \frac{\partial u}{\partial z} = ku + v\left[(\rho\mathbf{U})_{\tau,\tau} - \frac{\partial \mathbf{N}}{\partial t}\cdot\mathbf{J}\rho\right],$$

$$v\frac{\partial p}{\partial z}\mathbf{N} + \left(\frac{\partial}{\partial t} + w\frac{\partial}{\partial z}\right)\mathbf{U} = v\mathbf{U}\left[(\rho\mathbf{U})_{\tau,\tau} - \frac{\partial \mathbf{N}}{\partial t}\cdot\mathbf{J}\rho\right]$$
$$- v\left[(\rho\mathbf{U}\otimes\mathbf{U})_{\tau,\tau} - z\frac{\partial \mathbf{N}}{\partial t}\cdot\mathbf{J}\rho\mathbf{U} + \mathbf{J}\rho\right] + v\boldsymbol{\varphi},$$

$$\left(\frac{\partial}{\partial t} + w\frac{\partial}{\partial z}\right)s = \left(z\frac{\partial \mathbf{N}}{\partial t} - \mathbf{W}_\tau\right)\cdot\mathbf{J}s + \frac{v}{T}\psi.$$

$$(\mathrm{III.5})$$

1.3. Evolution of Chemical Species

Here, we follow all the authors who have contributed to the theoretical study of detonations, in admitting the existence of a *state of local thermodynamic equilibrium* $\mathbf{q}$ *at every point of the flow*. On the other hand, where the evolution of chemical species is concerned, we take a less restrictive position.

Whether implicit, as in Von Neumann [32] or Hayes [23], or explicit as in Friedrichs [20] or Hirschfelder and Curtiss [24], all authors assume that in a detonation the chemical transformation is a zero-order reaction of the type $\mathscr{E}_0 \to \mathscr{E}$. This hypothesis, which is obviously very simplistic if not completely unrealistic (see Chap. IV), nevertheless has the advantage of leading to an analytically simple law of the type

$$\frac{dx}{dt} = (1 - x)S(\mathbf{q}, x),$$

where $x \in [0, 1]$ denotes the mass fraction of $\mathscr{E}$ at every point of space at the instant t.

Starting from this outline, it is possible—without too much complication—to cover mechanisms more general than $\mathscr{E}_0 \to \mathscr{E}$ by admitting the existence of a state variable $m \in [0, 1]$:

(a) such that the chemical composition N is determined as a function of $\mathbf{q}$ and m by the relation

$$N = N_0 + m[\tilde{\tilde{N}}(\mathbf{q}) - N_0], \tag{III.6}$$

just as in the perfect fluid model (see §II.2.1); and
(b) governed by a law of the form

$$\frac{dm}{dt} = (1 - m)S(\mathbf{q}, m), \tag{III.7}$$

where the reactivity S is specified below.

The stability of the substance before any external stress and the irreversibility of its transformation, once *initiated*, imply that the function S confirms the following two properties:

- there exists a neighborhood V of $\mathbf{q}_0$ such that $S(\mathbf{q}, 0) = 0$ for $q \in V$; and
- $S(\mathbf{q}, m) > 0$ for $m \in]0, 1]$.

Studying deflagrations and detonations together in the framework of the mechanism $\mathscr{E}_0 \to \mathscr{E}$, Friedrichs [20] achieved these conditions by considering a function $S(T)$ of the absolute temperature T and a temperature $T^f > T_0$ such that

$$\left.\begin{array}{l} S(T \le T^f) = 0, \\ S(T > T^f) > 0. \end{array}\right\}$$

The choice is perfectly suited to flames: T^f is then an image of the ignition temperature. It is less suited to detonations insofar as it is a shock which raises the medium to a state where reactions are triggered (see Dixon [16], Vieille [38], Von Neumann [32], etc.), and insofar as the state reached downstream of this shock may not be characterized by the temperature, whereas it is always characterized by the specific entropy s (see Behe [4]). That is why we finally prefer to consider here an *ignition entropy* s^i and a function $S(s, m)$ null for $s \leq s^i$ and increasing positive for $s > s^i$ when $m = 0$

$$\left.\begin{array}{l} S(s \leq s^i, 0) = 0, \\ S(s > s^i, 0) > 0. \end{array}\right\} \tag{III.8}$$

Finally, it should be noted that, with the help of local variables, (III.7) is written

$$\left(\frac{\partial}{\partial t} + w\frac{\partial}{\partial z}\right)m + \left(\mathbf{W}_\tau - z\frac{\partial \mathbf{N}}{\partial t}\right)\cdot \mathbf{J}m = (1 - m)S. \tag{III.9}$$

1.4. Characteristic Quantities and External Variables

The use of an asymptotic expansion method—a developed form of reasoning on the orders of magnitude—calls for the equations to which it is applied to bear on dimensionless and near-to-unity independent and dependent variables. Given the quantities in question in equations (III.1) and (III.7), we must *a priori* introduce (see Kaplun [28], Lagerstrom and Cole [30]) three *primary characteristics*:

- a length $\tilde{L}$,
- a velocity $\tilde{D}$, and
- a volume mass $\tilde{\rho}$,

and two *secondary characteristics*:

- a length L, and
- a length L',

which "measure" the range of perturbations, by dissipation of thermomechanical origin and by deviation from the chemical equilibrium, respectively.

Let us agree as a general rule to indicate by a bar over the initial symbol the operation which consists of calculating in dimensionless form a variable (independent or dependent) using the above characteristics; let us also agree, in order to simplify the expressions of subsequent expansions, to make an exception to this rule for lengths counted according to the normal N to $\Sigma(t)$.

These two conventions are rendered by the following notations:

$$Z = \frac{z}{\tilde{L}}, \quad R' = \frac{r'}{\tilde{L}}, \quad R'' = \frac{r''}{\tilde{L}}, \quad K = \frac{1}{Z - R'} + \frac{1}{Z - R''},$$

$$\bar{\xi}' = \frac{\xi'}{\tilde{L}}, \quad \bar{\xi}'' = \frac{\xi}{\tilde{L}}, \quad \tilde{t} = \frac{\tilde{D}}{\tilde{L}} t, \quad \mathbf{J} = \frac{N'}{h'} \cdot \frac{\partial}{\partial \bar{\xi}'} + \frac{N''}{h''} \frac{\partial}{\partial \bar{\xi}''},$$

$$\bar{U} = \frac{U}{\tilde{D}}, \quad \bar{W} = \frac{W}{\tilde{D}}, \quad \bar{v} = \tilde{\rho}v, \qquad\qquad\qquad (III.10)$$

$$\bar{e} = \frac{e}{\tilde{D}^2}, \quad \bar{s} = \frac{s}{\tilde{D}^2}, \quad \bar{p} = \frac{p}{\tilde{\rho}\tilde{D}^2},$$

$$\bar{\mu}^{(n)} = \frac{\mu^{(n)}}{L\tilde{\rho}\tilde{D}}, \quad \bar{\lambda} = \frac{\lambda}{L\tilde{\rho}\tilde{D}^3}, \quad \bar{S} = \frac{L'}{\tilde{D}} S.$$

So, relative to the *external variables* $(Z, \bar{\xi}', \bar{\xi}'', \tilde{t})$, equations (III.5) and (III.9) are written

$$\left(\frac{\partial}{\partial \tilde{t}} + \bar{w}\frac{\partial}{\partial Z}\right) \text{Log } \bar{v} - \frac{\partial \bar{u}}{\partial Z} = K\bar{u} + \bar{v}\left[(\bar{\rho}\bar{U})_{\tilde{t},\bar{\xi}} - Z\frac{\partial N}{\partial \tilde{t}} \cdot \mathbf{J}\tilde{\rho}\right],$$

$$\bar{v}\frac{\partial \bar{p}}{\partial Z}\mathbf{N} + \left(\frac{\partial}{\partial \tilde{t}} + \bar{w}\frac{\partial}{\partial Z}\right)\bar{U} = \bar{v}\bar{U}\left[(\bar{\rho}\bar{U})_{\tilde{t},\bar{\xi}} - Z\frac{\partial N}{\partial \tilde{t}} \cdot \mathbf{J}\tilde{\rho}\right]$$

$$\qquad - \bar{v}\left[(\bar{\rho}\bar{U} \otimes \bar{U})_{\tilde{t},\bar{\xi}} - Z\frac{\partial N}{\partial \tilde{t}} \cdot \mathbf{J}\tilde{\rho}\bar{U} + \mathbf{J}\bar{p}\right] + \bar{v}\frac{L}{\tilde{L}}\Phi, \qquad (III.11)$$

$$\left(\frac{\partial}{\partial \tilde{t}} + \bar{w}\frac{\partial}{\partial Z}\right)\bar{s} = \left(Z\frac{\partial N}{\partial \tilde{t}} - \bar{W}_{\tilde{t}}\right) \cdot \mathbf{J}\bar{s} + \frac{\bar{v}}{T}\frac{L}{\tilde{L}}\Psi,$$

$$\frac{1}{L'/\tilde{L}}(1 - m)\bar{S} = \left(\frac{\partial}{\partial \tilde{t}} + \bar{w}\frac{\partial}{\partial Z}\right)m + \left(\bar{W}_{\tilde{t}} - Z\frac{\partial N}{\partial \tilde{t}}\right) \cdot \mathbf{J}m.$$

where the quantities Φ and Ψ are henceforth defined in terms of the dimensionless variables and the differential operators on the external variables by

$$\Phi = \bar{\mathbf{V}} \cdot \bar{\mathbf{\Pi}},$$

$$\Psi = \bar{\mathbf{V}} \cdot (\bar{\lambda}\bar{\nabla}T) + \text{Tr}(\bar{\mathbf{\Pi}} \cdot \bar{\nabla}\bar{U}),$$

$$\bar{\mathbf{V}} = \mathbf{N}\frac{\partial}{\partial Z} + \mathbf{J}, \qquad\qquad\qquad (III.12)$$

$$\bar{\mathbf{\Pi}} = \bar{\mu}'(\bar{\mathbf{V}} \cdot \bar{U})\mathbf{1} + \bar{\mu}[\bar{\nabla}\bar{U} + (\bar{\nabla}\bar{U})^{\mathsf{T}}].$$

The first important result of rewriting these equations in dimensionless form is to bring out—as a factor of the terms of dissipation Φ, Ψ, $\bar{S}$—the ratios $L/\tilde{L}$ and L'/L which measure the relative range of the respective perturbation domains. The following step consists of putting to use the experimental knowledge acquired about the detonation phenomenon to arrange the perturbations in order and retain—at first—only those with the greatest range.

1.5. Perturbation Parameter and Internal Variables

In the above analysis, three characteristic lengths appear: a primary length $\tilde{L}$ and two secondary lengths L and L'. *A priori*, the perturbation parameter in the sense of Van Dyke [36] can be either $L/\tilde{L}$, or $L'/\tilde{L}$; experiments enable us to decide between the two.

In fact, detailed observation of detonations showed long ago (Dixon, 1893 [16]; Chapman, 1899 [8]; and Paul Vieille, 1900 [38]) that it is a shock— today associated with the work of Zeldovich, 1940 [41]; Von Neumann, 1942 [32]; and Doring, 1943 [17]!—which raises the medium into a state where chemical reactions are triggered. More precisely, it shows that a transitional propagation, called *build-up*, where the temperature is too low behind the shock to lead immediately to a strong reactivity, most often precedes the main propagation, characterized by a rapidly increasing temperature immediately after the passage of the shock as a result of the energy released by the rearrangement of the atoms. *A contrario*, for a given particle, *built-up detonation* means a transformation where the time constant linked to the realization of chemical equilibrium in a given physical state decreases rapidly from the moment the reactions start, and only rises again later as the system relaxes.

This analysis shows that the inequality $L' \gg L$ can be applied at the transitional phase whereas the other inequality $L' \ll L$ prevails during the main phase; the perturbation parameter to choose for the latter phase is therefore $\varepsilon = L/\tilde{L}$.

The last step consists in choosing the *internal* independent variables. The Van Dyke principle of minimum degeneration [36] leads to the adoption of a linear extension along **N**

$$\zeta = \frac{Z}{\varepsilon}. \tag{III.13}$$

Relative to the variables $(\zeta, \bar{\xi}', \bar{\xi}'', \bar{t})$ the equations (III.11) are written

$$
\left.
\begin{aligned}
&\bar{w}\frac{\partial \log \bar{v}}{\partial \zeta} - \frac{\partial \bar{u}}{\partial \zeta} = 0(\varepsilon), \\[2mm]
&\bar{v}\mathbf{N}\frac{\partial}{\partial \zeta}\left(\bar{p} - \bar{\mu}''\frac{\partial \bar{u}}{\partial \zeta}\right) + \bar{w}\frac{\partial \bar{U}}{\partial \zeta} - \bar{v}\frac{\partial}{\partial \zeta}\left(\bar{\mu}\frac{\partial \bar{W}_\tau}{\partial \zeta}\right) = \mathbf{0}(\varepsilon), \\[2mm]
&\bar{w}\frac{\partial \bar{s}}{\partial \zeta} - \frac{\bar{v}}{T}\frac{\partial}{\partial \zeta}\left(\bar{\lambda}\frac{\partial T}{\partial \zeta}\right) + \bar{\mu}''\left(\frac{\partial \bar{u}}{\partial \zeta}\right)^2 + \bar{\mu}\left|\frac{\partial \bar{W}_\tau}{\partial \zeta}\right|^2 = 0(\varepsilon), \\[2mm]
&\bar{w}\frac{\partial m}{\partial \zeta} - \frac{L}{L'}(1 - m)\bar{S} = 0(\varepsilon),
\end{aligned}
\right\} \tag{III.14}
$$

where it can be seen that thermomechanical dissipation terms survive well in the first three equations when we make $\varepsilon = 0$, which justifies a posteriori the choice made above for the internal variables.

Taking account of the equalities $w = u - \delta$, $\partial\delta/\partial\zeta = 0$, and

$$\frac{\partial e}{\partial \zeta} = -p\frac{\partial v}{\partial \zeta} + T\frac{\partial s}{\partial \zeta},$$

we obtain

$$\left.\begin{array}{r}
\dfrac{\partial}{\partial \zeta}(\overline{\rho}\overline{w}) = 0(\varepsilon), \\[2mm]
\mathbf{N} \cdot \dfrac{\partial}{\partial \zeta}\left(\overline{p} - \overline{\mu}''\dfrac{\partial \overline{w}}{\partial \zeta}\right) + \overline{\rho}\overline{w}\dfrac{\partial \overline{\mathbf{W}}}{\partial \zeta} - \dfrac{\partial}{\partial \zeta}\left(\overline{\mu}\dfrac{\partial \overline{\mathbf{W}}_\tau}{\partial \zeta}\right) = \mathbf{0}(\varepsilon), \\[2mm]
\overline{\rho}\overline{w}\left(\dfrac{\partial \overline{e}}{\partial \zeta} + \overline{p}\dfrac{\partial \overline{v}}{\partial \zeta}\right) - \left[\dfrac{\partial}{\partial \zeta}\left(\overline{\lambda}\dfrac{\partial T}{\partial \zeta}\right) + \overline{\mu}''\left(\dfrac{\partial \overline{w}}{\partial \zeta}\right)^2 + \overline{\mu}\left|\dfrac{\partial \overline{\mathbf{W}}_\tau}{\partial \zeta}\right|^2\right] = 0(\varepsilon), \\[2mm]
\overline{w}\dfrac{\partial m}{\partial \zeta} - \dfrac{L}{L'}(1 - m)\overline{S} = 0(\varepsilon),
\end{array}\right\} \quad \text{(III.15)}$$

where it is interesting to note that velocity no longer occurs except through the components w and $\mathbf{W}_\tau$ of the relative velocity-vector $\mathbf{W}$.

1.6. Zero-Order Structures: Notion of Quasi C–J Detonation

With the notions and definitions introduced above, the construction through the method of matched asymptotic expansions (see Van Dyke [36]) of an approximate solution within $0(\varepsilon)$ and of the equations (III.1) and (III.7) in the neighborhood of a detonation wave can be summed up as follows.

Denoting the state of local thermodynamic equilibrium by $\mathbf{q}$ (see §1.3) and the vector $(\mathbf{U}, \mathbf{q}, m)$ by $\mathbf{Y}$, we are seeking:

— a solution $\mathbf{Y}^{(\circ)}(\zeta, \overline{\xi}', \overline{\xi}'', \hat{t})$ of the equations (III.15) where $\varepsilon = 0$;
— a solution $^{(\circ)}\mathbf{Y}^-(Z, \xi', \xi'', t)$ of the equations (III.11) where $\varepsilon = 0$, $Z \leq 0$;
— a solution $^{(\circ)}\mathbf{Y}^+(Z, \xi', \xi'', t)$ of the equations (III.11) where $\varepsilon = 0$, $Z \geq 0$;

such that

$$\mathbf{Y}^{(\circ)}(-\infty, \overline{\xi}', \overline{\xi}'', \hat{t}) = {}^{(\circ)}\mathbf{Y}^-(-0, \xi', \xi'', \hat{t}) = \mathbf{Y}_0(\overline{\xi}', \overline{\xi}'', \hat{t}), \quad \text{(III.16}^-\text{)}$$

$$\mathbf{Y}^{(\circ)}(+\infty, \overline{\xi}', \overline{\xi}'', \hat{t}) = {}^{(\circ)}\mathbf{Y}^+(+0, \xi', \xi'', \hat{t}) = \mathbf{Y}_1(\overline{\xi}', \overline{\xi}'', \hat{t}), \quad \text{(III.16}^+\text{)}$$

where $\mathbf{Y}_0(\overline{\xi}', \overline{\xi}'', \hat{t})$ is the upstream state of the explosive substance in the perfect fluid model, and where $\mathbf{Y}_1$ denotes a state of the detonation arc (H_+) of the Crussard curve of origin $\mathbf{q}_0$. Subject to extending:

$^{(\circ)}\mathbf{Y}^-$ for $Z > 0$ uniformly by $\mathbf{Y}_0$; and
$^{(\circ)}\mathbf{Y}^+$ for $Z < 0$ uniformly by $\mathbf{Y}_1$;

$$\mathbf{Y} = \mathbf{Y}^{(\circ)} + {}^{(\circ)}\mathbf{Y}^+ + {}^{(\circ)}\mathbf{Y}^- - \mathbf{Y}_0 - \mathbf{Y}_1 \quad \text{(III.17)}$$

constitutes the desired approximation. The vectors $\mathbf{Y}^{(\circ)}$, $^{(\circ)}\mathbf{Y}^-$, $^{(\circ)}\mathbf{Y}^+$, $\mathbf{Y}$ are called, respectively, zero-order internal, upstream external, downstream external, and composite structures.

Until now, apart from the chemical evolution equation, the procedure is in every way identical to that which would be used for the study of a shock considered as a layer of steep gradients in a Navier–Fourier dissipative fluid. Henceforward, it differs fundamentally from the latter procedure in that the state $\mathbf{Y}_1$ is unambiguously determined on the Hugoniot by the datum w_0, whilst it does not exist on the Crussard curve unless $w_0(\xi', \xi'', t) \geq D_*[\mathbf{q}_0(\xi', \xi'', t)]$.

Just as in the last subsection, it is experiment which directs the continuation of the model.

As a general rule, the normal upstream relative velocity w_0 varies with the position of the surface $\Sigma(t)$. This dependence results *a priori* from both:

(i) the global effect of the boundary conditions of the flow of detonation products; and

(ii) the local effect of the upstream state $\mathbf{q}_0(\xi', \xi'', t)$.

In a uniform medium, only the global dependence survives; experiment shows then—at least for the built-up detonations which we shall consider from now on—that the less the average curvature, the less w_0 varies. This experimental fact leads us to allow that $w_0(\xi', \xi'', t)$ can generally be approximated by

$$w_0(\cdot) = D^{(\circ)}[\mathbf{q}_0(\cdot)] + 0(\varepsilon)D^{(1)}[\mathbf{q}_0(\cdot), \cdot]$$

with   (III.18)

$$D^{(\circ)}[\mathbf{q}_0(\cdot)] \geq D_*[\mathbf{q}_0(\cdot)].$$

Under these conditions, the state (states) $\mathbf{Y}_1$ to be considered in order to find the zero-order structures is (are) defined on (H_+) by $w_0 = D^{(\circ)}$.

It must be stressed that the case $D^{(\circ)} = D_*$ can occur whether w_0 is greater than, equal to, or less than D_*; this case, which is comprehensively treated in what follows, is called *quasi C–J detonation* to show that it contains C–J detonation in the strict sense of ($w_0 = D_*$) but is not limited to such a detonation.

2. Zero-Order Internal Structure

2.1. Equations

By making $\varepsilon = 0$ in the equations (III.15) and returning to the physical dependent variables (with no line above them), it can be seen that $\mathbf{Y}^{(\circ)}$ should confirm the equations

$$\frac{\partial}{\partial \zeta}(\rho w) = 0,$$

$$\frac{\partial}{\partial \zeta}\left(p - \frac{\mu''}{L}\frac{\partial w}{\partial \zeta} + \rho w^2\right) = 0,$$

$$\frac{\partial}{\partial \zeta}\left(\rho w \mathbf{W}_\tau - \frac{\mu}{L}\frac{\partial \mathbf{W}_\tau}{\partial \zeta}\right) = 0,$$

$$\frac{\partial}{\partial \zeta}\left(\rho w\left(e + \frac{|\mathbf{W}|^2}{2}\right)\right) + \left(p - \frac{\mu''}{L}\frac{\partial w}{\partial \zeta}\right)w - \frac{\lambda}{L}\frac{\partial T}{\partial \zeta} - \frac{\mu}{L}\mathbf{W}_\tau \cdot \frac{\partial \mathbf{W}_\tau}{\partial \zeta} = 0,$$

$$\frac{\partial m}{\partial \zeta} = \frac{L}{w}(1 - m)S.$$

(III.19)

By introducing the functions M, K, K', K" of $(\bar{\xi}', \bar{\xi}'', \bar{t})$, the integrals of the differential system (III.19) can be written

$$\rho w = M, \tag{a}$$

$$\frac{\mu''}{L}\frac{\partial w}{\partial \zeta} = p - K + Mw, \tag{b}$$

$$\frac{\mu}{L}\frac{\partial \mathbf{W}_\tau}{\partial \zeta} = M\mathbf{W}_\tau - \mathbf{K}', \tag{c}$$

$$\frac{\lambda}{L}\frac{\partial T}{\partial \zeta} = M\left(e + \frac{|\mathbf{W}|^2}{2}\right) - K'' + (K - Mw)w + (\mathbf{K}' - M\mathbf{W}_\tau)\cdot \mathbf{W}_\tau, \tag{d}$$

$$\frac{w}{L}\frac{\partial m}{\partial \zeta} = (1 - m)S. \tag{e}$$

(III.20)

According to the conditions (III.16)$^-$ and (III.16)$^+$, a solution $\mathbf{Y}^{(o)}(\zeta; \cdot)$ must be found for the system $\mathscr{S}$ formed by the equations (III.20b–e) such that

$$\left.\begin{array}{l} w(-\infty; \cdot) = D^{(o)}(\cdot), \\ \mathbf{W}_\tau(-\infty; \cdot) = \mathbf{W}_{\tau 0}(\cdot), \\ T(-\infty; \cdot) = T_0(\cdot), \\ m(-\infty; \cdot) = 0, \end{array}\right\} \tag{III.20$^-$}$$

$$\left.\begin{array}{l} w(+\infty; \cdot) = w_1(\cdot), \\ \mathbf{W}_\tau(+\infty; \cdot) = \mathbf{W}_{\tau 1}(\cdot), \\ T(+\infty; \cdot) = T_1, \\ m(+\infty; \cdot) = 1. \end{array}\right\} \tag{III.20$^+$}$$

Now such a solution cannot exist (Sansone and Conti [34], p. 31) unless the points $(D^{(\circ)}, \mathbf{W}_{\tau 0}, T_0, 0)$ and $(w_1, \mathbf{W}_{\tau 1}, T_1, 1)$ are singular points of $\mathscr{S}$, i.e., unless the second members of $\mathscr{S}$ cancel there. This happens if, and only if

$$
\left.
\begin{aligned}
M &= \rho_0 D^{(\circ)}, \\
K &= p_0 + M^2 v_0, \\
K' &= M\mathbf{W}_{\tau 0}, \\
K'' &= M\left(e_0 + \frac{|\mathbf{W}_0|^2}{2} + p_0 v_0\right).
\end{aligned}
\right\}
\tag{III.21}
$$

Taking account of (III.21), the equations which $\mathbf{Y}^{(\circ)}$ should confirm are written

$$
\left.
\begin{aligned}
0 &= w - Mv, \\
\frac{M\mu''}{L}\frac{\partial v}{\partial \zeta} &= p - p_0 + M^2(v - v_0), \\
\frac{\mu}{ML}\frac{\partial \mathbf{W}_\tau}{\partial \zeta} &= \mathbf{W}_\tau - \mathbf{W}_{\tau 0}, \\
\frac{\lambda}{LM}\frac{\partial T}{\partial \zeta} &= (e - e_0) + (v - v_0)\left[p_0 + \frac{M^2}{2}(v_0 - v)\right] - \frac{|\mathbf{W}_\tau - \mathbf{W}_{\tau 0}|^2}{2}, \\
\frac{w}{L}\frac{\partial m}{\partial \zeta} &= (1 - m)S.
\end{aligned}
\right\}
\tag{III.22}
$$

Given that $\mathbf{W}_{\tau 1} = \mathbf{W}_{\tau 0}$ according to (I.29), the solution in $\mathbf{W}_\tau$ can only be

$$
\mathbf{W}_\tau = \mathbf{W}_{\tau 0}.
\tag{III.23}
$$

So, in fact, it is a question of finding for the system

$$
\left.
\begin{aligned}
\frac{M\mu''}{L}\frac{\partial v}{\partial \zeta} &= p - p_0 + M^2(v - v_0), \\
\frac{\lambda}{LM}\frac{\partial T}{\partial \zeta} &= e - e_0 + (v - v_0)\left[p_0 + \frac{M^2}{2}(v_0 - v)\right], \\
\frac{w}{L}\frac{\partial m}{\partial \zeta} &= (1 - m)S,
\end{aligned}
\right\}
\tag{III.24}
$$

where M equals $\rho_0 D^{(\circ)}$, a solution such that

$$
\left.
\begin{aligned}
v(-\infty; \cdot) &= v_0(\cdot), \\
T(-\infty; \cdot) &= T_0(\cdot), \\
m(-\infty; \cdot) &= 0,
\end{aligned}
\right\}
\tag{III.24$^-$}
$$

$$
\left.\begin{aligned}
v(+\infty; \cdot) &= v_1(\cdot), \\
T(+\infty; \cdot) &= T_1(\cdot), \\
m(+\infty; \cdot) &= 1.
\end{aligned}\right\} \tag{III.24}^+
$$

2.2. Existence and Behavior at Infinity

The existence of the zero-order internal structure and its behavior at infinity were analyzed by Friedrichs [20] in the case where $w_1 \neq a_1$ assuming that both the explosive and its detonation products are an ideal gas of constant specific heats c_p and c_v. We were able to generalize his results to the hypotheses of §1.3 and solve the case $w_1 = a_1$ under the same hypotheses (see Appendix B). The results thus obtained are summed up in Table III.1 assuming, for simplicity, that ζ is chosen so that $m\,(\zeta > 0) > 0$.

These results lead to two basic observations.

(a) In a strong detonation, $(v, T, m)^{(\circ)}$ reaches exponentially the downstream state tangential to:

- the direction $(v_1 P_1', T_1 Q_1', -v_1)$ if $-r_1 > v_1$,
- the direction $(v_1, -T_1 \mathscr{G}_1, 0)$ if $-r_1 < v_1$.

In the quasi C–J detonation, $(v, T, m)^{(\circ)}$ reaches the downstream state hyperbolically for v and T, exponentially for m, the overall approach being made according to the direction $(v_1, -T_1 \mathscr{G}_1, 0)$. These two behaviors, analytically distinct, are also physically distinct insofar as the downstream approximation of the strong detonation depends on the kinetic law $S(q, m)$ owing to P_1', Q_1', and k_1' whilst the downstream approximation of the quasi C–J detonation does not depend on it. For these reasons, we must exclude the idea—however natural—according to which the C–J detonation would be the limiting case of the strong detonation when $D^{(\circ)}$ tends towards $D_* + 0$.

(b) Because weak detonations are impossible (see Appendix B *in fine*) it appears that strong ($w_1 < a_1$) and quasi C–J ($w_1 = a_1$) detonations are equivalent, respectively, to the cases $D^{(\circ)} > D_*$ and $D^{(\circ)} = D_*$. As a rule it would suffice to know $D^{(\circ)}[q_0(\xi', \xi'', t)]$ and to compare it to $D_*[q_0(\xi', \xi'', t)]$ in order to say that such a detonation is locally strong or quasi C–J. But, in reality, $D^{(\circ)}$ is not a quantity which we can reach through experiment; experiment can give it only an approximate value $w_0(\xi', \xi'', t)$ which differs from it by an unknown quantity of the order of ε. In other words, it is impossible to distinguish a strong detonation from a quasi C–J detonation by a simple measurement of normal upstream velocity. We shall see later that, on the other hand, observation of the motion of detonation products makes this distinction possible.

Table III.1

$w_1 > a_1$	Weak detonations are parasitic solutions
$w_1 = a_1$	$Y^{(o)}$ "exists" and confirms systems (0) and (1)*
$w_1 < a_1$	$Y^{(o)}$ "exists" and confirms systems (0) and (1)' or (1)"

Behavior of $Y^{(o)}$ at infinity	Definition of functions $\ell, r, v, \mathcal{G}, \alpha$
(0) $\begin{cases} \dfrac{v - v_0}{v_0} = k_0 e^{r_0\zeta} + 0(e^{r_0\zeta}), \\[2mm] \dfrac{T - T_0}{T_0} = -k_0 \mathcal{G}_0 e^{r_0\zeta} + 0(e^{r_0\zeta}), \\[2mm] m = 0 \quad \text{for } \zeta = 0. \end{cases}$	$G = v\dfrac{\partial p}{\partial e}(v, e, m); \ \omega = \dfrac{\mu'' c_p}{\lambda}; \ \mathfrak{R} = \dfrac{a^2}{w^2};$
$v_1 > -r_1 = 0:$	$\mathcal{F} = \mathfrak{R}^2 + 2[(\gamma - 1)\omega + \gamma(\omega - 1)]\mathfrak{R} + (\omega - \gamma)^2.$
(1)$_*$ $\begin{cases} \dfrac{v - v_1}{v_1} = -\alpha_1 \dfrac{\ell_1}{L\zeta} + 0\left(\dfrac{1}{\zeta}\right), \\[2mm] \dfrac{T - T_1}{T_1} = \mathcal{G}_1 \alpha_1 \dfrac{\ell_1}{L\zeta} + 0\left(\dfrac{1}{\zeta}\right), \\[2mm] m - 1 = -v_1 k_1 e^{-v_1\zeta} + 0(e^{-v_1\zeta}). \end{cases}$	$\ell = \begin{cases} \dfrac{\mu''}{M}\dfrac{2\gamma}{\sqrt{\mathcal{F}} + \mathfrak{R} - \gamma - \omega} & \text{if } \mathfrak{R} \neq 1, \\[3mm] \dfrac{\lambda v}{a c_p}(\omega + \gamma - 1) & \text{if } \mathfrak{R} = 1. \end{cases}$
$v_1 > -r_1 > 0:$	$r = -\dfrac{L}{\ell},$
(1)' $\begin{cases} \dfrac{v - v_1}{v_1} = k_1' e^{r_1\zeta} + 0(e^{r_1\zeta}), \\[2mm] \dfrac{T - T_1}{T_1} = -k_1' \mathcal{G}_1 e^{r_1\zeta} + 0(e^{r_1\zeta}), \\[2mm] m - 1 = -v_1 k_1 e^{-v_1\zeta} + 0(e^{-v_1\zeta}). \end{cases}$	$v = \dfrac{LS}{w},$ $\mathcal{G} = \dfrac{2G\omega}{\sqrt{\mathcal{F}} + \mathfrak{R} - \gamma + \omega},$
$-r_1 > v_1 > 0:$	$\alpha = \dfrac{\rho^3 w^2}{\dfrac{\partial^2 p}{\partial v^2}(v, s, m)},$
(1)" $\begin{cases} \dfrac{v - v_1}{v_1} = k_1 P_1' e^{-v_1\zeta} + 0(e^{-v_1\zeta}), \\[2mm] \dfrac{T - T_1}{T_1} = k_1 Q_1' e^{-v_1\zeta} + 0(e^{-v_1\zeta}), \\[2mm] m - 1 = -v_1 k_1 e^{-v_1\zeta} + 0(e^{-v_1\zeta}). \end{cases}$	N.B. (1) $r_0 > 0, r_1 < 0, \alpha_1 > 0;$ (2) $\mathcal{G}G > 0;$ (3) $\mathfrak{R} = 1 \Rightarrow \mathcal{G} = G.$
P_1' and Q_1' are defined in Appendix B. $k_0, k_1, k_1',$ are constants peculiar to the solution $Y^{(o)}$.	

3. Zero-Order External Downstream Structure

3.1. Equations

By making $\varepsilon = 0$ in the equations (III.11) then coming back to the physical dependent variables using (III.10), we see that $^{(\circ)}Y^+$ confirms the equations

$$
\left.
\begin{aligned}
\frac{\partial u}{\partial Z} - w\frac{\partial \text{ Log } v}{\partial Z} &= -uK + vZ\frac{\partial N}{\partial \tilde{t}} \cdot \bar{J}\rho - (\rho U)_{\tau,\tilde{t}} + \tilde{D}\frac{\partial \text{ Log } v}{\partial \tilde{t}}, \\
w\frac{\partial U}{\partial Z} + v\frac{\partial \rho}{\partial Z}N &= v[U(\rho U)_{\tau,\tilde{t}} - (\rho U \otimes U)_{\tau,\tilde{t}} - \bar{J}p] - \tilde{D}\frac{\partial U}{\partial \tilde{t}} \\
&\quad - v\tilde{D}Z\left[U\left(\frac{\partial N}{\partial \tilde{t}} \cdot \bar{J}\rho\right) - \frac{\partial N}{\partial \tilde{t}} \cdot \bar{J}\rho U\right], \\
w\frac{\partial s}{\partial Z} &= \left(\tilde{D}Z\frac{\partial N}{\partial \tilde{t}} - W_\tau\right) \cdot \bar{J}s - \tilde{D}\frac{\partial s}{\partial \tilde{t}}, \\
(1 - m)S &= 0.
\end{aligned}
\right\} \quad \text{(III.25)}
$$

According to the condition (III.16)$^+$ a solution $^{(\circ)}Y^+(Z, \cdot)$ of (III.25) must be sought, such that

$$\langle v, U, s; m\rangle(+0, \cdot) = \langle v_1, U_1, s_1(\cdot); 1\rangle.$$

Let us note here and now the consequences of two hypotheses satisfied in every current experimental case:

• *the upstream state is uniform*;

then $D^{(\circ)}$ does not depend on (ξ', ξ'', t) over $\Sigma(t)$ with the result that the state Y_1 defined as shown in §1.6 *in fine*, confirms

$$\frac{\partial Y_1}{\partial \xi'} = \frac{\partial Y_1}{\partial \xi''} = \frac{\partial Y_1}{\partial t} = 0; \quad \text{(III.26a)}$$

• *the upstream state is at rest*;

then according to (I.29)

$$W_{\tau 1} = 0. \quad \text{(III.26b)}$$

To shorten the discussion, a detonation which takes place following the two hypotheses stated above will be called *simple*.

3.2. Behavior in Downstream State of a Simple Detonation

For a simple detonation, the nullity of the transverse gradients and indifference to time in the downstream state ((see (III.26)) make it possible to write the equations (III.25) in the very simple form

$$
\left.
\begin{array}{r}
\dfrac{\partial u}{\partial Z} - w\dfrac{\partial \operatorname{Log} v}{\partial Z} = \dfrac{2u}{R_m} + \cdots, \\[2ex]
w\dfrac{\partial u}{\partial Z} - a^2\dfrac{\partial \operatorname{Log} v}{\partial Z} = \cdots, \\[2ex]
\dfrac{\partial \mathbf{W}_\tau}{\partial Z} = \cdots, \\[2ex]
\dfrac{\partial s}{\partial Z} = \cdots,
\end{array}
\right\}
\tag{III.27}
$$

which involves the average curvature $1/R_m = \frac{1}{2}(1/R' + 1/R'')$, and where the ellipses denote quantities which tend towards 0 when Z tends toward $+0$. An equivalent form of the first two equations is

$$
\left.
\begin{array}{r}
(a^2 - w^2)\dfrac{\partial u}{\partial Z} = 2\dfrac{ua^2}{R_m} + \cdots, \\[2ex]
(a^2 - w^2)\dfrac{\partial \operatorname{Log} v}{\partial Z} = 2\dfrac{uw}{R_m} + \cdots,
\end{array}
\right\}
\tag{III.28}
$$

where we see that the behavior of $^{(\circ)}Y^+$ in $Z = 0$ is different according to whether the detonation is strong ($w_1 < a_1$) or quasi C–J ($w_1 = a_1$). In the first case, the result is obtained immediately, whereas in the second, the conclusion is not obvious. In Appendix C we show the results given in Table III.2, and illustrated by Figure III.1.

N.B. This analysis is, of course, faulty if $R_m^{-1} = 0$. But, in view of the significance of the surface $\Sigma(t)$—the geometric approximation of the steep gradients layer in the perfect fluid model—this case is of physical interest only insofar as it corresponds to a uniaxial flow ($|r'| = |r''| = 0$). This circumstance will be considered favorably in Section 5 at the same time as the other one-dimensional propagations (cylindrical and spherical).

Table III.2. α is defined in Table III.1, and $\beta = -((w/u)\alpha)$.

Detonation	Strong ($w_1 < a_1$)	Quasi C–J ($w_1 = a_1$)
$\dfrac{u - u_1}{u} =$	$\dfrac{2a_1^2}{a_1^2 - w_1^2}\dfrac{Z}{R_m} + 0\left(\dfrac{Z}{R_m}\right)$	$\pm\sqrt{\beta_1 \dfrac{Z}{R_m}} + 0\left(\sqrt{\dfrac{Z}{R_m}}\right)$
$\dfrac{v - v_1}{v_1} =$	$\dfrac{2u_1 w_1}{a_1^2 - w_1^2}\dfrac{Z}{R_m} + 0\left(\dfrac{Z}{R_m}\right)$	$\pm 2\dfrac{u_1}{a_1}\sqrt{\beta_1 \dfrac{Z}{R_m}} + 0\left(\sqrt{\dfrac{Z}{R_m}}\right)$

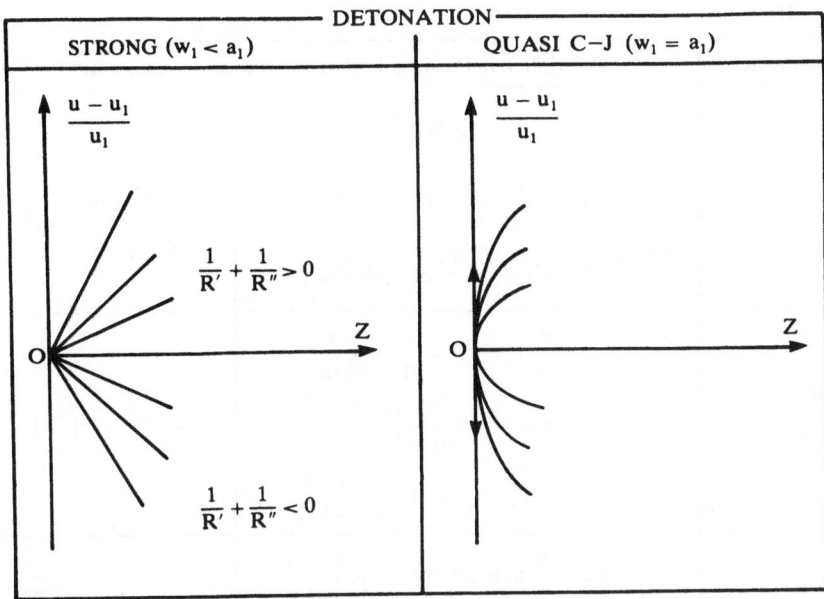

Figure III.1. Sketch of the zero-order field of external downstream velocity in the neighborhood of the downstream state.

4. Zero-Order Composite Structure of a Simple Detonation

4.1. Description in the Neighborhood of the Downstream State

The zero-order composite structure of a simple detonation, as defined by (III.17), amounts to

$$\mathbf{Y}(z, \xi', \xi'', t) = \mathbf{Y}^{(o)}\left(\frac{z}{L}\right) + {}^{(o)}\mathbf{Y}^+\left(\frac{\xi'}{\tilde{L}}, \frac{\xi''}{\tilde{L}}, \frac{\tilde{D}t}{\tilde{L}}\right) - \mathbf{Y}_1,$$

and the results summarized in Tables III.1 and III.2 make it possible to give a description of a simple detonation in the neighborhood of the downstream state. The three possible cases are given in Table III.3.

These results have a great wealth of consequences. We comment only on the four points which appear most noteworthy:

(a) In the neighborhood of the downstream state, the description can contain terms affected by thermomechanical dissipation: those which exhibit the length ℓ_1 (defined in Table III.1), which itself depends on the longitudinal Prandtl number as well as on one of the two coefficients of transport λ and μ''.

(b) We meet once more, in a way which is quite striking, something which had already appeared separately on each of the zero-order internal and

Table III.3

Detonation	Description in the neighborhood of the downstream state	
Quasi C–J	$\dfrac{u(z,\cdot)}{u_1} = 1 + \beta_1\dfrac{\ell_1}{z} \pm 2\sqrt{\beta_1 z/r_m},$ $\dfrac{v(z,\cdot)}{v_1} = 1 + \dfrac{u_1}{a_1}\left(\beta_1\dfrac{\ell_1}{z} \pm 2\sqrt{\beta_1 z/r_m}\right),$ $\dfrac{T(z,\cdot)}{T_1} = 1 - G_1\dfrac{u_1}{a_1}\left(\beta_1\dfrac{\ell_1}{z} \pm 2\sqrt{\beta_1 z/r_m}\right).$	$+0\left(\dfrac{\ell_1}{z}\right) + 0\left(\sqrt{\dfrac{z}{r_m}}\right)$
Strong $\ell_1 > \dfrac{w_1}{S_1}$	$\dfrac{u(z,\cdot)}{u_1} = 1 + \dfrac{w_1}{u_1}k_1' e^{-z/\ell_1} + \dfrac{2a_1^2}{a_1^2 - w_1^2}\dfrac{z}{r_m},$ $\dfrac{v(z,\cdot)}{v_1} = 1 + k_1' e^{-z/\ell_1} + \dfrac{2u_1 w_1}{a_1^2 - w_1^2}\dfrac{z}{r_m},$ $\dfrac{T(z,\cdot)}{T_1} = 1 - \mathcal{G}_1 k_1' e^{-z/\ell_1} - G_1\dfrac{2u_1 w_1}{a_1^2 - w_1^2}\dfrac{z}{r_m}.$	$+0(e^{-z/\ell_1}) + 0\left(\dfrac{z}{r_m}\right)$
Strong $\ell_1 < \dfrac{w_1}{S_1}$	$\dfrac{u(z,\cdot)}{u_1} = 1 + \dfrac{w_1}{u_1}k_1 P_1' e^{-S_1 z/w_1} + \dfrac{2a_1^2}{a_1^2 - w_1^2}\dfrac{z}{r_m},$ $\dfrac{v(z,\cdot)}{v_1} = 1 + k_1 P_1' e^{-S_1 z/w_1} + \dfrac{2u_1 w_1}{a_1^2 - w_1^2}\dfrac{z}{r_m},$ $\dfrac{T(z,\cdot)}{T_1} = 1 + k_1 Q_1' e^{-S_1 z/w_1} - G_1\dfrac{2u_1 w_1}{a_1^2 - w_1^2}\dfrac{z}{r_m}.$	$+0(e^{-S_1 z/w_1}) + 0\left(\dfrac{z}{r_m}\right)$

external structures: quasi C–J detonation cannot in any way be considered as the limit of strong detonation when $D^{(o)}$ tends towards $D_* + 0$.

(c) In quasi C–J detonation, in the neighborhood of the downstream state, the evolution of the kinematic u and physical v and T dependent variables does not depend on the chemical reactivity $S(s, m)$; it depends only on the values of state variables and dissipation coefficients calculated in the C–J state of the detonation products, a state which is a thermodynamic characteristic of the medium considered.

(d) The ambiguity of sign which remains in the development relative to quasi C–J detonation cannot be removed without an appropriate discussion (see the following subsection).

N.B. For the interpretation of the results summarized in Table III.3, it is essential to note that

$$u_1 = D^{(o)}\left(\frac{v_1}{v_0} - 1\right) < 0,$$

in accordance with the upstream rest hypothesis, with $w_0 = D^{(o)}$ and with the relations (III.20a) and (III.21a).

4.2. Rules of Propagation; Notion of Autonomous Detonation

In §2.2 we saw the necessity of distinguishing weak detonation ($w_1 > a_1$), strong detonation ($w_1 < a_1$), and quasi C–J detonation ($w_1 = a_1$) in order to be more specific about the existence of $\mathbf{Y}^{(o)}$ and its behavior at $\zeta = +\infty$; we then concluded that the first was impossible, and so decided on the equivalence relations

$$w_1 < a_1 \quad \Leftrightarrow \quad D^{(o)} > D_*,$$
$$w_1 = a_1 \quad \Leftrightarrow \quad D^{(o)} = D_*.$$

To be able to say whether a detonation is strong or quasi C–J, it would suffice, in principle, to know the velocity $D^{(o)}$ and to compare its value to that of D_* calculated *a priori*. But, in fact, $D^{(o)}$ is not a quantity which can be found by experiment, which can only give an approximate value $w_0(\xi', \xi'', t)$ for it, differing from it by a quantity which is generally small, certainly, but unknown. So it is impossible to determine the nature of the wave solely from a measurement of the velocity w_0. On the other hand, some valuable conclusions result from consideration of the curvature of the wave, starting from the results given in Table III.3.

Two of these are immediate:

(i) a quasi C–J detonation necessarily has a positive curvature; and
(ii) a detonation with negative curvature is necessarily strong.

Other, even more precise, conclusions are valid for the case of *autonomous* detonations where propagation is assured simply by releasing chemical energy to the exclusion of any compression wave traveling through the flow of products up to the ignitor shock. In this case, the ignitor shock is followed by a decompression, which means—amongst other things—that the field of absolute normal velocities $u(z)$ confirms

$$u(z)/u_1 < 1 \qquad \text{for} \quad z \gg \ell_1. \tag{III.29}$$

Referring to Table III.3, we see that (III.29) makes it possible to:

(iii) remove the ambiguity of sign in front of the term in $\sqrt{z/r_m}$ in the case of autonomous quasi C–J detonation; and
(iv) conclude that an autonomous detonation with positive curvature is necessarily quasi C–J.

The four conclusions above are essential for the interpretation of experimental results where the curvature is known, whether it results directly from the initial stress conditions or is observed in the course of propagation. Thus, as an example, we can state that:

- a convergent spherical detonation is strong;
- an autonomous divergent spherical detonation is quasi C–J;

- a concave axisymmetric detonation is strong; and
- an autonomous convex axisymmetric detonation is quasi C–J.

The following section, which deals with one-dimensional detonations (uniaxial, cylindrical, spherical), highlights the differences with separate autonomous and nonautonomous divergent propagations and from another point of view justifies the name chosen.

5. One-Dimensional Detonations

5.1. Zero-Order External Structure Equations

The symmetry of one-dimensional configurations makes the "transverse" terms disappear from equations (III.25) which take the *exact* form

$$\left.\begin{aligned}
\frac{\partial u}{\partial Z} - w\frac{\partial \operatorname{Log} v}{\partial Z} &= \frac{Nu}{R - Z} + \tilde{D}\frac{\partial \operatorname{Log} v}{\partial \bar{t}}, \\
w\frac{\partial u}{\partial Z} + v\frac{\partial p}{\partial Z} &= -\tilde{D}\frac{\partial u}{\partial \bar{t}}, \\
w\frac{\partial s}{\partial Z} &= -\tilde{D}\frac{\partial s}{\partial \bar{t}}, \\
m &= 1,
\end{aligned}\right\} \tag{III.30}$$

where N is 0, 1, or 2 according to whether the wave is plane, cylindrical, or spherical.

As in §3.2, let us solve in $\partial u/\partial Z$, $(\partial \operatorname{Log} v)/\partial Z$, $\partial s/\partial Z$, by taking account of

$$\left.\begin{aligned}
\frac{\partial p}{\partial Z} &= -\frac{a^2}{v}\frac{\partial \operatorname{Log} v}{\partial Z} + \frac{\partial p}{\partial s}\frac{\partial s}{\partial Z}, \\
\frac{\partial p}{\partial \bar{t}} &= -\frac{a^2}{v}\frac{\partial \operatorname{Log} v}{\partial \bar{t}} + \frac{\partial p}{\partial s}\frac{\partial s}{\partial \bar{t}},
\end{aligned}\right\} \tag{III.31}$$

we obtain

$$\left.\begin{aligned}
(a^2 - w^2)\frac{\partial u}{\partial Z} &= \frac{Nua^2}{R - Z} + \tilde{D}\left(w\frac{\partial u}{\partial \bar{t}} - v\frac{\partial p}{\partial \bar{t}}\right), \\
(a^2 - w^2)\frac{\partial \operatorname{Log} v}{\partial Z} &= \frac{Nuw}{R - Z} + \tilde{D}\left(\frac{\partial u}{\partial \bar{t}} - \frac{v}{w}\frac{\partial p}{\partial \bar{t}} + \frac{w^2 - a^2}{w}\frac{\partial \operatorname{Log} v}{\partial \bar{t}}\right), \\
w\frac{\partial s}{\partial Z} &= -\tilde{D}\frac{\partial s}{\partial \bar{t}}.
\end{aligned}\right\} \tag{III.32}$$

At this stage, it is convenient to return to the independent physical variables

$$(a^2 - w^2)\frac{\partial \, \text{Log} \, |u|}{\partial z} = +a^2 \left[\frac{N}{r-z} + \frac{w}{ua^2}\left(\frac{\partial u}{\partial t} - \frac{1}{\rho w}\frac{\partial p}{\partial t}\right) \right],$$

$$(a^2 - w^2)\frac{\partial \, \text{Log} \, v}{\partial z} = uw \left[\frac{N}{r-z} + \frac{1}{uw}\left(\frac{\partial u}{\partial t} - \frac{1}{\rho w}\frac{\partial p}{\partial t} + \frac{w^2 - a^2}{w}\frac{\partial \, \text{Log} \, v}{\partial t}\right) \right],$$

$$w\frac{\partial s}{\partial z} = -\frac{\partial s}{\partial t},$$

$$\text{(III.33)}$$

and to introduce the acceleration A(t) by

$$\frac{\partial u}{\partial t} - \frac{1}{\rho w}\frac{\partial p}{\partial t} = \frac{u}{w}A(t). \tag{III.34}$$

Proceeding as in §3.2, we obtain:

- if the detonation is strong ($w_1 < a_1$)

$$\frac{u - u_1}{u_1} = z\frac{a_1^2}{a_1^2 - w_1^2}\left[\frac{N}{r} + \frac{A_1}{a_1^2}\right] + 0\left(\frac{z}{r}\right),$$

$$\frac{v - v_1}{v_1} = z\frac{u_1 w_1}{a_1^2 - w_1^2}\left[\frac{N}{r} + \frac{A_1}{w_1^2} + \frac{w_1^2 - a_1^2}{u_1 w_1^2}\frac{\partial \, \text{Log} \, v_1}{\partial t}\right] + 0\left(\frac{z}{r}\right), \tag{III.35}$$

$$s - s_1 = \frac{1}{w_1}\left(\frac{\partial s}{\partial t}\right)_1 z + 0\left(\frac{z}{r}\right).$$

- if the detonation is quasi C–J ($w_1 = a_1$)

$$\frac{u - u_*}{u_*} = \pm\sqrt{2\beta_*\left(\frac{N}{r} + \frac{A_*}{a_*^2}\right)} + 0\left(\frac{z}{r}\right)^{1/2},$$

$$\frac{v - v_*}{v_*} = \pm\frac{u_*}{a_*}\sqrt{2\beta_*\left(\frac{N}{r} + \frac{A_*}{a_*^2}\right)}z + 0\left(\frac{z}{r}\right)^{1/2}, \tag{III.36}$$

$$s - s_* = \frac{1}{w_*}\left(\frac{\partial s}{\partial t}\right)_* z + 0\left(\frac{z}{r}\right).$$

N.B. It is possible to transform A_1 taking the jump relations into account; however, the expression obtained is not simple except in the hypothesis $u_0 = 0$ which leads to

$$A_1 = w_1 \frac{a}{\partial t} \, \text{Log}(\rho_0 D^{(o)} u_1^2). \tag{III.37}$$

The main significance of the expansions (III.35) and (III.36) is that they show the essential role of the relations (III.26a and b)—which express the hypothesis of a uniform upstream rest state—in the demonstration of the propagation rules set out in §4.2. Thus, as an example, and taking account of $\beta_* > 0$ according to (II.16a), it can be seen that a nonsimple convergent detonation could be quasi C–J, provided that A_* is sufficiently large so that

$$\frac{A_*}{a_*^2} + \frac{N}{r} > 0. \tag{III.38}$$

It can also be seen that, except for the factor a_*^2, the quantity which appears on the left in (III.38) is the one which is written λ which occurs in Brun's article [6] on sonic shock waves; the above calculations, carried out within the context of the analysis of zero-order structures for both strong and quasi C–J detonations give it a different emphasis.

The case of quasi C–J divergent detonation calls for special attention. Indeed, the characteristics $\mathfrak{C}^+$ of a solution flow of the equations (III.33) have as an envelope—in the plane (r, t)—the line $\mathscr{L}$ of local slope D_*. Since two distinct $\mathfrak{C}^+$ cannot pass through a point next to $\mathscr{L}$, one and the same $\mathfrak{C}^+$ cannot both end at $\mathscr{L}$ and start off again from the same side. So two possible configurations remain; in the first (1) they move away from $\mathscr{L}$ when t increases; in the second (2) they come to an end at $\mathscr{L}$.

Brun's study [6], reinterpreted in the context of zero-order structures, makes it possible   in formulas (III.36)—to link:

— the *minus* sign to configuration (1), therefore a release for $z \gg \ell_*$, and an autonomous detonation in the sense of §4.2,
— the *plus* sign to configuration (2), therefore a compression for $z \gg \ell_*$, and a nonautonomous detonation.

Formulas (III.36) thus specified make it possible, quite obviously, to initialize the calculation of the solution of (III.33) starting from an arc of $\mathscr{L}$ insofar as they make it possible to substitute for the latter an infinitely close arc which carries new data from which the $\mathfrak{C}^+$ do not start tangentially, and thus come back to a normal Cauchy problem.

Regarding the gradual construction of the solution, Brun also makes interesting observations from which we borrow the following. Consider a point P next to $\mathscr{L}$ "linked" by a $\mathfrak{C}^+$, a $\mathfrak{C}^-$, and a trajectory, respectively, to the points L^+, L^-, and L (see Fig. III.2). In the configuration (1), $t_{L^+} < t_L < t_{L^-}$ with the result that the domain of dependence of P contains its own trajectory. On the other hand, in the configuration (2), $t_L < t_{L^-} < t_{L^+}$ with the result that the domain of dependence of P does not contain its own trajectory. In the first case only, a part of the downstream zero-order external structure is determined at a given instant by the succession of the states which the wave has encountered in the course of its propagation, independently of the conditions at the downstream boundary (by "independently" it must be understood that the conditions at the downstream boundary are involved in the determination of the structure as a whole, but that this involvement is limited to the determination of the extent of an *autonomous zero-order domain* and the solution outside this domain). In the second case,

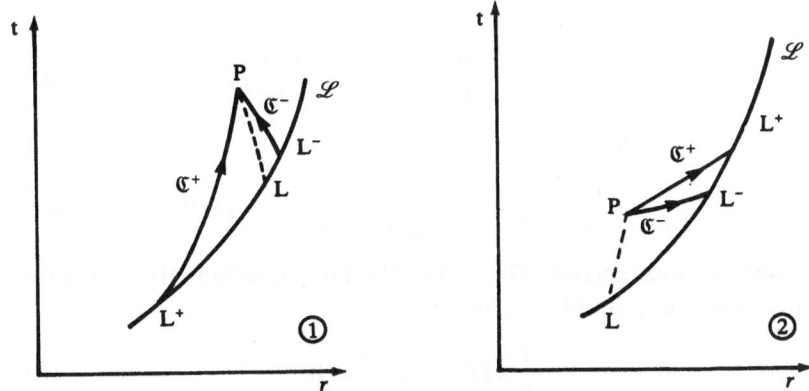

Figure III.2. Sketch of characteristics in the neighborhood of the wave.

on the other hand, the situation is identical to that which we meet down-stream of a shock. Thus for a quasi C–J detonation, the notion of *autonomous* detonation and its significance, which is as much physical (no compression coming from the boundaries) as mechanical (existence of a zero-order domain of autonomy) are greatly clarified.

5.2. Divergent Simple Detonation (N = 1 or 2)

Because of its geometric ($r = r' = r'' > 0$) and physical simplicity (uniform upstream state of rest in a frame linked to the symmetry element), divergent simple detonation has received the attention of many authors, from Jouguet [27] to Taylor [35] and Zeldovich [42], and, more recently, Chéret [11] and Brun [6]. Notations suited to this simplicity are essential: the distance $x = r - z$ to the element of symmetry, as well as the dimensionless variable $\eta = x/r$. Moreover, it is essential to note that, in accordance with (III.8), divergent simple detonation is necessarily propagated with a *constant velocity to within* $0(\varepsilon)$; in other words, in the zero-order approximation in ε accepted below, everything occurs as though the detonation were being propagated with a constant velocity $D^{(o)}$.

Taking up again *from this standpoint* (i.e., in the framework of zero-order structures) the question put by Brun in [6], we examine below the nature of detonation when the boundary B of the downstream domain moves away from the symmetry element (point, axis) with a constant velocity $\dot{x}_B$ selected over the interval $[0, +\infty[$; in such a situation, the variable η—which by definition equals 1 over the wave—also maintains a constant value $\dot{x}_B/D^{(o)}$ over B. This observation leads us to look closely at the solutions of (III.33) which depend only on η, i.e., on z/r, therefore self-similar in a frame linked to the wave.

For every quantity occurring in (III.33) we have

$$\frac{\partial}{\partial z} = -\frac{1}{r}\frac{d}{d\eta}, \qquad \frac{\partial}{\partial t} = \frac{1}{r}\frac{z}{t}\frac{d}{d\eta}, \qquad \text{(III.39)}$$

so also, by introducing $\dot{x} = -u > 0$,

$$w\frac{\partial}{\partial z} + \frac{\partial}{\partial t} = \frac{1}{r}\left(-w + \frac{z}{t}\right)\frac{d}{d\eta} = \frac{1}{r}(\dot{x} - \eta D^{(\circ)})\frac{d}{d\eta}. \qquad \text{(III.39')}$$

First, we take into account (III.39') in the last equation (III.33) relative to entropy; before any simplification, we get

$$\frac{1}{r}(\eta D^{(\circ)} - \dot{x})\frac{\partial s}{\partial \eta} = 0. \qquad \text{(III.40)}$$

Now, any solution of (III.33) satisfies, *outside the boundary* B,

$$\eta D^{(\circ)} - \dot{x} > 0 \qquad \text{(III.41)}$$

such as results from an extension of Brun's observation in [6]. What we have, indeed, in general, is

$$\eta D^{(\circ)} - \dot{x} \equiv \frac{x}{t} - \frac{dx}{dt} \equiv -t\frac{d}{dt}\left(\frac{x}{t}\right).$$

For $\eta = 1$, i.e., on the wave, $(\eta D^{(\circ)} - \dot{x})$ equals $(D^{(\circ)} - \dot{x}_1)$ which is no other than w_1, the downstream material velocity relative to the wave measured in the local frame, and always positive. It is then sufficient to show that x/t is a monotonic function of time (necessarily decreasing according to the preceding sign) along any trajectory. If x/t were to go through an extreme value $(x/t)_m$, on a trajectory $\mathcal{T}_m$, the latter would be locally on the same side of the straight line $x = (x/t)_m t$. Because all the trajectories can be generated from $\mathcal{T}_m$ by homothetic transformations of the center O, two trajectories would pass through any point adjacent to the line $x = (x/t)_m t$ taken on the appropriate side, which is impossible.

Thus, equation (III.40) is equivalent to

$$\frac{ds}{d\eta} = 0, \qquad \text{(III.42a)}$$

which means that the entropy $s(\eta)$ is uniform over every interval where it is continuous and differentiable; in other words, the entropy can possibly vary over a trajectory only in proportion to the passage of a shock for a value $\eta_c \in [\dot{x}_B/D^{(\circ)}, 1]$.

This preliminary result allows us, in the first two equations (III.33), to express $\partial p/\partial t$ according to the velocity of sound and $\partial v/\partial t$. So the application of the formulas (III.39) leads to a homogeneous linear system in $d \operatorname{Log} \dot{x}$, $d \operatorname{Log} \rho$, and $d \operatorname{Log} \eta$

$$\frac{d \operatorname{Log} \dot{x}}{a^2} = \frac{d \operatorname{Log} \rho}{\dot{x}(\eta D^{(\circ)} - \dot{x})} = N\frac{d \operatorname{Log} \eta}{(\eta D^{(\circ)} - \dot{x})^2 - a^2}. \qquad \text{(III.42b)}$$

N.B. The calculations which lead to (III.42b) are simpler if we go back to the form (III.30) from which (III.33) came.

Let us leave the detonation context for the moment and turn to the integration of the differential system (III.42) starting from ($\eta = 1$; $\dot{x} = \dot{x}_1$, $\rho = \rho_1$, $s = s_1$) by *decreasing values of* η. Three cases appear, whose nature as stated below justifies the choice of reference letters.

($\underline{S}$) The state (x_1, p_1, s_1) is *subsonic relative to the wave*

$$D^{(o)} - \dot{x}_1 < a_1,$$

so the integration of (III.42) is immediate (as soon as the velocity of sound $a(\rho, s)$ is known) and provides—according to (III.41)—a series of states which are *subsonic relative to the wave*. It can be seen, in particular, that $\dot{x}(\eta; \cdot)$ grows from $\dot{x}_1 < D^{(o)}$ when η decreases from 1, which justifies the introduction of the value $\dot{x}_B$ defined by

$$\dot{x}\left(\frac{\dot{x}_B}{D^{(o)}}; \cdot\right) = \dot{x}_B.$$

($\underline{S}^k$) The state ($\dot{x}_1$, ρ_1, s_1) is *sonic relative to the wave*

$$D^{(o)} - \dot{x}_1 = a_1$$

and the integration of (III.42) is initialized by the formulas of Table III.2 with the $+$ sign. This takes us back to the preceding case with just the one difference: the slope of $\dot{x}(\eta; \cdot)$ is infinite in $\eta = 1$ instead of finite. We note $\dot{x}_B^k$ the particular value taken here by the velocity $\dot{x}_B$ defined above.

($\bar{S}$) The state ($\dot{x}_1$, ρ_1, s_1) is *sonic relative to the wave*

$$D^{(o)} - \dot{x}_1 = a_1$$

and the integration of (III.42) is initialized by the formulas of Table III.2 with the $-$ sign. Thus integration leads to a series of states which are *supersonic relative to the wave*. We note, in particular, that $\dot{x}(\eta; \cdot)$ decreases from $\dot{x}_1$ until it reaches the value O for $\eta = \eta_0$ tangentially to $\dot{x} = 0$ (see [31], p. 609). The only feature common to the solutions ($\underline{S}^k$) and ($\bar{S}$) is the infinite value of the slope $\dot{x}(\eta; \cdot)$ in $\eta = 1$.

The different features mentioned above for the cases ($\underline{S}$), ($\underline{S}^k$), and ($\bar{S}$) appear in Figure III.3.

Let us now return to the problem raised at the start of this subsection. Consideration of the extreme cases $\dot{x}_B = 0$ and $\dot{x}_B \gg D^{(o)}$ lets us assume the existence, downstream of detonation, of

— an expansion for fairly small $\dot{x}_B$ (the detonation is then autonomous in the sense of §4.2); and
— a compression for fairly strong $\dot{x}_B$;

and therefore the existence of a critical value for the boundary velocity which separates the quasi C–J detonation regime from the strong detonation regime. In order to specify this assumption and the critical value, we shall show

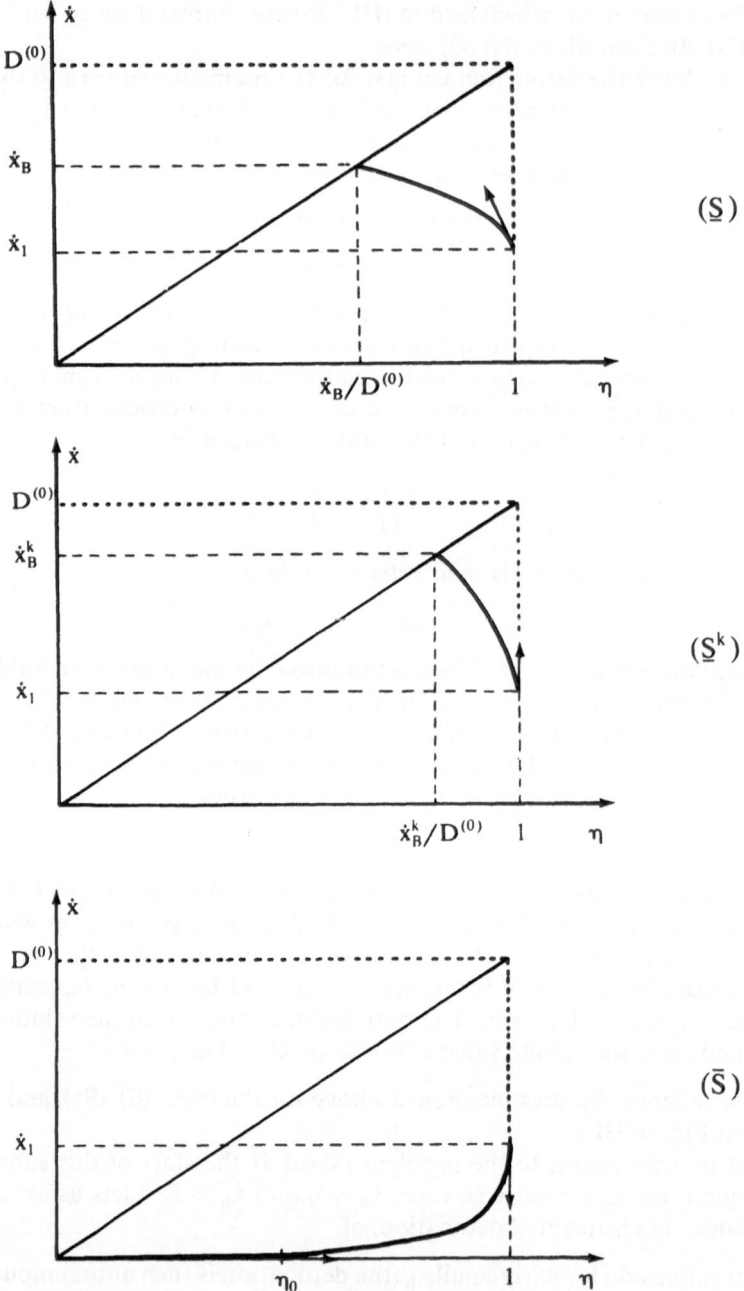

Figure III.3. Integration of the differential system (III.42) based on ($\eta = 1$; $x = \dot{x}_1$, $\rho = \rho_1$, $s = s_1$). The profiles $\underline{S}$ ($\eta \leq 1$) and $\underline{S}^k$ ($\eta < 1$) are formed from subsonic states relative to the wave; the profile $\bar{S}$ ($\eta < 1$) is formed from supersonic states relative to the wave; the states $\underline{S}^k$ ($\eta = 1$) and $\bar{S}$ ($\eta = 1$) are sonic.

that it is possible, for every value of $\dot{x}_B$ in the interval $[0, +\infty[$, to construct $^{(\circ)}Y^+$ starting from the locally self-similar solutions of (III.42).

To do this we shall consider the solution obtained by juxtaposing upstream and downstream, respectively, of a shock $\mathfrak{C}$ with absolute velocity $D_*\eta_C$, the solution $(\bar{S})$ and the solution $(\underline{S})$ relative to the downstream state of the shock $\mathfrak{C}$. Using the properties of $(\bar{S})$ and $(\underline{S})$ which have just been analyzed, and recalling the jump relations, we can establish that, when η_C increases from $(\eta_0 + 0)$ to $(1 - 0)$:

- the absolute velocity of $\mathfrak{C}$ increases from $(\eta_0 D_* + 0)$ to $(D_* - 0)$;
- the absolute velocity upstream of $\mathfrak{C}$ increases from $+0$ to $(|u_*| - 0)$;
- the relative velocity upstream of $\mathfrak{C}$ increases from $(\eta_0 D_* + 0)$ to $(a_* - 0)$; and
- the boundary velocity varies from $+0$ to $\dot{x}_B^k - 0$.

The group of solutions thus defined, completed in :

- $\eta_C = \eta_0$ by the solution $(\bar{S})$ itself extended by a uniform rest from η_C to 0; and
- $\eta_C = 1$ by the solution $(\underline{S}^k)$;

makes it possible to satisfy every boundary velocity over $[0, \dot{x}_B^k]$.

Moreover, the group of solutions $(\underline{S})$ relative to the strong detonation states makes it possible to satisfy every boundary velocity on $]\dot{x}_B^k, +\infty[$.

Thus it appears that $\dot{x}_B^k$ is the critical value of the boundary velocity below which the detonation is quasi C–J and above which detonation is strong, below which a zero-order domain of autonomy exists and above which such a domain no longer exists. For $\dot{x}_B = \dot{x}_B^k$ we have a bifurcating configuration which obviously merits the name of *critical detonation*.

In [6], Brun gives the value of $\dot{x}_B^k/|u_*|$ and of η_0 in the hypothesis of a $\bar{\gamma}$-polytropic behavior of the detonation products for different values of the coefficient $\bar{\gamma}$. We give below those values relative to $\bar{\gamma} = 3$, nicely representative of explosives in the condensed phase.

N	1	2
$\dot{x}_B^k/|u_*|$	2.40	2.88
η_0	0.49	0.45

Respecting the orders of magnitude, we illustrate synthetically the set of results above in Table III.4 plotted on the plane $(\eta D^{(\circ)}, \dot{x})$, relative to the spherical case: the velocity profiles are represented by unbroken lines, whilst the locus $C_0 C' JD$ of downstream states of shock $\mathfrak{C}$ when the boundary velocity varies from 0 to $+\infty$ is shown by dotted lines. Its bearing in C_0 and in J is not evident and results from the following observations:

- in the neighborhood of $J + 0$, the infinite slope results from the minimum property of the velocity of detonation compared with the jump in material velocity when we move along the H_+ arc of the Crussard curve in the neighborhood of the C–J point;

- in the neighborhood of $J - 0$, the shock $\mathfrak{C}$ is close to a sound wave of velocity a_*; the infinite slope results from the infinite slope of $(\bar{S})$ and the proportionality in the neighborhood of J between the velocity difference $\eta D_* - a_*$ and the jump in material velocity $\dot{x}_{C'} - \dot{x}_C$; and
- in the neighborhood of $C_0 + 0$, the shock $\mathfrak{C}$ is close to a sound wave with the speed $\eta_0 D_*$; the nonnull finite slope results from the null finite slope of $(\bar{S})$ in C_0 (see above) and the proportionality in the neighborhood of B between velocity difference $\eta D_* - \eta_0 D_*$ and jump in material velocity $\dot{x}_{C'} - \dot{x}_C$.

These observations enable us to correct the erroneous representation given by Brun in [6].

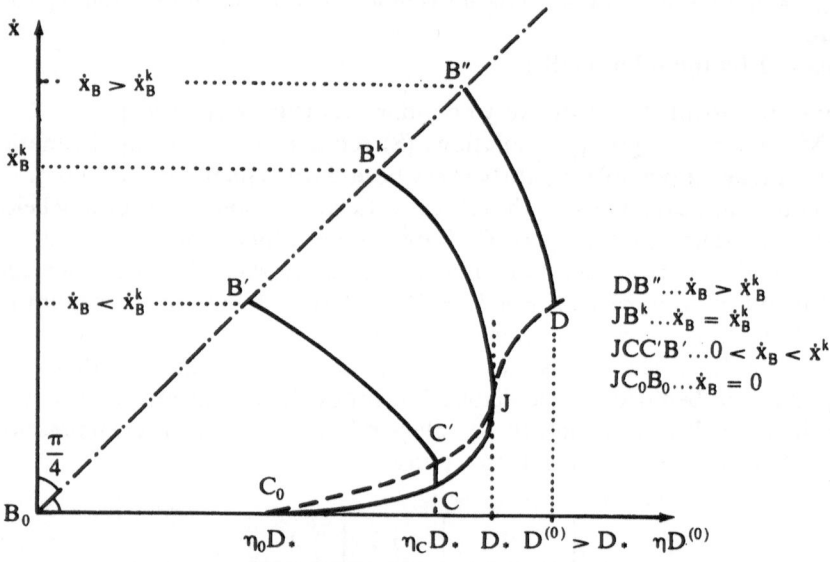

Figure III.4. Profile of material velocity $\dot{x}$ in terms of $\eta D^{(\circ)}$ (with $D^{(\circ)} \geq D_*$) for various values of the boundary velocity $\dot{x}_B$.

5.3. Convergent Simple Detonation

Although endowed with the same attractions ($r \equiv r' = r'' < 0$, uniform upstream state of rest), convergent simple detonation received less attention from theoreticians—if not from practitioners—than divergent simple detonation. Compared with preceding expansions, this relative disinterest can be explained by the particular difficulty which is raised by the existence of a supplementary characteristic length: the radius where detonation is *built-up* after the transitional phase. So the expansion (III.18) of $w_0(t)$ in ε is no longer relevant and it is advisable—the consequences prove that this is fully justified —to substitute for that expansion the following hypothesis: a radius r^* and an instant t^* exist, such that

$$\begin{aligned} |r|(t^*) = |r^*|, & \quad w_0(t^*) = D_*, \\ t > t^* \quad \Leftrightarrow \quad |r|(t) < |r^*|, & \quad w_0(t) > D_*. \end{aligned} \tag{III.43}$$

First, we look at the two extreme solutions: that where the velocity w_0 barely exceeds the C–J velocity D_*, and the solution where detonation no longer differs from a strong shock. Then we propose an interpolation formula which will be compared with the experiment in Chapter XI.

Velocity just Exceeds Velocity D_*

The acceleration mechanism of the wave was first raised in 1959 by Zeldovich [43] and was taken up again by Damamme [15]. But the gaps which remain in these papers (see observations *in fine*) lead us to reconsider the question on the basis of the zero-order external structure equations (III.33) stated above and recalled here

$$\begin{aligned} (a^2 - w^2)\frac{\partial u}{\partial z} &= ua^2\,\frac{N}{r-z} + w\left(\frac{\partial u}{\partial t} - \frac{1}{\rho w}\frac{\partial p}{\partial t}\right), \\ (a^2 - w^2)\frac{\partial \operatorname{Log} v}{\partial z} &= uw\,\frac{N}{r-z} + \left(\frac{\partial u}{\partial t} - \frac{1}{\rho w}\frac{\partial p}{\partial t} + \frac{w^2 - a^2}{w}\frac{\partial \operatorname{Log} v}{\partial t}\right), \\ w\frac{\partial s}{\partial z} &= -\frac{\partial s}{\partial t}. \end{aligned}$$

When z tends towards $+0$, the differentials $\partial/\partial t$ become differentials taken along the Crussard curve; in particular, as a result of the jump relation $p - p_0 = M(u_0 - u)$ and the extremum properties of M and s at point J, the right-hand parentheses have $-2a_*(\partial u/\partial t)_*$ and $-2(\partial u/\partial t)_*$, respectively, as limits when $t \to t^* + 0$. Equations (III.43) therefore entail

$$\begin{aligned} \lim_{t\to t^*+0}\lim_{z\to+0}\left[(a^2 - w^2)\frac{\partial \operatorname{Log}|u|}{\partial z}\right] - 2a_*\left(\frac{\partial \operatorname{Log}|u|}{\partial t}\right)_* &= a_*^2\frac{N}{r^*}, \\ \lim_{t\to t^*+0}\lim_{z\to+0}\frac{v - v_*}{u - u_*} &= \frac{v_*}{a_*}, \\ \lim_{t\to t^*+0}\lim_{z\to+0}\frac{\partial s}{\partial z} &= 0. \end{aligned} \tag{III.44}$$

It is clearly seen from the first of the above equations that establishing a correlation between curvature r and acceleration of the wave starting from time t^* involves the evaluation of the limit on the left, and therefore involves modeling the incipient propagation.

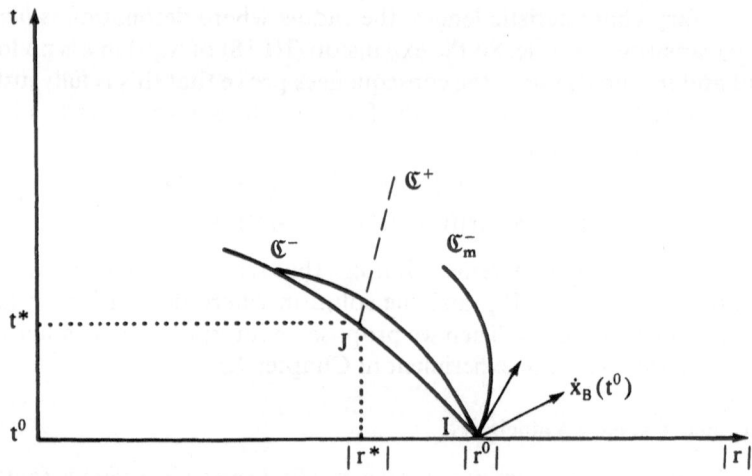

Figure III.5. Convergent simple detonation. Sketch of the characteristics in the neighborhood of priming boundary.

The experiments actually carried out [13] lead us to accept that propagation:

- results from the application of a velocity $\dot{x}_B$ ($t \geq t^0$) on the boundary of the explosive structure (radius $|r^0|$ at rest; t^0 as close to t^* as desired);
- takes place with C–J velocity D_* between the point $I(|r^0|, t^0)$ and the point $J(|r^*|, t^*)$.

In these conditions, let us examine (see Fig. III.5) the family of characteristic curves in the plane ($|r|$, t)

- the characteristics $\mathbb{C}^+$ leave the segment IJ with a uniform slope $u_* + a_* = 2a_* - D_*$;
- from point I, there starts a bundle of characteristics $\mathbb{C}^-$ whose slope is bounded by the slope $-D_*$ of the segment IJ and the slope $\mathbb{C}_m^-$ imposed by the initial boundary velocity $\dot{x}_B(t^0)$.

The zone $d(t^0)$ of the plane bounded by the characteristic $\mathbb{C}^+$ starting from J and the two extreme characteristics $\mathbb{C}^-$ starting from I (segment IJ and $\mathbb{C}_m^-$) depends only on the boundary condition at $t = t^0$. In this zone, the isentropic flow equations can be written

$$(u + \sigma) - (u + \sigma)_* = N \cdot 0(t^* - t^0) \quad \text{on } \mathbb{C}^+,$$

$$(u - \sigma) - (u - \sigma)_1 = N \cdot 0(t^* - t^0) \quad \text{on } \mathbb{C}^-,$$

(where σ is the Riemann integral defined by $\sigma = -\int a(dv/v)$). This written form shows that, apart from the terms $0(t^* - t^0)$, the flow in $d(t^0)$ is that of the "uniaxial expansion centered on I," i.e.,

$$\left.\begin{aligned} \frac{z}{t - t^0} &= u - u_* + (a - a_*), \\ u - u_* &= -\int_{v_*}^{v} a \frac{dv}{v}, \end{aligned}\right\}$$

(for a description of how the above equatons were established, see Berger and Viard [3], p. 39 *et seq.*, p. 164), from which we can deduce, on a line t = constant,

$$\frac{1}{t - t^0} = \frac{\partial u}{\partial z} + \frac{\partial a}{\partial z} = \frac{\partial u}{\partial z} + \frac{\partial a}{\partial v}\frac{\partial v}{\partial z} = \frac{\partial u}{\partial z} - \frac{v}{a}\frac{\partial a}{\partial v}\frac{\partial u}{\partial z}.$$

Taking account of the thermodynamic identity

$$\frac{\partial a}{\partial v}(v, s) = \frac{a}{v} - \frac{v^2}{2a}\frac{\partial^2 p}{\partial v^2}(v, s),$$

which results from the actual definition of the velocity of sound a, for flow inside d(t⁰) we obtain

$$\frac{\partial u}{\partial z} = \frac{2a^2}{v^3 \frac{\partial^2 p}{\partial v^2}(v, s)} \cdot \frac{1}{t - t^0} + 0(t^* - t^0).$$

At the limit, when $t^0 = t^*$, we obtain

$$\frac{\partial u}{\partial z}(z, t) = \frac{2a^2}{v^3 \left(\frac{\partial^2 p}{\partial v^2}\right)(v, s)} \cdot \frac{1}{t - t^*} \qquad \text{for} \quad (z, t) \in d(t^*). \qquad \text{(III.45)}$$

Going back to (III.44) and the limit which interests us, we see that an equivalent still has to be found for $a^2 - w^2$ in the neighborhood of (z = 0, t = t*). This equivalent can be established by writing

$$w^2 - a^2 \sim 2a_*(w - a),$$

noting that

$$w - w_* \sim (u - u_*),$$

$$a - a_* \sim (v - v_*)\left[\frac{a}{v} - \frac{v^2}{2a}\frac{\partial^2 p}{\partial v^2}(v, s)\right]_*,$$

and using the second equation (III.44)

$$v - v_* \sim \frac{v_*}{a_*}(u - u_*). \qquad \text{(III.46)}$$

This procedure leads to

$$w^2 - a^2 \sim \left[\frac{v^3}{a} \frac{\partial^2 p}{\partial v^2}(v, s) \right]_* (u - u_*). \tag{III.47}$$

Taking (III.45) and (III.47) into account simultaneously for an arbitrarily small fixed value of z and an arbitrarily selected value of t on the corresponding segment of $d(t^*)$ gives

$$\left[(a^2 - w^2) \frac{\partial u}{\partial z} \right]_{z=+0} \sim -2a \left[\left(\frac{v^3}{a} \frac{\partial^2 p}{\partial v^2} \right)_* \middle/ \left(\frac{v^3}{a} \frac{\partial^2 p}{\partial v^2} \right) \right] \cdot \left(\frac{\partial u}{\partial t} \right)_{z=+0},$$

with the end result

$$\lim_{t \to t^*} \lim_{z \to +0} (a^2 - w^2) \frac{\partial \operatorname{Log} |u|}{\partial z} = -2a_* \left(\frac{\partial \operatorname{Log} |u|}{\partial t} \right)_{\substack{z=+0 \\ t=t^*}}.$$

So, in the first equation (III.44), there remains only the derivative along (H_+) in J, for which dt can be replaced by $D_* dr$ (algebraic r). From which

$$\left(\frac{\partial \operatorname{Log} |u|}{\partial r} \right)_* = -\frac{N}{4} \frac{a_*}{D_*} \cdot \frac{1}{r^*},$$

or again

$$\frac{|u| - |u_*|}{u_*} = -\frac{N}{4} \frac{a_*}{D_*} \cdot \frac{|r| - |r^*|}{|r^*|} + \cdots,$$

which, thanks to (III.46), is equivalent to

$$\frac{v - v_*}{v_*} = \frac{N}{4} \frac{|u_*|}{D_*} \frac{|r| - |r_*|}{|r_*|} + \cdots. \tag{III.48}$$

The above relation gives the variation of the mass volume in the downstream state according to the radius of curvature $|r|$. A further calculation will make it possible to correlate the variations in detonation velocity and curvature using the jump relations and the properties of point J on (H_+).

Considering the extremum property of the velocity w_0 in J, the jump relation (II.15) where w_0 is replaced by D

$$D^2 = v_0^2 M^2$$

leads to the limit expansion

$$\frac{D - D_*}{D_*} = \frac{v_0^2}{4 D_*^2} \left(\frac{d^2 M^2}{dv^2} \right)_* (v - v_*)^2 + \cdots, \tag{III.49}$$

where d^2/dv^2 represents a derivative along (H_+).

Going back to the definition of M^2, i.e.,

$$M^2 = \frac{p - p_0}{v_0 - v},$$

we establish, at a current point of (H_+), the expression

$$\frac{d^2M^2}{dv^2} = \frac{1}{v_0 - v}\frac{d^2p}{dv^2} + \frac{2}{(v_0 - v)^2}\frac{dp}{dv} + 2\frac{p - p_0}{(v_0 - v)^3}.$$

At point J, using the expressions of $(dp/dv)_*$ and $(d^2p/dv^2)_*$ established in Chapter II ((II.22) and (II.23), respectively), we obtain

$$\left(\frac{d^2M^2}{dv^2}\right)_* = \frac{1}{v_0 - v_*} \cdot \frac{1}{1 - \frac{1}{2}\left(\frac{v_0}{v_*} - 1\right)G_*}\left[\frac{\partial^2 p}{\partial v^2}(v, s)\right]_*. \qquad \text{(III.50)}$$

Simply inserting (III.48) and (III.50) in (III.49) leads to the expression

$$\frac{D - D_*}{D_*} = \frac{v_*^2}{4\rho_0^2 D_*^2(v_0 - v_*)} \cdot \left(\frac{u_*}{D_*}\right)^2 \cdot \frac{\left[\frac{\partial^2 p}{\partial v^2}(v, s)\right]_*}{1 - \frac{1}{2}\left(\frac{v_0}{v_*} - 1\right)G_*}$$

$$\cdot \left[\frac{N}{4}\cdot\frac{|r| - |r^*|}{r^*}\right]^2 + \cdots,$$

which can be transformed by the jump relations $p_* - p_0 = -\rho_0 D_* u_*$ and $u_* = \rho_0 D_*(v_* - v_0)$ to reveal the dimensionless groupings

$$\frac{D - D_*}{D_*} = \frac{1}{2}\left(\frac{u_*}{D_*}\right)^2\left[\frac{v^2}{p - p_0}\cdot\frac{\partial^2 p}{\partial v^2}(v, s)\right]_*\left[2 - \left(\frac{v_0}{v_*} - 1\right)G_*\right]^{-1}$$

$$\cdot \left[\frac{N}{4}\frac{|r| - |r^*|}{|r^*|}\right]^2 + \cdots. \qquad \text{(III.51)}$$

It is convenient for what follows to introduce C_* by

$$C_* = \frac{1}{2}\left(\frac{u_*}{D_*}\right)^2\frac{v_*^2}{p_* - p_0}\left[\frac{\partial^2 p}{\partial v^2}(v, s)\right]_*\left[2 - \left(\frac{v_0}{v_*} - 1\right)G_*\right]^{-1}, \qquad \text{(III.52)}$$

and to note that this quantity is expressed simply when:

(i) it is accepted that the isentrope of the detonation products originating from the C–J state can be approximated, in the neighborhood of this state, by the representation

$$\frac{p}{p_*} = \left(\frac{v}{v_*}\right)^{-\Gamma_*};$$

(ii) p_0 may be neglected in comparison with p_*.

In fact, the jump relations then show immediately that

$$\frac{u_*}{D_*} = -\frac{1}{\Gamma_* + 1}; \qquad \frac{a_*}{D_*} = \frac{v_*}{v_0} = \frac{\Gamma_*}{\Gamma_* + 1}; \qquad \left[\frac{\partial^2 p}{\partial v^2}(v, s)\right]_* = \Gamma_*(\Gamma_* + 1)\frac{p_*}{v_*^2};$$

with the result that C_* takes the form

$$C_* = \frac{\Gamma_*^2}{2(\Gamma_* + 1)(2\Gamma_* - G_*)} \tag{III.52bis}$$

The Velocity is Close to ∞

For low values of $|r|$, convergent detonation is no different from a strong shock. In the hypothesis—not very restrictive in view of the states concerned —of a $\bar{\gamma}$-polytropic behavior of the detonation products, the flow becomes asymptotically self-similar (Guderley, 1942; Landau and Stanyukovich, 1944) so that

$$D \underset{|r|\to 0}{\sim} |r|^{-n} \tag{III.53}$$

where n depends on $\bar{\gamma}$ and N (as in (III.49), notation D is substituted for notation w_0 with the sole aim of simplifying future equations). A precise analysis of the ratio $m = n/N$ can be found in Witham [39]; completed for the highest values of $\bar{\gamma}$, it shows that m does not depend greatly on N and is very appropriately approximated (see the table below) by Witham's formula

$$m(\bar{\gamma}) = \left(1 + \frac{2}{\bar{\gamma}} + \sqrt{\frac{2\bar{\gamma}}{\bar{\gamma} - 1}}\right)^{-1} \tag{III.54}$$

in the physically interesting variation interval of $\bar{\gamma}$ ($1 < \bar{\gamma} \simeq 3$).

	6/5	4/3	7/5	5/3	3
$m(\bar{\gamma}, 1)$	0.161220	0.1868	0.197294	0.226054	0.2892
$m(\bar{\gamma}, 2)$	0.160376	0.1865	0.197182	0.226346	0.2857
$m(\bar{\gamma})$	0.163	0.188	0.197	0.225	0.294

Interpolated Functional Relation

If the velocity and curvature of a convergent simple detonation are functionally linked, such a relation is necessarily compatible with the expansions (III.51) and (III.53). The possibilities of interpolation are infinite. Lacking a physical argument, we rely on an argument of simplicity to propose

$$\frac{D - D_*}{D_*} = \frac{C_*}{4m^2}\left[\left(\frac{|r^*|}{|r|}\right)^{Nm/2} - 1\right]^2, \tag{III.55}$$

where m and C_* are given, respectively, by (III.54) and (III.52) or (III.52bis).

The validity of (III.55) will be examined in Chapter XI based on the experimental results described in [13].

Observations

We mentioned earlier the gaps in the papers written by Zeldovich [43] and Damamme [15]. The above results explain them more clearly.

In Zeldovich's paper we find the following sentence: "... at the beginning of the process, it is not difficult to obtain expansion of the solution in a series" followed by a formula compatible with (III.48) in the cylindrical case $N = 1$ and in the framework of the simplifying hypotheses (i) and (ii), which are, moreover, restricted to the case $\Gamma_* = 3$. If it is true that we can hardly question the ability of the author to carry out the calculations which lead to (III.48), the fact remains that these calculations are not displayed and that the form of the general result ($N = 1$ or 2, with no simplifying hypotheses or restriction on the value of Γ_*) has not been put forward.

In [15] we find an attempt—which follows from [7]—to examine in a general way the problem of the incipient propagation of a convergent detonation. However, the demonstration of the expansion (III.48), or its equivalent, is lacking. Moreover, the stated validity condition—boundary pressure less than C–J pressure at the initial instant—is not the right one. Indeed, the demonstration of (III.48) detailed above shows clearly that the validity condition is the existence of a nonempty domain $d(t^*)$; this condition is assured if and only if, in J, the slope of $\mathfrak{C}_m^-$ is greater than that of $\mathfrak{C}^+$, i.e.,

$$\dot{x}_B - a_B > u_* + a_* = 2a_* - D_*, \qquad (III.56)$$

which has no reason for being identified with the condition $p_B < p_*$.

A simple expression of (III.56) can be given in the Γ_*-polytropic hypothesis already considered above. Indeed, the theory of the centered simple wave (see e.g., [3]) shows that

$$a_B = a_* - \frac{\Gamma_* - 1}{2} \dot{x}_B,$$

with the result that (III.56) is equivalent to

$$\dot{x}_B > 2 \frac{2\Gamma_* - 1}{(\Gamma_* + 1)^2} D_*.$$

6. The Beginnings and Limits of Detonation

6.1. Priming Boundary and Free Boundaries

In Subsection 1.5, we have already stressed that the actual *detonation* is never established instantaneously. Whatever the intensity and the nature of the stress to which the *priming boundary* B^a of the explosive structure is subjected, a transitional propagation occurs where the temperature behind the shock wave is too low to entail an immediate strong reactivity. Moreover, researchers have often found circumstances where, for the same explosive structure, the existence of *free boundaries* B^1 which are initially unstressed (i.e., where e.g., $P \in B^1 \Rightarrow u(P) = 0$ and $p(P) \approx 0$ as long as $\Sigma(t)$ has not reached P) breaks the uniformity of propagation on which the simplicity of one-dimensional detonations is grounded. On and in the neighborhood of these boundaries, pressure and temperature result from an interaction of the explosive with the adjacent medium, governed by the jump relations (I.26b) applied to a contact surface $M = 0$. And we cannot exclude the possibility that this interaction leads—in a more or less extended domain—to a temperature of the explosive material too low to entail a strong reactivity.

Thus, for reasons which may be attached either to the microscopic processes of chemical reactions or to the existence of free boundaries, or to both, the study of explosives cannot ignore the *transitional phases* and the *marginal zones* of propagation whose modeling eludes that developed above (Sections 1–5) for the actual detonation. We can however draw upon it, as is shown below, to extract the essential features.

6.2. Birth of a Simple Detonation

Three characteristic lengths were introduced in Subsection 1.4: a principal length $\bar{L}$ and two secondary lengths L and L'. The modeling of built-up detonation as a layer of steep gradients in a dissipative fluid rests on the inequality $L' \ll L$ which leads to the choice of $\varepsilon = L/\bar{L}$ as a perturbation parameter. Obviously, the transitional circumstances described in Subsection 1.5 come under the opposite inequality $L' \gg L$, which this time leads to the choice of $\varepsilon' = L'/\bar{L}$ as perturbation parameter.

The process is thus self-imposed. The Van Dyke minimum degeneration principle [36] leads us to adopt a linear extension along $\mathbf{N}$

$$\zeta' = \frac{Z}{\varepsilon'}.$$

Relative to the variables $(\zeta', \bar{\xi}', \bar{\xi}'', \bar{t})$, equations (III.11) are written

$$\left.\begin{aligned}
\bar{w}\frac{\partial \operatorname{Log} \bar{v}}{\partial \zeta'} - \frac{\partial \bar{u}}{\partial \zeta'} &= 0(\varepsilon'), \\
v\mathbf{N}\frac{\partial \bar{p}}{\partial \zeta'} + w\frac{\partial \bar{U}}{\partial \zeta'} &= \mathbf{0}(\varepsilon'), \\
\bar{w}\frac{\partial \bar{s}}{\partial \zeta'} &= 0(\varepsilon'), \\
\bar{w}\frac{\partial m}{\partial \zeta'} - (1 - m)\bar{S} &= 0(\varepsilon'),
\end{aligned}\right\}$$

which differs from (III.15) only in the absence of thermomechanic dissipation terms and the substitution $(\zeta \to \zeta', \varepsilon \to \varepsilon', L \to L')$.

Taking account of $\bar{w} = \bar{u} - \bar{\delta}$ and $\partial\bar{\delta}/\partial\zeta' = 0$, we obtain

$$
\left.
\begin{aligned}
\frac{\partial}{\partial\zeta'}(\bar{\rho}\bar{w}) &= 0(\varepsilon'), \\[1mm]
N\frac{\partial\bar{p}}{\partial\zeta'} + \bar{\rho}\bar{w}\frac{\partial\mathbf{W}}{\partial\zeta'} &= 0(\varepsilon'), \\[1mm]
\frac{\partial\bar{s}}{\partial\zeta'} &= 0(\varepsilon'), \\[1mm]
\bar{w}\frac{\partial m}{\partial\zeta'} - (1-m)\bar{S} &= 0(\varepsilon').
\end{aligned}
\right\}
\qquad\text{(III.57)}
$$

By making $\varepsilon' = 0$ in equations (III.57), and returning to the physical dependent variables (with no overbar), we see that the internal structure $Y'^{(\circ)}$ of a transient system should confirm the equations

$$
\left.
\begin{aligned}
\frac{\partial}{\partial\zeta'}(\rho w) &= 0, & \text{(a)} \\[1mm]
\frac{\partial}{\partial\zeta'}(p + \rho w^2) &= 0, & \text{(b)} \\[1mm]
\frac{\partial}{\partial\zeta'}(\rho w W_\tau) &= 0, & \text{(c)} \\[1mm]
\frac{\partial s}{\partial\zeta'} &= 0, & \text{(d)} \\[1mm]
\frac{\partial m}{\partial\zeta'} &= \frac{L'}{w}(1-m)S. & \text{(e)}
\end{aligned}
\right\}
\qquad\text{(III.58)}
$$

Using again the observations made at the start of the century and taken up by Von Neumann (see Subsection 1.5), consider that a shock carries the explosive medium from the uniform state Y_0 to a state $\hat{Y}(\xi', \xi'', t)$ where chemical reactions are released. By hypothesis this state is such that

$$
m = 0, \qquad \left(\frac{dm}{d\zeta'}\right)_{\zeta'=\zeta} \neq 0.
$$

Moreover, in accordance with the results established in Chapter II on the subject of shocks, this state is perfectly defined—locally in space and time— when giving the normal relative velocity upstream w_0; in particular, this is the case with the specific energy $\hat{s}$ and the normal flow of mass $M = \rho_0 w_0$. The evolution of chemical reactions downstream of a shock is described by the solution of (III.58) which equals $\hat{Y}$ when $\zeta' = \zeta$. It is clear that the actual solving passes through the formulation of the pressure of the mixture born of decomposition, as a function of variable m, specific entropy $\hat{s}$, as well as the characteristics of the initial medium and the products formed. But it is also clear that, once this relation $\mathscr{P}$ is chosen, the integration of equations (a), (b), and (d) with

$$v = \frac{w}{M},$$

$$p = p_0 + M(w_0 - w),$$

$$s = \hat{s},$$

makes it possible to express w as a function of m, and to reach an equation of local chemical evolution

$$\frac{\partial m}{\partial \zeta'} = L'(1 - m)\frac{S(\hat{s}, m)}{w[M(\hat{s}), \mathscr{P}(\hat{s}, m)]}, \qquad (III.59)$$

where $\hat{s}$ and M are two linked parameters.

Thus the expansion arising from the chemical reactions downstream of a shock depends at the same time on:

— chemical mechanisms by S;
— the thermodynamics of the mixture by $\mathscr{P}$; and
— the local conditions of shock by $\hat{s}$.

This shows the complexity and variety of the possible situations.

6.3. Extinction or Bifurcation of an Automomous Simple Detonation

Experimenters and users of explosives know very well that:

• detonation which builds up in a cartridge primed at one end displays a "front" whose curvature is more pronounced and whose velocity is lower, the smaller the cartridge diameter; and
• propagation is no longer observed below a diameter called the *critical diameter*.

Generally speaking, the facts mentioned above and others besides prove that effective propagation of a quasi C–J detonation is incompatible with too great a curvature of the wave surface. We attempt in what follows to understand the mechanism of this correlation by developing [12].

In the neighborhood of the critical propagation conditions, the relative order of the three characteristic lengths of the problem is $L \ll \tilde{L} \ll L'$. In other words, the primary length becomes L' and the secondary length $\tilde{L}$, whilst L can be ignored. The relevant dimensionless variables are no longer those defined in (III.10) but those deduced from (III.10) by replacing $\tilde{L}$ by L', *without, however, modifying the definition of* $\bar{S}$. Moreover, the perturbation parameter to choose becomes $\tilde{\varepsilon} = \tilde{L}/L'$. Except for the definition of the variables marked with a bar, other than $\bar{S}$, and subject to replacing $\tilde{L}$ by L', equations (III.11) remain valid. Then by making $L/L' = 0$, and returning to the physical dependent variables, we find the first three equations (III.25), whilst the fourth should be replaced by

$$L'(1 - m)S = \left(\tilde{D}\frac{\partial}{\partial \bar{t}} + w\frac{\partial}{\partial Z}\right)m + \left(\mathbf{W}_\tau - \tilde{D}Z\frac{\partial \mathbf{N}}{\partial \bar{t}}\right)\cdot\bar{\mathbf{J}}m. \quad \text{(III.60)}$$

Taking up the so-called Von Neumann hypothesis once more and therefore the existence of a state $\hat{\mathbf{Y}}(\xi', \xi'', t)$ such as

$$m = 0, \qquad \left(\frac{\partial m}{\partial t}\right)_{Z=+0} \neq 0,$$

and taking advantage of the definitions (III.26a and b) of simple detonation, we arrive at the equations which govern the near downstream flow: these are the equations (III.27) which must be associated with the equation of near chemical evolution deduced from (III.60)

$$\frac{w}{L'}\frac{\partial m}{\partial Z} = (1 - m)S - \frac{\partial m}{\partial t} + \cdots,$$

where the ellipses designate the quantities which tend to zero when Z tends to $+0$.

Finally, taking account of (III.28), we can say that the near downstream flow is governed by

$$\left.\begin{array}{l}(a^2 - w^2)\dfrac{\partial u}{\partial Z} = 2\dfrac{u^2}{R_m} + \cdots, \\[2mm] (a^2 - w^2)\dfrac{\partial \text{Log } v}{\partial Z} = 2\dfrac{uw}{R_m} + \cdots, \\[2mm] \dfrac{\partial \mathbf{W}_\tau}{\partial Z} = \cdots, \\[2mm] \dfrac{\partial s}{\partial Z} = \cdots, \\[2mm] \dfrac{\hat{w}}{L'}\dfrac{\partial m}{\partial Z} = S(\hat{s}, 0) - \left(\dfrac{\partial m}{\partial t}\right)_{Z=+0} + \cdots.\end{array}\right\} \quad \text{(III.61)}$$

The first four equations show that the near downstream flow is an isentropic perfect fluid flow. Furthermore, the last equation shows that effective propagation

$$\left(\frac{\partial m}{\partial Z}\right)_{Z=+0} \gg 0 \quad \text{(III.62)}$$

is possible if and only if

$$S(\hat{s}, 0) \gg \left(\frac{\partial m}{\partial t}\right)_{Z=+0}, \quad \text{(III.63)}$$

that is to say if the initial reactivity $S(\hat{s}, 0)$ exceeds the local variation rate of the composition.

Consider now the typical case of simple detonation in a cartridge with a diameter $\varnothing$: axisymmetric, and stationary in a frame moving at velocity $\mathbf{W}_0$ in an absolute frame of reference. Then $w_0 = |\mathbf{W}_0| \cos \psi$, where ψ is the angle formed by the normals to the wave Σ_0 respective to the summit A and the

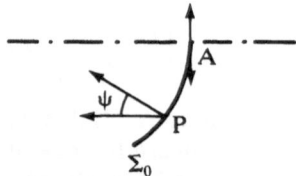

current point P. Without the need for an exact calculation, it is clear that—for a convex surface Σ_0—ψ increases with the curvature. Consequently, the more the wave is curved, the less the values of w_0 and therefore $\hat{s}$ and the initial reactivity $S(\hat{s}, 0)$. In these conditions, it is understood that stationary axisymmetric propagation which necessarily confirms

$$\left(\frac{\partial m}{\partial t}\right)_{z=+0} = 0$$

ceases to exist when the limiting conditions on the boundaries of the cartridge impose too great a curvature on the detonation front. We attempt below to argue more closely and add some refinements to these semi-quantitative considerations on simple detonation in a cartridge.

Consider the relative flow at a reference point made to move with velocity $\mathbf{W}_0 = |\mathbf{W}_0|\mathbf{i}$ in an absolute frame of reference. As the preceding study suggests, we are concerned with reactive fluid flow in the near downstream of the ignitor shock represented by a surface Σ_0 of summit A. As throughout the preceding subsections the normal $\mathbf{N}$ to Σ_0 is oriented from upstream to downstream.

As appears in the expansions which follow, it is convenient to make use of the intrinsic coordinates associated with the flow lines and their orthogonal trajectories. Let us agree to orient the unit vector $\mathbf{1}$ on the flow line by the velocity-vector $\mathbf{W}$ and adopt the following definitions:

- the unit vector τ on Σ: $(\mathbf{N}, \tau) = +\pi/2$;
- the unit vector $\mathbf{n}$ on the orthogonal trajectories: $(\mathbf{1}, \mathbf{n}) = +\pi/2$;
- the deflection of $\mathbf{W}$ compared with the upstream state: $(\mathbf{i}, \mathbf{1}) = \theta$; and
- the angle of Σ and the upstream flow line: $(\mathbf{i}, \tau) = \eta$.

At the heart of the flow, we designate the downstream state of Σ_0 by the superscript $\hat{}$; the deflection of $\mathbf{W}$ compared with upstream is therefore $\hat{\theta}$ at the point of Σ_0 whose distance from the axis is $\hat{r}$; by extension, the angle $\hat{\eta}$ is written $(\hat{\mathbf{1}}, \tau) = \eta - \hat{\theta}$. Finally, the curvature of the flow line is written $1/R$, whilst the curvature of the meridian of Σ_0 is written $1/R_\Sigma$; by the definition of

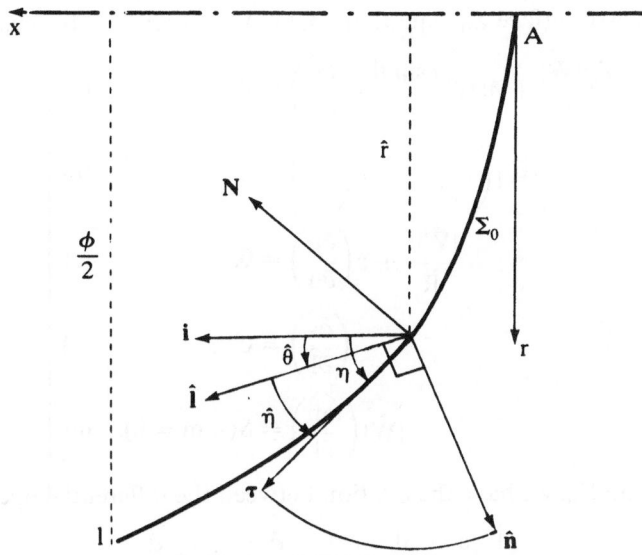

Figure III.6. Diagram and notations for studying the relative flow of detonation products.

η and $\hat{r}$, the latter is

$$\frac{1}{R_\Sigma} = \sin \eta \frac{d\eta}{d\hat{r}}. \qquad (III.64)$$

Figure III.6 amalgamates the majority of the notations defined above.

By the interaction of the differential operators

$$\left.\begin{aligned}
\frac{\partial}{\partial l} &= \cos \theta \frac{\partial}{\partial x} + \sin \theta \frac{\partial}{\partial r}, \\
\frac{\partial}{\partial n} &= -\sin \theta \frac{\partial}{\partial x} + \cos \theta \frac{\partial}{\partial r},
\end{aligned}\right\}$$

and taking account of $\partial/\partial t = 0$, equations (I.26a)—which approximate the first four equations (III.61) in the neighborhood of Σ_0—become

$$\left.\begin{aligned}
\frac{\partial \rho |\mathbf{W}|}{\partial l} + \rho |\mathbf{W}| \left(\frac{\sin \theta}{r} + \frac{\partial \theta}{\partial n} \right) &= 0, \\
|\mathbf{W}| \frac{\partial |\mathbf{W}|}{\partial l} + v \frac{\partial p}{\partial l} &= 0, \\
\frac{|\mathbf{W}|^2}{R} + v \frac{\partial p}{\partial n} &= 0, \\
\frac{\partial s}{\partial l} &= 0,
\end{aligned}\right\}$$

On the wave surface Σ_0, equations (III.61) can therefore be replaced by

$$
\left.
\begin{array}{rl}
\widehat{\dfrac{\partial \rho |\mathbf{W}|}{\partial l}} + \rho |\widehat{\mathbf{W}}| \left(\dfrac{\sin \theta}{\hat{r}} + \widehat{\dfrac{\partial \theta}{\partial n}} \right) = 0, & \text{(a)} \\[3mm]
|\widehat{\mathbf{W}}| \left(\widehat{\dfrac{\partial |\mathbf{W}|}{\partial l}} \right) + \widehat{v} \left(\widehat{\dfrac{\partial p}{\partial l}} \right) = 0, & \text{(b)} \\[3mm]
\dfrac{|\widehat{\mathbf{W}}|^2}{\hat{R}} + \widehat{v} \left(\widehat{\dfrac{\partial p}{\partial n}} \right) = 0, & \text{(c)} \\[3mm]
\left(\widehat{\dfrac{\partial s}{\partial l}} \right) = 0, & \text{(d)} \\[3mm]
|\widehat{\mathbf{W}}| \left(\widehat{\dfrac{\partial m}{\partial l}} \right) = S(\hat{s},\, m = 0). & \text{(e)}
\end{array}
\right\} \quad \text{(III.65)}
$$

Moreover, on Σ_0, we have the relation between the differential operators

$$
\frac{1}{R_\Sigma} \frac{d}{d\eta} \equiv \frac{d}{d\tau} = \cos \hat{\eta} \frac{\partial}{\partial l} + \sin \hat{\eta} \frac{\partial}{\partial n},
$$

which, applied to the pressure $\hat{p}$ and the deflexion $\hat{\theta}$, gives

$$
\left.
\begin{array}{rl}
\dfrac{1}{R_\Sigma} \dfrac{d\hat{p}}{d\eta} = \cos \hat{\eta} \, \widehat{\dfrac{\partial p}{\partial l}} + \sin \hat{\eta} \, \widehat{\dfrac{\partial p}{\partial n}}, & \text{(f)} \\[3mm]
\dfrac{1}{R_\Sigma} \dfrac{d\hat{\theta}}{d\eta} = \cos \hat{\eta} \, \widehat{\dfrac{\partial \theta}{\partial l}} + \sin \hat{\eta} \, \widehat{\dfrac{\partial \theta}{\partial n}}. & \text{(g)}
\end{array}
\right\} \quad \text{(III.65)}
$$

Finally, the dependence of p in (v, s, m) gives on Σ_0

$$
\left(\widehat{\frac{\partial p}{\partial l}} \right) = \hat{c}^2 \left(\widehat{\frac{\partial p}{\partial l}} \right) + \left(\widehat{\frac{\partial p}{\partial s}} \right) \left(\widehat{\frac{\partial s}{\partial l}} \right) + \left(\widehat{\frac{\partial p}{\partial m}} \right) \left(\widehat{\frac{\partial m}{\partial l}} \right), \quad \text{(h)} \qquad \text{(III.65)}
$$

where $\hat{c}^2$ is the velocity of sound in the nondecomposed explosive, brought to the state produced downstream of the ignitor shock.

The elimination of the six differentials

$$
\widehat{\frac{\partial \theta}{\partial n}},\ \widehat{\frac{\partial p}{\partial n}},\ \widehat{\frac{\partial \rho}{\partial l}},\ \widehat{\frac{\partial p}{\partial l}},\ \widehat{\frac{\partial s}{\partial l}},\ \widehat{\frac{\partial m}{\partial l}},
$$

taken in the downstream state of Σ_0, from the eight linear equations (III.65) leads to the two relations

$$
\left.
\begin{array}{l}
\dfrac{1}{R_\Sigma} = \dfrac{\widehat{\dfrac{\partial p}{\partial m}} \dfrac{\hat{S}}{\rho |\widehat{\mathbf{W}}|} - \dfrac{1}{\hat{R}} \dfrac{\widehat{\mathbf{W}}^2 \sin^2 \hat{\eta} - \hat{c}^2}{\sin \hat{\eta} \cos \hat{\eta}} - \hat{c}^2 \dfrac{\sin \theta}{\hat{r}}}{\left(1 - \dfrac{\hat{c}^2}{|\widehat{\mathbf{W}}|^2} \right) \dfrac{\hat{v}}{\cos \hat{\eta}} \dfrac{d\hat{p}}{d\eta} + \dfrac{\hat{c}^2}{\sin \hat{\eta}} \dfrac{d\hat{\theta}}{d\eta}}, \\[8mm]
-\dfrac{1}{\hat{R}} = \cotan \hat{\eta} \, \widehat{\dfrac{\partial \operatorname{Log} |\mathbf{W}|}{\partial l}} + \dfrac{1}{R_\Sigma} \cdot \dfrac{\hat{v}}{|\widehat{\mathbf{W}}|^2 \sin \hat{\eta}} \dfrac{d\hat{p}}{d\eta}.
\end{array}
\right\} \quad \text{(III.66)}
$$

In the second equation of (III.66), with reference to the quantities marked $\wedge$ other than $(\partial \log |\mathbf{W}|)/\partial l$, it is convenient to:

- note the presence of the rate of pressure variation at the start of decomposition $\widehat{\partial p/\partial m}$ (which depends on the mixture law between initial compressed medium and decomposition products) and the reactivity at the start of decomposition $\hat{S}$; and
- stress that the other dimensions marked $\wedge$, as well as their derivatives in η, are perfectly defined based on $|\mathbf{W}_0|$ and η by the jump relations (I.29) put in equivalent form

$$
\left.
\begin{aligned}
\hat{e} - e_0 &= \tfrac{1}{2}(\hat{p} + p_0)(v_0 - \hat{v}), && \text{(a)} \\
\hat{p} - p_0 &= \rho_0^2 |\mathbf{W}_0|^2 (v_0 - \hat{v}) \sin^2 \eta, && \text{(b)} \\
\cotan \hat{\theta} &= (\tan \eta)\left(\frac{\rho_0 |\mathbf{W}_0|^2}{\hat{p} - p_0} - 1\right), && \text{(c)} \\
|\hat{\mathbf{W}}|^2 &= |\mathbf{W}_0|^2 - (\hat{p} - p_0)(v_0 + \hat{v}). && \text{(d)}
\end{aligned}
\right\} \qquad \text{(III.67)}
$$

In the case of an *autonomous simple detonation* defined above by (III.29), the velocity-vector $\mathbf{W}$, having decreased in modulus from $|\mathbf{W}_0|$ to $|\hat{\mathbf{W}}|$ across the ignitor shock Σ_0, increases again when we move away from Σ_0 on the flow line so that we can admit

$$
\frac{\overline{\partial \log |\mathbf{W}|}}{\partial l} = 0. \qquad \text{(III.68)}
$$

Taking account of the definition (III.64) of R_Σ, the two equations (III.66) lead to the differential equation in $\eta(\hat{r})$

$$
\frac{\sin \eta}{\sin \hat{\eta}} \frac{d\eta}{d\hat{r}} \left[\frac{d\hat{\theta}}{d\eta} + \frac{\hat{v}}{|\hat{\mathbf{W}}|^2} \frac{d\hat{p}}{d\eta} \cotan \hat{\eta}\right] = \frac{1}{\hat{c}^2} \frac{\widehat{\partial p}}{\partial m} \frac{\hat{S}}{\rho |\hat{\mathbf{W}}|} - \frac{\sin \hat{\theta}}{\hat{r}}, \qquad \text{(III.69)}
$$

the behavior of which will now be examined at the limits $\hat{r} = 0$ and $\hat{r} = \varnothing/2$ of the integration interval.

At point A, where Σ_0 intersects the symmetry axis, η_A and $\hat{\eta}_A$ coincide and equal $\pi/2$, the deflection $\hat{\theta}_A$ is 0, with the result that the jump relations have the form, directly deduced from (I.29) and (I.30)

$$
\left.
\begin{aligned}
\hat{e}_A - e_0 &= \tfrac{1}{2}(\hat{p}_A + p_0)(v_0 - \hat{v}_A), \\
\hat{p}_A - p_0 &= \rho_0^2 |\mathbf{W}_0|^2 (v_0 - \hat{v}_A), \\
\rho_0 |\mathbf{W}_0| &= \hat{\rho}_A |\hat{\mathbf{W}}|_A, \\
|\mathbf{W}_0|^2 - |\mathbf{W}_A|^2 &= (\hat{p}_A - p_0)(v_0 + \hat{v}_A).
\end{aligned}
\right\} \qquad \text{(III.70)}
$$

In the neighborhood of A, let $\eta = -\pi/2 - \varepsilon$ with $\varepsilon > 0$; the expansion of

(III.67)—taking account of (III.70)—leads to

$$
\left.
\begin{aligned}
&\hat{\theta} \sim \varepsilon\left(\frac{v_0}{\hat{v}_A} - 1\right), \\[2ex]
&\frac{\pi}{2} - \hat{\eta} \sim \frac{v_0}{\hat{v}_A}, \\[2ex]
&\frac{d\hat{\theta}}{d\eta} \sim 1 - \frac{v_0}{\hat{v}_A}, \\[2ex]
&\left(\frac{d\hat{p}}{d\eta}\right) = 2(\cotan \eta)(\hat{p} - p_0)\left[1 - \left(\frac{\hat{p} - p_0}{\hat{v} - v_0}\right)\bigg/\left(\frac{d\hat{p}}{d\hat{v}}\right)\right]^{-1}.
\end{aligned}
\right\} \tag{III.71}
$$

Noting that the definition (III.64) of the curvature $1/R_\Sigma$ leads to

$$
\frac{\varepsilon}{\hat{r}} \underset{\varepsilon \to 0}{\sim} -\left(\frac{1}{R_\Sigma}\right)_A, \tag{III.72}
$$

the equation (III.69) gives at $\hat{r} = 0$

$$
\left(\frac{d\eta}{d\hat{r}}\right)_A \equiv \frac{1}{(R_\Sigma)_A} = \frac{\hat{S}_A}{2|W_0|}\cdot\left[\frac{\partial}{\partial m} \mathrm{Log}\,(\rho - \rho_0)\right]_{(p=\hat{p}_A,\,s=\hat{s}_A,\,m=0)} \neq 0.
$$

Let us now consider the flow in the neighborhood of the point of Σ where $\hat{r} = \varnothing/2$. Experiment shows that, in the adjacent medium "m" a shock wave Σ_m builds up which accompanies Σ_0 in the absolute frame of reference. The relative flow in "m" is thus characterized upstream of Σ_m by a relative velocity-vector $W_m = W_0$. Downstream of Σ_m, a state is established which should satisfy both the shock relations across Σ_m and the compatibility relations with the explosive undergoing decomposition along the "partition" which is constituted by the flow line passing through I.

Under these conditions, discussion of the existence of a solution of (III.69) on $\hat{r} \in [0, \varnothing/2]$ leads to the introduction of two critical values for $|W_0|$:

• one, D_c, is independent of $\varnothing$ but dependent on the adjacent medium such that $]D_c, D_*]$ is the interval for which the state in I downstream of Σ_0 confirms $\hat{s}_I > s^i$; let $\eta_I(|W_0|, \text{"m"})$ be the co-incidence in I resulting from this adaptation to "m"; and

• the other, D_c', is dependent on $\varnothing$ but independent of "m" so that $]D_c', D_*]$ is the interval of variation of $|W_0|$ for which (III.69) is regularly integrable from $\hat{r} = 0$ to $\hat{r} = \varnothing/2$; let $\eta_I'(|W_0|, \varnothing)$ be the co-incidence in I resulting from this integration.

So, for a cartridge with a given radius $\varnothing/2$ surrounded by a medium "m," two situations can arise, depending on whether the representative curves of the variations of η_I and η_I' according to $|W_0|$ do or do not have a common point.

If they do, there exists a value of $|W_0|$ (whose uniqueness is borne out by current experiment) linked to the diameter $\varnothing$ of the cartridge. Note that the measurement of $|W_0|$ is a well-established measurement and is essential in the dynamic characterization of an explosive substance.

If they do not, we must conclude that the explosive considered, enclosed in a cartridge of diameter $\varnothing$, and surrounded by a medium "m" cannot be the seat of a detonation of the type postulated at the start: a shock wave Σ_0 followed by a reactive laminar flow. But it is useful to distinguish the two possible origins of such an impossibility:

(a) the greatest possible value of $\varnothing$ of $\varnothing_c$ such that

$$\lim_{\varnothing \searrow \varnothing_c} |W_0|(\varnothing) = D_c(\text{"m"}),$$

then the impossibility of propagation for $\varnothing \le \varnothing_c$ is linked to the non-existence in the explosive, *in the vicinity of* "m," of a shocked state guaranteeing ignition on the boundary $(s_1 > s^i)$; and

(b) the greatest possible value of $\varnothing$ is $\varnothing'_c$ such that

$$\lim_{\varnothing \searrow \varnothing'_c} |W_0|(\varnothing) = D'_c(\varnothing'_c),$$

then the impossibility of propagation for $\varnothing \le \varnothing'_c$ is linked to the non-existence in the explosive, *in its entirety*, of a regular flow of the type postulated. We can assume that the values D'_c and $\varnothing'_c$ are critical values starting from which the flow bifurcates towards more complex forms (Spinning detonations).

It is also useful to note that the presence of the quantities $\widehat{\partial p/\partial m}$ and $\hat{S}$ in the second member of the differential equation (III.69) explains why the form of the curve $|W_0|(\varnothing)$ can depend very considerably (see [1], [2], for example) on the chemical nature of the explosive substance, at the molecular scale as well as the mesoscopic scale.

The preceding reasoning brings in the differential equation (III.69) without however necessitating its effective solution. Nevertheless, this solution becomes useful in interpreting the experimental curve $|W_0|(\varnothing)$. A calculation given in Appendix D shows how, based on the last three equations (III.67), (III.69) can be put in the form

$$\left. \begin{array}{l} g\dfrac{d\eta}{d\hat{r}} + \dfrac{\sin 2\eta}{2\hat{r}} + \dfrac{\hat{S}}{|W_0|} \cdot \left[\dfrac{\partial}{\partial m} \text{Log}\,(\rho - \rho_0) \right]_{(p=\hat{p}, s=\hat{s}, m=0)} = 0, \\[4mm] 1 + \dfrac{g}{\dfrac{v_0}{\hat{v}}\cos^2 \eta + \sin^2 \eta} = \dfrac{4\left[1 - \dfrac{\hat{p} - p_0}{\hat{v} - v_0} \Big/ \dfrac{d\hat{p}}{d\hat{v}} \right]^{-1}}{\left[1 + \dfrac{\hat{v}}{v_0}\tan^2 \eta + \dfrac{v_0}{\hat{v}}\left(1 + \dfrac{v_0}{\hat{v}}\cotan^2 \eta \right) \right]}. \end{array} \right\}$$

$$(III.73)$$

When we substitute for (III.67a) a linear relation between the relative normal upstream velocity U and the absolute value u of the jump of normal material velocity of the form $U = A + Bu$, then explicit writing in g of the quantities marked $\char`\^{}$ becomes possible (see Appendix D) as a function of $|W_0|$, η, A, and B by

$$\left.\begin{aligned}
\left[1 - \frac{\hat{p} - p_0}{\hat{v} - v_0}\Big/\frac{d\hat{p}}{d\hat{v}}\right]^{-1} &= 1 + \frac{1}{2}\left(\frac{|W_0|}{A}\sin\eta - 1\right)^{-1}, \\
B\left(1 - \frac{\hat{v}}{v_0}\right) = 1 - \frac{A}{|W_0|\sin\eta} &\quad\Leftrightarrow\quad \frac{\hat{v}}{v_0} = \frac{B-1}{B} + \frac{A}{B|W_0|\sin\eta}.
\end{aligned}\right\} \quad\text{(III.74)}$$

Since the values of A and B are accessible through experiment, we can consider studying the extinction (or bifurcation) according to the *kinetic* parameter

$$\hat{S} \times \left[\frac{\partial}{\partial m}\text{Log}(\rho - \rho_0)\right] \qquad (p = \hat{p},\ s = \hat{s},\ m = 0),$$

and in particular as a function of the parameters of the aggregate state of a composite solid explosive substance.

Since the work of Jones [26], Eyring *et al.* [19], and Wood and Kirkwood [40] over thirty years ago, the existence of a critical diameter, and more generally the correlation between velocity and curvature or velocity and diameter, hold the attention of many authors seeking a theoretical explanation of the experimental laws. Outstanding among these works are those of Dremin, Trofimov, and Savrov (see [18]), who foresaw the alternative between simple extinction and bifurcation towards a complex flow, but who— for want of introducing the notion of autonomous detonation—stopped short of equation (III.69) and the possibilities it offers of an interpretation of the influence of the physico–chemical parameters of the explosive on the propagation itself.

References to Part One

[1] AVEILLE, J., BACONIN, J., CHERET, R. et al. Célérités de détonation et profondeurs d'amorçage de deux compositions explosives à base d'octogène et de TATB. *Proc. Colloque de Pyrotechnie Fond. et Ap.*, Arcachon/France (1982), p. 385.

[2] AVEILLLE, J., BACONIN, J., CARION, N., ZOE, J. Experimental study of spherically diverging detonation waves. *Proc. 8th Symposium on Detonation*, Albuquerque/NM (1985), p. 523.

[3] BERGER, J., VIARD, J. *Physique des Explosifs Solides*. Dunod, Paris (1962).

[4] BETHE, H.A. On the theory of shock waves for an arbitrary equation of state. O.S.R.D. Report no. 544, 1942.

[5] BROCHET, C., BROSSARD, J., CHERET, R., MANSON, N., VERDES, G. A comparison of spherical, cylindrical and plane detonation velocities in some condensed and gaseous explosives. *Proc. 5th Symposium on Detonation*, Pasadena/CA (1970), p. 41.

[6] BRUN, L. Sur l'autonomie des ondes de choc à l'état aval sonique. Cas de la détonation. *J. Méc. Théor. Appl.*, 1, no. 4 (1982), pp. 623–646.

[7] BRUN, L., CHERET, R., VACELLIER, J. Considérations sur les détonations fortes. *Symposium H.D.P.*, Paris/France (September 1978), pp. 269–279.

[8] CHAPMAN, D.L. On the rate of explosions in gases. *Phil. Mag.* (1899).

[9] CHERET, R. Sur la nature singulière du couplage entre écoulement et réactions chimiques dans l'onde explosive idéale. *C. R. Acad. Sci. Paris*, 269 (1969), p. 603.

[10] CHERET, R. Sur la structure externe aval de rang O d'une onde de détonation. *C. R. Acad. Sci. Paris*, 270B (1970), p. 1517.

[11] CHERET, R. Contribution à l'étude des détonations sphériques divergentes dans les explosifs solides. Thèse de doctorat ès sciences. Poitiers, 1971, Rapport CEA no. 4283.

[12] CHERET, R. Extinction ou bifurcation d'une détonation autonome. *C. R. Acad. Sci. Paris*, 301, series II (1985), p. 961.

[13] CHERET, R., CHAISSE, F., ZOE, J. Some results on the converging spherical detonation in a solid explosive. *Proc. 7th Symposium on Detonation*, Annapolis/MD (1981), p. 602.

[14] CRUSSARD, J. Ondes de choc et onde explosive. Bulletin de la Société de l'Industrie Minérale, 4th series, vol. VI (1907).

[15] DAMAMME, G. Mouvement d'une onde de détonation sonique. *C. R. Acad. Sci. Paris*, 292, series II (1981), p. 381.

[16] DIXON, H.B. The rate of explosions in gases. *Philos. Trans. Roy. Soc. London, Ser. A*, 184 (1893), pp. 97–188.

[17] DÖRING, W. Uber den detonations vorgang in gasen. *Ann. Physik*, 6, **43** (1943), pp. 421–436.

[18] DREMIN, A.N., TROFIMOW, V.S. Nature of the critical detonation diameter of condensed explosives. *Fizika Gorenya i Vzryva*, **5**, no. 3 (1969), pp. 304–311.

[19] EYRING, H., POWELL, R.E., DUFFEY, G.H., PARLIN, R.B. Stability of detonation. *Chem. Rev.*, **45** (1949), pp. 69–181.

[20] FRIEDRICHS, K.O. On the mathematical theory of the deflagrations and detonations. Navord Report 79–46, 1946.

[21] GERMAIN, P. *Mécanique des Milieux Continus*. Masson, Paris (1962).

[22] GERMAIN, P., GUIRAUD, J.P. Conditions de choc et structure des ondes de choc dans un écoulement non stationnaire de fluide dissipatif. *J. Math. Pures Appl.*, **45** (1966), p. 311.

[23] HAYES, W.D. The basic theory of gasdynamic discontinuities. *High Speed Aerodynamics and Jet Propulsion*, vol. III, Part D (H.W. Emmons, editor), Princeton University Press, Princeton, NJ (1958).

[24] HIRSCHFELDER, J.O., CURTISS, C.F., BYRON BIRD, R. *Molecular Theory of Gases and Liquids*. Wiley, New York (1954).

[25] HUGONIOT, P.H. *J. l'École Polytechnique* (1887–1889), 57th and 58th cahiers.

[26] JONES, H. A theory of the dependance of the rate of detonation of solid explosives on the diameter of the charge. *Proc. Roy. Soc. London, Ser. A*, **189** (1947), p. 415.

[27] JOUGUET, E. *Mécanique des Explosifs*. Octave Doin, Paris (1917).

[28] KAPLUN, S. The role of coordinate systems in boundary-layer theory. *Z. Angew. Math. Phys.*, **5** (1954), p. 111.

[29] KEIL, K.A. Das qualitative Verhalten der Integralkurven einer gewöhnlichen differentialgleichung erster ordnung in der umgebung eines singulären punktes. *Jahresber. Deutsch. Matn.-Verein.*, **57** (1955), p. 111.

[30] LAGERSTROM, P.A., COLE, J.D. Examples illustrating expansion procedures for the Navier–Stokes equations. *J. Rat. Mech. Anal.*, **4** (1955), p. 817.

[31] LANDAU, L., LIFSCHITZ, E. *Fluid Mechanics*. Pergamon Press, London (1959).

[32] NEUMANN, J. VON Theory of detonation waves. O.S.R.D. Report no. 549, 1942. *Von Neumann's Collected Works* (A.H. Taub, general editor), Pergamon Press, London (1963).

[33] RAYLEIGH, J. Aerial plane waves of finite amplitude. *Proc. Roy. Soc. London*, **84** (1910), pp. 247–284.

[34] SANSONE, G., CONTI, R. *Nonlinear Differential Equations*. Pergamon Press, London (1964).

[35] TAYLOR, G.I. The dynamics of the combustion products behind plane and spherical detonation fronts in explosives. *Proc. Roy. Soc. London, Ser. A*, **200** (1950), p. 235.

[36] VAN DYKE, M. *Perturbation Methods in Fluid Mechanics*. Academic Press, New York (1964).

[37] WEYL, H. Shock waves in arbitrary fluids. *Comm. Pure Appl. Math.*, II (1949).

[38] VIEILLE, P. Étude sur le rôle des discontinuités dans les phénomènes de propagation. *Memorial des Poudres et Salpêtres*, **10** (1899–1900), p. 177.

[39] WITHAM, J.G.B. *Linear and Nonlinear Waves*. Wiley, New York (1974).

[40] WOOD, W.W. and KIRKWOOD, J.G. Diameter effects in condensed explosives. The relation between velocity and radius of curvature of the detonation wave. *J. Chem. Phys.*, **22** (1954), p. 1920.

[41] ZELDOVICH, I.B. *J. Exptl. Theoret. Phys (USSR)*, **10** (1940), p. 542.
[42] ZELDOVICH, I.B. Distribution de pression et de vitesse dans les produits de déto-
     nation; cas d'une onde sphérique divergente. *J. Exptl. Theoret. Phys (USSR)*, **12**
     (1942), p. 389.
[43] ZELDOVICH, I.B. Converging cylindrical detonation wave. *J. Exptl. Theoret.
     Phys. (USSR)*, **36** (1959), pp. 782–792.

Part Two
Molecular Mechanisms of Explosive Decomposition

CHAPTER IV

Sensitivity to "Shock" and Molecular Structure

1. Introduction

The study of the properties of explosives in relation to their structure assumes several aspects; one of these consists of the examination of explosive decomposition at the level of the molecule. The first process of this kind was proposed by Van't Hoff [1] at the end of the last century. Observing that it was the nitration of certain compounds which conferred explosive properties on obtained derivatives, properties moreover which were all the more pronounced because the density in NO_2 groups was high, he stated: "It is atomic bonds of a specific nature which generate the explosive character."

Thus the role of structural parameters of the molecule in the phenomenon of detonation was introduced. It was very soon clarified by Wieland, who defined the notion of explosophore groups.

The influence of molecular parameters and, in particular, of these groups was then considered at the level of two properties: the detonation velocity and sensitivity to "shock." (N.B. In Chapters IV–VI, the inverted commas round the word shock underline the fact that it does not refer to shock in the strict sense of Chapter II, but rather to any form of sudden mechanical stress.) It seems however that until the 1960s, this consideration was limited to the thermodynamic characteristics (energy of formation, etc.) of the molecule, the explosophore groups often only occurring through the mass, or the total formula of the compound.

It was not until the development of experimental—and in particular, spectroscopic—techniques, and the theoretical support (quantum mechanics, molecular shocks, exciton theory, etc.) which is indispensable in the analysis of the data provided by these techniques that there was a return, in a practical way, to an examination of the relationship between molecular structure and explosive properties.

Today it is clear that the essential difficulty lay in the method of tackling the problem, that is to say, in the choice of a measurable quantity which is unequivocally linked to the first step in the generation of detonation. The chosen parameter in the majority of studies was the sensitivity to "shock" of the substance, represented by the energy of the "shock" capable of inducing a certain probability of reaction in a specific explosive.

First, then, it is a question of extracting from the abundant literature the data which pertain to sensitivity to "shock." In fact, while the existence of a detonation velocity which is characteristic of every explosive composition has been established for a long time, sensitivity remains a feature which is both relative and ambiguous. The numerical value assigned to it depends almost as much on the explosive itself as on the conditions (physical state, operating protocol, external environment) in which it is found. Moreover, if the term "pyrotechnic sensitivity" means the more or less pronounced ability of an explosive to decompose through the action of external stresses, the very diversity of the nature of those stresses leads to consideration of different types of sensitivity. Indeed, there is one sensitivity defined by its type of aggression (friction, "shock," heat flow, shock wave, etc.).

Next, an experimental scale must be defined which can serve as a reference to different studies. In fact, simply describing the different "shock" hammers [2], [3], [4] reveals the extent to which, when a laboratory publishes a value of sensitivity without stating, if not the machine used, at least the chosen percussive mass or the operative characteristics of the test (glass-paper, confinement of the explosive, encapsulation, etc.), this result is difficult to situate in the collection of existing values. Moreover, when this information is known, it can be seen that the way in which results are analyzed is a determining factor: the number of trials carried out, the sequence used to define successive heights [5]. In addition, knowing the percentage of the tests judged to be positive which have been taken into account when determining this sensitivity is very important. It is, in fact, deceptive to compare sensitivities which correspond to different percentages of reaction [H (50%) ≠ H (30%)].

In the case of the nitro secondary explosives studied in this chapter, we give in Table IV.1(a) a scale which is the result of an analysis of the works referenced from [6] to [11]. In establishing this scale, account was taken of the aforementioned difficulties.

The explosives whose developed formulas are given in Table IV.1(b) are characterized by their figure of insensitivity (F.I.), which was first demonstrated by Rotter. This quantity contains implicitly the comparison of the test result with the result of the same test carried out on an explosive used as a control. It is in fact defined as follows:

$$\text{F.I. } (n\%) = 80 \frac{H\,(n\%)}{H_R(n\%)}.$$

In this expression, H is the height of fall which gives a certain n% probability of decomposition, independently of the rate of that decomposition. $[H_R]$ characterizes a certain batch of cyclonite (RDX), tested in the same conditions, and serving as a reference.

Table IV.1(a). Experimental scale of "sensitivity to shock" of the main explosive molecules, divided into three groups. F.I. is the shape of insensitivity; expanded formulas are given in Table IV.1(b).

Sensitivity to "shock"	Nictric esters	Nitramines	Nitroaromatics
F.I.			
10	Nitroglycerine		
	Nitrocellulose Dinitroglycerine Nitroglycide		
50			
	PETN		
		Cyclonite (RDX) Octogene	
100	Nitrometriol Nitroglycol	Tetryl Haleite	Tetryl Tetranitroaniline
	Dimethylolnitro- ethane dinitrate		
			Picric acid
			TNX
150			TNT Trinitrophenetole
			s-trinitrobenzene
			m-dinitrobenzene
200			
250		Nitroguanidine	
			TATB
300			

Table IV.1 (b). Expanded formulas of explosive molecules referred to in Table IV.1(a).

2. Concept of Explosophore Group

The most direct way of considering the influence of the molecular structure of the explosive on sensitivity consists of taking into account its atomic composition. In fact, at the 6th International Symposium on Detonation held in 1976 a correlation between the sensitivity to "shock" of a substance and its atomic composition was effectively formulated for a large number of explosives.

Due to Kamlet [12], the study presented rests on the notion of families arbitrarily defined in such a way that they bring together explosives which are characterized by some type of decomposition mechanism. It underscores, within each family, a linear relationship, which is statistically significant taking account of the number of explosives tested, between the "50% height of fall" logarithm and the values of an oxygen balance defined as follows:

$$\text{B.O.} = \frac{100(2\alpha_O - \alpha_H - 2\alpha_C - 2\alpha_{COO})}{M},$$

M being the molecular mass of the product, α_ℓ the number of ℓ atoms present in the molecule, and α_{COO} the number of carboxyl groups.

In Table IV.2 we show the linear relationships obtained for four families of compounds, [12], [13], as well as the number of explosives considered in each family in order to establish these relationships.

Note that in this work, the authors are sketching out an analysis of the causes which are capable of explaining, at a molecular level, the differences observed between the sensitivities of compounds which are neighbors by their structure, without formulating any simple interpretation of the observed correlations. The sole conclusion is that differences of sensitivity which have been observed between aromatic compounds are attributed to inter- and intramolecular oxidizing actions which are selective at the level of certain groups.

Table IV.2. Relationship between oxygen balance and height of fall entailing decomposition in 50 cases out of 100: H (50%).

	Log H (50%)	Number of compounds studied
Nitramines	1.372–0.168 B.O.	45
Nitric esters	1.753–0.233 B.O.	28
Nitroaromatics without CH group in position α of one NO_2	1.33–0.26 B.O.	11
Nitroaromatics with CH group in position α of one NO_2	1.73–0.32 B.O.	21

It was the observation that the nitration of compounds such as aromatic carbides, alcohols, or amines conferred explosive properties on obtained derivatives, and that these properties are more pronounced, the greater the density in NO_2 groups, that led Van't Hoff [1] to situate the origin of the explosive nature of nitro compounds in the NO_2 group. By generalizing to other groups, he stated: "It is atomic bonds of a specific nature which give rise to the explosive character." These bonds, thirteen in number at the time of Van't Hoff, belong to the chemical groups which, in 1909, Wieland named explosophore groups. Among these groups, for example, there are the nitro (NO_2), nitroso (NO), azido (N_3), perchloryl (ClO_3), and azo (—N=N—) chromophores, etc.

The prominence given to the special role of these groups under high pressure is however quite recent. It will be illustrated by the work of Owens and Sharma [14], [15] who show that, in the case of nitro explosives, it is always the bonds belonging to the $R—NO_2$ group which are preferentially perturbed by "shock."

The two techniques used to this end are X-ray photoelectron spectrometry (XPS) and electron paramagnetic resonance (EPR).

Results Obtained by XPS

The principle of XPS spectroscopy consists of bombarding the sample with X-rays of a known $h\nu$ energy and measuring the distribution of kinetic energy of the emitted photoelectrons.

Taking account of the equation of conservation of energy

$$E_L = h\nu - \tfrac{1}{2}mV^2,$$

an XPS spectrum reveals E_L energies of all the electronic levels of the different atoms of the molecule as well as the number of electrons occupying these levels. Moreover, for a given electronic level of an atom, the value of E_L

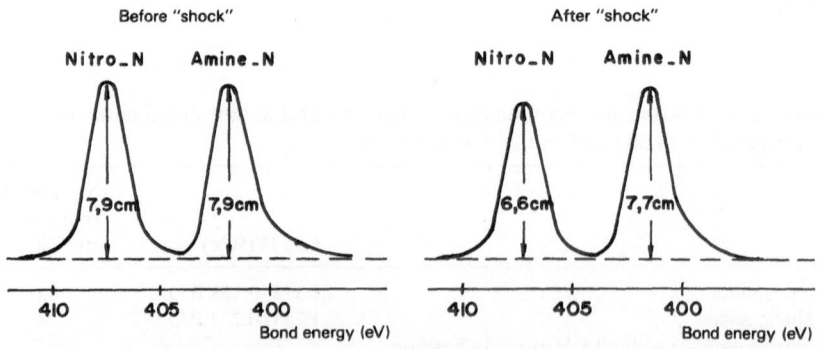

Figure IV.1. Changes under "shock" of the XPS spectrum of cyclonite. Under "shock," a decrease in intensity is observed of the peak relative to the nitrogen of the nitro group in comparison with that of cyclic nitrogen, Ref. [18].

depends on the chemical environment of that atom. Thus, for example, in the case of cyclonite (RDX), the bonding energy of the 1s electron of the nitrogen atom carried by the NO_2 group is greater by 5.5 eV to that of the 1s electron of the nitrogen atom which belongs to the cycle.

The principle of the experiments carried out by Owens and Sharma consists of comparing the intensity of the peaks relative to each nitrogen atom before and after "shock." An illustration is given in Figure IV.1 of the results obtained in the case of the 1s electrons belonging to the two types of nitrogen in cyclonite. From such experiments it emerges that the only modifications are observed at the level of the peak relative to the nitrogen of the NO_2 group. In fact, the intensity of this peak, which is equal at rest to that of the other nitrogen, diminishes by 14% after "shock" in comparison with the latter.

This result makes evident the splitting under high pressure of the bond linking the explosophore group to the rest of the molecule, the number of molecules affected being in the order of 14% of the total molecular population.

Results Obtained by EPR

Owens and Sharma completed their investigation by giving prominence to the appearance of paramagnetic species at the time of "shock." The nature of these species is determined with the help of EPR. The signal obtained is, as much by its form as by its position and the number of its constituent elements, identical to that of the NO_2 radical which was theoretically reconstituted by Schaafsma [16], [17]. This result thus brings a second confirmation of the release of the nitro group under "shock" in the case of cyclonite (RDX).

3. The Case of Nitro Explosives

The special role of these explosophore groups under high pressure having been perceived, the next stage consisted of linking, through sensitivity to "shock," the conditions of generation of detonation to the release of these groups. In the case of nitro explosives, this came down to establishing a correlation between this sensitivity and the more or less pronounced facility which these molecules may have in releasing NO_2.

3.1. Sensitivity to "Shock" and Amount of NO_2 Released

A first approach, proposed by Cherville et al. [20] consists of comparing this sensitivity to the quantity of NO_2 formed without judging what mechanism might be at the origin of this release. This quantity is assessed by the radiochemical yield of production of NO_2 (G_{NO_2}) which measures the number of NO_2 molecules formed per one hundred radiative electron volts absorbed in

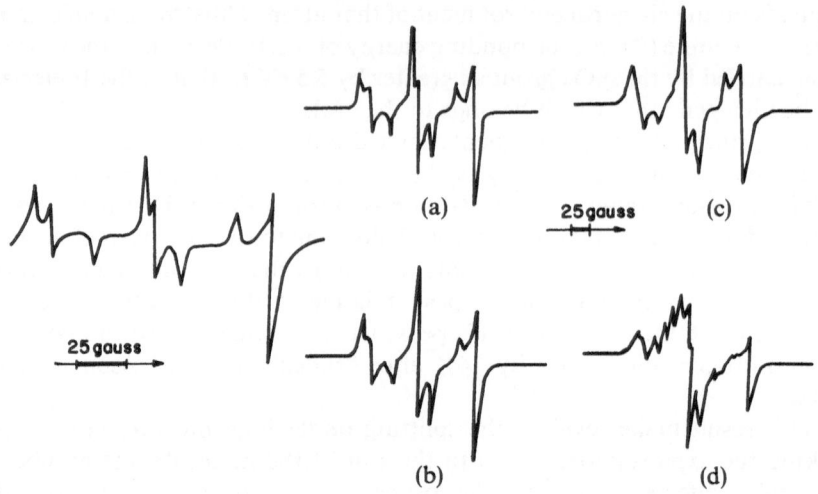

Figure IV.2. EPR spectra of: (a) penta-erythritol, (b) cyclonite (RDX), (c) octogen, (d) tetryl, irradiated and studied at 77 K, Ref. [2]. Comparison with the EPR spectrum of NO_2 in a crystalline medium theoretically reconstructed by Shaafsma, Ref. [19].

one gram of explosive. It should be stressed that the yield thus defined is independent of the mass dose absorbed, but that its value nevertheless depends on a possible distribution of the absorbed energy between the phenomenon of NO_2 release and other phenomena.

Since the work of Leverd [21] on the one hand and Darnez and Paviot [22] on the other, have shown that the nature of the final products of the radiolysis of cyclonite (RDX) is independent of the nature of incident radiation, radicals are obtained by irradiation in a vacuum at 77 K of explosives in a polycrystalline form, with the help of gamma rays produced by a ^{60}Co source with a dose rate of 1 Mrad/h.

The radicals formed are characterized by EPR. The choice of this method is explained by the fact that the concentration of the species formed is directly proportional to the surface of the signal. Figure IV.2 brings together the EPR spectrum of NO_2 theoretically reconstituted by Schaafsma, as well as those obtained by Cherville and his colleagues in the case of penta-erythritol, cyclonite, octogen, and tetryl.

If N represents the number of spins per gram of irradiated substance for a deposited radiation dose D (expressed by electron volt), G_{NO_2} is given by

$$G_{NO_2} = \frac{N \times 100}{D\,(eV)}.$$

The results obtained for eleven explosives belonging to the three secondary nitro families are given in Table IV.3. Examination of this table, in which the explosives are classified in order of decreasing sensitivity to "shock," leads to

Table IV.3. Number of molecules of NO_2 formed by irradiation of one gram of explosive with a dose of 1 Mrad; corresponding radio-chemical yield G_{NO_2}. The explosives are classified by order of decreasing sensitivity to "shock." Refs. [25], [29].

Explosive	No. of NO_2 molecules formed per gram of irradiated explosive	G_{NO_2}
Nitrocelulose (10.8)	16.9×10^{17}	2.70
Penta-erythritol	23.8×10^{17}	3.80
Cyclonite (RDX)	5.6×10^{17}	0.90
Octogen	5.0×10^{17}	0.80
Haleite	1.9×10^{17}	0.30
Nitroguanidine	0	0
Tetryl	3.7×10^{15}	0.006
Tetranitroaniline	$<10^{15}$	<0.001
TATB	6.2×10^{15}	0.010
Trinitrophenol	10^{15}	<0.001
TNT	$<10^{15}$	<0.001

two conclusions:

(a) the three explosive families considered differ significantly in value, indeed in the order of the magnitude of the G_{NO_2}; thus the very sensitive explosives (nitric esters) have a $G_{NO_2} > 1$, the explosives which are barely sensitive (nitroaromatics) have a $G_{NO_2} < 10^{-2}$, and nitramines occupy an intermediate position;

(b) the only consideration of the quantity of NO_2 formed is not sufficient to account for the scale of sensitivities to "shock", as shown below:

haleite, whose sensitivity is of the order of that of tetryl ($G_{NO_2} = 0.006$), has a G_{NO_2} of 0.3;

tetranitroaniline, whose sensitivity is similar to that of tetryl, has an almost nil G_{NO_2} (like that of TNT, for example); and

TATB, which is less sensitive than all these explosives, has a G_{NO_2} of 0.01, that is to say, higher than that of tetryl.

The conclusion of this work then is that the assessment alone of molecular decomposition, examined at the level of the explosophore group, is not sufficient to explain the sensitivity to "shock" of the substance.

3.2. Sensitivity to "Shock" and Conditions of Release of NO_2

The preceding considerations focused on the ability of the molecule to release the explosophore group globally, without prejudice to the factors which work towards it at the level of this molecule. This power is looked at in more detail below:

- by considering the tendency to separate this group from the rest of the molecule, taking into account the values of the bond energies $R—NO_2$; and
- by evaluating the ability of the molecule to harness the energy necessary for this decomposition.

Linking sensitivity to "shock" to the value of molecular bond energies is a logical step which has been the object of many studies. In the hypothesis of the preferential release of the explosophore group, consideration of this energy should, in the case of nitro explosives, be made at the level of the bond linking NO_2 to the rest of the molecule. By keeping to the big families, the differences in sensitivity of the nitric esters, nitramines, and nitro-aromatics can in fact be attributed to the differences observed between the values of the bond energies $O—NO_2$, $N—NO_2$, and $C—NO_2$ [24].

In order to verify whether in fact this relationship exists in the case of molecules belonging to the same family, the nitramine bond energy of cyclonite (RDX), octogen, and tetryl has been determined by mass spectrometry with an electronic impact source [25]. The threshold energy of an NO_2^+ ion produced by electronic bombardment is, in fact, equal to the sum of the dissociation energy E of the bond $N—NO_2$ and the ionization energy of NO_2 in NO_2^+ [26].

The ionization energy of NO_2 being constant [27], the determination of the energy of these bonds by that technique comes down to the measuring of the threshold energies of the NO_2^+ ion. Examination of the values obtained:

- cyclonite $E (N—NO_2) = 75.6$ kcal/mole;
- octogen $E (N—NO_2) = 54.9$ kcal/mole; and
- tetryl $E (N—NO_2) = 47.9$ kcal/mole;

is sufficient to illustrate that this energy cannot be correlated on a one-to-one basis with the sensitivity of the explosive. Paradoxically, in fact, it can be seen that it is the most sensitive explosive which has the strongest bond energy $N—NO_2$.

The above values of G_{NO_2} show the ability of the molecule to release a specific group. In the works [23], [25], [28], [29], this ability has been compared to that of the substance to harness the incident energy necessary for this release. This comparison leads to a definition of the idea of "selective utilization" of the incident energy at the breaking of a particular type of bond.

The ability of the molecule to harness incident energy is assessed by determining the probability of reaching the electronic levels contained in a given energy interval (E_0, E_ω). This potential is measured by the yield G_{EX} of excited molecules for a certain amount of energy deposited per molecule. Where this energy is taken to equal 100 eV, this yield is expressed in the form [18], [19]

$$G_{EX}(E_0, E_\omega) = \frac{100}{Z} \int_{E_0}^{E_\omega} \frac{df}{E - E_o},$$

where Z is the number of valence electrons in the molecule, and df is the differential oscillator force associated with each electronic transition of amplitude $E - E_0$.

It has been evaluated experimentally by the study of UV absorption spectra and, theoretically, by the use of the methods of quantum chemistry. However, the theoretical determination has the advantage of considering the molecules in their crystalline form and allowing every excitation level to be taken into account. In this case, the values of $G_{EX}(E_0, E_\omega)$ are determined by carrying out a summation on each of the levels E_n, ω being the number of levels considered

$$G_{EX}(E_0, E_\omega) = \frac{100}{Z} \sum_{n=1}^{n=\omega} \frac{f_n}{E_n - E_o}.$$

In this expression f_n is the oscillator force of an electronic transition corresponding to the excitation of an electron from the fundamental state E_0 to a given electronic state n.

The theoretical method used by Leverd [21] in the case of cyclonite and of octogen, and by Delpuech [25] in the case of 15 molecules belonging to three families of nitro secondary explosives, was the semiempirical method CNDO–2S/CI [30]. In the first-mentioned work, the ω number of excited

Table IV.4. Theoretical yield of excited molecules G_{EX} computed by the CNDO–2S/CI method.

Explosive	G_{EX}	G_{EX}/G_{EX} (hex)
Nitroglycerine	0.318	0.80
Dinitroglycerine α	0.298	0.75
Dinitroglycerine β	0.287	0.72
Penta-erythritol	0.346	0.87
Cyclonite (RDX)	0.398	1.00
δ Octogen	0.404	1.02
β Octogen	0.410	1.03
α Octogen	0.409	1.03
Haleite	0.422	1.06
Nitroguanidine	0.538	1.35
Tetryl	0.615	1.55
Tetranitroaniline	0.622	1.56
TATB	0.800	2.01
s. trinitrobenzene	0.832	2.09
m. dinitrobenzene	0.902	2.27

levels under consideration is limited to 20. In the second, every level whose difference of energy from the fundamental state is lower than or equal to 10 eV is considered; the results thus obtained are given in Table IV.4.

Upon examination of these results, it is seen that the yield of excited molecules varies markedly from one explosive family to another. If cyclonite (RDX) is taken as a reference, the molecules which belong to each of the three nitro families have fairly similar G_{EX} values:

$0.70 < G_{EX}/G_{EX}$ (RDX) < 0.90 in the case of nitric esters;
$1.00 < G_{EX}/G_{EX}$ (RDX) < 1.35 in the case of nitramines; and
$1.50 < G_{EX}/G_{EX}$ (RDX) < 2.30 in the case of nitroaromatics.

It can also be seen that the molecules which have a low G_{EX} have a high G_{NO_2} and conversely, the more sensitive the family, the greater the G_{NO_2}/G_{EX} ratio. Thus the explosives which are not very sensitive (with a low G_{NO_2}) have a higher G_{EX} which represents, for example, 90% of the excited molecules in nitrobenzene. On the other hand, the very sensitive nitric esters (with a higher G_{NO_2}) have a low G_{EX}. Finally, the nitramines occupy an intermediate position in both cases.

Altogether these results lead to the conclusion that sensitivity may depend not on the quantity of energy absorbed by the molecule, but rather on the effect of this absorption on a particular type of bond. This hypothesis leads on to an examination of the influence of electronic structure on sensitivity to "shock."

CHAPTER V

Sensitivity to "Shock" and Electronic Structure

1. Electronic Structure in the Fundamental State

1.1. Introduction

The influence of the electronic structure of explosives on the birth of the detonation regime is introduced quite naturally if it is noted that the rapid conversion of the energy stored in chemical bonds into the kinetic energy of decomposition fragments depends on the changes which must be brought about in the molecular electron levels in order to produce this conversion.

The study of these changes requires first of all a definition of the parameters able to account for them, i.e., among the electronic characteristics of the explosive molecule, those which are definitive for moving from state 1 to state 2 under the effect of an external energy supply (see Fig. V.1).

The first type of approach consists of studying the electronic structure of these compounds in the fundamental state. This structure may, in fact, be thought to play a role, not only in the ability of the molecule to harness energy, but also in its response to excitation. It is from this perspective that certain authors have studied the electronic structure of molecular explosives and others have established the band-structure of azides.

Section V.1 is devoted to the approaches used to correlate this electronic structure to sensitivity to "shock." It will be seen as we go along that simply considering this structure in its fundamental state does not explain the scale of sensitivity to "shock," whatever the parameter considered.

1.2. Distribution of Electrons in the Isolated Molecule

Determining the electronic structure of explosive molecules means employing the methods of quantum chemistry, which aim at solving the Schrödinger equation of the system. Due to the dimensions of the molecules under consideration (the cyclonite (RDX) molecule, for example, is made up of 24 atoms and 114 electrons), such a solving makes it necessary to set up approximations. Since the Hamiltonian is fixed, the problem consists first of defining the

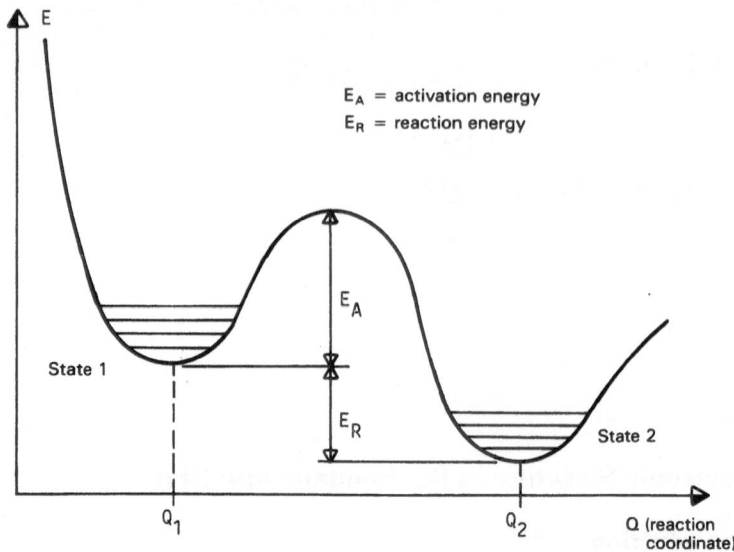

Figure V.1. Reaction path allowing passage from state 1 to a more energetically stable state 2.

type of wave function likely to represent the system. The Born and Oppenheimer approximation leads to a division of this wave function into two parts, one relating to nuclei, the other to electrons. The polyelectronic wave function is then constructed by considering monoelectronic orbitals as linear combinations of atomic orbitals—the LCAO (Linear Combination of Atomic Orbitals) approximation [32].

Calculating the energies associated with each molecular orbital requires the evaluation of a large number of terms. Therefore, the aim of semiempirical methods is to introduce devices which are capable of reducing the volume of these calculations and minimizing the secondary effects of these devices, by attributing numerical values to a certain number of integrals so that the results of the calculations are adjusted to fit the experimental data. However, the choice of any semiempirical method should take into account the type of structural property studied [33], since the parametrizations appropriate to each one actually make it more or less suited to the study of some characteristic or other (energy, geometry, dipole moment, etc.).

Thus it appears that, apart from some studies which call on the extended methods of Huckel [34], the majority of works use INDO (Intermediate Neglect of Differential Overlap) or CNDO (Complete Neglect of Differential Overlap) methods. In fact, these methods, proposed by Pople *et al.* [35], [36], have the advantage over other methods which look at orbitals including all the valence electrons, of being better suited to the determination of electronic densities.

It is noteworthy in the literature that this type of approach has been used frequently only on a limited number of explosives [21], [37]. Thus, taking into account the approximations which govern them, the results obtained by applying these methods have no real meaning except in a relative sense. That is why only the studies which looked at a consequent number of molecules will be considered here, i.e., those of Delpuech and Cherville on nitramines, nitroaromatics [23], nitric esters [38], tetrazoles [25], and picrylazoles [39], those of Haskins on tetrazoles [40], and those of Schroeder on tetrazoles [41] and picrylazoles [42].

The majority of these studies consider the isolated molecule. In these conditions, the only data necessary for calculation are the coordinates of all the component atoms of the molecule in the configuration studied. Depending on the explosive group concerned and according to the authors, the molecular configuration was treated in different ways. In the case of nitroaromatics and nitramines, Delpuech and Cherville examined the molecular configuration as it appeared in the crystal; from a practical point of view, this limits the study to molecules whose crystalline structure is known. In the case of liquid nitric esters, they determined, using the INDO method, the most likely configurations in energy terms by varying bond lengths and angles, with the effect of rotations around bonds C—C and C—ONO_2 being given particular consideration [38]. In all the studies relating to tetrazoles, the authors had recourse to the standard bond lengths and angles used in the semiempirical methods [43]. In the case of picrylazoles, Schroeder [42] examined standard structural parameters and limited his study to the nitrogen ring, the influence of the picryl ring being ignored. On the other hand, this effect is considered by Delpuech et al. [39], notably at the level of the relative position of the two rings and their resultant consequences on the orientation of the nitro groups.

Whatever the parameter considered by the different authors to link the electronic structure of a molecule to the sensitivity of the substance, all the studies consisted at first of establishing the electron distribution affecting all the constituent atoms of a molecule. Figure V.2, taken from [25], illustrates the charge distribution thus obtained using the INDO method in the case of nitro explosives. The charge density q_A carried by an atom A is defined as the difference between the core charge Z_A of this atom and the total electron population P_A which is localized there

$$q_A = Z_A - P_A.$$

The P_A population is equal to the sum of all the electron populations of the atomic orbitals "centered" on the atom. The values of Z_A are equal, respectively, to 1, 4, 5, and 6 in the case of the atoms of hydrogen, carbon, nitrogen, and oxygen.

1.3. Influence of the Crystalline Environment

However, the molecule is indissociable from the rest of the sample. Delpuech and Cherville have assessed the effect of crystalline environment on the

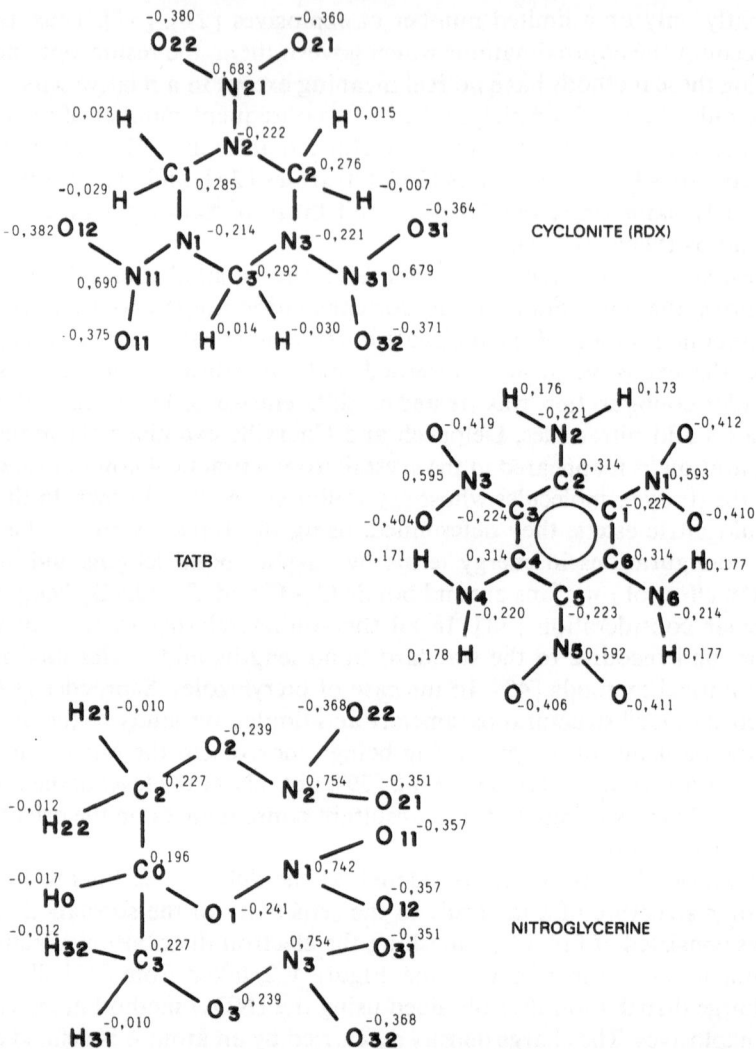

Figure V.2. Electron charge distribution in the fundamental state obtained by application of the INDO method, Ref. [25].

charge distributions explained above [44]. As a result of the authors' being more concerned with the changes brought about in the molecule rather than the global properties of the crystal, the method chosen is that developed by Bacon and Santry [45]. This method consists of performing a self-consistent-field (SCF) perturbation treatment of semi-infinite molecular crystals by localizing the orbitals on the molecules which form the network and by treating the delocalization on neighboring molecules with a technique based on the theory of third-order perturbations.

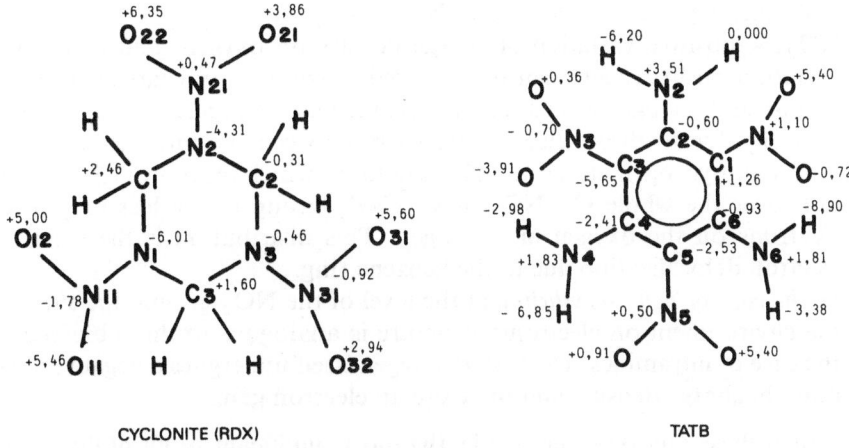

CYCLONITE (RDX)   TATB

Figure V.3. Effect in percentages of the crystalline environment on electron distribution in the molecule, Ref. [44].

However, the study is limited to molecules situated in a regular crystalline environment, and focused on crystals of cyclonite (RDX), β-octogen (β-HMX), haleite, triaminotrinitrobenzene (TATB), and pentaerythritol (PETN) whose characteristics are given in Refs. [46]–[49]. The relative variations of charge density due to crystalline interactions were established for all the constituent atoms of the molecule. Figure V.3, taken from [44], gives percentage values obtained in the case of cyclonite (RDX) and (TATB). It should be noted that, in the case of cyclonite (RDX), given the weak charge densities of the hydrogen atoms, the values of these variations have been omitted; in this case, the relative values actually lose all significance.

The influence of crystalline environment on the electron distribution of the compounds can be shown schematically as follows:

- *In the case of nitramines (haleite, cyclonite (RDX), octogen)*, there can be seen:
  — a positive variation on the oxygens which expresses an electron loss (this phenomenon is due to the fact that the strong electron density localized on oxygen induces a tendency to create hydrogen bonds and charge transfers);
  — a negative variation on nitrogen of the nitro group, that is to say, an electron gain resulting from the interaction with oxygens of neighboring groups; and
  — a greater negative variation on the other nitrogen which, in spite of its negative charge, drains a part of the surplus electrons on the oxygen atoms.
- *In the case of TATB*, variations in charge densities due to the environment express the different electronic characters of the two groups present on the

molecule, acceptor in the case of NO_2 and proton–donor in the case of NH_2. A positive variation of charge density on oxygens and a negative variation on hydrogens can be observed. Variations in charge density of carbon and nitrogen atoms can be explained as the superposition of modifications due to direct electron transfers between the nitrogen atoms of NO_2 and hydrogen atoms of NH_2, and those which result from a distribution over the whole $C—NO_2$ or $C—NH_2$ group of the loss or gain of electrons on the oxygen or hydrogen. This distribution is the result of electron delocalization due to the benzene ring.

- *In the case of penta-erythritol*, at the level of the NO_2 groups, the effect of the environment on electronic structure is analogous to that observed in the case of nitramines. The "ester" oxygen itself undergoes a negative variation in charge density, and therefore an electron gain.

When these analyses were made, the most significant result of this study was that the effect of crystalline interactions on charge density carried by the different atoms is very slight. From this point of view, then, it is possible to assert that at the level of electron distribution, this effect is considered when calculations are carried out on an isolated molecule looked at in the configuration which it occupies in the crystal.

1.4. Electronic Structure Before Excitation

The effect of molecular electron distribution in the fundamental state on sensitivity to "shock" can be seen from two angles:

- the ability of the molecule to absorb incident energy; and
- the ability of this molecule to focus energy on particular types of groups or bonds.

Electron Distribution and Energy Absorption

This aspect, in particular, was studied by Haskins in the case of six tetrazoles substituted in position 1 [40]. He looked for a correlation between the total electron population of the nitrogen cycle and the sensitivity of these compounds. The deficient or excess nature of this population was assessed through the electronegative character of the substitute, characterized by its charge. The results obtained by applying the EHMO method are quoted in column 1 of Table V.1. In this table, the explosives are classified in order of decreasing sensitivity to "shock." On examining these values, there appears to be no direct correlation between the two scales. Thus, for example, whereas the presence of the nitro group leads to an electron deficiency in the nitrogen ring greater than that which is due to the azide group, all the experimental results lead to the conclusion that nitrotetrazole has a lower sensitivity than azidotetrazole. In the course of this work, Haskins studied the nitrogen–nitrogen bond energies in β position of carbon. These bonds

Table V.1. Developed formula and electronic characteristics of six tetrazoles substituted in position 1. Ref. [40].

Explosive	Electronegative character of the substitute	Overlap population of N—N bonds in β of the substitute
Azidotetrazole	−0.197	0.829
Nitrotetrazole	−0.368	0.856
Nitraminotetrazole	−0.362	0.831
Tetrazole	+0.043	0.840
Aminotetrazole	−0.077	0.824
Methyltetrazole	+0.047	0.832

Azidotetrazole

Benzotetrazole

Methyltetrazole

Nitrotetrazole

Nitraminotetrazole

Bitetrazole

are, in fact, those whose rupture was first observed at the time of the decomposition phenomena studied by slow analysis methods. In order to evaluate the strength of these bonds, he determined their degree of covalence through the overlap population. The value of this population, in fact, expresses the contribution in the actual bond of the electrons belonging to the two atoms which make up that bond. This is expressed as

$$P(A - B) = \sum_{j \in A} \sum_{k \in B} n(A - B) C_j C_k S_{jk},$$

(n(A − B) is the number of electrons belonging to the molecular orbital associated with the bond; C_j and C_k are the coefficients of expansion of this orbital on the base of the atomic orbitals j and k which belong to the two atoms concerned; S_{jk} are the overlap integrals of the atomic orbitals). It can be seen—column 2 of Table V.1—that the results obtained confirm the conclusions of §IV.3.2, i.e., there is no simple relation between bond energy and sensitivity.

Electron Distribution and Focusing of Energy

The effect of electron distribution on the ability of a molecule to focus on a particular group was studied by Delpuech and Cherville [23], [25], [29]. The authors attempted, in the case of secondary nitro explosives to link the release ability of an NO_2^+ ion, to the probability of energy absorption by R—NO_2 bonds. Since this probability is greater the higher the polarity of the bonds, they established, for a large number of molecules belonging to nitroaromatic, nitramine, and nitric ester groups, the charge asymmetry carried by these bonds.

This asymmetry is defined as the difference between the charges carried by the R and N atoms, respectively, of the R—NO_2 bonds considered

$$\Delta C° \, (R—NO_2) = q_N - q_R.$$

The results obtained by applying the INDO method (Table V.2), allow two observations to be made:

(a) Each of the three groups considered is characterized by a polarity value of the nitro bonds. This value varies from one group to another, in the same direction as the scale of NO_2 yields (see §IV.3.1) and sensitivi-

Table V.2. Maximum charge asymmetries affecting R—NO_2 bonds of nitramines and nitroaromatics in the fundamental state classified by order of decreasing sensitivity to "shock," Ref. [25].

Explosive	$\Delta C_m°$	$\Delta C_m°/\ell$
Cyclonite (RDX)	0.905	0.660
δ-Octogen	0.937	0.673
β-Octogen	0.889	0.650
α-Octogen	0.902	0.662
Tetryl	0.841	0.624
Haleite	0.880	0.676
Nitroguanidine	0.867	0.642
Tetranitroaniline	0.673	0.461
TNX	0.579	0.391
Trinitrophenetol	0.644	0.435
s-TNB	0.575	0.391
m-DNB	0.539	0.364
TATB	0.820	0.579

ties to "shock." Thus $O—NO_2$ bonds have a polarity value of the order of one corresponding to a $+0.75$ charge carried by nitrogen and a -0.25 charge carried by oxygen. The $N—NO_2$ ($0.84 < \Delta C° < 0.94$) bonds are more strongly polarized than the $C—NO_2$ ($\Delta C° < 0.82$) bonds all stabilized by a resonant cycle.

(b) On the other hand, if the results concerning molecules belonging to a single group are taken into consideration, this correlation no longer holds. Table V.2, gives, as an example, the maximum polarities affecting $R—NO_2$ bonds of 13 molecules belonging to the nitroaromatic and nitramine groups. Consideration of this polarity leads to the following classifications:

— in the case of nitramines: δ-octogen (δ-HMX), cyclonite (RDX), α-octogen (α-HMX), β-octogen (β-HMX), haleite, nitroguanidine, tetryl,
— in the case of nitroaromatics: TATB, tetranitro-aniline, trinitrophenetol, TNX, s-TNB, mDNB.

It can be seen that neither classification allows for an explanation of the scale of sensitivity to "shock" commonly acknowledged for the molecules under consideration. It is established, moreover, that taking into account the polarity values per bond length unit ($\Delta C°/\ell$) does not reproduce this scale in any way.

The general conclusion of all the studies on the electronic structure of explosives in the fundamental state is that no simple relationship is apparent between this structure and the sensitivity to "shock" of the substance. This can be explained by the fact that if this structure correctly expresses the energy absorption capacity, it does not, on the other hand, explain its distribution. To evaluate this last parameter, a knowledge of the electronic structure in decomposition conditions, i.e., after excitation, is essential.

2. Electronic Structure After Excitation

2.1. Introduction

This section relies on studies carried out since 1976 by Delpuech and Cherville on changes in electron distribution in the molecule after energy is applied. These studies led to the suggestion that the sensitivity to "shock" of an explosive should be considered as a molecular property. This suggestion resulted from the prominence given, in the case of 40 molecules belonging to five different groups, to a very good correlation between the electronic structure of a molecule after energy was applied and the response to the hammer of the molecule in question. It is based, with the help of quantum chemistry methods, on the study of the changes in electron distribution after excitation.

One possible cause of the lack of correlation between sensitivity to "shock" and electronic structure of a molecule in the fundamental state may lie in the fact that this structure is not necessarily that of the molecule after "shock," that is to say in decomposition conditions. In order to approach these conditions, the authors looked closely at the different excited states likely to exist after energy absorption.

These states were arrived at by applying the CNDO–2.S/CI method developed by Jaffe [30] which theoretically simulates the electron transition spectra by determining the energy levels through "interaction" of configurations.

The excited states considered are singlet states whose energy difference with the fundamental state is less than 7 eV, i.e., are likely to exist in privileged ignition sites, taking account of the temperatures attained.

In order to make the best possible evaluation of the different parameters involved in this type of method, the results obtained in the case of simple molecules (nitramines, dimethylnitramine) were compared to experimental electron transition spectra [25]. From such a comparison it emerges that the most suitable parametrization is Pariser's [50]. Calculations were carried out by considering an interaction affecting 60 configurations. It was shown that this number, which may seem small in relation to the number of configurations which can be looked at theoretically for the molecules studied (1008 in the case of cyclonite (RDX)), does nevertheless explain the wave function accurately. On the other hand, it is probable that the results obtained by Orloff et al. [51] in a study using the same method in the case of cyclonite (RDX) and in which the authors, for programming reasons, considered only 25 configurations, would be different if this number were greater.

In order to find the polarity changes between the fundamental state and the different excited states, each bond $A - B$ is characterized by a relative polarity variation in the form

$$\delta(A - B) = \frac{\Delta C^* - \Delta C^\circ}{\Delta C^\circ}.$$

In this expression ΔC^* and ΔC° are, respectively, the charge asymmetries borne by this bond in the excited state considered and in the fundamental state.

2.2. Molecular Parameter of Sensitivity to "Shock"

At first, the authors considered molecules belonging to the following three groups: nitroaromatics, nitramines, and nitric esters. The study of polarity variations of each $R-NO_2$ bond, between the fundamental state and all excited states considered, led to the following conclusions [25], [29]:

- these variations are very different according to the excited state considered;
- in the majority of excited states the $R-NO_2$ bonds show a weaker polarity than in the fundamental state; and
- for a given molecule, the different nitro bonds undergo very different maximum polarity variations.

Table V.3. Maximum polarity variations during excitation. Resulting minimum charge asymmetries, related for each group to a reference compound [25], [39].

Explosive	δ_m (%)	ΔC^*	$\dfrac{\Delta C^*}{\Delta C^* \text{(N.G.)}}$	$\dfrac{\Delta C^*}{\Delta C^* \text{(RDX)}}$	$\dfrac{\Delta C^*}{\Delta C^* \text{(T.N.B.)}}$
Nitroglycerine	−49.15	0.625	1.00		
Dinitroglycerine β	−48.73	0.648	1.04		
Dinitroglycerine α	−47.80	0.651	1.04		
Nitroglycide	−47.45	0.659	1.05		
Penta-erythritol	−45.36	0.670	1.07		
Nitrometriol	−45.16	0.691	1.11		
Dimethylolnitro-ethane	−39.31	0.730	1.17		
dinitrate	to	to	to		
	−41.41	0.753	1.20		
Nitroglycol	−39.47	0.753	1.20		
Cyclonite (RDX)	−55.10	0.511		1.00	
δ-octogen (δ-HMX)	−46.72	0.602		1.18	
β-octogen (β-HMX)	−36.00	0.727		1.42	
α-octogen (α-HMX)	−35.30	0.735		1.44	
Tetryl	−23.42	0.786		1.54	
Haleite	−26.40	0.846		1.66	
Nitroguanidine	−22.08	0.962		1.88	
Tetranitroaniline	−16.02	0.518			0.85
TNX	−28.43	0.526			0.86
Trinitrophenetol	−23.72	0.565			0.93
s. trinitrobenzene	−14.77	0.609			1.00
m. dinitrobenzene	−14.85	0.612			1.01
TATB	−34.00	0.620			1.02

Table V.3 brings together for various molecules the δ_m value of the *maximum* polarity variation as well as the *minimum* charge asymmetry ΔC^* which results from this variation. In the case of the nitric esters, the values were obtained by considering all the rotation isomers around the C—O and C—C bonds as being probable in terms of energy after excitation [38]. On examining these values, it will be seen that the maximum polarity variation differs considerably from one explosive to another.

Thus, for example, in the case of nitramines, illustrated by Figure V.4:

• Cyclonite (RDX) with high $\Delta C°$ has a negative δ_m whose absolute value is high; thus excitation can confer a weak polarity on the bond in question.
• α-, β-, and δ-octogen (HMX), which have neighboring $\Delta C°$ only slightly lower than that of cyclonite (RDX), have δ_m whose absolute value is not nearly so high and is more variable from one form to another. The value of ΔC^* for the δ form is distinctly lower than that which relates to the α and β forms.
• Tetryl, haleite, and nitroguanidine are characterized by a very weak polarity variation during excitation. The resulting ΔC^* are therefore high; those

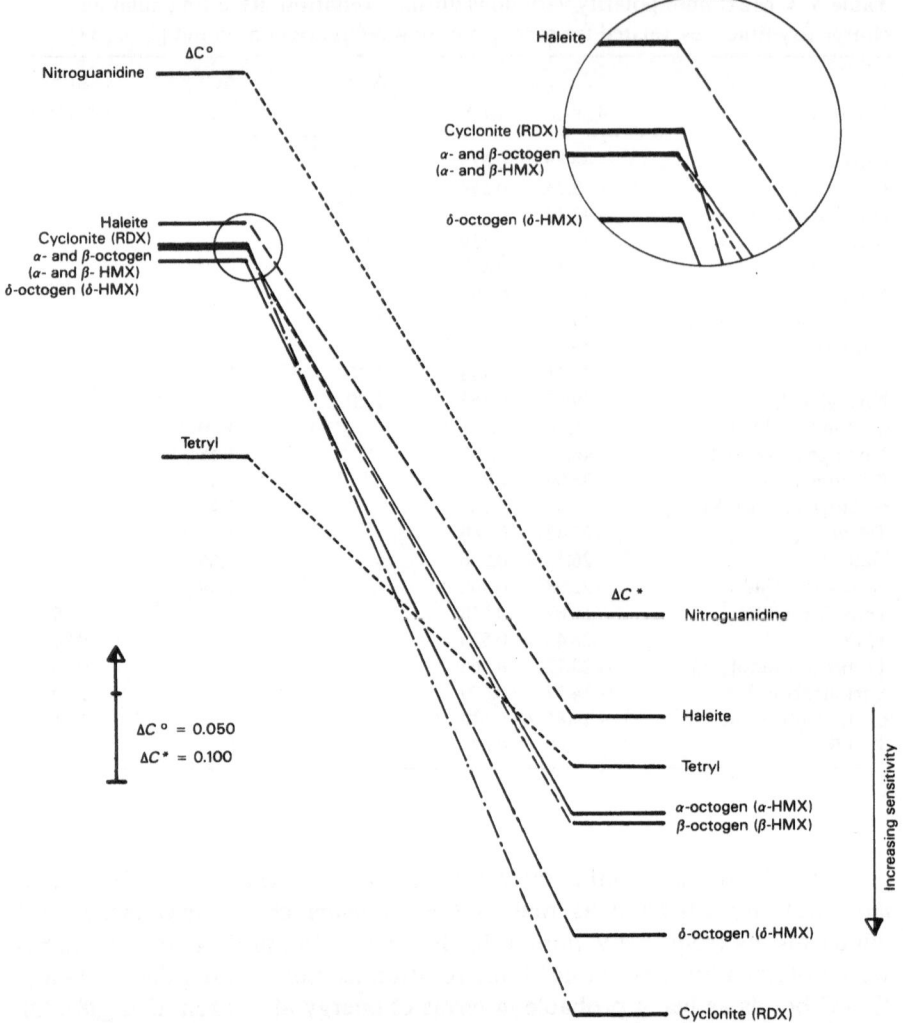

Figure V.4. Maximum polarity variation of nitramine bonds during excitation.

of tetryl and haleite, although neighboring, are nevertheless clearly lower than that of nitroguanidine which has the highest ΔC^*.

Thus it appears that if, within each group, the molecules are classified according to the minimum polarity which affects one of their nitro bonds after excitation, an analogous classification to that of sensitivity to "shock" is obtained. The most sensitive explosive has the smallest ΔC^*. This result leads to the idea that it is the molecular parameter ΔC^* which determines sensitivity to "shock" and which, in particular, makes it possible [38] to:

- explain the influence of molecular arrangement on the sensitivity of the explosive;
- establish a single sensitivity scale for liquid and solid compounds; and
- explain the difference in sensitivity observed in the case of a single explosive, according to its physical state.

This result also leads to the idea that the explosive molecule goes through an excited state and that there exists in this molecule a privileged bond whose polarity after excitation determines its sensitivity.

Before examining an extension of these conclusions to molecules without a nitro grouping, the following question must be answered: how, using a criterion which is solely molecular, can a property which depends largely on random factors (grain size, humidity, firing conditions, etc.) be explained? Two elements of the answer to this question have been given in [29].

(a) The scale resulting from the consideration of this molecular parameter coincides with the majority of sensitivity to "shock" tests, but particularly with a more pragmatic scale, to which specialists refer when relating the sensitivity of one explosive to that of another. Whilst apparently not very rigorous, this conventional scale must be taken into consideration, as it does in fact result from smoothing through experiment of every random factor such as those enumerated above. Thus the language retains only the intrinsic value of sensitivity which, not surprisingly, is seen to coincide with a molecular criterion.

(b) The tests considered are those of sensitivity to "shock," in which every chemical reaction that produces noise or smoke is counted as positive. So these tests merely record a decomposition reaction index, which explains why they coincide well with theoretical hypotheses which take into account precisely only the absorption of energy by the explosive molecule. It would be quite different had it been decided to refer to the sensitivity to *the shock wave* which is generally measured by the appearance of a deflagration or a detonation. The test in that case integrates, besides the generation mechanism, the entire process appropriate to the crystal studied, to its state, and to the ambient conditions which, starting from an identical stress, leads to combustion, deflagration, or detonation. In this case the existence of a purely molecular criterion is very unlikely.

2.3. Notion of Explosophore Bond

We now return to an extension of the foregoing results to explosive molecules belonging to other groups. The tetrazole and picrylazole groups are all particularly interesting for the following reasons:

- in the sensitivity scale, the tetrazoles occupy a place between primary and secondary explosives (this position, moreover, explains the interest which these compounds have aroused in various countries, which is expressed in the large number of studies, theoretical, in particular, [42] [53]); and

- among picrylazoles, molecules which are closely related from a structural point of view have sensitivities which are sometimes close to those of penta-erythritol, sometimes of the order of that of TATB [41].

As in the case of nitro explosives, no simple relationship to the fundamental state between sensitivity to "shock" and bond polarity in particular, or to electron distribution in general, has appeared in the referenced works [39]. This conclusion confirms those previously formulated by Haskins for tetrazoles and Schroeder for picrylazoles. On the other hand, the study of electronic variations during excitation brings to the fore some interesting correlations [29], [39], [54], [55]:

(a) For the two groups concerned, the maximum polarity variations undergone by N—N bonds in the nitrogen ring during excitation are very great. As a result, the minimum polarities of these bonds after excitation are very weak: less than 0.13 for the nitrogen–nitrogen bonds of picrylazoles, less than 0.10 in the case of tetrazoles.

(b) In the case of tetrazoles, the minimum polarity which is then shown by one of the bonds in the β-position of the substitute, i.e., one of the bonds which breaks first when decomposition phenomena occur as observed by slow analysis methods, as in the case of nitro explosives, correlates perfectly with the scale of sensitivities to "shock."

(c) In the case of picrylazoles (see Table V.4), consideration of nitrogen ring bond polarities does not offer such a correlation. Analysis of the electronic modifications shows that during excitation the nitrogen ring plays, with regard to the benzene ring, the role of an electron reservoir. This inter-ring electron transfer, whose intensity depends on the relative position of the two rings, is dominant with regard to the intra-ring electron modifications. As a result it modifies to a very great extent the electronic structure of the nitro groups of the picryl ring and, in particular, the polarities of C—N bonds of the chromophore in the ortho position. Consideration of the minimum polarity affecting these bonds after excitation confirms the validity for these compounds of the ΔC^* parameter as a sensitivity criterion.

Taken together these studies bring to the fore the existence, in each explosive molecule, *of a privileged bond whose electronic structure conditions sensitivity to "shock."* The nature of this bond is characteristic of the group concerned. This is, respectively:

- one of the $C—NO_2$ bonds in the case of nitroaromatics;
- one of the $N—NO_2$ bonds in the case of nitramines;
- one of the $O—NO_2$ bonds in the case of nitric esters;
- one of the $C—NO_2$ bonds of the picryl ring in the ortho position of the nitrogen ring in the case of picrylazoles; and
- one of the N—N bonds in β of the substitute in the case of tetrazoles.

Study of molecules which possess several types of explosophore bonds seems, moreover, to indicate that there exists a bond "hierarchy." It can be

Table V.4. Developed formula of nine 1-picrylazoles.

① 1 Picryl - 4 nitro-1,2,3 triazole
② 1 Picryl - 1,2,3 triazole
③ 1 Picryl - 4 nitro-1,2,5 triazole
④ 1 Picryl - 4 nitro-pyrazole
⑤ 1 Picryl - 1,2,5 triazole
⑥ 1 Picryl- 2 nitro-imidazole
⑦ 1 Picryl - 1,2,4, triazole
⑧ 1 Picryl - imidazole
⑨ 1 Picryl pyrazole

seen, for example, that in the case of tetryl, it is the electronic structure of the $N-NO_2$ bond which makes it possible to account for its position in the sensitivity scale [25].

2.4. Sensitivity Scale

The foregoing studies have shown, moreover, that during the fundamental state → excited state transition, the relative polarity variation of the exploso-phore bonds is very different according to the group concerned. Thus, $|\delta_m|$ is taken to be:

- between 15% and 35% in the case of nitroaromatics;
- between 25% and 55% in the case of nitramines;
- between 40% and 50% in the case of nitric esters;
- between 10% and 65% in the case of picrylazoles; and
- between 75% and 95% in the case of tetrazoles.

It can be seen that, in the nitramine group where sensitivity varies greatly from one compound to another, $|\delta_m|$ covers a wide range of values; on the other hand, it can also be seen that, in nitroaromatics and nitric esters where

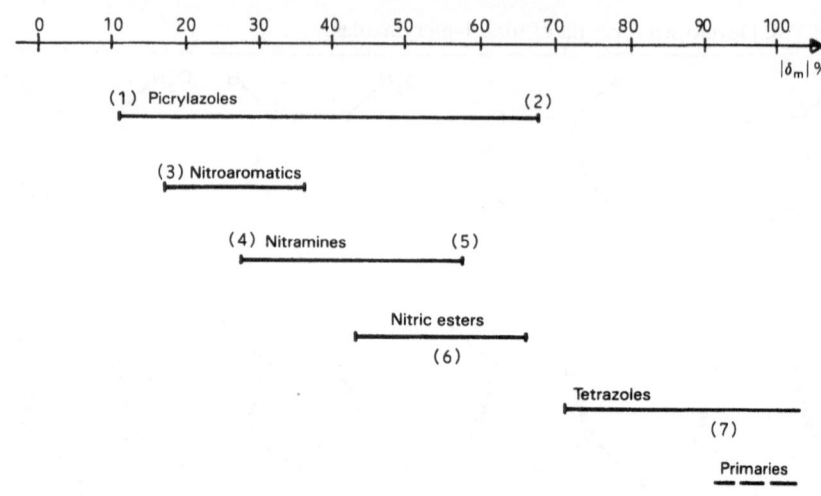

(1) 1 picrylpyrazole
(2) 1 picryl-4 nitro-1,2,3 triazole
(3) T.N.T.
(4) Nitroguanidine
(5) Cyclonite (RDX)
(6) Penta-erythritol
(7) Nitrotetrazole

Figure V.5. Classification of major explosive families according to sensitivity using parameter $|\delta_m|$. Refs. [39] and [54].

sensitivity varies little from one compound to another, the difference between the extreme values of $|\delta_m|$ is slight.

If it is then considered that, in the case of primary explosives, the first stage of decomposition is the removal of the valence electron, which is equivalent to $|\delta_m| = 100\%$, it can be seen that this parameter, which characterizes the ability of an explosive to decompose, allows exhaustive and continuous classification of explosive groups, from primary to the most insensitive secondary (see Fig. V.5).

So it seems possible to define a single molecular criterion of sensitivity ΔC^* for all explosives, which implies that the same process precedes the decomposition of any explosive, the nature of the explosophore bond being dependent on the group concerned. This conclusion is very similar to Shaw's [56], where the difference between primary and secondary explosives is one of degree rather than nature.

3. Conclusion

The recent developments referred to in Sections V.1 and V.2 demonstrate the very good correlation between the electronic structure of a molecule and the behavior of a sample under shock hammer. More precisely, adopting the minimum polarity which affects a particular bond in excited states of the molecule as the criterion of sensitivity to "shock" makes it possible to:

- account for the order commonly accepted in practice;
- establish a single scale for liquid and solid substances; and
- explain the effects of molecular arrangement and physical state.

It must be stressed that the very capacity of the model to provide a coherent explanation of the various experimental findings justifies, *a posteriori*, the two essential hypotheses:

- the first stage of decomposition of the molecule is the appearance of an excited state; and
- for every molecular species, there exists a privileged bond which conditions sensitivity to "shock."

Thus, at the end of this chapter, we appear to be justified in accepting the proposition, already well established, that sensitivity to "shock" is a molecular property. However, many questions still remain, including: What are the subsequent stages which lead effectively to explosive decomposition? Chapter VI repeats some recent, promising results which shed light on this debate, without necessarily bringing it to a final conclusion.

Explosive Decomposition

1. Reaction Mechanisms

The question which arises is to know how the modifications in electronic structure of the molecule, considered under "shock" in the last chapter, may be fitted into a mechanism of explosive decomposition. In other words, the problem to be solved is that of the effect of these modifications on the conditions of birth of the detonation regime. This effect may be considered on two levels: rupture of intramolecular bonds, and intermolecular energy transfers. In fact, the conditions of molecular decomposition are linked directly to the electronic state in which it is supposed that the molecule will be found after excitation. Moreover, electron distribution over the different bonds is one of the principal quantities influencing the transfer of energy between molecules [38].

Every phenomenon which brings into play the breaking of a bond can be described with the help of potential energy surfaces. Knowledge of these surfaces tells us, on the one hand, about the nature (dissociative or not) of the different electronic states, and, on the other, about the activation energy leading to possible decompositions.

We shall determine first the nature of these states by considering the isolated molecule. Next we shall emphasize the works of Bardo [57] which show the need, in the study of molecular decomposition methods, to look at models which consider the molecule in its environment. Only by considering this environment, in fact, is it possible to look at the effect of high pressures on the nature of electronic states and on activation energies.

1.1. Monomolecular Decomposition

Determining the dissociative character of the different electronic states of a molecule taken as isolated, consists of studying the energy variation of each state according to the lengthening of the different bonds. For a given bond, a state is dissociative when its energy curve in terms of the bond length presents no potential well (antibonding state); it is predissociative when the potential well exists (bonding state), but when the energy curve crosses that of an antibonding state in the energy neighborhood of the well. In the case of simple molecules (NO, NO_2, nitromethane), the energy variation can be determined using *ab initio* methods, by solving, for each electronic state, the characteristic Schrödinger equation [57]. In the case of more complex molecules (TATB, cyclonite (RDX), etc.), the use of semiempirical methods no longer allows such an approach; determination of the dissociative character of different states is based on a study of variations of electron transition energies according to bond length [39], [54], [57].

Figure VI.1 shows the results obtained by applying the CNDO/2-S/CI method in the case of the nitro bond of cyclonite (RDX) [39], whose polarity was correlated with the scale of sensitivities to "shock" by Delpuech [25]. From these studies it emerges that:

(a) The polarity of a bond is not linked on a one-to-one basis to its dissociative character. This result is in keeping with the observation according to which the effect of the value of the polarity of a bond on its energy is weak [43], [58]. It can be seen, however, that the electronic state characterized by a minimum bond polarity is always, in the case of nitro bonds, a dissociative or predissociative state and leads to the emission of NO_2 [39]. In the case of Figure VI.1, for example, if the transitions between the first vibrational levels of the fundamental state and the excited state

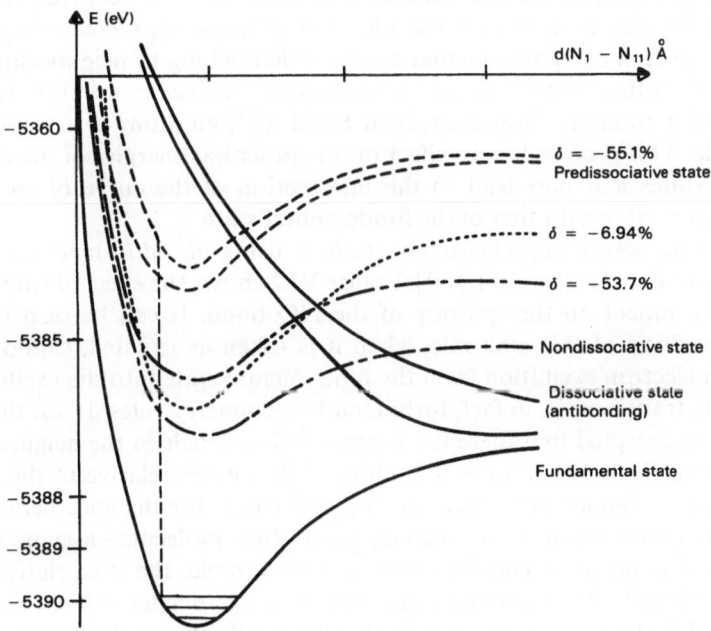

Figure VI.1. Potential energy curves of the N_1—N_{11} bond of cyclonite (RDX), Ref. [39].

characterized by a maximum polarity variation are considered, it can be seen that they lead effectively to the decomposition of the molecule through the intersection of the energy curve of this excited state and that of an antibonding state.

(b) Inside the molecule, excited electronic states of a nondissociative nature can exist.

1.2. Polymolecular Decomposition

The results which we have just set out for isolated molecules suppose that the motion of nuclei is negligible in comparison to the motion of electrons. In fact, there is no reason for such an approximation to exist whenever reactions occur in a propagation phenomenon. Bardo [59] has shown the effect of motion of nuclei in intermolecular splitting processes which involve the NO and NO_2 groups. Furthermore, he calculated the effect of the molecular environment on the nature of the different excited states under high pressure. This effect is explained by the formation of strong intermolecular bonds which, on the one hand, result in the appearance of states of a predissociative type, and on the other hand, in a reduction in activation energy.

Formation of Predissociative States

The effect of pressure can be considered at the level of the relative position of atoms belonging to different molecules. For example, an increase in pressure has the effect of bringing together atoms which belong to neighboring molecules. Under these conditions, an atom linked to a given molecule may, in the course of a rotation, form a covalent bond with an atom of a neighboring molecule. This process has an effect on the potential energies of the different excited states and may lead to the intersection of the curve of an excited dissociative state with that of the fundamental state.

This has been comprehensively studied, using *ab initio* methods, in the case of small molecules [57], [60]. Figure VI.2 shows the effect of considering the environment on the splitting of the NO bond. It can be seen that the decomposition of this molecule, when it is taken as isolated, cannot occur through electron excitation from the fundamental state into the excited state $^2\Sigma^+$; this transition is, in fact, forbidden by symmetry rules. If, on the other hand, it is accepted that there is a second NO molecule in the neighborhood of the first, then there is an intersection of the curves relative to the excited state and the fundamental state. In the case where the distance between the nitrogen and oxygen atoms belonging to two molecules assumed to be connected is taken as equal to 1.64 Å, for example, the dissociative curve associated with the $^3\pi$ excited state crosses the associated curve of the fundamental state at the point 100 kcal. This result shows the importance of the forbidden transitions, which are ruled out when the monomolecular decomposition mechanisms are considered.

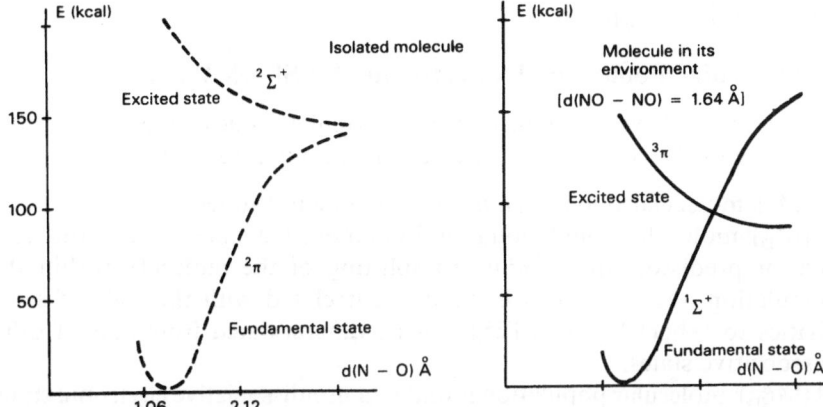

Figure VI.2. Effect of consideration of the environment on the splitting mechanism of the NO bond, Ref. [57].

This type of study is very complicated due to the effect on the results of the relative orientation of the molecules. Bardo points out, however, that the studies on nitrogen peroxide NO_2, nitromethane CH_3NO_2, nitrobenzene $C_6H_5NO_2$, nitroaniline $(NH_2)C_6H_4NO_2$, and triaminotrinitrobenzene $C_6(NO_2)_3(NH_2)_3$ show a similar effect of pressure on the electronic states considered. Such a similarity suggests to Bardo that the effect of pressure on the rate of decomposition reactions is identical in the case of molecules containing the NO_2 chromophore.

Reduction of Activation Energy

We shall limit ourselves here to stressing the unsuitability of values of the activation parameters determined in the gaseous, or even liquid, phase at ordinary pressure (weak interactions between molecules), to explain the decomposition mechanisms occurring in the detonation phenomenon in the condensed phase. In fact, it can be seen that the activation parameters, determined by considering the effect of pressure [57], are very different from those calculated when the molecule is taken as isolated. This difference is explained by the fact that the formation of a covalent bond under high pressure between atoms belonging to different molecules results in a weakening of the intramolecular bonds to which these atoms belong. This reduction has been calculated by Bardo using *ab initio* methods in the case of nitromethane CH_3NO_2 and nitric oxide NO. For nitromethane, the calculations show that the transition energy between the fundamental state and the first excited state, which is 2.45 eV when the molecule is considered isolated, changes to 1.18 eV for the compound $(CH_3NO_2)_2$. For NO, considering a bimolecular decomposition mechanism lowers the energy activation value for splitting the N—O bond from 6.5 eV to 4 eV.

2. Theoretical Approach

2.1. Molecular Population Downstream of a Shock Wave

Previous studies have led to the molecular population downstream of a shock wave being broken down into a total of three populations [39]:

- $P(M°)$: molecular population in the fundamental state.
- $P(M_D^*)$: molecular population found in an excited state, whether dissociative or predissociative, leading to splitting of the molecule; within this population, the one which is directly correlated with the scale of sensitivities to "shock" in Ref. [25] can be differentiated from that of other dissociative states.
- $P(M_{ND}^*)$: molecular population found in a nondissociative electronic state.

Thus, in the case of nitro explosives, the total population is the sum

$$P(M°) + P'(M_D^*) + P''(M_D^*) + P(M_{ND}^*),$$

where the second term designates the molecular population in an excited state leading to the release of NO_2.

2.2. Birth of Explosive Decomposition

Taking into account this division of excited molecules into three populations, Delpuech et al. [39] proposed that the initial stage of the propagation phase should be considered to be the appearance in certain areas of the explosive of zones where, as a result of electronic changes in the molecules, the following two phenomena occur *simultaneously*:

- the conditions of energy transfer between molecules are greatly modified as a result of the existence of the population $P(M_{ND}^*)$ where the polarity of bonds is very different from that of the fundamental state; and
- certain molecules are raised into a dissociative state and decompose [population $P(M_D^*)$].

The molecular parameters characteristic of this last population have, moreover, been correlated with sensitivity to "shock" (see §V.2.2). This is in perfect agreement with the fact that this type of sensitivity and measurement expresses only an observation of the decomposition of the explosive without predicting the nature of the propagation regime: detonation, deflagration, combustion, etc.

On the contrary, the authors take the view that the conditions of birth of the detonation, linked to the conditions for achieving a cooperative regime, are a direct function of the characteristics of the population $P(M_{ND}^*)$. This proposition is based on the fact that, if the splitting mechanism is instantaneous, it is the parameters of excited states characterized by a significant lifetime which make it possible to attain a cooperative regime.

In other words, according to this model, the generation of detonation implies the existence in a confined space of *spots* where the electronic structure of the majority of molecules is modified for a sufficient length of time, and where simultaneously decomposition of some molecules occurs [54].

A study of the structural modification, which affects an explosive molecule downstream of a shock wave and before decomposition, was undertaken in [39], [61], [62], [63] and provides experimental confirmation of these hypotheses.

3. Experimental Approach

3.1. The Principles of High-Speed Raman Spectrometry

The technique used is high-speed Raman spectrometry whose experimental device is described in Refs. [55], [62], and [63]. This technique, perfected by Bridoux and Delahaye [64], is, in fact, the only method which makes it possible to obtain data about molecular structure at a very localized point and in a sufficiently short time. The principle of the experiment consists in comparing the Raman spectrum emitted by an explosive crystal to that produced by the same zone of the same crystal, but immediately downstream of the shock wave [62].

Recording the spectra meets two goals:

(a) characterizing, downstream of the shock wave, the population P (M_{ND}^*), i.e., the molecules which display a modified electronic structure for a sufficient length of time (the presence of these molecules should be expressed by a modification in the intensity of the spectral lines); and

(b) observing the splitting of bonds, notably $R-NO_2$ (these splits should lead to spectrum modifications following the disappearance of certain lines and the appearance of new bands characteristic of the radicals formed).

In fact, the intensity of a Raman line relative to a transition between two levels p and q is proportional to the elements P_{ij} of the diffusion tensor of the molecule [65]. These elements are dependent on the wave functions ψ_p and ψ_q of the states considered in the molecule

$$(P_{ij})_{pq} = \frac{1}{h} \sum_\ell \frac{(M_i)_{p\ell}(M_j)_{\ell q}}{v_{\ell p} - v} + \frac{(M_i)_{p\ell}(M_j)_{\ell q}}{v_{\ell q} + v}.$$

(M is the dipole moment operator of the molecule; $i, j = 1, 2, 3$; $M_{p\ell} = \int \psi_\ell M \psi_p \, dv$; $v_{\ell p}$ is the frequency associated with the transition $\ell \to p$; v is the frequency of the activating electric field; index ℓ is relative to the energy levels of the molecule.)

3.2. Experimental Set-Up

The explosives studied were cyclonite (RDX) [63], [39] and penta-erythritol [61] in the monocrystalline state. The zone examined is situated 1 mm from the entrance face of the shock in the case of penta-erythritol, and 2 mm in the case of cyclonite (RDX). The shock pressure deduced from Doppler laser interferometry measurements is 140 kbar for penta-erythritol and 60 kbar for cyclonite (RDX) at the point of entry of the monocrystal. In these conditions, penta-erythritol detonates in every case, whereas cyclonite (RDX) does not (for the latter, the pressure value was chosen to avoid phase transitions in the crystal). The phenomenon was observed for 10 ns in a time slice of a few tens of nanoseconds after the passage of the shock wave.

The set-up (Fig. VI.3) consists essentially of a monocrystal sample, a light source and a detection system.

(a) The monocrystal is firmly attached to the end of a hollow copper piston which contains the priming system. This assembly is placed in a cell three-quarters full of a silicon oil of the same refractive index as the crystal studied in order to limit reflections due to the incidence or irregularities of the faces. The location of the observed zone of the monocrystal is set using a continuous laser beam colinear with a pulsed laser described below.

(b) The source of the energizing light consists of a two stage YAG laser, set off by Pockels cell. The flash, emitted at a wave length of 1064 nm, represents an energy of 1 Joule and lasts for 10 ns. A frequency doubler (KPD) makes it possible to obtain between 100 mJ and 400 mJ at 532 nm.

(c) The detection system used is a spectrometer consisting of two monochromators in series. The spectrum of rays emerging from the second monochromator enters a TSN 503 slit scanning camera whose shutter is controlled so as to mask the light emitted after detonation. The spectrum of lines is amplified, processed in an optical spectrometer and recorded.

The trigger circuit regulates the delay between the instant of emission of the laser flash and the instant when the shock enters the target zone of the monocrystal; it also sets the shuttering time of the camera, after recording the spectrometer data but before any light emission following the decomposition of the explosive. The firing order controls, on the one hand, the pumping of laser bars and sends, on the other hand, an impulse which reaches the modulator after a period which is equal to the pumping time. At this instant, two pulses occur:

- one initiates the detonator and creates a shock wave which, after a time t_1, moves to the position of the crystal to be studied; and
- the other delays—according to the chosen time t_1—both the camera aperture and the Pockels cell aperture which activates the laser.

The main cause of uncertainty lies in the operating time of the detonator. This uncertainty has been estimated at 30 ns [61].

3.3. Results and Discussion

The most immediately accessible result is that, for several tens of nanoseconds downstream of the shock wave, the spectrum obtained contains all

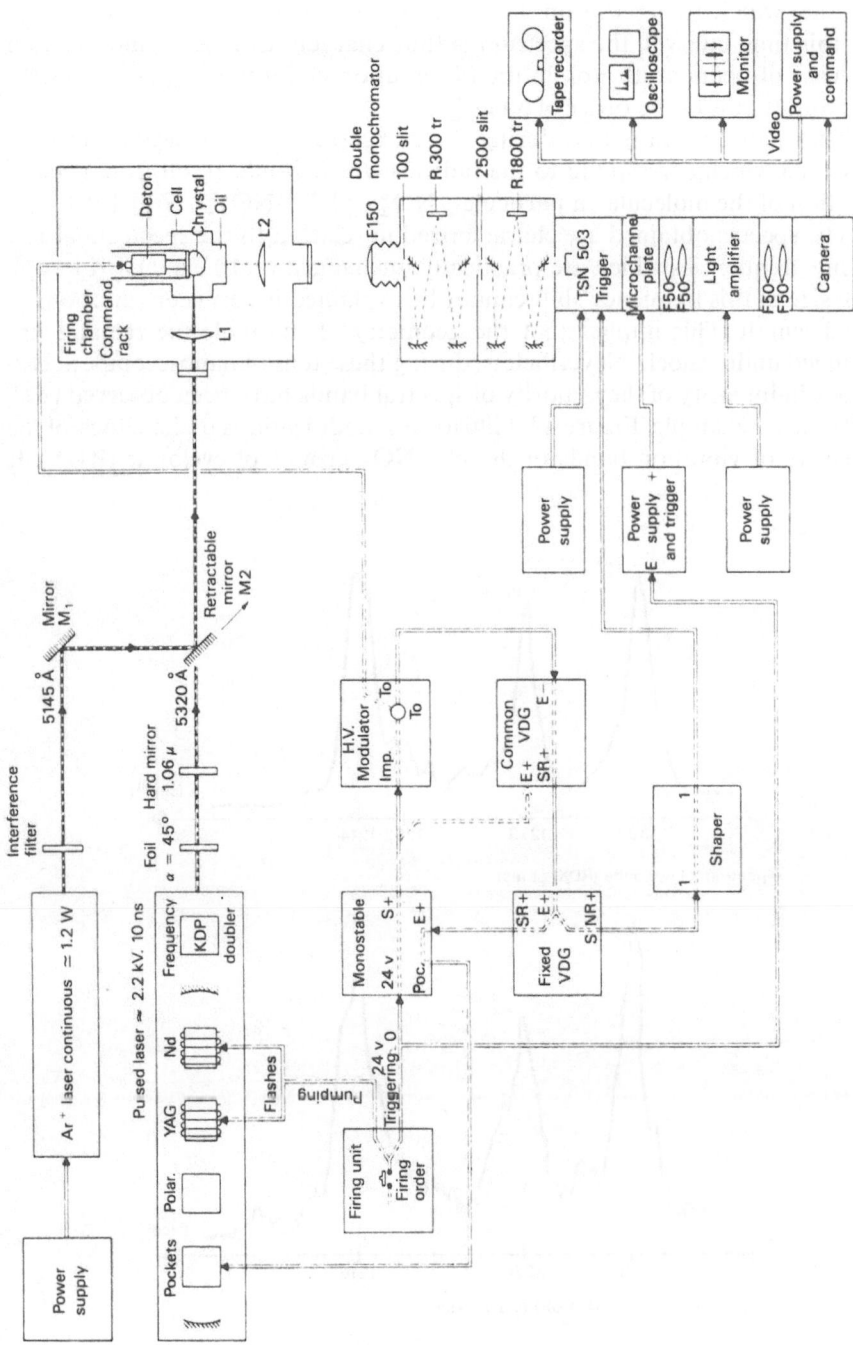

Figure VI.3. Diagram of high-speed Raman spectroscopy assembly, Ref. [61].

the lines which are present on the spectrum recorded at rest [61], [62]. During this time interval, the spectrum is thus characteristic of the molecule. In the case of penta-erythritol, in the observation conditions explained above, this duration is of the order of 50 ns [62].

The second result is that during these experiments lines have never been observed which correspond to the radicals which would result from decomposition of the molecule, in particular, $NO_2^\bullet$ or $[R\!-\!NO_2]^\bullet$, [61], [39].

The spectra obtained are characterized in relation to the spectrum at rest by line frequencies which are practically unchanged (< 10 cm^{-1}), [61], [63]. The size of this frequency shift cannot be explained in terms of variations in bond length. This implies that the geometry of the molecule remains unchanged under shock. Nevertheless, during these tens of nanoseconds, fluctuations in intensity of the majority of spectral bands have been observed [62], [39]. As an example, Figure VI.4 illustrates modifications under shock of the intensity of vibration bands of the $N\!-\!NO_2$ groups of cyclonite (RDX). It

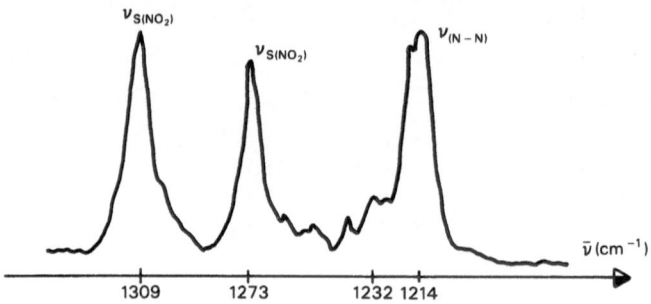

Spectrum of cyclonite (RDX) at rest.

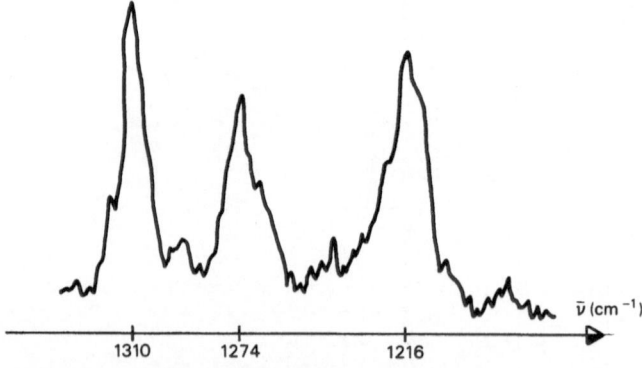

Spectrum of cyclonite (RDX) under shock

Figure VI.4. High-speed Raman spectrometry study of structural modifications of cyclonite (RDX) under shock. Examination of the vibration frequencies of the $N\!-\!NO_2$ groups, Ref. [39].

can be seen that the relationship of the intensities of the v_{NN} bands (centered at 1214 cm^{-1}) to those of the $v_{S(NO)_2}$ bands (centered at 1273 cm^{-1} 1309 cm^{-1}) which is 1.00 at rest increases to 1.40 downstream of the shock wave.

The first conclusion which emerges from analysis of the above experimental results is that decomposition of a group of molecules does not occur immediately after the passage of the shock front. Moreover, all observations agree with the hypothesis of the existence, behind this front, before any explosive decomposition, of a molecular population for which the only modification would consist of a reorganization of the electronic structure. In fact, since examination of the frequencies failed to show an intermolecular reorientation under shock, the intensity fluctuations of the Raman lines appear to be explicable only by modifications in the number and nature of the excited levels occupied (electronic and vibrational).

These modifications depend in particular on the conditions of pressure, temperature, and lifetime of the levels involved.

On the other hand, these studies have not been able to show the presence during this period of molecules in the process of decomposing [59], [61]. The authors explain this result by the fact that the P' (M$_D^*$) population is small relative to the whole population, and that the species formed are not sufficiently stable to be observed by this method [22].

4. Conclusion

The study of explosive decomposition at the molecular scale has been marked in recent years by the consideration of the effect of the electronic characteristics of a molecule on the ignition conditions. This observation is largely explained by the contribution of quantum chemistry to the knowledge of these characteristics and their modifications after energy has been supplied. The first conclusion lies in establishing a correlation between sensitivity to "shock" of a substance and the electronic structure of a molecule.

Thus, this correlation substantiates the long-standing hypothesis according to which explosive decomposition originates from a sudden structural modification. However, it concerns only the start of the decomposition reaction and not the subsequent production of combustion, deflagration, or detonation.

In order to offer a representative model of this second phase of the phenomenon, recent studies suggest that the energy communicated by a shock is transmitted preferentially to particular molecules. Thus, behind the shock front, several distinct molecular populations would exist at the level of their electronic characteristics. The parameters of particular populations would make it possible to account for the sensitivity of the substance, whilst those of other populations would influence the conditions of birth of a cooperative decomposition process through exchange phenomena and energy transfers.

The results of an experimental study carried out by high-speed Raman spectrometry of structural modifications of a molecule downstream of a shock wave are in keeping with the existence of molecules which have actually undergone a modification in their electron distribution.

These propositions are compatible with all the detonation generation mechanisms considered at the macroscopic scale. In spite of their great diversity, the majority of these mechanisms do, in fact, allow for, as we shall see, thermal or mechanical heterogeneity of the substance downstream of the shock. The most important of these mechanisms are discussed in the following chapter.

References to Part Two

[1] VAN'T HOFF, J.H. *Leçons de Chimie Physique*, 3rd vol., p. 104, Hermann, Paris (1898).

[2] KECKER, K.R., MASON, C.M., WATSON, R.W. U.S. Bureau of Mines Report no. 7670 (1972).

[2] MACEK, A. *Chem. Rev.*, **62** (1962), 41.

[2] POPOLATO, A., ABLARD, J.E., JAMES, E. Discussion Sessions 5, 6. *Proc. International Conference on Sensitivity and Hazards of Explosives*, London (1963). Picatinny Arsenal Report no. 3278 (1970).

[2] ROBERTSON, R. *J. Amer. Chem. Soc.*, **119** (1921), 16.

[2] MARSHALL, A. *Explosives*, Vol. 3, Churchill, London (1932).

[2] MORTLOCK, H.N., WILBY, J. *Explosivestoffe*, **14** (1966), 49.

[2] MURAOUR, H. *Mémorial de l'Artillerie Française*, **22** (1933), 559.

[2] MEDARD, L. *Mémorial de la Poudres*, **33** (1951), 330.

[2] BIGOURD, J., MICHOT, C., CARBONEL, P. *Symposium International sur le Comportement des Milieux Denses sous Hautes Pressions Dynamiques*, Paris (August 1978).

[2] FEDEROF, B.T., SHEFFIELD, D.E. *Encyclopedia of Explosives and Related Items*, Vol. 7, I.35, Picatinny Arsenal, Dover, New Jersey (1975).

[3] TOMLINSON, W.R., Jr. Picatinny Arsenal Technical Report no. 1401 (1950).

[3] TOMLINSON, W.R., Jr., SHEFFIELD, O.E. Picatinny Arsenal Technical Report no. 1740 (1958).

[4] SOREL, J., BROSSE, J.M., LUCAS, G., POULARD, S. *Colloque sur la Sécurité en Pyrotechnie*, Bourges (1977).

[5] BROWNLEE, K.A., HODGES, J.L., Jr., ROSENBLATT, M. *J. Amer. Statist. Assoc.*, **43** (1953), 262. Principles of Explosives Behavior. U.S. Army Material Command. 706–108, Washington (1972).

[6] *Manuel Technique*, Explosifs Militaires T.M. **9**, (1958), 1910.

[7] URBANSKY, T. Vols. 1–3, Pergamon Press, New York (1965).

[8] CALZIA, J. *Les Substances Explosives et leurs Nuisances*. Dunod, Paris (1969).

[9] DOBRATZ, B.M. Lawrence Livermore Laboratory, U.C.R.L. 51319.

[10] FEDEROFF, B.T., SHEFFIELD, D.E. *Encyclopedia of Explosives and Related Items*, Vols. 1–9 Picatinny Arsenal, Dover, New Jersey (1975).

[11] MEYER, R. *Explosives*. Verlag-Chemie, New York (1977).

[12] KAMLET, M.J. *Proc. 6th Symposium on Detonation*, San Diego ICA (1976).

[13] KAMLET, M.J., ADOLPH, H.G. *Propellants and Explosives*, **4** (1979), 30.

[14] Owens, F.J., Sharma, J. *J. Appl. Phys.* **51** (3), (1980), 1494.

[15] Sharma, J., Gora, T., Bulusu, S., Wiegand, D.A. Rapport AD.785.670, Picatinny Arsenal, Dover, NJ (1974).

[16] Shaafsma, T.J. *Chem. Phys. Lett.*, **1**, (1967), 16.

[17] Shaafsma, T.J., Kommandeur, J. *Molecular Phys.*, **14** (1968), 517.

[18] Santar, I., Bednar, J. *Collect. Czecholov. Chem. Commun.*, **32** (1967), 953.

[19] Santar, I., Bednar, J. *Radiat. Chem.*, **1** (1968), 523.

[20] Cherville, J., Linares, B., Poulard, S., Schulz, C. *Third Symposium on Chemical Problems Connected with the Stability of Explosives*, Ystad/Sweden (May 1973).

[21] Leverd, L.J. Thèse de spécialité, Université de Toulouse (1974).

[22] Darnez, C., Paviot, J. *Internat. J. Radiat. Phys. Chem.*, **4** (1972), 11.

[23] Delpuech, A., Cherville, J. *Propellants and Explosives*, **3** (6) (1978), 169.

[24] Feuer, H. *The Chemistry of the Nitro and Nitroso Groups*. Interscience, New York (1969).

[24] Gray, P., Williams, A. *Chem. Rev.*, **59** (1959), 239.

[24] Gowenlock, B., Jones, P., Mater, J.R. *Trans. Faraday Soc.*, **57** (1961), 23.

[24] Vedeneyev, V.I. et Coll. *Bond Energy, Ionisation Potentials and Electron Affinities*. Edward Arnold, London, (1966).

[25] Delpuech, A. Thèse Doctorat ès Sciences no. 656, Université de Bordeaux (1980).

[26] Field, F.H., Franklin, J.L. *Electron Impact Phenomena*. Academic Press, New York (1957).

[27] Killgoar, P.C., Leroi, G.F., Berkowitz, J., Chupka, W. *163th American Chemical Society Meeting*, Boston (1972).

[28] Delpuech, A., Cherville, J. *Fourth Symposium on Chemical Problems Connected with the Stability of Explosives*, Molle/Sweden (May 1976).

[29] Delpuech, A., Cherville, J. *Proc. Symposium H.D.P.*, Paris/France (1978), p. 21.

[30] Del Bene, J., Jaffe, H.H. *J. Chem. Phys.*, **48**, (1968) 1807; **48**, (1968) 4050; **49**, (1968) 1221; **50**, (1969) 1126.

[30] Ellis, R.L., Kuenhlenz, G., Jaffe, H.H. *Theoret. Chem. Acta*, **26** (1972), 131.

[31] Gora, T., Dows, D.S., Kemmey, P.J., Sharma, J. *Energetic Materials*, Vol. 1, p. 193 (H.D. Fair, F.R. Walker, editors), Plenum Press, New York (1977).

[32] Roothaan, C.C.J. *Rev. Modern Phys.*, **23** (1951) 68; **32** (1960) 179.

[33] Jaffe, H.H. *Acc. Chem. Res.*, **2** (5) (1969), 136.

[34] Hoffman, R. *J. Chem. Phys.*, **39** (1963), 1397.

[34] Rein, R., Fukuda, N., Win, H., Clarke, G.A. *J. Chem. Phys.*, **45** (1966), 4743.

[35] Pople, J.A., Bederidge, D.L., Dobosh, P.A. *J. Chem. Phys.*, **47** (1967), 2026.

[36] Pople, J.A., Santry, D.P., Segal, G.A. *J. Chem. Phys.*, **43** (1965), S130.

[36] Pople, J.A., Segal, G.A. *J. Chem. Phys.*, **43** (1965), 5136; **44** (1966) 3289.

[37] Stals, J., Barraclough, C.G., Buchanan, A.S. *Trans. Faraday Soc.*, **65** (1969), 904.

[37] Stals, J. Austral. *J. Chem.*, **22** (1969) 2505; **22** (1969) 2515.

[37] Stals, J. *Trans. Faraday Soc.*, **67** (1971), 1739.

[38] Delpuech, A., Cherville, J. *Propellants and Explosives*, **4** (6) (1979), 121.

[39] Delpuech, A., Cherville, J., Michaud, C. *Proc. 7th Symposium on Detonation*, Annapolis/MD (1981).

[40] HASKINS, P.J. *International Conference on Research in Primary Explosives Communication* no. 6, Waltham Abbey/England, March 1975.

[41] SCHROEDER, M.A. Ballistic Research Laboratories, Report no. 2340 (1973).

[42] SCHROEDER, M.S. Ballistic Research Laboratories, Report no. 1348 (1975).

[43] POPLE, J.A., BEVERIDGE, D. *Approximate Molecular Orbital Theory.* McGraw-Hill, New York (1971).

[44] DELPUECH, A., CHERVILLE, J. *Propellants and Explosives,* 4 (2) (1979), 61.

[45] BARON, J., SANTRY, D.P. *J. Chem. Phys.,* 56 (1972), 2011.

[46] FILHOL, A. Thèse de Docteur Ingénieur, Université de Bordeaux (1971).

[47] CADY, H., LARSON, A.C. *Acta Crystallogr.,* 18 (1965), 485.

[48] BOOTH, D., LLEWELLYN, J.F. *J. Chem. Soc.* (1947), 837.

[49] TURLEY, J.W. *Acta Crystallogr.,* B24, (1968), 942.

[50] PARISER, R., PARR, R.G. *J. Chem. Phys.,* 21 (1963), 767.

[51] ORLOFF, M.K., MULLEN, P.A., RAUCH, F.C. *J. Phys. Chem.,* 74 (1970), 2189.

[52] HEAVENS, S.N., FIELD, J.E. *Proc. Roy. Soc. London, Ser. A,* 338, (1974), 77.

[53] BLAY, N.J. and colleagues. Rapport ERDE TR 163 (1975).

[53] FARNCOMB, R.E., CHANG, M. N.S.W.C. Technical Report 77–82 (1976).

[53] SCOTT, S.L. *International Conference on Research in Primary Explosives.* Communication no. 15, Waltham Abbey/England (March 1975).

[53] WHITE, J.R., WILLIAMS, R.J. *International Conference on Research in Primary Explosives.* Communication no. 16, Waltham Abbey/England (March 1975).

[54] DELPUECH, A., CHERVILLE, J. *Colloque International de Pyrotechnie Fondamentale et Appliquée,* Arcachon/France (1982).

[55] DELPUECH, A., CHERVILLE, J. *Colloque 1977 sur la Sécurité en Pyrotechnie,* Bourges/France (1977).

[55] DELPUECH, A., CHERVILLE, J. *Sciences et Techniques de l'Armement* 55 (1) (1981).

[56] SHAW, R. *International Conference on Research in Primary Explosives.* Communication no. 5, Waltham Abbey/England (March 1975).

[57] BARDO, R.D. *Proc. 7th Symposium on Detonation,* Annapolis/MD (1981).

[58] LEROY, G., MARIN, P., PEETERS, D. *J. Chimie Physique,* 71 (3) (1974), 319.

[59] BARDO, R.D., WOLFSBERG, M. *J. Chem. Phys.,* 67 (2) (1977), 593.

[60] BARDO, R.D. Rapport NSWC TR 79-175 (1980).

[61] TAILLEUR, M.-H., CHERVILLE, J. *Propellants, Explosives, Pyrotechnics,* 7, (1982), 22.

[62] TAILLEUR, M.-H. Thèse de spécialité no. 1549, Université de Bordeaux I (1980).

[63] BOISARD, F., LINARES, B., DELPUECH, A., CHERVILLE, J. *Proc. Symposium H.D.P.,* Paris/France (1978), p. 33.

[64] BRIDOUX, M., DELHAYE, M. *Nouv. Rev. Opt. Appl.,* 1 (1970), 123.

[65] PLACZEK, G. UCRL Trans. no. 526 (L), March 22 (1959).

Part Three
Macroscopic Mechanisms of Generation of Detonation

CHAPTER VII

Cooperative Mechanisms

1. Ignition

This section is based on a preliminary report by Robert Belmas.

1.1. The Hot Spot Concept

The theory of shock sensitivity developed in Chapters IV–VI is based on a hypothesis which states that the energy of an appropriate thermomechanical stress is distributed between the molecules in such a way that some of them dissociate. This hypothesis of localized reactions at the molecular scale evolving in a propagation to the macroscopic scale, has been outlined by Muraour [27] and Garner [13] in 1938, made more explicit by Mampel [23] in 1940, and developed by Bowden and Yoffe [2] in 1952, to whom we owe the term "*hot spot.*"

Since then, the literature has grown considerably. Hot spots are now considered as original loci of discontinuities or heterogeneities which affect an explosive substance at the mesoscopic level: impurities in a liquid, stacking faults of a crystal, interfaces within a compressed or bonded aggregate. Indeed, under particular experimental conditions, hot spots have been observed [10] and the subsequent phase—evolution of hot zones—has been analyzed by IR emission [36]. However, by the very nature of their properties —small dimension, short life, high temperature, high pressure environment, destructive testing—hot spots are difficult to describe, so that a number of questions remain unanswered with respect to their correlation with the mesoscopic properties of the initial state.

Thus, due to a lack of incontrovertible experimental evidence, it is difficult to avoid giving a provisional presentation, where *a priori* modeling is the rule. For the convenience of this account, the models are divided into two classes, according to whether they invoke the *pore collapse* mechanism or the *mesoscopic shear* mechanism.

The former rest on publications which take as their starting point the very notion of hot spot: their importance lies in the fact that their domain of validity encompasses both homogeneous substances (a monocrystal with empty interstices, nondegassed liquids) as well as inhomogeneous substances (compressed, cast or bonded aggregates).

The latter, by contrast, are more recent: their development is linked partly to the setting up of an area of the physics of solids dealing with defects, and partly to the emergence of experimental facts such as the observation by Howe [15] of shear zones in TNT under shock, or the increase of sensitivity to "shock" of penta-erythritol by the addition of polymers that fracture by shear (see Swallowe and Field [33]). Their domain of validity is apparently limited only to initially solid substances: but their importance is in the role they play in practical applications due to their high chemical energy mass density.

In the next part of the account, the word *ignition* will be used to designate the appearance of hot spots, whatever mechanism is responsible.

1.2. Ignition by Pore Collapse

Whether it be liquid or solid, an explosive substance can be considered "porous" owing to the fact that gas bubbles are occluded at different manufacturing or handling stages: *compactness* α_0, defined as the ratio of the real volume mass ρ_0 to the maximum theoretical volume mass, can vary from 100% in a perfect monocrystal to 50% in a compressed aggregate. This is why "pores" are prime candidates for potential hot spot sites, several mechanisms being invoked to support this point of view.

The first rests on Bowden's famous experiments—which showed that sensitivity to "shock" of nitroglycerin is greatly increased by the presence of gas bubbles—and attributes the formation of a hot spot to compressive heating of the gas. The most successful form of this model is due to Partom [28] who proposed that the gas undergoes a spherical adiabatic compression, during which it obeys a polytropic law. These hypotheses are debatable, just as are the conclusions of experiments by Seay and Seely [29] and their successors [6], [24]. In fact, a good part of the debate arises from the stress rate and the nature of the associated compression (adiabatic, isothermal, etc.), as Starkenberg showed [31], where an explosive endowed with a large cavity (5–10 mm) is stressed by a pressure (a few kilobars) maintained for a long time (several milliseconds). Whatever the case may be, some experiments (see [9], for example) cast doubt on the effect of the nature and pressure of the interstitial gas on ignition, therefore also on the preponderance of pore collapse.

It is largely because of this conviction that many authors [3], [4], [22], [16], [25], [38] have proposed a second mechanism dominated by dissipative work of viscous stresses in the deformation of the material around the pore. The reader is referred to the very detailed work of Maiden [22] where stress has, among other effects, the ability to create a viscous melt zone around the pore, which acts as a heat source when collapse follows on: a hot spot appears if thermal conduction is low relative to thermomechanical dissipation. A comparison of the results of the model with those of experimental generation of detonation of PBX 9404 by shock is satisfactory; it is less so for compounds based on TATB or penta-erythritol, but we cannot say whether we should implicate the mechanism itself, or the supposed direct link between the appearance of a hot spot in the model and the appearance of detonation in the experiment.

The controversy is not limited to the above two mechanisms. In fact, Mader, for his part, proposes a purely hydrodynamic mechanism where he applies numerical modeling with a high degree of generality [19], [20], [21]. His approach consists of causing a shock wave to interact with isolated pores or a group of pores arranged in a cubic grid. During compaction, flow is convergent on the neighborhood of the edge of attack of the pore; the material increases in density there and accelerates until it comes into contact with the opposite edge, thus creating momentarily a more compressed and hotter restricted zone (see Fig. VII.1(a–d), reproduced from [35]). When inserted into a three-dimensional hydrodynamic code, this model allows isolation of certain numerical properties of the transition from shock to detonation:

- the stress of an isolated pore is not sufficient to ensure the transition; on the other hand, a cooperative mechanism resulting from interactions between incident and reflected shocks during local impacts allows this to occur;
- for constant porosity, smaller pore size implies larger numbers; this effect can be "sensitizing" by increasing the number of contributions to the transition, but "desensitizing" by reducing each individual contribution (N.B. another account of this effect is presented in §IX.3.1); and
- desensitization by a weak fore-shock results from pore collapse by a stress insufficient to entail ignition, but sufficient to cause an after-shock to meet with a material lacking sources of interaction (see §IX.3.1).

Mader's model, originally applied to liquid explosives [19], has been generalized [20], [21] to solid explosives (PETN, HMX, TATB, NQ, HNS, etc.) under one and the same hypothesis, that of ideal fluid behavior of the material. However, with regard to solids, it is difficult not to envisage the consequences of elastic–plastic behavior whether or not it is associated with viscosity; this is the position adopted by Taylor [36] who, without going beyond the elastic-perfectly plastic behavior stage, proposes an interesting attempt to establish a correlation between real and numerical porosity.

In each of the above three models, the same criticism can be leveled: that of giving preference to one mechanism and ignoring the other two. This "cold war" state (nonconfrontation by mutual ignorance) has not escaped Frey [12], who has undertaken—though in the restricted framework of compaction by implosion—a comparison of several factors, rate of pressure increase:

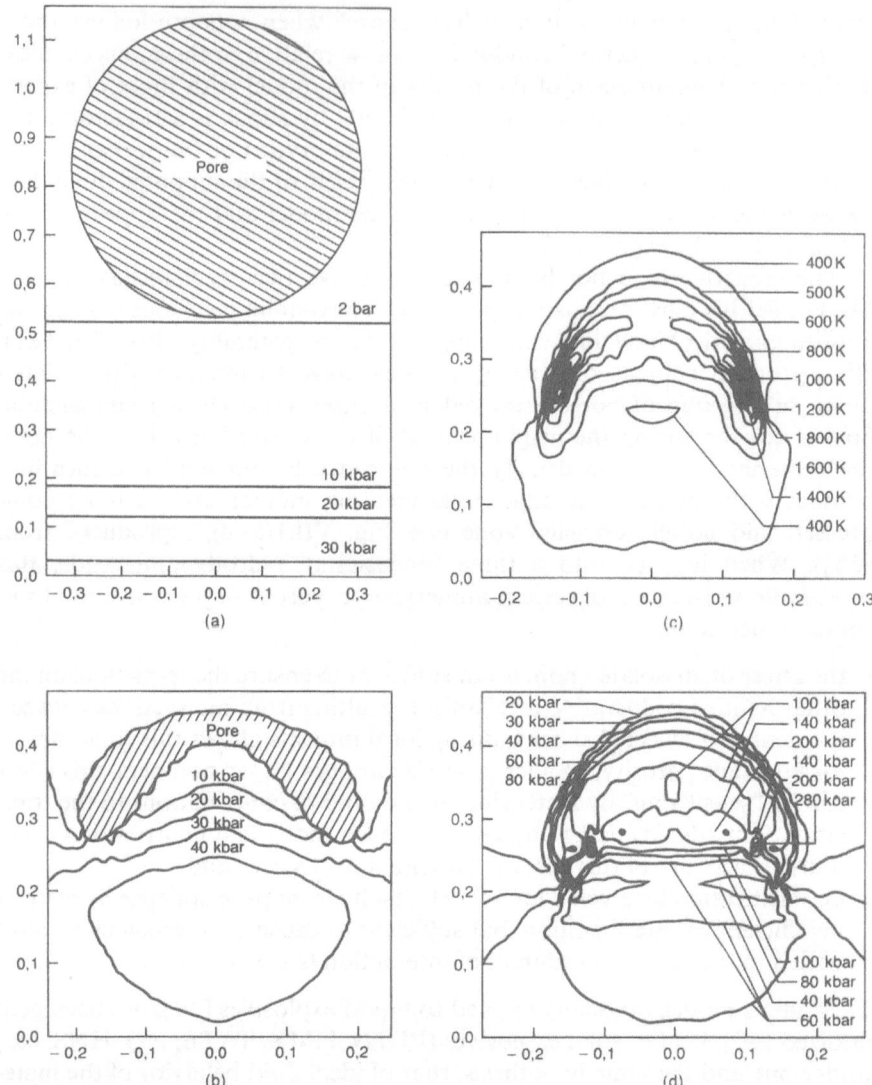

Figure VII.1. Compaction of a cylindrical pore by a plane shock parallel to the generating lines of the cylinder. (a) A 30 kbar shock propagating vertically upward is situated in the lower neighborhood of the pore. (b) Partial implosion of pore: the lower part of the pore is accelerated towards the upper part. (c) and (d) Isotherms and isobars (from [35]). Scales on both axes are in micrometers.

pore diameter, viscosity, elastic limit, and melting temperature. His moderated conclusions—"under appropriate conditions, each mechanism may be dominant"—need to be consulted in the original: we content ourselves by drawing attention to the last sentence of [12]: "These conclusions apply only within the framework of pore implosion and may not be extrapolated to other situations," which shows that the debate is far from over, more especially as another class of ignition mechanisms is necessary to account for the hexogen grain size influence (see Moulard et al. [26]) in compositions where the compaction α_0 equals 1 to within at most a thousandth.

1.3. Ignition by Mesoscopic Shear

Frey [11], already referred to in the previous subsection, was the first to formulate ideas governing the formation of hot spots by shearing:

- the velocity field is heterogeneous at the mesoscopic scale, as is the shock impedance field, due to the existence of voids, or to the existence of rigid grains immersed in a flexible matrix, or to internal defects in the grains;
- in a shear zone, the temperature results from competition between heating from the work done by viscoplastic forces and cooling by thermal conduction;
- the viscosity coefficient is an increasing exponential function of p and $1/T$; and
- fusion of material nullifies the limiting shear tension but not viscous tensions.

But it was Walker [37] who first attempted to link shear ignition to observations on the generation of detonation by shock. However, this attempt is limited in scope by the nature of the explosive concerned (composition B: HMX as grains dispersed in a matrix of TNT) and by very restrictive hypotheses: localization of shear internally within the matrix alone, perfect adhesion of the grains to the matrix. Furthermore, its conclusive character is greatly undermined by the author's concern to attach the validation of the model to the establishment of the $p^2\tau = Cte$ detonation criterion by one-dimensional impact (see §IX.2.1), though experiments prove amply that it is not universal. It could nevertheless open the way to an attractive modeling of bonded aggregates, subject to—as Belmas [1] emphasizes—the thorough study of the thermomechanical behavior of the bonding agent (as a function of pressure, temperature, and rate of shear) and the formulation of a realistic condition of contact between grains and bonding agent.

This attractive prospect should not cause us to neglect another possible shear localization—within the grains themselves—by analogy with well-established results on metals and alloys. In this field, where research is still at an early stage, we refer the reader to Coffey [7], Kipp [17], and Grady and Kipp [14].

2. Induction

This section is based on a preliminary report by Christian Michaud.

2.1. From Ignition to Propagation

Section VII.1 dealt with mechanisms proposed to explain ignition, i.e., the appearance of preferential sites where the temperature conditions are such as to involve rupture of chemical bonds. But ignition does not mean decomposition at the macroscopic level. For a stress that is too brief, too weak, or too diffuse, hot spots can cool by thermal conduction and die out. According to Setchell [30], that is the case for fine hexanitrostilbene compressed at $\rho_0 = 1.6$ g/cm^3 and subjected to a sustained 34 kbar shock. On the other hand, if the stress conditions (duration, intensity, distribution) permit, the temperature field may evolve—under the combined influence of thermal conduction and release of chemical energy—such that decomposition progresses, as if "propagating" beyond ignition sites. This initial propagation is a step which is difficult to define other than by the occurrence of an unusual event (often called "explosion") at the end of a finite time t^l (called *induction* duration or explosion time). Whatever event is chosen to mark this outcome, an experimental fact has to be taken into account: the value of t^l can spread over at least ten orders of magnitude according to the stress itself, and also according to the dimensions of the explosive mass. Thus t^l is of the order 10^{-6} s in the case of experiments to generate detonation by plane shock (see Chap. IX), but of the order of 10^4 s in the case of Zinn's experiment [19, p. 147]. The exceptional dynamics reveal an immense variety of situations, which is the source not only of dramatic errors in assessing the risk caused by a particular stress on a particular explosive mass, but is also the source of great difficulties in modeling.

All attempts at taking account of the initial decomposition of an explosive have a common base: the energy conservation equation associated with relative rest at constant volume of the stressed material:

$$\rho c_v \frac{\partial T}{\partial t} - (\lambda T_{,i})_{,i} = \rho \frac{\partial \Delta e}{\partial t}, \tag{VII.1}$$

where $-\Delta e$ is the variation of specific internal energy at the time of decomposition of the explosive. For the given boundary conditions, equation (VII.1) governs a temperature field which is coupled to a decomposed mass fraction field by $\partial \Delta e/\partial t$ and according to a more or less complex reaction scheme (see §2.2). The difficulties encountered in counting and clarifying with confidence the stages of this scheme explain why all authors have first of all considered zero-order kinetics

$$\frac{\partial \Delta e}{\partial t} = QZ \exp(-E^{\neq}/\Re T), \tag{VII.2}$$

where Q is the mass heat given out at constant volume, $E^{\neq}$ is the activation energy, and Z is the frequency factor. Moreover, taking care to give as simple a numerical resolution as possible leads to one's attention being limited to one-dimensional geometries, where (VII.1) takes the form

$$\rho c_v \frac{\partial T}{\partial t} - \lambda \left(\frac{\partial^2 T}{\partial x^2} + \frac{N}{x} \frac{\partial T}{\partial x} \right) = \rho Q Z \exp(-E^{\neq}/\mathfrak{R}T). \tag{VII.3}$$

Below, a stands for the radius of a sphere ($N = 2$), or the radius of a cylinder ($N = 1$) or the half-thickness of a slab of explosive ($N = 0$).

Mader [19] presents the results of the numerical solution of (VII.3) carried out in 1960 using as an approximation constant values for ρ, c_v, and λ and with the following boundary conditions:

$$\begin{aligned} T(|x| \leq a) = T_0 \qquad &\text{for} \quad t \leq 0, \\ T(|x| = a) = T^a > T_0 \qquad &\text{for} \quad t > 0. \end{aligned} \tag{$*$}$$

For the reader who is not familiar with this work [19, §3A], we recall the essential results here, rewritten with the notation used in this book.

- There exists a critical value T^i of T^a such that at the end of a finite time $t^l(T^a)$, $T^a \geq T^i$ induces an *explosion*, defined as the existence of a region of a mass of explosive where the temperature increases indefinitely while t tends towards t^l.
- Geometry has an effect on the values of T^i and $t^l(T^a)$:

$$\begin{aligned} T^i(N = 0) < T^i(N = 1) < T^i(N = 2), \\ t^l(N = 0) > t^l(N = 1) > t^l(N = 2), \end{aligned} \tag{$**$}$$

by the surface-to-volume ratio $(N + 1)/a$.

- For $T^a = T^i$, the initial explosion region is localized on the symmetry element (median plane, axis, center). For $T^a > T^i$, this region is closer to the external surface (where T^a is applied) the higher T^a is. For $T^a \gg T^i$, this region is identical to the external surface: thus, time t^l is identical to the time calculated by Cook [8] for a semi-infinite slab of solid explosive.
- Induction time t^l is always greater than adiabatic explosion time t^l_{ad}, defined by

$$t^l_{ad} = \frac{c_v}{QZ} \int_{T^a}^{T^i} \exp(E^{\neq}/\mathfrak{R}T) \, dT \simeq \frac{c_v \mathfrak{R}(T^a)^2}{QZE^{\neq}} \exp(E^{\neq}/\mathfrak{R}T^a), \tag{VII.4}$$

which represents the time taken by the mass of explosive to reach T^i if, at $t = 0$, it was raised to temperature T^a in totality.

Most of these conclusions are immediately apparent from an examination of Figure VII.2.

Too rapid a reading of the above results can lead to the acceptance of the idea that very high temperatures occur in effect. In reality, once the temperature is greatly in excess of T^a (e.g., $T > 2T^a$), equation (VII.1) for heat

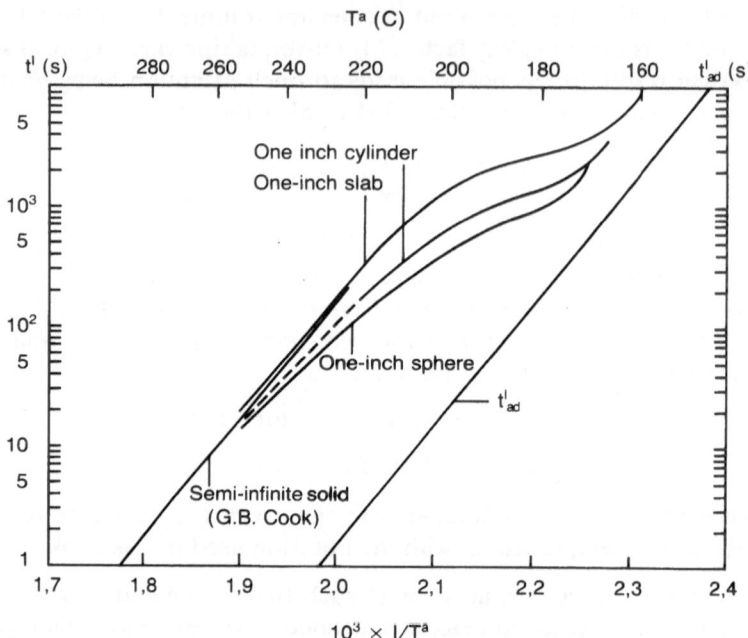

Figure VII.2. Cyclonite (RDX) at initial temperature = 25° C. Adiabatic explosion times t^l_{ad} and induction times t^l versus T^a in various one-dimensional geometries (after Mader [19], p. 145).

propagation in a medium at relative rest becomes inappropriate. In other words, the occurrence of the explosion—in the sense given above—should be regarded as the sign that propagation has begun, where motion and decomposition are henceforth linked. An examination of this linkage is the subject of Chapter VIII. However, before tackling this we devote the following subsection to another linkage, that of induction and the reaction process at the molecular level.

2.2. Coupling to the Reaction Scheme

For several pure explosives (HMX, RDX, etc.), there is excellent agreement between calculated and experimental T^i values in a given geometry [19, p. 146], subject to possible adjustments of zero-order $E^{\neq}$ and Z kinetic data [5], and/or experimental precautions [34]. (N.B. This well-established fact justifies keeping zero-order kinetics in Chapter VIII for dealing with the propagation phase proper.) On the other hand, the comparison of calculated and experimental t^l values is less satisfactory (see [5]), an obvious result of the nonreal chronology of a zero-order rate scheme, from which the only way

out is by returning to a less simplistic view of decomposition, i.e., a superposition of elementary reactions. This point of view is still scarcely found in
the literature. However, one can gain an idea of possible developments from
[18].

For TNT and TATB, MacGuire *et al.* [18] propose a three-stage auto-
catalytic mechanism

$$\text{①} \qquad \mathscr{E}_0 \rightarrow \mathscr{E}',$$

$$\text{②} \quad \mathscr{E}' + \mathscr{E}_0 \rightarrow \mathscr{E}, \qquad\qquad *$$

$$\text{③} \quad \mathscr{E}' + \mathscr{E}' \rightarrow \mathscr{E},$$

where the first is endothermic and the other two are exothermic. The mass
rate $\partial \Delta e/\partial t$ of heat released in the aggregate is then, with the usual notation,

$$\partial \Delta e/\partial t = Q_1 Z_1 X_0 \exp(-E_1^{\neq}/\mathfrak{R}T) + Q_2 Z_2 X_0 X' \exp(-E_2^{\neq}/\mathfrak{R}T)$$

$$+ Q_3 Z_3 X'^2 \exp(-E_3^{\neq}/\mathfrak{R}T). \tag{VII.5}$$

For cyclonite and octogen, they propose another three stage sequence,

$$\text{①} \quad \mathscr{E}_0 \rightarrow \mathscr{E}' \quad (H_2 C = N - NO_2),$$

$$\text{②} \quad \mathscr{E}' \rightarrow 2\mathscr{E}'' \quad (HNO_2 + HCN \text{ or } CH_2O + N_2O),$$

$$\text{③} \quad 2\mathscr{E}'' \rightarrow \mathscr{E},$$

where stage ① is endothermic, stage ② is weakly exothermic and stage ③
is strongly exothermic. The rate $\partial \Delta e/\partial t$ is then

$$\frac{\partial \Delta e}{\partial t} = Q_1 Z_1 X_0 \exp(-E_1^{\neq}/\mathfrak{R}T) + Q_2 Z_2 X' \exp(-E_2^{\neq}/\mathfrak{R}T)$$

$$+ Q_3 Z_3 (X'')^2 \exp(-E_3^{\neq}/\mathfrak{R}T). \tag{VII.6}$$

Thus, expressions (VII.5) or (VII.6) can be substituted for (VII.2) in the heat
propagation equation (VII.1).

However, this substitution will only be a quite unnecessary complication
if the hypothesis that ρ, λ, and c_v are constant is not also revised at the same
time. This is implicit in [18] to which we refer the reader who is interested in
this specifically thermal aspect of the problem.

The completed model leads to some remarkable results where the ratio
between observed and measured values does not exceed 2 for either the pure
explosives mentioned above or bonded compositions based on HMX or
TATB. One might think that the problem was solved. In reality, much more
remains to be written on this subject; further progress augurs well, whether in
the field of stress (uniform heating by electron beam [32]), or from observation (high-speed spectroscopy applied to Raman diffusion or stimulated
Raman diffusion).

References

[1] BELMAS, R. Phénoménologie et modélisation des points chauds. Rapport Interne CEA (1987).

[2] BOWDEN, F.P., YOFFE, A.D. *Initiation and Growth of Explosions in Liquids and Solids.* Cambridge University Press, Cambridge, UK (1952).

[3] CARROL, M.M., HOLT, A.C. Static and dynamic pore collapse relations for ductile porous materials. *J. Appl. Phys.*, **43**, no.4 (1972), p. 1626.

[4] CARROL, M.M., KIM, K.T. The effect of temperature of viscoplastic pore collapse. *J. Appl. Phys.*, **59**, no. 6 (1986), p. 1962.

[5] CATALANO, E. *et al.* The thermal decomposition and reaction of confined explosives. *Proc. 6th Symposium on Detonation*, Coronado/CA (1976), p. 214.

[6] CHICK, M.C. The effect of intersticial gas on the sensitivity of low density explosives. *Proc. 4th Symposium on Detonation*, White Oak/MD (1965), p. 349.

[7] COFFEY, C.S. The formation of hot spots and the initiation of explosive crystals by shock and impact. *Proc. Atelier sur la Détonation*, Megève/France (1987), p. 253.

[8] COOK, G.B. Initiation of explosion in solid secondary explosives. *Proc. Roy. Soc.*, **246** (1958), p. 154.

[9] DREMIN, A.N., SHVEDOV, K.K. On shock wave explosive decomposition. *Proc. 6th Symposium on Detonation*, Coronado/CA (1976), p. 29.

[10] FIELD, J.E., SWALLOWE, G.M., HEAYENS, S.N. Ignition mechanisms of explosives during mechanical deformation. *Proc. Roy. Soc.* **A382** (1982), p. 231.

[11] FREY, R.B. The initiation of explosives charges by rapid shear. *Proc. 7th Symposium on Detonation*, Annapolis/MD (1981), p. 36.

[12] FREY, R.B. Cavity collapse in energetic materials. *Proc. 8th Symposium on Detonation*, Albuquerque/NM (1985), p. 68.

[13] GARNER, W.E. Detonation or explosion arising out of thermal decomposition. *Trans. Faraday Soc.* **34** (1938), p. 985.

[14] GRADY, D.E., KIPP, M.E. The growth of unstable, thermoplastic shear with application to steady-wave shock compression in solids. *J. Mech. Phys. Solids*, **35**, no. 1 (1987), p. 95.

[15] HOWE, P.M. *et al.* An experimental investigation of the role of shear in initiation of detonation by impact. *Proc. 8th Symposium on Detonation*, Albuquerque/NM (1985), p. 294.

[16] KHASAINOV, B.A., BORISOV, A.A., ERMOLAYEV, B.S. Shock wave predetonation processes in porous high explosives. *Progr. Astronaut. Aeronaut.*, **87** (1981), p. 492.

[17] KIPP, M.E. Modeling granular explosive detonation with shear band concepts. *Proc. 8th Symposium on Detonation*, Albuquerque/NM (1985), p. 35.

[18] MAC GUIRE, R.R., TARVER, C.M. Chemical decomposition models for the thermal explosion of confined explosives. *Proc. 7th Symposium on Detonation*, Annapolis/MD (1981), p. 56.

[19] MADER, C.L. *Numerical Modeling of Detonations.* University of California Press, Berkeley (1979).

[20] MADER, C.L., KERSCHNER, J.D. Three dimensional modeling of shock initiation of heterogeneous explosives. *19th Symposium of the Comb. Inst.*, Haïfa/Israël (1982), p. 685.

[21] MADER, C.L., KERSCHNER, J.D. The three-dimensional hot spot model applied to PETN, HMX, TATB, NQ. LANL Report 10203 (1984).

[22] MAIDEN, D.E. A hot spot model for calculating the threshold for shock initiation of pyrotechnics and explosives. 3e Congrès de Pyrotechnie Spatiale, Juan-les-Pins/France (1987), p. 17.

[23] MAMPEL, K.L. The time dependant conversion formulas for heterogeneous reactions at the interface of solid bodies (transl. by R. McElroy Co, Austin/Texas, 1970). Z. Phys. Chem., A187 (1940), p. 43.

[24] MARSHALL, W.W. The role of interstitial gas in the detonation build-up characteristics of low density granular HMX.

[25] MERZHANOV, A.G., BARZIKIN, V.V., GONTKOVSKAYA, V.T. The problem of hot spot thermal explosion. Dokl. Akad. Nauk. SSSR, 148 no. 2 (1963), p. 380.

[26] MOULARD, H., KURY, J.W., DELCLOS, A. The effect of RDX particle size on the shock sensitivity of cast PBX formulation. Proc. 8th Symposium on Detonation, Albuquerque/NM (1985), p. 902.

[27] MURAOUR, H. Note on the theory of explosive reaction. Trans. Faraday Soc., 34 (1938), p. 989.

[28] PARTOM, Y. A void collapse model for shock initiation. Proc. 7th Symposium on Detonation, Annapolis/MD (1981), p. 506.

[29] SEAY, G.E., SEELY, L.B. Initiation of a low-density PETN pressing by a plane shock wave. J. Appl. Phys., 32 (1961), p. 1092.

[30] SETCHELL, R.E. Experimental studies of chemical reactivity during shock initiation of HNS. Proc. 8th Symposium Detonation, Albuquerque/NM (1985), p. 15.

[31] STARKENBERG, I. Ignition of solid high explosive by the rapid compression of an adjacent gas layer. Proc. 7th Symposium on Detonation, Annapolis/MD (1981), p. 3.

[32] STOLOVY, A. et al. Electron beam initiation of high explosives. Proc. 7th Symposium on Detonation, Annapolis/MD (1981), p. 50.

[33] SWALLOWE, G.M., FIELD, J.E. Effects of polymers on the drop weight sensitiveness of explosives. Proc. 7th Symposium on Detonation, Annapolis/MD (1981), p. 24.

[34] TARVER, C.M. et al. The thermal decomposition of explosives with full containment on 1-D geometries. Proc. 17th Symposium of Comb. Inst., Leeds U.K. (1978), p. 1407.

[35] TAYLOR, P.A. The effect of material structure on the shock sensitivity of porous granular explosives. Proc. 8th Symposium Detonation, Albuquerque/NM (1985), p. 26.

[36] VON HOLLE, W.G., TARVER, C.M. Temperature measurement of shocked explosives by time-resolved infrared radiometry. A new technique to measure shock induced reactions. Proc. 7th Symposium on Detonation, Annapolis/MD (1981), p. 993.

[37] WALKER, E.H. Derivation of the $p^2\tau$ detonation criterion. Proc. 8th Symposium on Detonation, Albuquerque/NM (1985), p. 1119.

[38] ZABABAKHIN, E.I. Collapse of bubbles in a viscous liquid. PMM 24, no. 6 (1960), p. 1129.

CHAPTER VIII

Coupling of Decomposition and Motion

1. Introduction

The first part (Chapters I–III) dealt with the major macroscopic laws governing the propagation of detonations, starting from fundamental principles and some essential observations. After a long digression (Chapters IV–VII), in which we examined various aspects of microscopic mechanisms, we resume our treatment of the whole phenomenon, this time with the aim of arriving at a method of numerical evaluation of a flow with detonation.

The problems which arise by linking mechanical phenomena (movement) to chemical phenomena (evolution of species) have already appeared in Chapter I with regard to the molecular and continuum models of the reactive fluid. We have already dwelt (see Section I.4) on the complexity of the systems to which one is, respectively, led, and on the difficulty of finding realistic closure conditions for them. From 1960, and with the first computers, this complexity and this difficulty have been investigated by "numerical models." They all have the following in common:

(i) they reduce the reactive fluid to a two-component fluid: the initial substance and a "final substance" called "detonation products";
(ii) they take both substances separately to be in thermochemical equilibrium; and
(iii) they are based on the continuum model of the reactive fluid, while neglecting external forces ($f_i = 0$).

Putting such a view into practice supposes that two problems have been tackled. The first starts from a knowledge of the "final substance" and putting its equation of state into a form suitable for numerical calculations; it depends on chemical thermodynamics and on molecular fluid dynamics and can be treated in isolation, which is what we will do in Chapter XII.

The second relates to establishing a practical formulation of the continuum model for a two-component reactive fluid; it depends at the same time on the method employed in section I.3 and on numerical analysis and is dealt with in the next section, VIII.2. We ponder over one-dimensional flow within section VIII.3, endeavoring to arrange the variants in order and to discriminate between those resulting from the model and those resulting from the numerical solving algorithm. The chapter concludes with section VIII.4, which is devoted to finding a continuum law of decomposition based on the *hot spot* concept introduced in the previous chapter.

2. Modeling of Two-Component Reactive Fluid

2.1. Balance Equations for Each Component

The following procedure to establish the balance for each species is very close to that used in sections I.1 and I.3. The notations are identical and the superscript k used as a reference for the species takes only values $k = a$ and $k = b$. The differences in presentation result from the fact that summation over k is not carried out in the balances, and average velocity $\tilde{u}_i$ is not introduced. Furthermore, as in §I.3.1, we assume that no discontinuity surface crosses the reactive fluid.

The mass balance given by (I.47, k) can be rewritten in the form

$$\frac{d^k \rho^k}{dt} + \rho^k u^k_{j,j} = P^k, \qquad \text{(VIII.1, } k)$$

where P^k is the rate of volume production of species (k).

The momentum balance for species (k) appears in §I.3.4 before summing over k

$$\frac{d^k}{dt} \int_D \rho^k u^k_i \, DV = \int_{\partial D} \sigma^k_{ij} n_j \, DS, \qquad \text{(VIII.2, } k)$$

From lemma (I.4) on the particle derivative of a volume integral, the left-hand side of (VIII.2, k) can be put in the form

$$\int_D \left[\frac{d^k}{dt}(\rho^k u^k_i) + (\rho^k u^k_i)u^k_{j,j} \right] DV,$$

which can in turn be modified (from VIII.1, k) by eliminating the divergence term and writing

$$\int_D \left[\rho^k \frac{d^k u^k_i}{dt} + P^k u^k_i \right] DV.$$

As far as the right-hand member of (VIII.2, k) is concerned, it can be transformed by Green's theorem (I.1). Finally, we have as the balance equation (VIII.2, k) for momentum of species (k)

$$\rho^k \frac{d^k u^k_i}{dt} = \sigma^k_{ij,j} - P^k u^k_i, \qquad \text{(VIII.2', } k)$$

which differs from that (I.22) for a one-component fluid by the term $-P^k u_i^k$ inserted in the right-hand side.

The energy balance for species (k) appears in §I.3.5 before summing over k

$$\frac{d^k}{dt} \int_D \rho^k (e^k + \tfrac{1}{2} u_i^k u_i^k) \, DV = \int_{\partial D} (u_i^k \sigma_{ij}^k - q_j^k) n_j \, DS. \qquad (VIII.3, k)$$

Proceeding as above for (VIII.2, k), the left-hand side can first of all be transformed with the help of (I.4)

$$\int_D \left[\frac{d^k}{dt} \rho^k (e^k + \tfrac{1}{2} u_i^k u_i^k) \, DV + (e^k + \tfrac{1}{2} u_i^k u_i^k) \rho^k u_{j,j}^k \right] DV,$$

then from (VIII.1, k) by elimination of the divergence term

$$\int_D \left[\rho^k \frac{d^k}{dt} (e^k + \tfrac{1}{2} u_i^k u_i^k) + (e^k + \tfrac{1}{2} u_i^k u_i^k) P^k \right] DV.$$

As far as the right-hand term of (VIII.3, k) is concerned, it can be transformed by Green's theorem (I.1). Finally, after applying the fundamental lemma of §I.1.1, we obtain the energy balance equation for species (k)

$$\rho^k \frac{d^k}{dt} (e^k + \tfrac{1}{2} u_i^k u_i^k) = (u_i^k \sigma_{ij}^k - q_j^k)_{,j} - P^k (e^k + \tfrac{1}{2} u_i^k u_i^k),$$

which may be further simplified with the help of (VIII.2', k)

$$\rho^k \frac{d^k e^k}{dt} = u_{i,j}^k \sigma_{ij}^k - q_{j,j}^k + P^k (\tfrac{1}{2} u_i^k u_i^k - e^k). \qquad (VIII.3', k)$$

This last equation differs from that (I.22) for a one-component fluid by the P^k proportionality term which occurs on the right-hand side.

Since they are so important for the rest of the chapter, the balance equations (VIII.1, k), (VIII.2', k), and (VIII.3', k) are rewritten below for a perfect fluid ($\sigma_{ij}^k = -p^k \delta_{ij}$, $q_j^k = 0$) of mass volume $v^k = 1/\rho^k$

$$\frac{d^k \rho^k}{dt} = -\rho^k u_{j,j}^k + P^k, \qquad (a)$$

$$\rho^k \frac{d^k u_i^k}{dt} = -p_{,i}^k - P^k u_i^k, \qquad (b) \qquad \left. \right\} \; (VIII.4, k)$$

$$\frac{d^k e^k}{dt} = -p^k \frac{d^k v^k}{dt} - P^k v^k (e^k + p^* v^k - \tfrac{1}{2} u_i^k u_i^k). \qquad (c)$$

2.2. Balance Equations for the Mixture

Despite its educational value, the system of equations (VIII.4) where each component is identified in the flow is not the one which we meet in current practice of numerical modeling of flow with detonation. Two arguments— one utilitarian, the other historical—may be advanced to explain this gap.

"Current" uses of an explosive arrange for it to be generally associated with an adjoining medium which behaves as a chemically inactive fluid. This is why a system of partial differential equations which would be able to handle a flow with detonation and would yet be similar to that commonly used to deal with chemically inactive flow, has obvious practical interest. This remark allows us to understand one of the constants of research in this field: a code structure where the active medium is distinguished from inactive media only by the "branching of a decomposition law" on a hydrodynamic algorithm, termed "standard."

Whether from the angle of Riemann characteristic curves theory or from the angle of pseudoviscosity invented a century later by Richtmyer and Morton, the basic tools available for the numerical study of a flow are those associated with a system of four partial differential equations

$$
\left.
\begin{aligned}
\frac{d\rho}{dt} &= -\rho u_{i,i}, && \text{(a)} \\[2mm]
\rho \frac{du_i}{dt} &= -p_{,i}, && i = 1, 2, 3, \quad \text{(b)} \\[2mm]
\frac{de}{dt} &= -p \frac{dv}{dt}, && \text{(c)} \\[2mm]
e &= E(p, v), && \text{(d)}
\end{aligned}
\right\} \qquad \text{(VIII.5)}
$$

(one of which is vectorial) in four unknowns (one of which is vectorial). It is therefore not surprising that the natural tendency was to find:

(i) a fifth unknown which describes the variation in chemical composition of the mixture from the initial stage to the stage of detonation products in chemical equilibrium: this is the function m already introduced in §II.2.1 and §III.1.3; and

(ii) a fifth equation called (VIII.5, e) which expresses more or less empirically reactivity $(1 - m)^{-1} dm/dt$ as a function $S(m; a, b)$ of the state of the mixture and/or the initial substance (a) and/or the final substance (b)

$$
\frac{1}{1-m} \frac{dm}{dt} = S(a, b, m) \quad \text{(e)} \qquad \text{(VIII.5)}
$$

In order to complete this extension of the "standard" hydrodynamic scheme to flow with detonation, it remains to specify the thermodynamic conditions which govern the transformation of the explosive substance (referred to hereinafter by the index $k = a$) into detonation products (referred to hereinafter by the index $k = b$ as burnt). [N.B. Certain publications use indices s for solid and g for gas; we have dispensed with these unnecessarily restrictive symbols.] To do this, it is difficult to escape the introduction of six additional unknowns: internal mass energies e^a and e^b, mass volumes v^a and v^b, and pressures p^a and p^b. The seven equations generally chosen (see discussion in §3.3) to be added to equations (VIII.5(a) (b), (c), and (e)) and substituted for (VIII.5, d) are

$$v = (1 - m)v^a + mv^b \qquad \text{(f)}$$
definition of m,

$$e = (1 - m)e^a + me^b \qquad \text{(d')}$$
additivity of internal energy,

$$e^a = E^a(v^a, p^a) \qquad \text{(g)}$$
thermodynamic definition of the initial substance,

$$e^b = E^b(v^b, p^b) \qquad \text{(h)}$$
thermodynamic definition of the final substance,

$$p = p^a = p^b \qquad \text{(i, j)}$$
homobaric mixture hypothesis,

$$T^a = T^b \qquad \text{(k)}$$
homothermal mixture hypothesis.

(VIII.5)

We thus obtain a system of 11 equations with 11 unknowns.

2.3. Comparison

In view of systems (VIII.4, $k = a, b$) and (VIII.5), we may legitimately ask ourselves the question about the possibility of correlating the first with the second. To answer this question, we introduce "primed" unknowns ρ', v', e', p', u' through the following notations and hypotheses:

$$\rho' = \rho^a + \rho^b,$$
$$e' = e^a + e^b,$$
$$v' = v^a + v^b,$$
$$u'_i = u^a_i = u^b_i \quad \text{(homokinetic hypothesis)},$$
$$p' = p^a = p^b \quad \text{(homobaric hypothesis)}.$$

(VIII.6)

Note that the first four do not necessarily coincide with the "nonprimed" unknowns of system (VIII.5); note also that ρ' does not equal $1/v'$. Summing equations (VIII.4, k) over k ($k = a, b$) and bearing in mind that $P^a + P^b = 0$ (see I.3.1), we immediately obtain

$$\frac{d\rho'}{dt} = -\rho' u'_{j,j}, \qquad \text{(a)}$$

$$\rho' \frac{du'_i}{dt} = -p'_{,i}, \qquad \text{(b)}$$

$$\frac{de'}{dt} = -p' \frac{dv'}{dt} - \sum_k P^k v^k (e^k + p' v^k - \tfrac{1}{2} u'_i u'_i). \qquad \text{(c)}$$

(VIII.7)

These *three* equations involve *thirteen* unknowns:

- four for the initial substance: v^a, ρ^a, e^a, P^a;
- four for the final substance: v^b, ρ^b, e^b, P^b; and
- five associated with the "mixture": v', ρ', e', p', u_i'.

They are connected by the first *three* equations (VIII.6) rewritten below

$$\left.\begin{array}{ll} \rho' = \rho^a + \rho^b, & \text{(d)} \\[2mm] e' = e^a + e^b, & \text{(e)} \\[2mm] v' = v^a + v^b, & \text{(f)} \end{array}\right\} \qquad \text{(VIII.7)}$$

which obviously may be completed by the following *six*

$$\left.\begin{array}{ll} v^a = 1/\rho^a, & \text{(g)} \\[2mm] v^b = 1/\rho^b, & \text{(h)} \\[2mm] e^a = E^a(v^a, \text{p}'), & \text{(i)} \\[2mm] e^b = E^b(v^b, \text{p}'), & \text{(j)} \\[2mm] P^a = -P^b = P(a, b). & \text{(k), (l)} \end{array}\right\} \qquad \text{(VIII.7)}$$

and by *one* hypothesis on the temperatures T^a and T^b which may, for example, be that of the homothermal mixture as in (VIII.5, k)

$$T^a = T^b. \quad \text{(m)} \qquad\qquad \text{(VIII.7)}$$

Systems (VIII.5) and (VIII.7) have obvious formal analogies (mass balance and impulsion balance, homobaric and homothermal hypotheses), but also notable differences. The first concerns the homobaric hypothesis: its neglect in (VIII.7) considerably modifies equation (b) on conservation of momentum and equation (c) on conservation of energy, which is not the case for system (VIII.5) since the homobaric hypothesis is simply joined to the conservation equations. The second and more fundamental difference concerns the unknowns associated with the mixture: whereas u' and p' are identified with u and p, on the other hand, ρ', v', and e' are not identified with ρ, v, and e. In particular, the relation $v = 1/\rho$ implicit in (VIII.5) has no equivalent in (VIII.7). We shall postpone a discussion of the consequences of this lack to section VIII.3, while examining the "standard" resolution algorithm of (VIII.5) where $v = 1/\rho$ arises explicitly. Here it will suffice to bear in mind that it is to a "nonstandard" algorithm that we must address ourselves to solve (VIII.7). The third difference concerns functions S and P which determine in (VIII.5, e) and (VIII.7, k, l), respectively, the chemical evolution of the mixture: when the rates P^a and P^b have dimensions $ML^{-3}T^{-1}$ and belong to chemical kinetics, reactivity has dimension T^{-1} and is defined by a balance formula (m and $1 - m$ are dimensionless terms which balance the additive unknowns v^a, v^b, e^a, e^b in the expressions of the unknowns v and e of the mixture) so that the former are not reducible to the latter.

System (VIII.7) thus has the advantage of introducing real kinetics (by the expedient of P^a and P^b) but also has the major practical disadvantage of not being in a form amenable to a "standard" hydrodynamic treatment. This also explains why system (VIII.5), despite the intrinsic difficulty which the formulation of function S raises in (VIII.5, e) (see §4.3), is still widely used, and why the next section is devoted to this point. There we shall abandon the three-dimensional Eulerian system (VIII.5) in favor of a one-dimensional Lagrangian system by plane sections which has been more extensively studied, and is more appropriate for making apparent the linkages with what we have, on several occasions, called the "standard" hydrodynamic approach (for a simple introduction to this Lagrangian system, the reader is referred to the classic work by Landau and Lifshitz [9, p. 5—*in fine*).

3. One-Dimensional Rectilinear Reactive Flow

3.1. Model and Reference Algorithm

Consider one-dimensional rectilinear motion of a two-component perfect reactive fluid and the associated Lagrangian equations as written and numbered by Richtmyer and Morton [16]

$$\frac{\partial x}{\partial t} = u, \tag{1}$$

$$\frac{\partial u}{\partial t} = -v_0 \frac{\partial p}{\partial X}, \tag{2}$$

$$\frac{\partial e}{dt} = -p \frac{\partial v}{\partial t}, \tag{3}$$

$$\frac{\partial x}{\partial X} = \frac{v(X, t)}{v_0}, \tag{4}$$

$$e = me^b + (1 - m)e^a \equiv E(p, m, v^a, v^b), \tag{5}$$

$$\frac{\partial m}{\partial t} = (1 - m)S(p, T^a, v^b, m), \tag{6}$$

$$v = mv^b + (1 - m)v^a, \tag{7}$$

$$\theta^a(p, v^a) = \theta^b(p, v^b), \tag{8}$$

$$e^a = E^a(p, v^a), \tag{9}$$

$$e^b = E^b(p, v^b), \tag{10}$$

$$\text{(VIII.8)}$$

i.e., 10 equations with 10 unknowns

$$x, u, p, v, e, m, e^a, e^b, v^a, v^b.$$

Before moving on to the algorithm proper, we make the point that this model is implicitly homobaric and explicitly homothermal; these hypotheses will be discussed in §3.3.

The discretized equations are designated by the "starred" number of the associated equation with partial derivatives. The boundaries of the mesh and their velocities are referred to by the subscript k; the thermodynamic state is referred to at the center of the mesh by the subscript $k + \frac{1}{2}$; the instant of calculation is designated by the superscript n. At time t^n, we know for all k,

$$x_k^n, \quad p_{k+1/2}^n, \quad p_{k-1/2}^n, \quad e_{k+1/2}^n, \quad m_{k+1/2}^n,$$

$$u_k^{n-1}, \quad v_{k+1/2}^n, \quad (v^a)_{k+1/2}^n, \quad (v^b)_{k+1/2}^n.$$

Six calculation steps lead to the state at time t^{n+1}:

(I) Calculation of u_k^{n+1} by the "driving gradient"

$$\frac{u_k^{n+1} - u_k^{n-1}}{\Delta t} = v_0 \frac{p_{k+1/2}^n - p_{k-1/2}^n}{\Delta X}. \qquad \text{(VIII.8, 2)}*$$

(II) Calculation of x_k^{n+1} $(0 < k < K)$ from conditions at the limits u_0^{n+1} and u_K^{n+1}

$$\frac{x_k^{n+1} - x_k^n}{\Delta t} = u_k^{n+1}. \qquad \text{(VIII.8, 1)}*$$

(III) Calculation of $v_{k+1/2}^{n+1}$

$$v_{k+1/2}^{n+1} = v_0 \frac{x_{k+1}^{n+1} - x_k^{n+1}}{\Delta x}. \qquad \text{(VIII.8, 4)}*$$

(IV) Calculation of $e_{k+1/2}^{n+1}$ by conservation of energy

$$e_{k+1/2}^{n+1} - e_{k+1/2}^n = -p_{k+1/2}^n (v_{k+1/2}^{n+1} - v_{k+1/2}^n). \qquad \text{(VIII.8, 3)}*$$

(V) Calculation of $u_{k+1/2}^{n+1}$ by the decomposition law

$$\frac{m_{k+1/2}^{n+1} - m_{k+1/2}^n}{(1 - m_{k+1/2}^n)\Delta t} = S[p_{k+1/2}^n, (T^a)_{k+1/2}^n, (v^b)_{k+1/2}^n, m_{k+1/2}^n], \qquad \text{(VIII8, 6)}*$$

where $(T^a)_{k+1/2}^n = \theta^a[p_{k+1/2}^n, (v^a)_{k+1/2}^n]$.

(VI) End of calculation of pressures $p_{k+1/2}^{n+1}$ by the equations of state. We calculate $(v^a)_{k+1/2}^{n+1}, (v^b)_{k+1/2}^{n+1}, p_{k+1/2}^{n+1}$ by

$$e_{k+1/2}^{n+1} = m_{k+1/2}^{n+1} E^b[p_{k+1/2}^{n+1}, (v^b)_{k+1/2}^{n+1}]$$

$$+ (1 - m_{k+1/2}^{n+1}) E^a[p_{k+1/2}^{n+1}, (v^a)_{k+1/2}^{n+1}], \qquad \text{(VIII.8, 5)}*$$

$$(v)_{k+1/2}^{n+1} = m_{k+1/2}^{n+1}(v^b)_{k+1/2}^{n+1} + (1 - m_{k+1/2}^{n+1}) \cdot (v^a)_{k+1/2}^{n+1}, \qquad \text{(VIII.8, 7)}*$$

$$\theta^a[p_{k+1/2}^{n+1}, (v^a)_{k+1/2}^{n+1}] = \theta^b[p_{k+1/2}^{n+1}, (v^b)_{k+1/2}^{n+1}]. \qquad \text{(VIII.8, 8)}*$$

We thus complete the calculation cycle and we know the state at time t^{n+1}, i.e.,

$$x_k^{n+1}, \quad p_{k+1/2}^{n+1}, \quad p_{k-1/2}^{n+1} \quad e_{k+1/2}^{n+1}, \quad m_{k+1/2}^{n+1},$$
$$u_k^{n+1}, \quad v_{k+1/2}^{n+1}, \quad (v^a)_{k+1/2}^{n+1}, \quad (v^b)_{k+1/2}^{n+1}.$$

3.2. Modifications According to Source Term

The reference algorithm may be qualified as "pseudo-isentropic." This expression aims to recall in vivid fashion the "source" of the variation in velocity in the starred equations. It takes on its profound significance when we note that the use of (VIII.8, 3) implies that there is no release of energy in the mesh so that it is the variation of m which modifies the pressure (steps V and VI). Then an attractive modification of the reference algorithm is one called "energy release in the mesh." At stage IV, we calculate $e_{k+1/2}^{n+1}$ by

$$de = -p \, dv + \text{source},$$

the source depending in general on m; step V and step IV remain the same, but it is the source term which fixes the new pressure and no longer the variation in chemical composition induced by reactivity.

We shall attempt to clarify these differences with the help of the notations in this book. For a perfect fluid in the absence of a field of moments and external forces, but in the presence of a rate ℓ of volume heat supplied externally (see Section I.1), the first law of thermodynamics is written

$$de = -p \, dv + \frac{\ell}{\rho} \, dt. \tag{VIII.9}$$

In the reference algorithm, the term ℓ is taken equal to zero: there is no release of energy in the mesh. This is the chemical transformation (by the expedient of m) which modifies the pressure field and which "drives" the flow. In order to establish thoroughly this formalism, whose rigor is indisputable, we should recall that the internal mass energy e depends not only on v and s, but also on m. Thus, writing (VIII.9) with $\ell = 0$ shows that the transformation (dv, ds, dm) of the explosive is effected according to the law

$$\frac{\partial e}{\partial s}(v, s, m) \cdot ds + \frac{\partial e}{\partial m}(v, s, m) \, dm = 0, \tag{VIII.10}$$

with the result that the specific entropy s of a particle cannot remain constant during the evolution of the composition $(dm \neq 0)$. The qualification "pseudo-isentropic" is thus justified.

In the so-called variant "with energy release in the mesh," the term ℓ/ρ is strictly positive. Certain authors use it in the form

$$e_X \frac{dm}{dt}, \tag{VIII.11}$$

where e_X is a mass reaction energy to be defined (see §XII.4.2).

In spite of the nonequivalence of the base algorithm and its variant, a practical equivalence can exist for an *ad hoc* choice of e_X, as the following example shows. When the reference algorithm is used with functions E^a and E^b linked by

$$E^b(v, p) = E^a(v, p) - \Delta E,$$

then we can easily verify that it is sufficient to choose $e_X = \Delta E$ such that (VIII.9) is written in both cases

$$m \, dE^b + (1 - m) \, dE^a + (E^b - E^a) \, dm = -p \, dv.$$

But this example also demonstrates the weakness of the variant: the validity of its results is strictly linked to the physical reality of the relation postulated between E^a and E^b.

3.3. Modifications According to Exchange Terms

In order to determine the possibilities of variations about the reference model, it is useful to return to the general approach which underlies system (VIII.8):

- consider a perfect fluid flow with five unknowns x, u, v, p, e and the associated Richtmyer–Morton system;
- in order to simulate the chemical evolution of the fluid, we introduce a supplementary unknown m as well as an equation giving dm/dt; and
- in view of the need to substitute a family of surfaces of state for a unique surface of state, and in view of the introduction by this expedient of the individual states of the two components, we introduce all or part of a set of eight supplementary unknowns

$$e^a, \quad v^a, \quad p^a, \quad u^a,$$
$$e^b, \quad v^b, \quad p^a, \quad u^b,$$

and as many supplementary equations as there are really independent supplementary unknowns.

The current approach is that of the reference model, characterized by the homobaric and homothermal hypotheses; it is implemented in Mader's SIN code [12], Donguy's TCD code [4], and many more. However, other approaches are possible and have been effectively implemented.

Thus, in the KRAKATOA code [3], Damamme and Missonnier introduce u^a and u^b as supplementary independent unknowns, challenge the two homobaric and homothermal hypotheses, and close the system by five hypotheses: the isentropic evolution hypothesis quoted above, two local composition hypotheses concerning velocities and pressures, and two hypotheses of local conservation of mass and momentum fluxes. This approach avoids having to introduce temperatures T^a and T^b as independent unknowns: this has to be seen as "an easy way out" in a sense which will become more apparent in the following subsection.

But there is no unanimity in heterothermal approaches from which temperature is excluded, at least apparently. Thus, Nunziato [13] and Kipp [8] have developed a model where not only temperature varies from one component to another, but also occurs explicitly in the exchange terms. Since the corresponding formalism deviates considerably from system (VIII.8), we refer the reader to the relevant publications.

The reader might also refer to [10] for another treatment of the problems posed by writing the exchange terms.

3.4. The Use of Pseudopotential p(v, e)

Here we regroup the numerical problems posed by the solution of equations (VIII.8, 5)*, (VIII.8, 7)*, and (VIII.8, 8)* which arise from functions E^a and E^b, θ^a and θ^b.

In the classical algorithm of Richtmyer and Morton, modified by Wilkins [19], substances (a) and (b) do not occur individually: only the unknowns of the mixture p, v, and m occur. The variable $p_{k+1/2}^{n+1}$ is therefore obtained by simple recourse to the subroutine p(v, e) with called variables $v_{k+1/2}^{n+1}$, $e_{k+1/2}^{n+1}$, the latter being obtained with the help of (VIII.8, 3)*. In other words, in order to link state and movement, it is sufficient to make use of a "black box" which gives p as a function of e and v after eliminating temperature from the functions $e(v, T)$ and $p(v, T)$ calculated by thermodynamicists.

The convenience which this "blind" use of the surface of state offers the thermodynamicist, explains the tendency for models to replace the homothermal hypothesis with an isentropic evolution hypothesis. Indeed, starting from identity (e.1) in Appendix A and writing de as a linear form of dv and dp, we find the identity

$$T\, ds = \left[p + \frac{\partial e}{\partial v}(v, p) \right] dv + \left[\frac{\partial e}{\partial p}(v, p) \right] dp,$$

so that ds = 0 is written

$$\left[p - \frac{\frac{\partial p}{\partial v}(v, e)}{\frac{\partial p}{\partial e}(v, e)} \right] dv + \left[\frac{\partial p}{\partial e}(v, e) \right]^{-1} dp = 0, \qquad \text{(VIII.12)}$$

where the sufficient nature of the "black box" shows up well enough. In the numerical solution scheme of the (VIII.8) starred equations, the advantage is equally obvious. Suffice it to recall that ds = 0 is equivalent to

$$\Delta e = -p(v, e)\Delta v;$$

in the corresponding algorithm, the called variables are v, e, and Δv and, eventually, the constants which remain in the function p(v, e), such that the output variable is p(v, e − pΔv); since this algorithm is used in another context in stage IV (see §3.1), we see that the isentropic hypothesis allows the use of the same "subroutine" twice per time step: once for the conservation of energy (VIII.8, 3)* and once for the supplementary relation replacing the homothermal equation (VIII.8, 8)*.

On the other hand, when we renounce the benefits of this ease of numerical nature (which owes nothing to physics), then temperature has to be expressed as a function of the variables v and p. We may start from the identity (e.2) of Appendix A

$$T = \frac{\partial e}{\partial s}(v, s),$$

and, introducing variables v and p, write successively

$$T(v, p) = \frac{\partial e}{\partial p}(v, p) \cdot \frac{\partial p}{\partial s}(v, s)$$

$$= -\frac{\partial e}{\partial p}(v, p) \cdot \frac{\partial T}{\partial v}(v, s)$$

$$= -\frac{\partial e}{\partial p}(v, p)\left[\frac{\partial T}{\partial v}(v, p) + \frac{\partial T}{\partial p}(v, p) \cdot \frac{\partial p}{\partial v}(v, s)\right],$$

Taking account of (VIII.12), we can replace $\partial p/\partial v(v, s)$ by

$$-\left[p + \frac{\partial e}{\partial v}(v, p)\right]\bigg/\frac{\partial e}{\partial p}(v, p),$$

and obtain

$$T(v, p) = \left[p + \frac{\partial e}{\partial v}(v, p)\right]\frac{\partial T}{\partial p}(v, p) - \frac{\partial e}{\partial p}(v, p) \cdot \frac{\partial T}{\partial v}(v, p),$$

which can be further written as

$$T(v, p) = \left[p - \frac{\frac{\partial p}{\partial v}(v, e)}{\frac{\partial p}{\partial e}(v, e)}\right] \cdot \frac{\partial T}{\partial p}(v, p) - \frac{\frac{\partial T}{\partial v}(v, p)}{\frac{\partial p}{\partial e}(v, e)}. \tag{VIII.13}$$

There too, the "black box" is sufficient for calculating the coefficients of the temperature derivatives. For all that, however, all the difficulties are not overcome concerning the calculation of the temperature itself, as is apparent from the highly pertinent analysis made by Fickett and Davis [6, pp. 122–123]: "If e(v, p) is known in some region of space (v, p) and if the temperature is given along some nonisentrope (noncharacteristic) arc, then the temperature T(v, p) is determined over the band between the isentropes through the two end points of the arc, and within the domain of definition of e."

Concerning the use of the pseudopotential p(v, e), there is another difficulty, which is not trivial: that of a representation which is thermodynamically admissible in the entire domain of application. This problem will be dealt with in §XII.4.2.

4. Decomposition Law

4.1. Temporary Solutions

This section aims to provide a precise description of function S introduced in (VIII.5, e), to express the variation in reactivity as a function of the state of the mixture and/or the initial substance and/or the final substance. Because of the nature of m—a macroscopic unknown indicative of the degree of progress of the fictitious transformation (a) → (b) to which the chemical evolution of the explosive is reduced—this function cannot result from *ab initio* calculations which would take account of the succession of steps of the reaction process. It can result only from a phenomenological analysis on the one hand, and from an interplay between numerical and physical experiments on the other. Furthermore, both types of experiment must be well chosen. This implies that in the case of the first type, the results must be indicative of function S, to the exclusion of the other terms in the numerical model; this condition is but easily satisfied, as results from the coupling with the variants according to source term and exchange terms (see Section VIII.2), with the equations of state (see Chap. XII), and with the methods of numerical analysis proper. This means that, for the second type, the experimental apparatus and the measurements, must be sensitive to the microscopic parameters influencing S (see Chap. IX). Thus, because physical modeling, numerical analysis, dynamic instrumentation and, indeed, even the static characterization of samples overlap, the problem of determining function S remains one of the most controversial questions in publications devoted to explosives.

However, one thing at least is clear (see Section VII.1), the ignition of a condensed explosive takes place in privileged sites called hot spots, whose appearance relates to several local heating mechanisms and whose distribution depends on the initial heterogeneity of the explosive: texture in a solid explosive, impurities, and even initial temperature fluctuations about an average value (see Dremin [5]) in a liquid explosive.

Considering the easy terms of numerical nature which the impasse on temperature affords (see §3.4), we should not be surprised to see a method termed "C–J volume burn" appearing at an early date (since the first computers) (see Wilkins [19], Mader [11]), which consists of replacing equations (VIII.5, e, f, d', g, i, j, k) by

$$m_k^n = (v_0 - v_k^n)/(v_0 - v_*),$$
$$p_k^n = m_k^n p^b(v_k^n, e_k^n),$$

(VIII.14)

where only the pseudopotential $p^b(v, e)$ of (b) appears. With certain adjustments in the neighborhood of limits m = 0 and m = 1, this method appears remarkably simple and effective for one-dimensional rectilinear and one-dimensional convergent flows, but is unsuitable for one-dimensional divergent and two-dimensional flows which are found to be the commonest. The numerous attempts made to remedy this want find expression in algorithmic devices which clearly emphasize the physical limits of the method.

Diametrically opposed to "C–J volume burn," which ignores temperature, we have the "Arrhenius burn law"

$$\frac{dm}{dt} = (1 - m)Z^{\neq} \exp\left(-\frac{T^{\neq}}{T}\right), \qquad \text{(VIII.15)}$$

which is explained by the finite difference equation

$$\frac{m_k^{n+1} - m_k^n}{\Delta t} = (1 - m_k^n)Z^{\neq} \exp\left(-\frac{T^{\neq}}{T_k^n}\right). \qquad \text{(VIII.15)*}$$

This method makes the progress of the transformation $(a) \rightarrow (b)$ no longer rest on the volume compression of the mixture, but on the heating of (a). To take an example, we quote the constants used by Mader [11] for nitromethane and penta-erythritol, which will serve as a reference on many occasions in the following chapters,

	$Z^{\neq}$ (s^{-1})	$E^{\neq}$ (kcal/mole)	$T^{\neq} = E^{\neq}/\mathfrak{R}$
NM	$4 \cdot 10^{14}$	53.6	26,973
PETN	$6.3 \cdot 10^{19}$	47	23,651

To evaluate this method, we quote Mader [11, pp. 156–157]: "One can use Arrhenius kinetics to describe the gross features of the explosive thermal decomposition of nitromethane over a surprising range of temperatures, times, and pressures. Since it is well known that the dominant mechanism of decomposition varies with temperature and pressure ($\ldots$), it would be useful to investigate the compensating mechanisms which permit the use of such a simple description of the kinetics." Unfortunately, the qualities of law (VIII.15) vis-à-vis homogeneous macroscopic explosives (liquids and mono-crystals) are of no help when we wish to take account of the influence of parameters of aggregation or crystalline texture of an inhomogeneous solid explosive at the scale of one micron, which is what must be considered in a phenomenon of propagation where the velocity is of the order of millimeters per microsecond, and the time resolution of currently available instruments (see Chap. X) is of the order of one nanosecond. Hence the need to have recourse to other methods.

4.2. From Hot Spot to Reactivity

Models which deviate from both "C–J volume burn" and "Arrhenius burn" may as a first approach be classified into two groups.

In the first ("bulk burn") we suppose that the mixture is sufficiently intimate that dm/dt can be considered as resulting directly from a volume transformation. This hypothesis fits in well with the homobaric and homothermal hypotheses of the reference model, and also with a function S having as argument the state variables of the mixture. The "C–J volume burn" and "Arrhenius burn" laws belong to this group.

In the second ("surface burn") we suppose that during a nonnegligible phase, m tends to 1 by divergent expansion of flame fronts, each of these fronts having as origin a hot spot and separating the final hot component (b) in the interior from the cold component (a) at the exterior. This hypothesis may or may not be paired with the homobaric hypothesis (N.B. elementary flame theory neglects the pressure drop across the front); on the other hand, it is incompatible with the homothermal hypothesis for which a possible substitute is that of the isentropic evolution of (a) downstream of the ignitor shock.

It would be rather tedious to recall every attempt over 25 years to go from "hot spot to reactivity." With the risks which that entails, we shall limit ourselves to recalling in this subsection the attempts of three schools, which were either important landmarks for their time or which still have practical significance.

"Surface burn" first appeared in Bernier [2] who in 1964 envisaged "grain burn" and "hole burn" simultaneously, and proposed a flame front velocity of type $ap^\alpha + b$ (where a, b, and α are adjustable parameters), based on many experimental arguments. This idea of retaining pressure as a pilot quantity of the transformation $(a) \rightarrow (b)$ is slowly gaining ground. Thus, from 1976, Mader and Forest [12], while remaining faithful to "bulk burn," abandoned the Arrhenius law for a law of the type

$$\frac{dm}{dt} = (1 - m) \cdot \exp\left(\sum_{n=0}^{n=14} a_{(n)} p^n \right) \qquad \text{for} \quad p_{min} < p < p_*, \quad \text{(VIII.16)}$$

called "Forest fire," which certainly is based on a physical hypothesis (see §IX.1.3) but which suffers from not having been established independently of the thermodynamic treatment of the mixture $(a) + (b)$. A thorough phenomenological analysis was performed in 1985 by Tang [17] and Johnson [7] to overcome this difficulty: their model, called "explicit hot spot process" is innovative in many respects and arises from the linkage of two rate laws.

New approaches appeared in 1978. The works of Wackerlé [18] and Anderson [1] on the one hand and Nunziato [13] on the other are of the "bulk burn" type and consist of an extension of the Arrhenius law. The former allow the frequency factor Z to be proportional to the square of the ignitor shock pressure $\hat{p}$, and that it may be corrected by a factor $G(p, t)$ which acts as an induction time variable in inverse ratio to the pressure

$$\frac{dm}{dt} \approx (1 - n) \cdot \hat{p}^2 \cdot G(p, t) \cdot \exp\left(-\frac{1200}{T}\right). \qquad \text{(VIII.17)}$$

The latter allow a locally higher temperature to exist on the surface of the grains (almost homothermal hypothesis) and it is this that must be taken into account in the Arrhenius law. Their arguments remain phenomenological and the outcome rather qualitative, according to the authors themselves. Moreover, the stated inadequacies led Kipp and Nunziato [8] to give it a supplementary dimension:

- by distinguishing grains of explosive and bonding agent in the initial substance (a);
- by supposing the mixture to be heterobaric–heterothermal;
- by decomposing the bonding agent according to an Arrhenius law; and
- by making the grains regress by a flame front velocity proportional to a variable nth power of the pressure p^b.

The authors point out the "good general agreement with the experiment" obtained in the framework of Wondy's one-dimensional Lagrangian hydroreactive code. One might also add, however, that the complexity of the equations is in opposition to the goal sought: the applicability to two-dimensional geometries.

For their part, Lee and Tarver [10] in 1980 proposed yet another model of the "bulk burn" family with an additive law of the type

$$\frac{dm}{dt} = (1 - m)^{2/9} \left[I\left(\frac{v_0}{\hat{v}} - 1\right) + G \cdot m^{2/3} p^z \right], \qquad \text{(VIII.18)}$$

where I, G, and z are variable parameters which, for penta-erythritol, for example, have values

$$I = 20 \ \mu s^{-1},$$
$$G = 400 \ \mu s^{-1} \cdot (M \ bar)^{-z},$$
$$z = 1.4.$$

(N.B. The hat over v stands for the value of the mass volume of (a) downstream of the ignitor shock, according to the notation introduced in §III.6.3.) In (VIII.18), the first term is introduced in order to simulate the hot spot formation phase ("ignition"), while the second term is constructed to take account of the growth phase of the transformation (a) → (b) starting from hot spots ("growth"). Like the authors of other models, Lee and Tarver could take advantage of numerous examples of agreement between experimental and numerical results descended from a one-dimensional hydroreactive code; they could also, like Mader and Forest then Tang and Johnson—and for the same reason of closeness to the reference model—achieve the insertion in a two-dimensional reactive flow code.

4.3. Ingredients of a Unified Theory

(See Cheret, *Comptes Rendus Acad. Sciences, Paris,* vol. 306, Series II (1988), p. 863.)

The models summarized in the preceding subsection are very diverse. "Bulk burn" and the homothermal hypothesis figure alongside "surface burn" and the heterothermal hypothesis; the additive formulation has some adherents (e.g., Lee and Tarver), also the multiplicative formulation (e.g., Wackerlé and Anderson); the factor $1 - m$ is even questioned by certain researchers (Lee and Tarver). Within each team, the complexity and number of floating parameters continue to increase in proportion to the increase in the number and precision of the experimental results. Each team records successful results, though it is, nevertheless, difficult to discern what is owed to the choice of experiments, to the adjustments carried out, and to the physical value of the law itself. So much so that none of them enjoys unqualified support, whether from the point of view of the physical basis or from the point of view of the algorithm facility.

At the risk of adding to the reader's perplexity, we venture to lay the foundations of a unified model, starting from the following idea: just as the theory of sensitivity to "shock" was (see Chap. VI) due to including primary and secondary explosives, so also the formulation of reactivity S is due to including explosives termed homogeneous and those termed heterogeneous. This requirement has two consequences. It leads to adopting uniformly the homobaric–homothermal hypotheses which are indispensable for homogeneous explosives. It also leads to searching for S in the form of a sum of two terms:

- an Arrhenius type *molecular rate law* term S',

$$S' = Z \exp\left(-\frac{T^{\neq}}{T}\right); \quad \text{and}$$

- an *aggregate rate law* term S'' which is to be specified.

It must be emphasized that this description means the superposition of two rate laws, as in Partom [14], and not a coupling, as in Tang and Johnson [7], [17].

We note first that the contribution of S'' becomes all the more important as the time which the wave of velocity U takes to traverse the characteristic length $\underset{\approx}{L}$ of the aggregate diminishes, and approximates to the characteristic time associated with the rate law $\underset{\approx}{S'}$; this argument invites consideration that S'' is proportional to the ratio $U/\underset{\approx}{L}$. We then note that S'' must be, for $m = 0$, a function sensitive to temperature $\hat{T}$ attained at the surface of purely explosive domains, according to an Arrhenius law in $\exp(-T^{\neq}/\hat{T})$. We note finally that $(1 - m)S''$ must be an increasing function of m in the neighborhood of $m = +0$ (geometric increase of flame front issuing from each hot spot) and a decreasing function of m in the neighborhood of $m = 1 - 0$ (merging of flame fronts).

The simplest choice of S'' which corresponds to these conditions is

$$S'' = \frac{U}{\widetilde{\widetilde{L}}} \frac{1 - 2m_1 + m}{(1 - m_1)^2},$$

where m_1 is defined in $]0, \frac{1}{2}[$ by

$$\frac{1 - 2m_1}{(1 - m_1)^2} = \exp\left(-\frac{T^{\neq}}{\widehat{\widetilde{T}}}\right) \in]0, 1[.$$

Figures VIII.1(a) (b) lighten the variation of function $(1 - m) \cdot S''(m, m_1)$.

In order to formulate the aggregative rate law completely, in terms compatible with the continuum model of VIII.2, it remains to express $\widetilde{\widetilde{L}}$ and $\widehat{\widetilde{T}}$ as a function of the stress and of the aggregate state parameters. As far as $\widetilde{\widetilde{L}}$ is concerned, all the experimental knowledge points to a fundamental, if not unique, role for the degree of division; the initial idea was to choose the average grain size for $\widetilde{\widetilde{L}}$. However, owing to the practical difficulty which arises when attempting to estimate this size from a histogram, it seems preferable to introduce the specific surface σ_0 of the powder which serves as a base for the aggregate, and to let

$$\widehat{\widetilde{L}} = \frac{1}{\rho_0 \sigma_0}.$$

It remains to give a formulation of the temperature $\widehat{\widetilde{T}}$; this will be done in

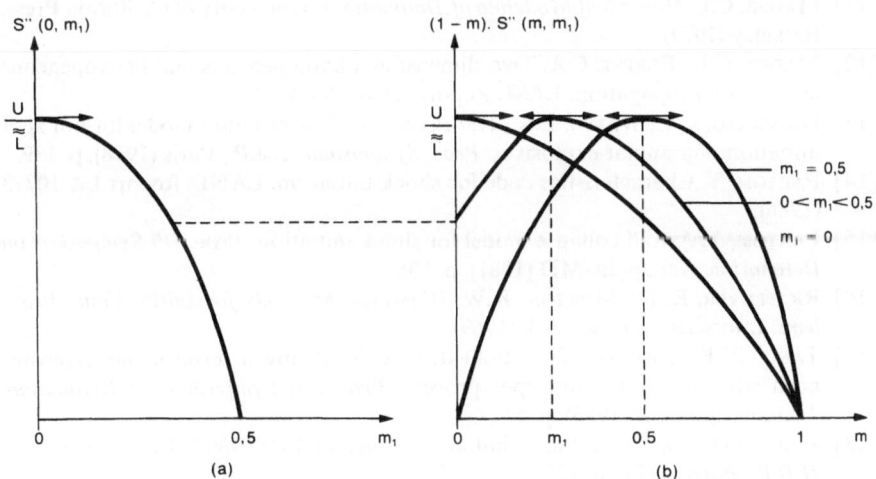

Figure VIII.1. Aggregative rate law: (a) $S''(0, m_1)$ function of m_1; (b) $(1 - m) \cdot S''(m, m_1)$ function of m.

§XI.3.1 in the following chapter on the generation of detonation by plane shock.

References

[1] ANDERSON, A.B. et. al. Shock initiation of porous TATB. Proc. 7th Symposium on Detonation, Annapolis/MD (1981), p. 385.

[2] BERNIER, H. Contribution à l'étude de la génération de la détonation par impact. Thèse de doctorat ès sciences. Paris (1964). Rapport CEA 2497.

[3] DAMAMME, G., MISSONNIER, M. Simulation of the reaction zone of heterogeneous explosives. Proc. 7th Symposium on Detonation, Annapolis/MD (1981), p. 641.

[4] DONGUY, P., LEGRAND, N. Numerical simulations of non-ideal detonations of a heterogeneous explosive with 2-D Eulerian code CEE. Proc. 7th Symposium on Detonation. Annapolis/MD (1981), p. 695.

[5] DREMIN, A.N. On condensed explosives detonation decomposition. Proc. Symposium H.D.P. Paris (1978), p. 175.

[6] FICKETT, W., DAVIS, W.C. Detonation. University of California Press, Berkeley (1979).

[7] JOHNSON, J.N., TANG, P.K., FOREST, C.A. Shock wave initiation of heterogeneous reactive solids. J. Appl. Phys., 57 (1985).

[8] KIPP, M.E., NUNZIATO, J.W., SETCHELL, R.E. Hot spot initiation of heterogeneous explosives. Proc. 7th Symposium on Detonation, Annapolis/MD (1981), p. 394.

[9] LANDAU, L. D. LIFSCHITZ, E.M. Fluid Mechanics. Pergamon Press, London (1959).

[10] LEE, E.L., TARVER, C.M. Phenomenological model of shock initiation in heterogeneous explosives. Phys. Fluids, 23 (1980), p. 2362.

[11] MADER, C.L. Numerical Modeling of Detonations. University of California Press, Berkeley (1979).

[12] MADER, C.L., FOREST, C.A. Two-dimensional homogeneous and heterogeneous detonation propagation. LASL Report LA 6959 (1976).

[13] NUNZIATO, J.W., WALSH, E.K., KENNEDY, J.E. A continuum model for hot spot initiation of granular explosives. Proc. Symposium H.D.P., Paris (1978), p. 139.

[14] PARTOM, Y. Characteristics code for shock initiation. LANL, Report LA 10773 (1986).

[15] PARTOM, Y. A void collapse model for shock initiation. Proc. 7th Symposium on Detonation, Annapolis/MD (1981), p. 506.

[16] RICHTMYER, R. D., MORTON, K.W. Difference Methods for Initial Value Problems. Interscience, New York (1969).

[17] TANG, P. K., JOHNSON, J.N., FOREST, C.A. Modeling heterogeneous explosive burn with an explicit hot spot process. Proc. 8th Symposium on Detonation, Albuquerque/NM (1985), p. 52.

[18] WACKERLÉ J. et al. A shock initiation study of PBX 9404. Proc. Symposium H.D.P., Paris (1978), p. 127.

[19] WILKINS, M.L. Calcul de détonations mono et bidimensionnelles. Proc. Colloque C.N.R.S. Ondes de Détonation, Gif-sur-Yvette/France (1961), p. 165.

CHAPTER IX

Generation of Detonation by Plane Shock

1. From Shock to Detonation

1.1. Field of Investigation

In a very particular sense, generation of detonation by plane shock has already been approached in §III.6.2, which dealt with the birth of a simple detonation. There we came to conclusions bearing on the *local* conditions of incipient development of chemical reactions downstream of a shock.

We take up this problem again here, as in [13], in a *global* form: we discuss the evolution of the entire flow from the instant when a shock type stress begins to reach the explosive to the instant when the detonation can be considered to be *built-up* in the sense of §III.6.2 (L' ≪ L). However, the exposition is intentionally restricted from two points of view. As far as its objective is concerned, the task is to extract from the voluminous literature the experimental laws which govern this evolution as a function of stress parameters and parameters proper to the explosive. As far as the field of investigation is concerned, we restrict ourselves to a consideration of geometries where:

- the stress is a "plane" shock;
- the priming boundary is a plane with which the incident shock "coincides" at instant 0; and
- the observation zone is sufficiently far away from the free boundaries so that the flow there is independent of lateral compressions, and therefore one-dimensional rectilinear in the direction normal to the incident shock.

The conception and realization of such experiments is far from trivial, as we have endeavored to show from the indications given below on the problems raised and the precautions taken.

The first of these concern the planarity of the incident shock. In the case of a shock induced by a projectile, the quality of the experiment depends on controlling the deformation of the projectile and/or taking account of possible deviations from planeness on the one hand, and parallelism with the priming boundary on the other. In the case of a shock induced by explosive generator, the quality of the experiment depends on controlling the wave shapers. In all cases, it is important to ensure good homogeneity of the receptor explosive, in terms of volume mass for a solid ($\Delta\rho_0 \leq 3$ mg/cm^3), temperature for a liquid ($\Delta T_0 \leq 0.1$ K), or composition for a compressed gaseous mixture. In a general manner, the time deviation from planeness must remain coherent with standard precision for the dimensions of the assembly (5×10^{-2} mm), related to the typical shock velocity (5 mm/μs), say 10 ns.

The second factors concern the one-dimensional rectilinear character of the flow, and derive from a well-known result from fluid mechanics: for an induced shock of velocity U, downstream of which the material velocity is $\hat{u}$ and the velocity of sound is $\hat{a}$, the depth of the zone perturbed by the lateral compression at a distance X from the entry surface is $X \tan(1/U)\sqrt{\hat{a}^2 - (U - \hat{u})^2}$. When $\hat{a}$ varies due to chemical evolution, this estimate should be raised by replacing, for example, $\hat{a}$ by a_*. Typical values,

$$U = 4.5 \text{ mm}/\mu\text{s},$$

$$\hat{u} = 1.5 \text{ mm}/\mu\text{s},$$

$$a_* = 6 \text{ mm}/\mu\text{s},$$

lead to an angle of perturbation of 50° which researchers are accustomed to make allowance for when they choose an "angle of observation" less than 40° (36° 5′ according to Campbell [6], 16° 45′ according to Bernier [4], and 20° according to Droux et al. [11]).

In other words, the distance over which a rectilinear flow may be observed is at most of the order of the smallest dimension of the priming boundary. Also, in other words, the effective observation of the build-up to a detonation in a rectilinear flow, when this build-up is slow, is conceivable only on a large-diameter structure. As, however, the observation of a rapid build-up is limited by the space–time resolution of the apparatus, it can be seen that the current experiment allows an exploration of only a restricted band of variation of the priming parameters. This established fact shows the interest which attaches to controlling the numerical modeling (Chap. VIII) and in particular that of the reactivity S.

Not to be "fooled" by decompressions is a permanent anxiety of the experimenter, which shows up in a number of details of the firing geometries and associated observation modes. Thus, for nitromethane, a transparent liquid explosive, a glass tube is generally chosen and two observations by slit scanning cameras (see §X.1.3): "lateral" for a cylinder generator, and "end on" for the diameter of the section opposite to the priming boundary. For a pentaerythritol (cyclonite (RDX), etc.) monocrystal, a face is made true to serve as a priming boundary; in the case of an observation by Raman spectroscopy (see Section VI.3) the other faces are left in the rough state but, in order to limit reflections due to irregularities of these faces, the crystal is immersed in a liquid of similar refractive index (silicone oil, zinc chloride, etc.) itself contained in a glass cell; in the case of an observation by slit camera, two other

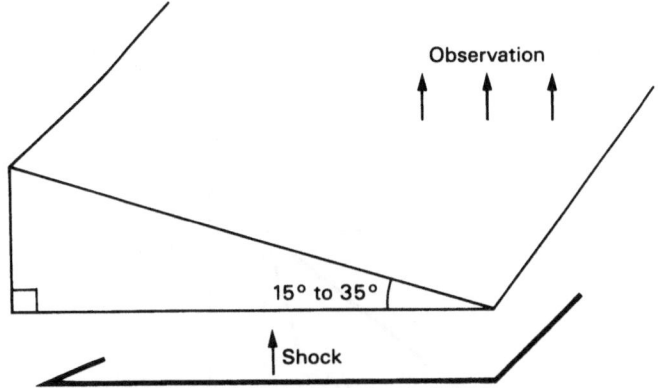

Figure IX.1. Diagram of a "wedge" arrangement.

faces are machined so as to form with the priming surface the lateral face of a regular prism whose section is a right-angled triangle (see Campbell *et al.* [5]), priming being achieved along the right-angle face gently inclined towards the hypotenuse (around 15°–35°) while the observation relates to the hypotenuse. This last geometry is known as a "wedge."

This arrangement has also been employed for bonded aggregate-type solid explosives, unchanged by the American authors, and in the improved "double wedge" form by Bernier [4] and then Droux [11].

1.2. General Aspects of Propagation

After the preliminary work of Chaiken [8] and Holland [17], two notable articles by Campbell *et al.* definitely established, in 1960, the main features of generation of a detonation by plane shock in the case of [5] homogeneous explosives (nitromethane, liquid TNT, penta-erythritol monocrystal, etc.) and in the case of [6] inhomogeneous explosives (free aggregates of TNT, cyclonite (RDX)-based bonded aggregate, nitromethane-impregnated carborundum, etc.).

The essential qualitative difference between the two groups appears in wave path and interface path diagrams in the (t, x)-plane, as shown on Figure IX.2(a) and (b). Along the trajectory OI, the priming boundary of the homogeneous explosive is moved with the material velocity û associated with the shock velocity U which precedes it along trajectory OI^0; a detonation is born in I and propagates with a velocity D_*^+—much greater than the Chapman–Jouguet (C–J) velocity D_* assignable to its initial state of rest—up to the "transition point" I^0 where it overtakes the induced shock, penetrates the explosive at rest, and decelerates rapidly until it attains a value close to D_* from $I^*(t^*, X^*)$. Along its trajectory, the shock induced in the inhomogeneous explosive at first propagates with velocity U; then an increasing

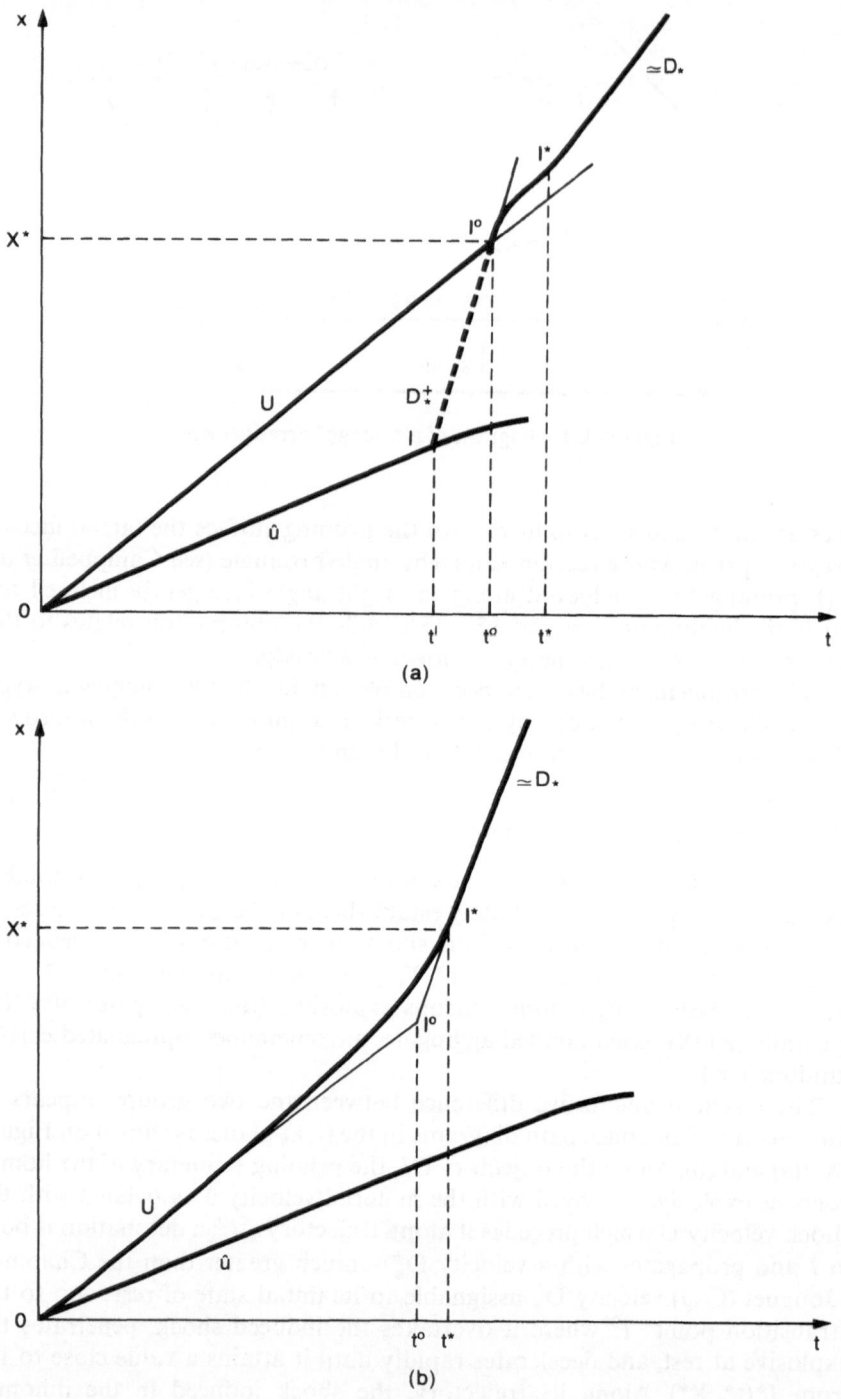

Figure IX.2. Sequence of events in the generation of a detonation by plane shock: (a) in a homogeneous explosive; and (b) in an inhomogeneous explosive.

acceleration phase becomes perceptible which is followed by a decreasing acceleration phase which itself comes to an end at a point $I^*(t^*, X^*)$ where the velocity stabilizes to a value close to D_* and where generally an intense luminous emission is observed.

In order to summarize these differences, one could scarcely do better than quote the actual words used by Campbell et al. [6] in the discussion to which the experiment led them:

... (in spite of certain imperfections) "the shock waves were of sufficient quality to permit the drawing of a number of conclusions from the experimental results.

It is of interest first to contrast the initiation behavior of homogeneous and inhomogeneous explosives (...). The contrasts are as follows:

(a) The initial shock wave in a homogeneous explosive shows a constant or slightly decaying velocity as a function of time; the corresponding wave in inhomogeneous explosive accelerates throughout its travel.
(b) The transition to high-order detonation is very abrupt in homogeneous explosive; the transition in inhomogeneous explosive is less so.
(c) The onset of high-order detonation in homogeneous explosive is accompanied by an overshoot in the velocity, amounting to about 10% in the case of nitromethane; no demonstrable overshoot has been recorded for inhomogeneous explosive in our experiments.
(d) Detonation is observed to originate at the shock attenuator–explosive interface in homogeneous explosive; at present, it is believed probable that detonation occurs at or near the shock front in inhomogeneous explosives. (...)
(e) The experiments with nitromethane–carborundum mixtures have shown that the mixtures are much more sensitive than the homogeneous liquid nitromethane. The inhomogeneities in the mixture cause shock interactions with resultant local heating. For initiation, the detailed structure of the shock properties of the explosive is more important than are the values of the thermochemical constants.
(f) The material behind the initial shock wave in homogeneous explosive is relatively nonconducting for electricity until the onset of detonation; in inhomogeneous explosive the material behind the initial shock front is quite conducting and becomes even more so as the transition to high-order detonation is approached.
(g) The initiation process in homogeneous explosive is much more sensitive to variation of the initial temperature or to variation of the shock pressure than it is in inhomogeneous explosive. It is logical to attribute the difference between the initiation behavior of inhomogeneous and homogeneous explosives to the voids and other defects in the former."

Another difference, this time quantitative, appears through the variation —as a function of the induced shock pressure—of two quantities to which frequent reference is made in the literature: *time to detonation* t* (or more

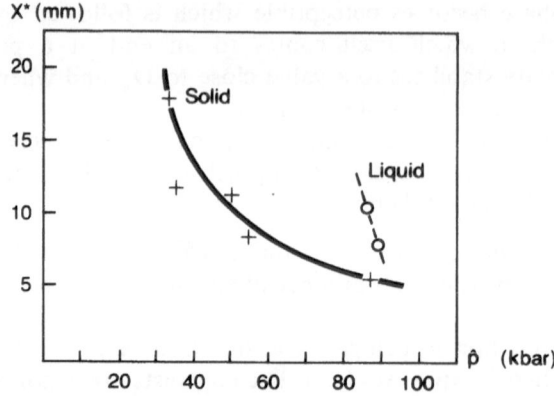

Figure IX.3. Run to detonation X* as a function of pressure $\hat{p}$ induced by the shock (after [6], Fig. 13).

rigorously, *build-up time*) and *run to detonation* X* (or more rigorously, *build-up distance*). As a suitable example, consider the variation in X* for liquid homogeneous nitromethane compared with inhomogeneous solid cyclotol (see Fig. IX.3, adapted from Figure 13 in [6]). Note, in particular, that the same distance X* of around 10 mm is obtained for $\hat{p} \approx 85$ kbar in nitromethane and for a pressure of about half that value in cyclotol. This point is elaborated in the next subsection.

1.3. Time and Run to Detonation

Consider first the case of a homogeneous explosive which, despite the apparent complexity of Figure IX.1(a), reveals itself as falling within the scope of a simple theoretical approach. In fact, as a function of induction time t^l and velocity D_*^+, the time to detonation t* and run to detonation X* are written

$$\begin{cases} t^* = t^l + \dfrac{X^* - \hat{u}t^l}{D_*^+}, \\ X^* = Ut^*, \end{cases}$$

which is easily transformed to

$$\begin{cases} t^* = t^l \dfrac{D_*^+ - \hat{u}}{D_*^+ - U}, \\ X^* = Ut^*. \end{cases}$$

Now one estimate of t^l is given by the thermal explosion model of Hubbard and Johnson [20]

$$t^l = \frac{\hat{T}^2}{ZT^{\neq}\Delta T} \exp(T^{\neq}/\hat{T}), \tag{IX.1}$$

Table IX.1. Calculation of t^l, D_*^+, t^*, and X^* for industrial nitromethane in the type III reference arrangement of Campbell et al. [5]. Values are calculated for $\Delta T = 3120$, and not $3120 + \hat{T} = 4260$ as Campbell did, by mistake.

Initial state of nitromethane	$\rho_0 = 1.125$ g/cm³, $T_0 = 25°$ C (purity $\simeq 97\%$)
Induced shock in nitromethane	$\left.\begin{array}{l} U = 4.5 \text{ mm/}\mu s \\ \hat{u} = 1.7 \text{ mm/}\mu s \end{array}\right\} \Rightarrow \hat{p} = \rho_0 U \hat{u} = 86$ kbar
C–J state of nitromethane	$D_* (\rho_0 = 1.125$ g/cm³$) = 6.30$ mm/μs $\Delta D_*/\Delta \rho_0 = 3.20$ m s^{-1}/g cm^{-3}
Molecular rate law constants	$Z = 10^{14.6}$ s$^{-1} = 4.10^{14}$ s^{-1} $\Re T^* = 53,600$ cal/g $\Leftrightarrow$ $T^* = 27,000$ K
$t^l = 0.75$ μs $D_*^+ = 10.18$ mm/μs	$t^* = 1.12$ μs $X_* = 5.04$ mm

which is none other than (VII.4) where $E^{\neq}$ is replaced by $\Re T^{\neq}$ and Q by $c_v \Delta T$. D_*^+ is estimated by another route by supposing that $D_*^+ - \hat{u}$ is little different from the C–J detonation velocity for an initial density equal to that of $\rho_0 U/(U - \hat{u})$ realized downstream of the induced shock. Table IX.1 presents Campbell's data for estimates of t^l, D_*^+, t^*, and X^* in one of the reference arrangements for nitromethane. Formula (IX.1) provides ample evidence for the extreme sensitivity of t^l and, consequently, of t^* and X^* at temperature $\hat{T}$; we shall return to this aspect in section IX.2 devoted to "sensitivity."

For an inhomogeneous explosive, the observation of wave and interface motions alone does not allow us to imagine the fate suffered by each slice of explosive and even less how this fate varies from one slice to another. This is why most of the studies subsequent to those of Campbell et al. relate to the *Lagrangian* analysis of the flow [1], [9], [12], [21], [22], [35], [36], [39]. Some of the methods of measurement developed in X.2 permit direct access to pressure profiles (piezoresistive gauges) and material velocity profiles (electromagnetic gauges).

Two typical diagrams (see Fig. IX.4(a) and (b)) derived from Erickson [12] and Wackerlé [36] show how the decomposition consecutive to the passage of the induced shock in the explosive is coupled to the flow: the pressure profile (material velocity) not increasing on the priming boundary ($X = 0$) acquires a progressively more pronounced peak ($X = 2$, $X = 4.1$, $X = 6.1$ on Fig. IX.4(a); $X = 1.04$, $X = 2.10$ mm; $X = 3.79$ mm on Fig. IX.4(b)) and tightens until reaching a maximum when $X = X^*$. The currently acceptable interpretation is as follows. In a slice adjacent to the priming boundary, decomposition develops slowly and thus is not manifest until some considerable time after the passage of the induced shock. This decomposition, even partial, is sufficient (see §VIII.3.1) to augment the "driving gradient" and thus progressively reinforce the induced shock, and increase the temperature

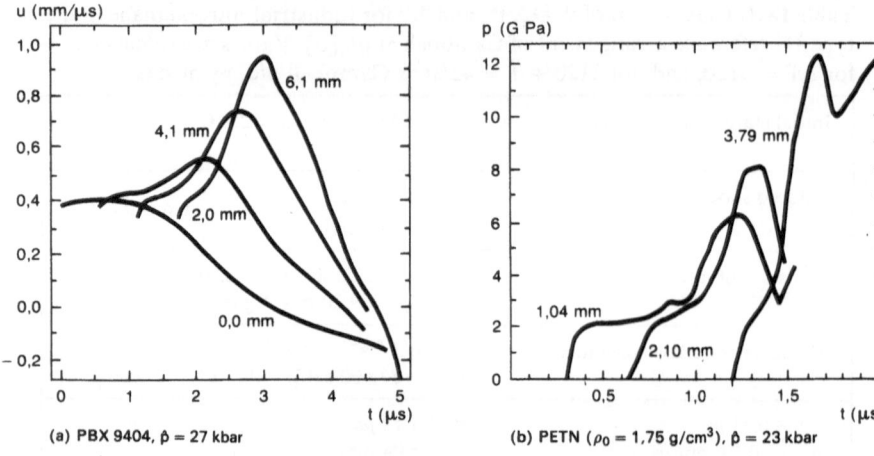

(a) PBX 9404, p̂ = 27 kbar   (b) PETN ($\rho_0 = 1,75$ g/cm³), p̂ = 23 kbar

Figure IX.4. Lagrangian analysis. Typical recordings of: (a) velocity by electromagnetic gauge (after [12]); and (b) pressure by piezoresistive gauge (after [36]).

attained downstream of the shock and at the same time magnify the reactivity. The pressure and velocity profiles become narrower; finally, an *autonomous* detonation (see §III.4.2) is built up *when the maximum velocity is attained immediately downstream of the ignitor shock.*

For each slice, we can define an induction time $t^I(X)$ by choosing as reference event (see §VII.2.1) the start of the decrease in pressure or material velocity; then, to say that a detonation builds up is equivalent to saying that there is a X^* such that $t^I(X)$ tends to zero when X tends to X^*.

Thus, for an inhomogeneous explosive, the time and run to detonation result in a coupling between decomposition and movement over the entire domain between the priming boundary and the induced wave as long as it has not become an autonomous detonation. Under these conditions, one may imagine that an estimate of t^* and X^* is not within the compass of elementary theory, but belongs on the other hand to the specialized domain of numerical investigations based on experimental laws. Among these, two assume a simple physical significance and are generally the subject of linear representations (see Table IX.2):

- $U = A + B\hat{u}$, giving the induced shock velocity U as a function of the material velocity $\hat{u}$ at the interface; and
- $\log X^* = a - b \log \hat{p}$, giving the run-to-detonation distance X^* as a function of pressure $\hat{p}$ downstream of the induced shock (often called "Popplot" after A. Popolato, one of its proponents).

These laws play an essential role in Forest-fire type decomposition laws (see §VIII.4.2) based on the so-called "single curve build-up principle" due to Mader [25] and Lindstrom [23]. According to this principle, everything

Table IX.2. Linear representation of diagrams (log $\hat{p}$, log X^*), ($\hat{u}$, U): (α) values obtained when linear smoothing takes account of an experimental estimate of the hydrodynamic velocity of sound when $\hat{u} = 0$; and (β) X^* in mm, $\hat{p}$ in kbar, U and $\hat{u}$ in mm/μs.

	log $X^* = a - b$ log $\hat{p}$ (β)		U = A + B$\hat{u}$ (β)		
	a	b	A	B	References
PETN $\rho_0 = 1.75$ g/cm^3	1.31	2.2	2.26 (α)	2.32	Ramsay and Popolato [30]
PBX 9404 $\rho_0 = 1.83$ g/cm^3	3.33	1.57	2.42 (α)	1.88	Kennedy [22] Mader [26]
comp. X/X_1 ($\rho_0 = 1.82$ g/cm^3)	3.33	1.37	2.89	2.00	Aveillé et al. [2]
comp. T/T_1 ($\rho_0 = 1.88$ g/cm^3)	6.71	2.72	2.50	2.10	Aveillé et al. [2]

happens as if

- in the time interval (0, t*) the propagation was that of a "reactive shock" of velocity u', pressure $\hat{p}'$, velocity $U' = A' + B'u'$ ($A' \approx A$, $B' \approx 4B/3$, see [26]); and
- pressure $\hat{p}'$ at X abscissa in itself determined the distance $X^* - X$ remaining to run to detonation, the corresponding functional relation coinciding with the pop-plot, considered as giving the distance X^* remaining to run when the pressure is $\hat{p}$.

This principle—for details of which we refer the reader to the original papers [26], [27]—has the advantage of dispensing with a direct formulation of the reactivity and yet leading to results in agreement with experiments in a fair number of arrangements. It has limitations, however, the most serious of which is not being able to account for the arrangement where a weak shock "desensitizes" the inhomogeneous explosive and confers to it, relative to any subsequent shock, the reactivity of a homogeneous explosive.

A final remark regarding the relation between X^* and $\hat{p}$ deserves attention. Certain authors (see Droux et al. [11]) propose a representation where X^* tends to infinity when $\hat{p}$ decreases towards a limiting value $\hat{p}_c$ ($\hat{p}_c \approx 26$ kbar for composition X/X_1 in Table IX.2). It is tempting to rely on this limit $\hat{p}_c$ to devise a scale of *sensitivity to shock* (without inverted commas), an explosive being termed more sensitive the lower the value of $\hat{p}_c$ with which it is associated. This approach must be considered with prudence, for two reasons. The first is because $\hat{p}_c$ is a practical limit associated with a firing configuration, increasing the dimensions of which is sufficient to reduce $\hat{p}_c$ and this is more notably the case the more "insensitive" the composition. The second reason is because it is possible to find by Lagrangian analysis a decomposition on the priming boundary or, indeed, at some distance from this boundary, even for pressures $\hat{p}$ less than $\hat{p}_c$ but still greater than a value $\hat{p}^i$. Concerning $\hat{p}^i$ and $\hat{p}_c \geq \hat{p}^i$, it is appropriate to employ a more precise language (*ignition threshold pressure* for the former and *critical pressure* for the latter), and to replace these quantities in the framework of an analysis of the "sensitivity" of an explosive substance. This is the subject of the next section, where we consider the group of impact-induced stresses, which lends itself to a parametric approach.

2. "Sensitivity" of an Explosive Substance

2.1. Detonability Thresholds by One-Dimensional Impact

In the general discussion of Section IX.1, it was implicitly supposed for the sake of simplicity that the stress on the priming boundary varies only slightly during the time to detonation t*. In fact, this condition is only rarely an automatic consequence of the precautions taken to ensure planeness of the stress and one-dimensional rectilinear character of the flow; additional precautions are necessary with regard to the thickness of the barrier which conveys the stress emitted by an explosive generator or that induced by the impact of a projectile. Some simple rules, based on the properties of one-dimensional decompression waves, well known since Riemann, provide suitable dimensioning. In the case of a projectile launched by powder gun or by exploded metal foil (see §XI.4.2), they provide the means of controlling the time τ during which pressure maintains a quasi-constant value on the priming boundary: the highest-performing equipment [38] justifies τ values from 10 nanoseconds to a few hundred nanoseconds. A possibility therefore opens up for exploring the detonability of an explosive, not only as a function of the induced shock pressure but also of the time for which it is maintained.

Taking account of the general observations reported in the previous section relating to build-up to detonation, it is obvious that a pressure $\hat{p}$ leads to the build-up of the detonation if the time τ during which it is maintained is at least equal to $t^*(\hat{p})$, and that it does not result in a detonation if τ is very much less than $t^*(\hat{p})$. There is therefore a threshold value $\tau(\hat{p})$ for which $t^*(\hat{p})$ is an over-estimate; the excess itself can be approximated by the time taken by the decompression, which originates behind the projectile, to rejoin the shock induced in the explosive. This simple observation is at the base of the experimental determination of a threshold relation between projectile thickness and velocity. Having chosen a thickness e for the projectile (therefore, the induced shock pressure $\hat{p}$ survives for a time τ), we proceed to a series of shots while making the velocity V of the projectile increase (therefore, the value of the induced shock pressure) until observing a luminous effect indicating an established detonation; this is repeated for different values of e. With reference to the critical energy model of Walker and Wasley [37], which supposed $\tau(\hat{p})$ to be proportional to $1/\hat{p}^2$, experimental results are

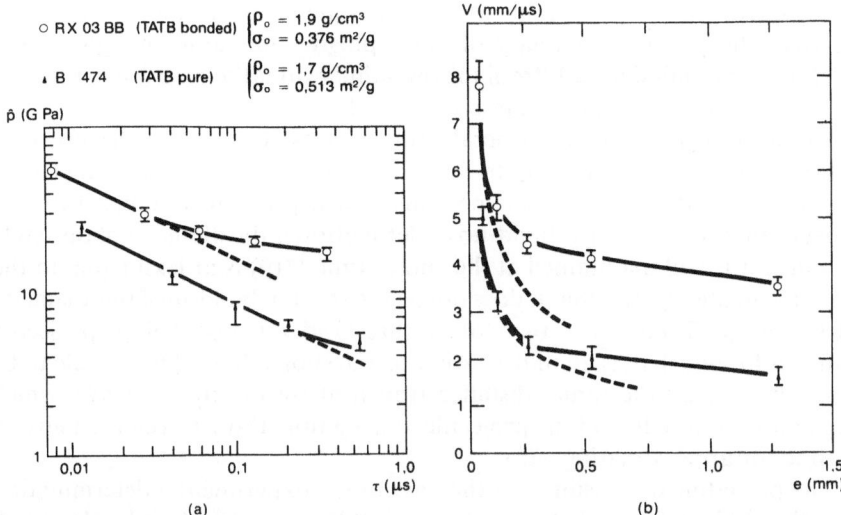

Figure IX.5. Detonability threshold by one-dimensional impact (after Monodel [18] for two TATB compounds): (a) relation between pressure $\hat{p}$ and duration τ; and (b) relation between velocity V and thickness e of the projectile. Solid lines join experimental points. Broken lines correspond to the law $\hat{p}^2\tau = $ constant.

often the subject of a linear representation of log $\hat{p}$ as a function of log $\tau(\hat{p})$. In fact, the numerous results accumulated by Weingart and co-workers [38], [39] show that this model has a domain of validity limited to brief pulses (up to at most 100 nanoseconds). It is therefore preferable to retain a representation in the plane (e, V) which lends itself to a simple physical interpretation in the entire range of experimental conditions (see Fig. IX.5(a), (b)).

The interpretation of this detonability threshold by one-dimensional impact has been the subject of many papers (see Hayes [16] and Longueville [24]) which we could not hope to summarize, so great is the diversity of approaches and explosives considered. On the other hand, it is easy to localize the origin of this diversity in the essential role played (see §2.1 second alinea) by the difference $t^*(\hat{p}) - \tau(\hat{p})$, the evaluation of which is linked to the model of the propagation of a decompression in a reactive medium.

2.2. Detonability Thresholds by Two-Dimensional Impact

The general discussion of §IX.1 contains another implicit hypothesis, knowing that the stress is maintained uniformly over the entire priming boundary during the time t^* necessary for build-up to detonation. In fact, due to release waves originating from the free surface of the projectile, this condition is never strictly satisfied, whatever precautions are taken to ensure planarity of the stress and the one-dimensional rectilinear character of the flow in some part of the explosive. This situation lends itself to a simple qualitative analysis, when the impact is due to an axisymmetric projectile of diameter ϕ less than the priming boundary of the explosive, but long enough for the effects called to mind in §2.1 (from the one-dimensional release wave originating from behind the projectile) to be negligible.

Taking into account the general observations reported in the previous section with regard to build-up to detonation and the conclusions of §III.6.3 on the critical diameter ϕ_c, it is obvious that a pressure $\hat{p}$ induced during impact can only lead to the build-up of detonation if the diameter of the circle on which it is still maintained at the end of time $t^*(\hat{p})$ is at least equal to the critical diameter ϕ_c, and that it does not lead to such a build-up if this diameter is less than ϕ_c. There exists, therefore, a threshold value $\varphi(\hat{p})$ of the projectile diameter for which ϕ_c is an under-estimate, the shortfall itself being able to be approximated by the radial distance traversed by the release wave, which originates on the edges of the projectile, during time $t^*(\hat{p})$. Green [15] gives a method for approximating $\varphi(\hat{p})$.

The preceding discussion is at the base of the experimental determination of a threshold relation between the projectile diameter ϕ and velocity V. Having chosen a projectile diameter φ, we proceed to a series of shots while making the velocity V of the projectile increase (therefore, the value of the induced shock pressure $\hat{p}$) until observing an effect (luminous or other) indicating a built-up detonation; this is repeated for different values of φ. An example of detonability threshold by two-dimensional impact is given in Figure IX.6(a) and (b), taken from [3] and [38]: we note that the excess $t(X) - (X/D_*)$ becomes negative when the pressure exceeds a value of the order of p_*, and becomes infinite positive when $\hat{p}$ drops below 25 kbar.

Based on experimental results, Moulard [28] generalized the concept of detonability threshold by two-dimensional impact to geometries where the impact area is not a circle but a nonaxisymmetric (rectangle) or non-connected (ring) area. He obtained a unique representation of all his results and those of [3] by substituting the initial impact area for the diameter φ. However, the justification which he gives (volume of explosive affected by the ignition) and the consequences which he derives from it ("sensitivity" to impact) are, at least, debatable: another approach would consist, for each configuration, of a preliminary investigation of the extinction/bifurcation conditions in the extension of that in §III.6.3, with regard to the circular area.

2.3. Nonmolecular Factors of "Sensitivity"

In Chapters I–VI, we relied mainly on molecular and continuum models for the definition of the explosive and the study of detonation and sensitivity to "shock." In Chapter VII, first with regard to hot spots, then in Chapter VIII with regard to reactivity, the gaps in and limits of these models have emerged from a consideration of experimental results. They recall, if need be, that an explosive structure—as the subject of physical experiments or practical uses —is a product of the chemical industry and therefore governed by laws which result not only from basic molecule(s), but also from others attributed to the

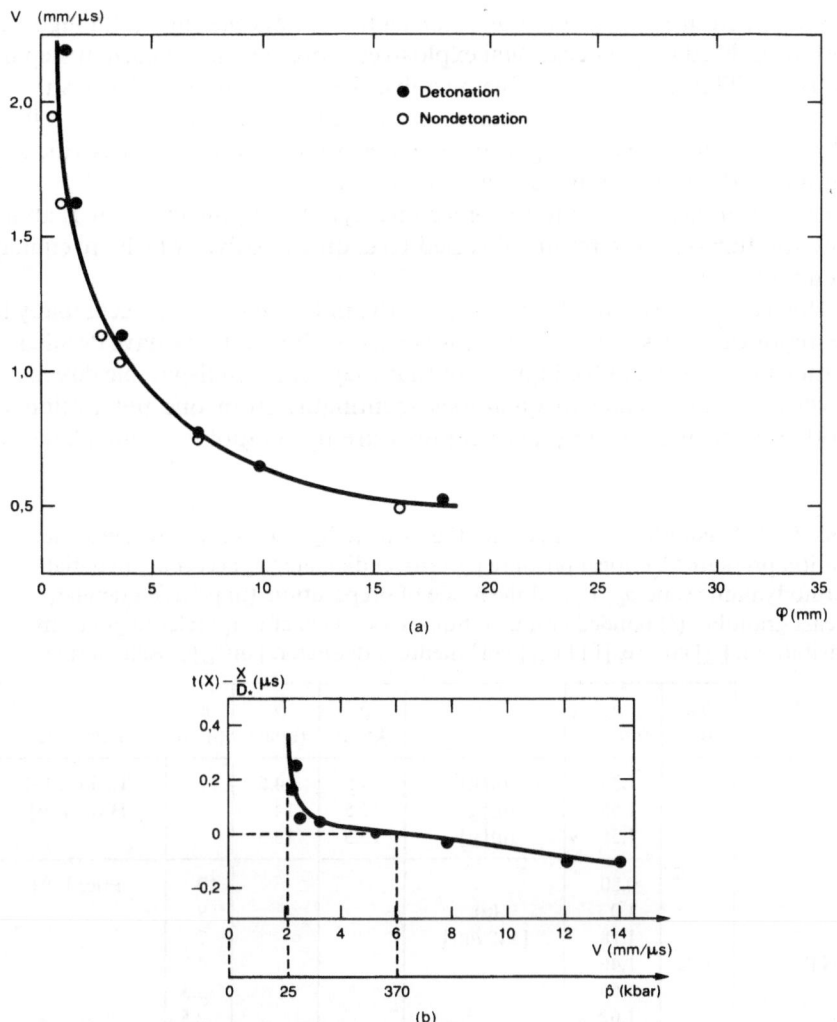

Figure IX.6. Plane impact of a PBX 9404 target by a projectile of velocity V and diameter φ: (a) Detonability threshold V(φ), after [3]; and (b) Excess transit time as a function of V, after [38]. • Detonation. ○ Nondetonation.

physical state and still others attributed to all the stages of the operational mode of preparation.

There is little need to search very far or deep for an illustration of this point. We know what Nobel's fortune owes to the difference between nitroglycerine and dynamite; likewise, the powder worker knows how much—during cold weather—his own safety depends on his care to avoid a speck of nitroglycerine slipping under his feet; further, gun merchants know what they owe to the low melting point of TNT? Nevertheless, in spite of the already long history of the explosives industry, it is illusory to claim to have drawn up a compete list of the nonmolecular factors of detonability (although the work of L. Medard on occasional explosives represents an important step on this road). This is why we confine ourselves here to a point of view restricted by two hypotheses: the initial macroscopic state is well described by a fluid behavior (variables T_0, ρ_0, p_0) while the mesoscopic state is well described by a granular distribution or a *specific surface* σ_0.

Even though restricted to the preceding hypothesis, the literature is abundant and furnishes the results of varied tests on explosives which are equally abundant.

This very variety, and the disparity of the initial data which accompany it, are major obstacles to synthesis. This is why at this point we have decided to extract some results judged typical in that they reflect indisputable directions of variation and relate to quantities identifiable from one publication to another: critical diameter ϕ_c, critical pressure $\hat{p}_c$, ignition pressure $\hat{p}^i$, induc-

Table IX.3. Quantities associated with the "sensitivity" of an explosive substance. Ignition pressure $\hat{p}^i$, critical pressure $\hat{p}_c$, critical diameter ϕ_c according to initial thermodynamic state ρ_0, T_0 and the mode of preparation: (p) pressed granular, (c) cast granular, (ℓ) bonded granular. Square brackets after (p) refer to grain size distribution: [g] coarse, [f] fine, [μm] medium diameter, [m^2/g] specific surface.

	T_0 (C)	ρ_0 (g/cm^3)		$\hat{p}^i$ (kbar)	$\hat{p}_c$ (kbar)	ϕ_c (mm)	References
TNT	$\simeq 25$	1.30	(p)[g]	4.5	4.5		Taylor [34]
		1.55	(p)[g]	11.5	14		Howe [19]
		1.55	(p)[g]	15.5	15.5		
		0.80				10	Price [29]
		1.0				9	
		1.18	(p)			7	
		1.46	[140 μm]			4.3	
		1.55				3.3	
		1.62				2.5	
		1.18	(p)			2.5	
		1.62	[30 μm]			2.5	
		1.62	(c)			14.5	Campbell [7]
		1.44	liquid			62.6	
HNS	25	1.57	(p)	17	23.2		Roth [31]
	25	1.39	[15 μm]	9	16		($\#$) quoted in
	260	1.46	[0.27 m^2/g]	9	13.2		[26] p. 266
PBX 9404	25	1.84	(ℓ)	15	25	1.2	[7] p. 647
	150	1.77	[94 % HMX]	15		($\#$)	
PBX 9502	−55		(ℓ)			10.5	Travis
	25	1.898	[95 % TATB]			7.6	(Private
	+75					6	communication)

Table IX.4. Quantities associated with "sensitivity" of industrial nitromethane at atmospheric pressure. Induction time t^l, critical diameter ϕ_c in a glass tube. (N.B. Only the relative variation is to be used in the last two lines.)

T_0 (C)	$\hat{p}$ (kbar)	t^l (μs)(+)	ϕ_c (mm)(++)	References
−20			34.8	(+) Campbell [5]
−10			29	(++) Campbell et al.
0			24.5	quoted by
1.7		5		Enig and Petrone
6.3		1.5		Proc. 5th Symp.
10	86		20.9	on Detonation,
20			18	Pasadena CA.
26.7		0.8		(1970), p. 99.
30			15.7	
36.7		0.57		
40			13.3	
45.5		0.45		
25	86	2.26		
	89	1.74		

tion time t^l. Tables IX.3 and IX.4 provide convincing proof of the validity of certain experimental laws displayed in Table IX.5 and give a physical framework to the design of explosive devices. Thus:

- a device operating at normal temperature cannot operate at a very much lower temperature (ϕ_c and $\hat{p}_c$ increase when T_0 decreases);
- the choice of conditions ρ_0 and σ_0 of a "sensitive" device consists of finding a small enough ϕ_c without at the same time increasing $\hat{p}_c$ too greatly, the

Table IX.5. Direction of variation of critical diameter ϕ_c; critical pressure $\hat{p}_c$, induction time t^l, as a function of initial temperature T_0, packing density ρ_0 and specific surface σ_0.

	Inhomogeneous explosive T_0, ρ_0, σ_0		Homogeneous explosive $T_0, p_0, \rho_0(T_0, p_0)$	
	ϕ_c	$\hat{p}_c$	ϕ_c	t^l
$\dfrac{\partial}{\partial T_0}$	−	−	−	−
$\dfrac{\partial}{\partial \rho_0}$	− if σ_0 low O if σ_0 high	+	+	+
$\dfrac{\partial}{\partial \sigma_0}$	− if ρ_0 low O if ρ_0 high	+		

compromise being generally obtained for a small enough value of ρ_0 and a value of σ_0 that is not too great; and

• the choice of conditions ρ_0 and σ_0 of an "insensitive" device consists of raising $\hat{p}_c$ to a suitable level without at the same time increasing ϕ_c inordinately, a compromise being generally obtained for a large enough value of σ_0 and a value of ρ_0 that is not too small.

Table IX.5 also allows us to measure how much the term "sensitivity"—no longer applied to a molecule as in Chapters IV–VI, but to an *explosive substance*—is unsuitable and ambiguous, despite its everyday nature. In fact, it is as natural to consider that an explosive is much more "sensitive" the smaller its critical diameter ϕ_c or its critical pressure $\hat{p}_c$. But, within one group of pressed explosives with the same ρ_0 and T_0, ϕ_c and $\hat{p}_c$ vary in opposite directions when σ_0 increases, so that those with a larger specific surface will be called "more sensitive" or "less sensitive" according to whether we rely on ϕ_c or $\hat{p}_c$ to judge the sensitivity! This apparent lack of coherence results from the fact that the parameters ϕ_c and $\hat{p}_c$ do not refer to the some phenomenon: the former results from the analysis of the propagation conditions of a built-up detonation (see §III.6.3) while the latter results from the analysis of the detonation build-up conditions (see Section IX.3). In fact, their value and their variation is the outcome of the coupling of the reactivity S to the flow under two different sets of boundary conditions. Hence, the importance attaching to a standardized formulation of S, already broached in §VIII.4.3 and brought to a conclusion in the following section.

3. Reactivity of an Explosive Substance

3.1. Unified Model

(See Chéret, *Comptes Rendus Acad. Sciences, Paris*, vol 306, Series II (1988), p. 863.)
In order for the aggregative rate law introduced in §VIII.4.3 to be completely formulated in terms compatible with the continuum model of §VIII.2, one further stage remains to be examined (see §VIII.4.3); this consists in expressing the temperature $\hat{T}$ as a function of quantities linked to the macroscopic and mesoscopic states which prevail downstream of the ignitor shock $\hat{\Sigma}$.

Let ∂D be the boundary, with external normal $\mathbf{n}$, of a connected explosive domain D in the downstream neighborhood of $\hat{\Sigma}$. The temperature $\hat{T}$ which prevails on this boundary is the result of thermal exchanges between the interstitial medium where hot sources ("hot spots") originate and domain D proper. Taking account of the fact (see §1.3) that the relative normal velocity $\mathbf{W}$ and the pressure remain stationary in the downstream neighborhood of $\hat{\Sigma}$ as long as the decomposition remains insignificant, and neglecting the components of grad T normal to $\mathbf{n}$ vis-à-vis the component along $\mathbf{n}$, one may, in each medium, write the energy equation (I.28c) of the relative movement in the form:

$$\rho(\mathbf{W} \cdot \mathbf{n}) c_p \frac{dT}{d\zeta} = -\lambda \frac{d^2 T}{d\zeta^2}. \qquad (IX.2)$$

Let us denote the neighborhood of ∂D by a length $\hat{L}$ defined later and introducing the reduced distance $\bar{\zeta} = \zeta/\hat{L}$ and the Péclet numbers $\hat{\mathscr{P}_e}$ and $\hat{\mathscr{P}_e'}$ respectively relative to the downstream state of the explosive medium and of the interstitial medium; then (IX.2) is written

$$\mathscr{P}_e \frac{dT}{d\bar{\zeta}} = -\frac{d^2 T}{d\bar{\zeta}}, \qquad \mathscr{P}_e = \hat{\mathscr{P}_e} \ \text{ or } \ \hat{\mathscr{P}_e'},$$

which, in the approximation $\mathscr{P}_e = $ constant, is integrated in each medium in the form

$$\frac{T(\bar{\zeta}) - T(0)}{\theta} = \frac{1}{\mathscr{P}_e}[1 - \exp(-\mathscr{P}_e \bar{\zeta})], \qquad \mathscr{P}_e = \hat{\mathscr{P}_e} \ \text{ or } \ \hat{\mathscr{P}_e'} \qquad (IX.3)$$

$$\theta = \dot{T}_0 \ \text{ or } \ \dot{T}_0'$$

It remains to choose $\hat{L}$.

The first idea consists of identifying $\hat{L}$ with $\hat{L}$ already introduced in §VIII.4.3. Such a choice would however be erroneous, as can be seen by taking into account the effects of comminution under shock in Graham's papers [14]. Consideration of these effects leads, *a contrario*, to constructing $\hat{L}$ from the specific surface $\hat{\sigma} \gg \sigma_0$ realized downstream of the ignitor shock; in another connection, the nullity of $\hat{L}$ required for the homogeneous state leads to choosing $\hat{L}$ proportional to the initial *porosity* π_0, defined as the difference $1 - Y_0 \alpha_0$ (α_0 compacity of the aggregate; Y_0 mass fraction of the explosive in the aggregate): finally, one is led to writing

$$\hat{L} = \pi_0/\hat{\rho}\hat{\sigma},$$

which may be interpreted as the average depth of the interstitial medium in the downstream neighborhood of $\hat{\Sigma}$. Taking this physical significance into account, it seems reasonable to admit that:

(i) everything happens as though the interstitial medium were the source of a hot spot $\tilde{T}$ situated at an infinite distance $\bar{\zeta}$ from ∂D; and
(ii) the temperature excess $\hat{T} - \hat{T}$ in the explosive medium affects only the zone of unit thickness: $-1 \le \bar{\zeta} \le 0$.

Then the integral (IX.3) gives

$$\hat{T} - \hat{\hat{T}} = \frac{\dot{T}(0)}{\hat{\mathscr{P}_e}}[1 - \exp(+\hat{\mathscr{P}_e})], \quad (a)$$

$$\tilde{T} - \hat{\hat{T}} = \frac{\dot{T}'(0)}{\hat{\mathscr{P}_e'}}. \qquad\qquad (b)$$

$$(IX.4)$$

In another connection, the continuity of the normal component of the heat flux vector across ∂D

$$\hat{\lambda}\dot{T}(0) = \hat{\lambda}'\dot{T}'(0)$$

allows us to eliminate the derivatives $\dot{T}(0)$ and $\dot{T}'(0)$ in (IX.4) and obtain

$$\hat{\bar{T}} = \hat{T}\frac{1 - \dfrac{\tilde{T}}{\hat{T}}\dfrac{\hat{\lambda}'\hat{\mathscr{P}}_e'}{\hat{\lambda}\hat{\mathscr{P}}_e}(1 - \exp\hat{\mathscr{P}}_e)}{1 - \dfrac{\hat{\lambda}'\hat{\mathscr{P}}_e'}{\hat{\lambda}\hat{\mathscr{P}}_e}(1 - \exp\hat{\mathscr{P}}_e)}. \tag{IX.5}$$

Making use of the continuity the normal mass flux across $\hat{\Sigma}$ and of the uniform state of rest upstream of $\hat{\Sigma}$, we can write (N normal to $\hat{\Sigma}$ oriented from upstream toward downstream)

$$(\widehat{\rho\mathbf{W}}) = (\widehat{\rho\mathbf{W}})' = \rho_0\,U\mathbf{N},$$

from which we have the equality

$$\frac{\hat{\lambda}'\hat{\mathscr{P}}_e'}{\hat{\lambda}\hat{\mathscr{P}}_e} = \frac{\hat{c}_p'}{\hat{c}_p}.$$

This relation, taking account of well-known laws on c_p, is scarcely different from unity, so that an approximate quantitative expression of $\hat{\bar{T}}$ is

$$\hat{\bar{T}} \simeq \hat{T}\frac{1 + \dfrac{\tilde{T}}{\hat{T}}[\exp\hat{\mathscr{P}}_e - 1]}{\exp\hat{\mathscr{P}}_e}, \tag{IX.6}$$

which may further be written as

$$\frac{\hat{\bar{T}}}{\hat{T}} = 1 + [1 - \exp(-\hat{\mathscr{P}}_e)]\left[\frac{\tilde{T}}{\hat{T}} - 1\right]. \tag{IX.6'}$$

Furthermore, an examination of the specialist literature (see Ref. [18] of Chap. V) and the exponential variation of λ with pressure (see Ref. [11] of Chap. VII) suggest the following orders of magnitude:

$$\hat{c}_p \simeq 0.3 \text{ cal/g} \cdot \text{K},$$

$$\hat{\lambda} \simeq 10\lambda_0 \simeq 0.02 \text{ cal/cm} \cdot \text{s} \cdot \text{K},$$

so that an approximate average value of $\hat{\mathscr{P}}_e$ is

$$\hat{\mathscr{P}}_e \simeq 150 \cdot \pi_0 \cdot \frac{U}{\hat{\sigma}} \cdot \langle\mathbf{n}\cdot\mathbf{N}\rangle \qquad (U \text{ in mm/}\mu\text{s}, \hat{\sigma} \text{ in m}^2/\text{g}), \tag{IX.7}$$

where $\langle\mathbf{n}\cdot\mathbf{N}\rangle$, is the average cosine of $\mathbf{n}$ relative to $\mathbf{N}$.

The five factors governing $\hat{\bar{T}}$ now clearly appear in the approximate formulas (IX.6) and (IX.7). In increasing order of complexity these are:

- initial porosity π_0;
- induced shock velocity U;

Table IX.6. Values of $\hat{\tilde{T}}/\hat{T}$ for realistic values of $\tilde{T}/\hat{T}$ and $\hat{\mathscr{P}}_e$ and for $\langle \mathbf{n} \cdot \mathbf{N} \rangle = 2/3$.

$\tilde{T}/\hat{T}$ \ $\hat{\mathscr{P}}_e$	0	1	2	∞
1	1	1	1	1
2	1	1.63	1.86	2
3	1	2.26	2.73	3

- specific surface $\hat{\sigma}$ after comminution by induced shock;
- the ratio $\tilde{T}/\hat{T}$ of hot spot temperature to shock temperature; and
- the average $\langle \mathbf{n} \cdot \mathbf{N} \rangle$ of the cosine of $\mathbf{n}$ relative to $\mathbf{N}$.

As much as the first two appear accessible, the last three factors appear outside the range of any measurement or theory, and must merit being treated as floating parameters. The few figures presented in Table IX.6 show the extreme sensitivity of possible variations, considering that S'' is proportional to $\exp(-\hat{\tilde{T}}/T^{\neq})$.

Allowing for the possibility that the state outside thermal equilibrium, which underlies the above analysis of the temperature field, ceases to exist from the initial decomposition, i.e., from when $m > 0$, then the preceding arguments added to those in §VIII.4.3 justify a unified formulation of the reactivity $S = (1 - m)^{-1} \, dm/dt$ of an explosive substance, whether it be homogenous or inhomogeneous. For the convenience of the discussion, we regroup them below:

$$\hat{\mathscr{P}}_e = 150 \frac{\pi_0}{\hat{\sigma}} U \langle \mathbf{n} \cdot \mathbf{N} \rangle \quad \text{(U in mm/\mu s, \hat{\sigma} in m}^2\text{/g),}$$

$$\hat{\tilde{T}}/\hat{T} = 1 + [1 - \exp(-\hat{\mathscr{P}}_e)] [(\tilde{T}/\hat{T}) - 1],$$

$$\frac{1 - 2m_1}{(1 - m_1)^2} = \exp(-T^{\neq}/\hat{\tilde{T}}) \ni \,]0, 1[,$$

$$S' = Z \exp(-T^{\neq}/\hat{T}),$$

$$S'' = \rho_0 \sigma_0 U \frac{1 - 2m_1 + m}{(1 - m_1)^2},$$

$$S = S' + S'',$$

(IX.8)

It is satisfactory to state that the total reactivity depends on:

- the induced shock velocity U;
- parameters inherent to the molecule (frequency factor Z and activation temperature $T^{\neq}$);

- parameters inherent to the initial macroscopic or mesoscopic state (volume mass ρ_0, porosity π_0, and specific surface σ_0); and
- parameters inherent to the downstream state of the induced shock (temperature $\tilde{T}$ of hot spot, shock temperature $\hat{T}$ of compact material, specific surface $\hat{\sigma}$, and orientation factor $\langle \mathbf{n} \cdot \mathbf{N} \rangle$).

It is equally satisfactory to state that the lack of porosity ($\pi_0 = 0$) leads to a molecular type of initial reactivity; such a limiting behavior explains simply the phenomenon of "desensitization" of a pressed explosive by a weak shock, established by Campbell [6] and by many subsequent authors: compaction by a weak shock suppresses the initial porosity, so that the compressed explosive reacts to a subsequent shock like a homogeneous explosive and, in particular, presents a higher critical pressure than that belonging to the initial state.

3.2. Lagrangian Analysis

A complete view of the consequences of the unified formulation (IX.8) can emerge only by a numerical investigation of the coupling between decomposition and motion in a broad spectrum of experimental arrangements. However, without going as far as the complete numerical solution of the system (VIII.8), one can have a semiquantitative view of it by applying it to a series of highly instrumented experiments reported by Setchell [32] on hexanitrostilbene powders compressed at $\rho_0 = 1.60 \pm 0.01$ g/cm^3 ($\alpha_0 = 0.92$; $Y_0 = 1$; $\pi_0 = 0.08$): each base powder (HNS-FP; HNS-I; HNS-II) has a known characteristic specific surface (8.1 m^2/g; 3.9 m^2/g 0.94 m^2/g) and has been tested under three stresses referenced by the pressure $\hat{p}$ of the induced shock (25 kbar; 30 kbar; 34 kbar) and the material induced velocity $\hat{u}$ (330 m/s; 390 m/s; 430 m/s), from which the velocity $U = \hat{p}/\rho_0 \hat{u}$ of the induced shock is deduced.

For each of the experimental arrangements, Table IX.7 gives the values of the ratio $\tilde{T}/\hat{T}$ and $\hat{T}$ (column 1), $\rho_0 \sigma_0 U s^{-1}$ (column 2), and initial aggregate reactivity (column 3) estimated from

$$
\begin{cases}
Z = 1.53 \cdot 10^9 & \text{(see [26]),} \\
T^{\neq} = 15{,}250 \text{ s}^{-1} & \text{(see [26]),} \\
\hat{T} = 400, 422, 440 \text{ K} & \text{(see [33]),} \\
\tilde{T}/\hat{T} = 1.6 & \text{(see [24]);} \\
\hat{\sigma} = 10\,\sigma_0 & \text{(see [14]),} \\
\langle \mathbf{n} \cdot \mathbf{N} \rangle \simeq 2/3 & \text{(spherical grains)}
\end{cases}
$$

It is important to note that the frequency factor Z in the molecular rate law (1.53×10^9 s^{-1}) is less than the pseudofrequency factor $\rho_0 \sigma_0 U$ of the aggregate rate law in the nine figured cases, with the result that the initial molecular reactivity is always less than the initial aggregate reactivity.

Table IX.7. For each experimental arrangement, values of $\hat{\hat{T}}/\hat{T}$ and $\hat{T}$ (column 1), $\rho_0\sigma_0 U \, 10^{-9} \, s^{-1}$ (column 2), and initial aggregate reactivity in s^{-1} (column 3).

	HNS-FP $\sigma_0 = 8.1 \, m^2/g$			HNS-I $\sigma_0 = 3.9 \, m^2/g$			HNS-II $\sigma_0 = 0.94 \, m^2/g$		
	(1)	(2)	(3)	(1)	(2)	(3)	(1)	(2)	(3)
$U = 4.7$ mm/μs $\hat{T} = 400$ K	1.223 / 489	60.9	0.003	1.371 / 548	29.3	0.024	1.589 / 636	7.1	0.274
$U = 4.8$ mm/μs $\hat{T} = 422$ K	1.226 / 547	62.2	0.010	1.376 / 581	30.0	0.112	1.590 / 671	7.2	0.973
$U = 4.9$ mm/μs $\hat{T} = 440$ K	1.230 / 541	63.5	0.036	1.380 / 607	30.6	0.376	1.591 / 700	7.4	2.561

Table IX.7 provides a simple interpretation of Setchell's recordings shown in Figure IX.7:

- in the five cases where ignition is followed by propagation ((I.30), (I.34), (II.25), (II.30), (II.34)), the initial value S'' is greater than 0.1 s^{-1} while it is less than 0.1 s$^{\pm 1}$ in the other four cases;
- in the above five cases, the induction time (defined in §1.3 as the time at the end of which the maximum velocity is registered) is shorter the greater the initial value of S''; an analogous correlation with $\hat{T}$ is not verified; and
- in the two cases of ignition (II.30) and (II.34) where induction comprises a stage of ultrafast increase (≈ 10 ns), the initial value of S'' is at least of the order of one, while it is considerably less in the three other cases of propagation.

The values of $S''(m = 0)$ thus determined illustrate the complex role played by the degree of division of the explosive substance: the dependence of $\hat{T}$ on $\hat{\sigma}$ means that too much initial division makes *ignition* more difficult (see Table IX.3: $\hat{p}^i$ increases when σ_0 increases), while the proportionality of S'' to σ_0 means that too little initial division makes *build-up* more difficult.

In order not to allow the reader to commit himself to a hasty generalization, it is necessary to dwell on the fact that hexanitrostilbene is an exception: for current explosives, Z is greater than the largest realistic values of $\rho_0 \sigma_0 U$ (10^{11} s^{-1}):

TNT $2.5 \cdot 10^{11}$,   RDX $2.015 \cdot 10^{18}$,

TATB $3.18 \cdot 10^{19}$,   HMX $5 \cdot 10^{19}$,   PETN $6.3 \cdot 10^{19}$.

Thus, for a given explosive prepared under $\langle \rho_0, \sigma_0, \pi_0 \rangle$ conditions, there generally exists a particular value of the shock pressure $\hat{p}$ such that the initial molecular and aggregate reactivities are equal. This value separates "low" pressures, where the initial reactivity is essentially aggregate, from "high" pressures where the initial reactivity is essentially molecular. For reasons which relate both to this partitioning role and the very notion of induction period (see Section VII.2 and §IX.1.3), we call this value *critical induction pressure* and denote it by $\hat{p}^l_c$; obviously, we have

$$\hat{p}^i < \hat{p}_c < \hat{p}^l_c.$$

The partition thus realized among the ignitor shocks explains the majority of the observations reported by Dremin and Shvedov [10] concerning TNT

		$\hat{p}^l_c$ (kbar)
liquid TNT		135
cast TNT		120
paraffined TNT	($\rho_0 = 1.55$–1.65 g/cm^3)	110
pressed TNT	($\rho_0 = 0.9$ g/cm^3)	70

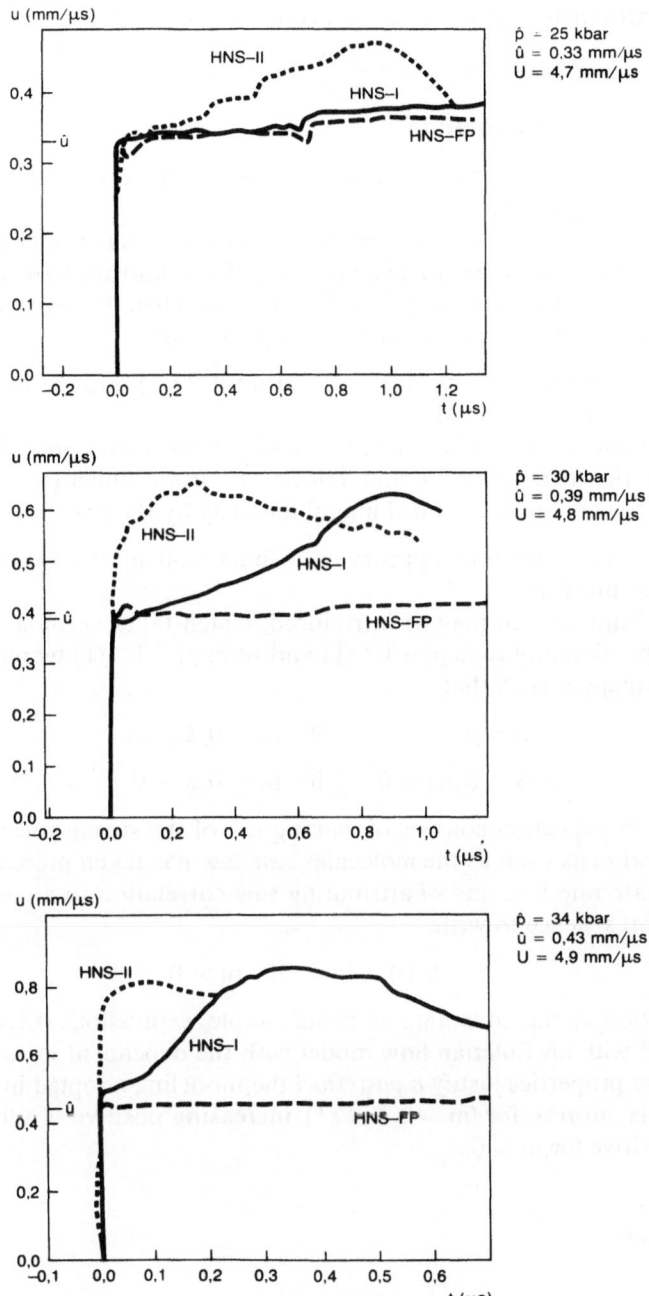

Figure IX.7. Evolution with time of velocity of rear face of a 2 mm thick HNS disc. (N.B. The zero instant corresponds to the arrival of $\hat{\Sigma}$ on this face.) After Setchell [32].

but moderates their conclusion as to the impossibility of "advancing a unified theory."

3.3. Eulerian Approximation

Fortunately, it is not always necessary to consider the total reactivity S in its explicit and complete form (IX.8).

First, the analysis of §3.2 shows that one may expect a reasonable approximation by assigning values to $\tilde{T}/\hat{T}$ and $\hat{\sigma}/\sigma_0$, if not universally valid then at least consistent within a given group of explosives. Then the reactivity $S(M, t)$ for the particle $M(\rho_0, \sigma_0, \alpha_0)$ at instant t depends on:

- the instantaneous state by the temperature $T(M, t)$ and the decomposed mass fraction $m(M, t)$; and
- the downstream state of the ignitor shock which M has gone through in terms of the temperature $\hat{T}$ and velocity U, from which (see §II.1.5) we know that they are determined unambiguously by the specific entropy $\hat{s}$.

Lagrangian reactivity thus appears as a functional of the variables $\hat{s}(M)$, $T(M, t)$, and $m(M, t)$.

A first simplification may be introduced, which takes account, at $m = 0$, of the proportionaiity to $\exp(-T^{*}/\tilde{T})$ and to $\exp(-T^{*}/\hat{T})$ by means of an ignition entropy s^i such that

$$S = 0 \qquad \text{if} \quad m = 0, \hat{s} \leq s^i,$$
$$S = S_0(s) > 0 \qquad \text{if} \quad m = 0, \hat{s} > s^i. \qquad \text{(IX.9a)}$$

A second simplification consists of making use of the strong correlation between T and m as soon as the molecular rate law has taken precedence over the aggregate rate law, and of attributing this correlation to a common dependence on s, hence to write

$$S = S_1(s) > 0 \qquad \text{if} \quad m \gg 0. \qquad \text{(IX.9b)}$$

In addition to the advantage of being simple, expressions (IX.9) are also compatible with an Eulerian flow model with the dependent variables (U, v, s, m). These properties justify *a posteriori* the modeling adopted in §III.1.3: a function $S(s, m)$ zero for $(m = 0, s \leq s^1)$, increasing positive of s for $(m = 0, s > s^i)$, positive for $m > 0$.

References

[1] ANDERSON, A.B. *et al.* Shock initiation of porous TATB. *Proc. 7th Symposium on Detonation*, Annapolis/MD (1981), p. 385.
[2] AVEILLÉ, J. *et al.* Célérité de détonation et profondeur d'amorçage de deux compositions explosives. *Proc. Colloque Pyrotechnie Fond. et Ap.*, Arcachon/France (1982), p. 396

[3] BAHL, K.L., VANTINE, H.C., WEINGART, R.C. Shock initiation of bare and covered explosives by projectile impact. *Proc. 7th Symposium on Detonation*, Annapolis/MD (1981), p. 325.

[4] BERNIER, H. Contribution à l'étude de la génération de la détonation provoquée par impact sur un explosif. Thèse de Doctorat ès Sciences, Paris (1964).

[5] CAMPBELL, A.W., DAVIS, W.C., TRAVIS, J.R. Shock initiation of detonation in liquid explosives. *Phys. Fluids*, 4 (1961), p. 498.

[6] CAMPBELL, A.W., DAVIS, W.C., RAMSAY, J.B., TRAVIS, J.R. Shock initiation of solid explosives. *Phys. Fluids*, 4 (1961), p. 511.

[7] CAMPBELL, A.W., ENGELKE, R. The diameter-effect in high density heterogeneous explosives. *Proc. 6th Symposium Detonation*, Coronado/CA (1976), p. 642.

[8] CHAIKEN, R.F. Comments on hypervelocity wave phenomena in condensed explosives. *J. Chem. Phys.*, 33 (1960), p. 760.

[9] COWPERTHWAITE, M., ROSENBERG, J.T. A multiple Lagrange gage study of the shock initiation process in cast TNT. *Proc. 6th Symposium on Detonation*, San Diego/CA (1976), p. 786.

[10] DREMIN, A.N., SHVEDOV, K.K. On shock wave explosive decomposition. *Proc. 6th Symposium on Detonation*, Coronado/CA (1976), p. 29.

[11] DROUX, R., MOUCHEL, C. Étude du comportement sous choc et de la génération de la détonation d'explosifs hétérogènes par la méthode du double coin. *Proc. Symposium H.D.P.*, Paris (1978), p. 103.

[12] ERICKSON, L.M. et al. The electromagnetic velocity gauge. *Proc. 7th Symposium on detonation*, Annapolis/MD (1981), p. 1062.

[13] FAUQUIGNON, C., CHÉRET, R. Generation of detonation in solid explosives. *Proc. 12th Symposium of the Combustion Institute*, Poitiers/France (1968), p. 745.

[14] GRAHAM, R.A. Shock-induced inorganic chemistry. *Proc. APS Shock Waves Meeting*, Menlo Park/CA (1981), p. 4.

[15] GREEN, L. Shock initiation of explosives by the impact of small diameter cylindrical projectiles. *Proc. 7th Symposium on Detonation*, Annapolis/MD (1981), p. 273.

[16] HAYES, D.B. A $p^n t$ detonation criterion from thermal explosion theory. *Proc. 6th Symposium Detonation*, San Diego/CA (1976), p. 76.

[17] HOLLAND, T.E., CAMPBELL, A.W., MALIN, M.E. Phenomena associated with detonation in large single crystals. *J. Appl. Phys.*, 28 (1957), p. 1212.

[18] HONODEL, C.A. et al. Shock initiation of TATB formulations. *Proc. 7th Symposium on Detonation*, Annapolis/MD (1981), p. 425.

[19] HOWE, P. et al. Shock initiation and the critical energy concept. *Proc. 6th Symposium on Detonation*, Coronado/CA (1976), p. 11.

[20] HUBBARD, H.W., JOHNSON, M.H. Initiation of detonations. *J. Appl. Phys.*, 30 (1959), p. 765.

[21] KANEL, G.I., DREMIN, A.N. Decomposition of cast trotyl in shock waves. *Combustion, Explosion and Shock Waves*, 13 (1977), p. 71.

[22] KENNEDY, J.E. Pressure-field in a shock-compressed high explosive. *Proc. 14th Symposium of the Combustion Institute*, Pittsburgh/PA (1972), p. 125.

[23] LINDSTROM, I. E. Plane shock initiation of an RDX plastic bonded explosive. *J. Appl. Phys.*, 37 (1966).

[24] LONGUEVILLE, Y. de et al. Initiation of several condensed explosives by a given duration shock wave. *Proc. 6th Symposium on Detonation*, San Diego/CA (1976), p. 105.

[25] MADER, C.L. The two-dimensional hydrodynamic hot spot. Vol. II. LASL Report LA 3235 (1965).

[26] MADER, C.L. *Numerical Modeling of Detonations.* University of California Press, Berkeley (1979).

[27] MADER, C.L., FOREST, C.A. Two-dimensional homogeneous and heterogeneous detonation propagation. LASL Report LA 6259 (1976).

[28] MOULARD, H. Critical conditions for shock initiation of detonation by small projectile impact. *Proc. 7th Symposium on Detonation*, Annapolis/MD (1981), p. 316.

[29] PRICE, D. Shock sensitivity, a property of many aspects. *Proc. 5th Symposium on Detonation*, Pasadena/CA (1970), p. 207.

[30] RAMSAY, J.B., POPOLATO, A. Analysis of shock wave and initiation data for solid explosives. *Proc. 4th Symposium on Detonation*, White Oak/MD (1965), p. 233.

[31] ROTH, J. Shock sensitivity and shock Hugoniots of high-density granular explosives. *Proc. 5th Symposium on Detonation*, Pasadena/CA (1976), p. 219.

[32] SETCHELL, R.E. Microstructural effects in shock initiation of granular explosives. *Proc. Symposium on Pyrotechnics and Explosives*, Beijing (1987), p. 635.

[33] SHEFFIELD, S.A., MITCHELL, D.E., HAYES, D.B. The equation of state and chemical kinetics of HNS explosive. *Proc. 6th Symposium Detonation*, Coronado/CA (1976), p. 748.

[34] TAYLOR, B.C., ERWIN, L.W. Separation of ignition and build-up to detonation in pressed TNT. *Proc. 6th Symposium on Detonation*, Coronado/CA (1976), p. 3.

[35] VORTHMAN, J., ANDREWS, G., WACKERLÉ, J. Reaction rates from electromagnetic gauge data. *Proc. 8th Symposium on Detonation*, Albuquerque/NM (1985), p. 99.

[36] WACKERLÉ, J., JOHNSON, J.O., HALLECK, P.M. Shock initiation of high density PETN. *Proc. 6th Symposium on Detonation*, San Diego/CA (1976), p. 20.

[37] WALKER, F.E., WASLEY, R.J. Critical energy for shock initiation of heterogeneous explosives. UCRL Report 70891 (1968).

[38] WEINGART, R.C. *et al.* Acceleration of thin flyers by exploding metal foils: application to initiation studies. *Proc. 6th Symposium on Detonation*, San Diego/Ca (1976), p. 653.

[39] WEINGART, R. C. *et al.* Magnetic stress gages in reacting high explosive environment. *Proc. Symposium H.D.P.*, Paris (1978), p. 451.

Part Four
The Dynamic Characterization
of Explosives

CHAPTER X

Experimental Methods

*The specificity of the experimental methods necessary for the dynamic charac-
terization of explosives basically depends on two of the properties of the physical phenomena involved: their transitory nature and their destructive character.
In fact, the total duration of an experiment rarely exceeds a few microseconds,
and the signals collected at the level of the assembly must be transmitted in real
time and over a distance of several meters before being recorded in an appropriate shed.*

*Furthermore, the orders of magnitude in such experiments are very unusual:
pressures of tens or hundreds of kilobars, speeds of hundreds or thousands of
meters per second, temperatures in hundreds or thousands of degrees. To meet
these needs, which are both diverse and specific, many techniques have been
devised, which are often used together. Because of the size of the subject, Chapter X is devoted to principles of instrumentation, with Chapter XI illustrating
them for elementary configurations of simple detonation. In our arrangement,
we have used classification into optical measurements (Section X.1), electronic
measurements (Section X.2), and radiographic measurements (Section X.3), and
devoted Section X.4 to stress generators. As our aim is more didactic than
practical, the reader in need of methods and details of equipment should refer
to the numerous specialized publications, in particular, to the* Proceedings of
the First European Congress on Cineradiography Using Photons and Particles, *published in 1981 by Jacques Marilleau under the aegis of the Society of
Photo-Optical Instrumentation Engineers (SPIE).*

*This synthesis would not have been possible without the friendly assistance
of Jean Aveillé, Claudine Bizeuil, Jean Bourguignon, Pierre Chapron, Philippe
Elias, Bertrand Laurent, Jacques Leyris, Roland Loichot, Jacques Montrosset,
Jean Perraud, Serge Roux, Bernard Udiment, and Patrick Vibert, who have
given us valuable knowledge gained from their practical daily work in the firing
area.*

1. Optical Techniques

Optical techniques allow high temporal resolutions to be reached, with easy
transfer of the collected signals over several meters, and measurements to be

taken without disturbing the flows being studied. These three reasons explain why considerable progress is being made with optical techniques in detonation experiments. Among the numerous techniques which have thus been developed, three main families can be distinguished at present:

— high-speed cinematography and photography;
— slit scanning cameras and associated optical pick-ups; and
— velocimetry by laser interferometry.

1.1. Light Sources

For every optical technique, there is the problem of choosing the light source. This problem is particularly characteristic in the case of measurements in detonics, because of the very short duration of the phenomena to be recorded, and the significant distance between the experimental assembly and the recording shed. These two factors thus require recourse to light sources which are very intense but which can function in a pulsed manner.

Here we will leave on one side the particular case of interferometry techniques for velocimetry, for which the necessity of coherent light indicates the use of lasers which can be considered as integral parts of the measuring equipment itself.

For photographic or cinematographic techniques—which may use integral images or slit scanning—the experimenter is often helped by the intense autoluminescence of the phenomena studied. However, in some cases one cannot count on this providential aid, and must then resort to auxiliary sources of lighting, which generally result from luminous phenomena linked to ionization of gases, either by shock (explosive flash technique) or by electrical discharge (electronic flash technique). According to circumstances, using some possible developments, they can be used with reflection from the object being studied, or with transmission (observation by transparency) or with occultation (observation by shadow).

(a) Explosive Flash

It has been known for a long time that the passage of an intense shock wave in a gas causes considerable heating of the latter, which can lead to its total or partial ionization behind the shock front. This ionization is accompanied, because of the free–free transitions (Bremsstrahlung) and because of the recombination radiation emitted at the time of the free–bound transitions, by the emission of a continuous spectrum of high luminosity. This phenomenon was very soon used to produce intense luminous sources [22].

As monoatomic rare gases have a very small number of degrees of freedom, they ionize very readily under shock. Among them, xenon and krypton have the weakest ionization potential and thus allow, for equal shocks, the most intense luminous emissions. However, argon is generally preferred, for

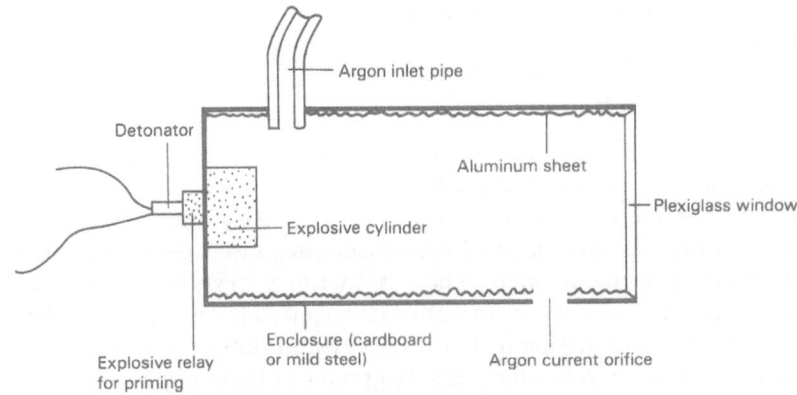

Figure X.1. Diagram of an explosive flash.

reasons of cost. To cause ionization, a block of solid explosive is generally used as shock generator, the duration of emission from the source being regulated by the length of the reservoir of argon in which the shock is propagated.

Diverse models of more or less sophisticated explosive flashes have been built on this principle. However, the most classically used source is that shown in Figure X.1. It is formed from an enclosure made from cardboard or mild steel sheet, closed on the front face by a transparent plexiglass window. A regular cylinder of fast explosive, usually weighing about 500 g, is fixed onto the opposite face, within the enclosure. This cylinder is primed across the wall of the enclosure by a small relay cylinder of the same explosive, which in turn is fired by a detonator. A sheet of aluminum foil covers the internal wall of the enclosure to reduce light absorption.

A few minutes before carrying out the experiment, the air contained in the enclosure is expelled by a permanent current of argon passing through two orifices cut in the enclosure. The synchronization of the luminous emission with the phenomenon studied is regulated by electronic delay generators on the firing of the experimental assembly and the explosive flash(es), respectively.

This device, placed some tens of centimeters from the object to be observed, provides a source of light which is broadly sufficient for the needs of high-speed cinematography and photography.

This source, while relatively cheap and very easy to use in the firing area, still presents some problems. In particular, because of the propagation and weakening of the shock transmitted through argon, the luminous emission is variable over time in position, intensity, and color temperature, which is sometimes a nuisance in cinematography. Furthermore, one is dealing with a nonuniform extended source. This last fault can, however, be reduced by the interposition of diffusing screens in front of the transparent window, which can, for example, provide uniform luminous backgrounds for observation by shadow.

(b) Electronic Flash

The first electronic flash devices used in high-speed photography to produce intense and very brief sources of light benefitted from an electric spark produced in the air between two electrodes [3]. At present, the commonest systems utilize the discharge of condensers in pulsed discharge tubes filled with a rare gas (most often xenon, sometimes krypton or argon). These systems function similar to the photographer's classic electronic flash, the principal difference arising from the energy and luminous (and thus electric) power, which have to be considerably greater in these applications.

The actual luminous source is of very simple design: an elongated tube of glass or quartz (straight, U, or spiral) filled with a rare gas at low pressure (a few bars) and bearing an electrode at each extremity. The use of an external electrode for activation is not necessary here: the sharp commutation of the high voltage on the electrodes is enough to autoactivate the tube.

To control the shape and duration of the luminous impulse produced, a simple capacitor is not used but, instead, a line formed of "inductance-capacitance" cells connected in a ladder, whose impedance is matched to that of the discharge circuit. An example of a possible assembly [20] is given in Figure X.2.

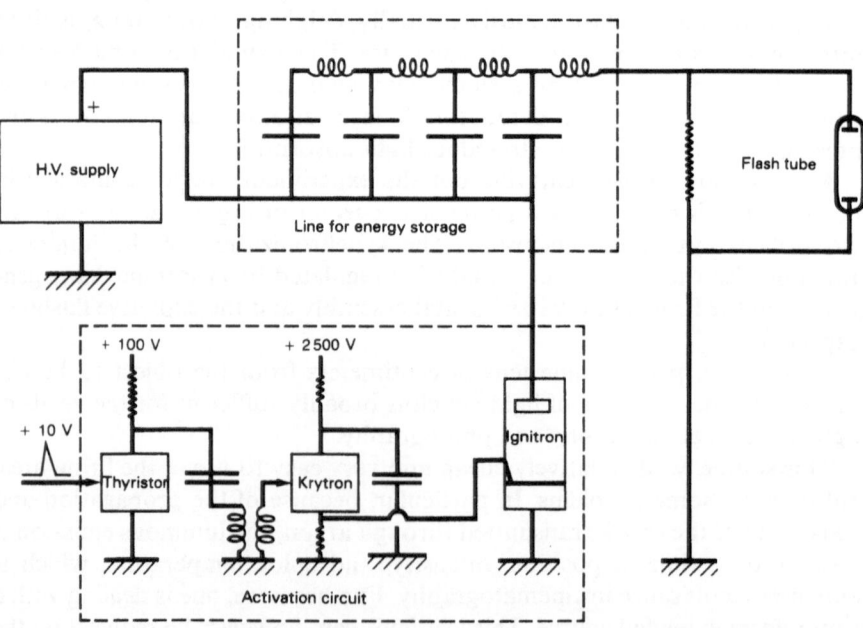

Figure X.2. Diagram of an electronic flash system.

Using a storage line with a capacity of a few microfarads it is possible, with a voltage of 10–20 kV, to store something of the order of 1 kJ of electrical energy. For a discharge circuit with matched impedance $Z = \sqrt{L/C}$, an intensity is obtained in the flash tube which has a form close to a pulse with duration $\mathscr{T} = 2\sqrt{L/C}$. Since the impedance of the tubes is several ohms, this fixes the value of the equivalent self-induction of the line to a few tens of microhenries and the duration of the discharge to a few tens of microseconds. Since the efficiency of the luminous conversion of the tube is of the order of 30%, one can therefore obtain a flash which is amply sufficient for a high-speed shot when the source tube is a few tens of centimeters from the observed experimental assembly.

1.2. High-Speed Cinematography and Photography

These techniques essentially correspond to the needs of global visualization of the detonation phenomena and of tracking the position and shape of material surfaces and wave surfaces.

The use of photography in detonics is characterized by the necessity of obtaining very short exposure times. In fact, the displacement velocities commonly encountered are of the order of several millimeters per microsecond. Since the size of the objects studied is usually about ten centimeters, the importance can be seen of the kinetic blur on the image if the exposure times used are not definitely less than a microsecond. Furthermore, since the total duration of the phenomena under study is most often of the order of several microseconds, the framing rate in cinematography must exceed one million images per second. Such performance is well outside the capacity of conventional cameras and photographic equipment, and has thus forced the development of specific materials.

(a) High-Speed Shutters

In order to succeed in taking shots of a detonation phenomenon, the first thought was to fit a conventional photographic chamber with a high-speed shutter. Since the performance of mechanical shutters was limited to a few tenths of a millisecond, in order to gain the thousandfold increase in exposure time, it was necessary to conceive new shutter systems by looking to different principles. However, it must be said that these shutter techniques are now of little more than historic interest, since cinematography has widely supplanted photography. We only mention them here because of the important role that they played in their time.

One of the first families of shutters used the *Kerr effect*: the birefringence of certain substances (nitrobenzene, in particular) subjected to an electric field involves a path difference between the component of polarized light which is parallel to the electric field and the component which is perpendicular. If the incident beam is polarized at 45° relative to the electric field, for a characteristic length covered in the electric field, then an output beam is obtained whose plane of polarization has been rotated by 90°. If an output polarizer (crossed) is placed at 90° relative to the input polarizer, then a shutter is obtained, since it only allows light to pass in the presence of an electric field. With this principle it is quite simple to obtain shutters with exposure times of the order of 10 ns.

Another way to obtain a high-speed shutter system consists of transforming the optical image into an electronic image by using an *image converter tube*, the commutation of the image being ensured by the fast electronics associated with the tube. The simplest tube which has been employed for this effect is the "bi-planar diode tube with uniform electric field" or "tube for proximity focusing" [13]. This tube has two flat parallel electrodes: a photocathode (usually of cesium) serving as entry converter (photon → electron) and an electroluminescent anode as exit converter (electron → photon).

The electrons emitted by the cathode are accelerated directly by the uniform electric field created between the two electrodes, without a deflection device. The shutter function is ensured by the simple commutation of the high voltage on the two electrodes, thus allowing exposure times of the order of one nanosecond, or even less, to be attained. The definition of the tube improves as the distance between the electrodes decreases and the voltage rises, and is therefore limited by the risk of breakdown. The luminous gain is proportional to the applied voltage, but generally remains low because of the limited aperture of the optics for returning the anodic image.

Other and more sophisticated tubes have been made by using electron multiplying stages and systems for moving the image by optical fibers and electron optics, but they can no longer be considered as simple shutters, and we will describe them in the chapter devoted to electronic cameras.

Furthermore, we will see that high-speed cameras with a rotating mirror risk recording superimposed images if the autoluminosity of the phenomenon studied lasts too long. It is then necessary to be able to shutter the field of view of the camera after the useful recording. The devices used for this effect can be compared to high-speed shutters which operate only on closing. Since the time available for their functioning is of the order of several microseconds, simple pyrotechnic artifices are commonly used for them, which are set out in the firing area and activated with a delay calculated to match the firing of the experimental assembly. Different methods are possible: opacifying a slab of glass on the optical path by shock waves created by two detonators placed on two opposite sides, or by destroying a mirror for returning the image by using a detonator placed behind the mirror, or by creating a smokescreen before the viewing window with two detonating fuses.

(b) Electronic Cameras

We have seen, when looking at the image converter tube, that it is possible to utilize the conversion of an optical image into an electronic image to provide

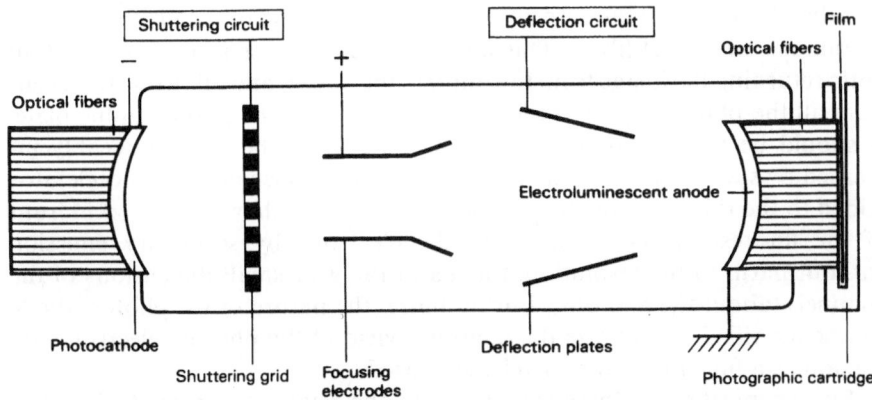

Figure X.3. Diagram showing the principle of a high-speed electronic camera.

a high-speed shutter function in front of a photographic chamber. But we can also, with this simple principle, make use of the possibilities offered by electron tubes (electron optics, amplification of brilliance, etc.) to produce a true high-speed electron camera which can record several images at rates higher than one million per second. Figure X.3 shows the principle of such a camera.

The camera tube is essentially composed of a photocathode, a shuttering grid, an electrostatic optic, and an electroluminescent anode (fluorescent screen). The photocathode is held at a constant high negative potential (some tens of kilovolts) with respect to the fluorescent screen, which is earthed. The strong acceleration thus communicated to the electrons allows amplification of the brilliance between the image entering the tube and the image leaving (of the order of several tens of times). The grid placed in front of the photocathode provides the shuttering function of the camera. Linked to a generator of square waves of very short duration, it ensures the very brief exposure times (from one microsecond down to a few nanoseconds) which are necessary for high-speed shots and thus the sequencing of the images in time (with rates reaching several tens of millions of images per second). The focusing electrodes, which are held at a fixed potential, form the electronic image of the photocathode on the fluorescent screen with the chosen magnification. Deflection plates, struck by voltage pulses synchronized to the opening instants of the shuttering grid, allow the images taken at different instants to be positioned in different places on the fluorescent screen. These juxtaposed images are thus recorded on the same film. It should be noted that, unlike television cameras, the image is not formed by line scanning, but deflected in its entirety.

Since the increase in the number of images acts to the detriment of their size, this type of camera generally leads to a small number of frames per recording.

The transfer of images between the entry screen and the photocathode, and then between the fluorescent screen and the plane of the film, is ensured by bundles of optical fibers. This device avoids the loss in luminosity of an optic returning the image from the fluorescent screen, and allows us to benefit from all the photonic gain of the tube at the level of the photographic plate. It should be noted that this gain can, if necessary, be very noticeably increased by such devices as amplifying the brilliance (second stage accelerator added to the tube) or a bundle of microchannels which, when placed in front of the fluorescent screen, multiply the electrons by secondary emission without harm to the resolution, thanks to the very small dimensions of the channels (about ten microns). Furthermore, the nature of the photocathode can be modified to optimize the quantum yield of the photon–electron conversion as a function of the wavelengths used.

Synchronization of these cameras with the phenomenon studied is very easy thanks to the incorporated electronic delay generators, which activate the shuttering with a delay preregulated to accord with an impulse from the firing.

The performance of the electronic cameras conceived in this way is of great interest in regard to exposure times and shooting speed, as well as luminous gain. However, despite the quality of the tubes and of the electronics, their definition remains somewhat limited, particularly because of the small size of the images obtained (at best some tens of pairs of lines per millimeter for an image of some 10 mm in diameter). The small number of available images, as well as the loss of chromatic information, are both serious handicaps in the use of these cameras.

(c) Rotating Mirror Cameras

In fact, the defects inherent in high-speed electron cameras—monochromatism, poor resolution, and above all the small number of images—have greatly limited their use in detonics, especially since mechano-optic cameras of high quality were soon developed which are perfectly adapted to the performances required for these experiments [5], [21].

In such a camera, a primary lens forms an image of the object studied in a plane passing through the axis of rotation of a plane mirror. The reflected image is not affected by any translation when the mirror rotates—but the bundle of rays which has formed the image and which is reflected by the mirror suffers a rotation double that of the mirror. A crown of lenses placed around the mirror serves to recapture the reflected image, for its different angular positions, and to reform it on a photographic film. The principle of such a device is shown in Figure X.4.

The rotation of the mirror, in projecting the reflected bundle in each lens in turn, ensures the separation of the images, both on the film and in time. The framing speed is fixed by the angular separation of the lenses and the speed of rotation of the mirror, while the exposure time is defined as the time during which the bundle penetrates into each lens. During the time of expo-

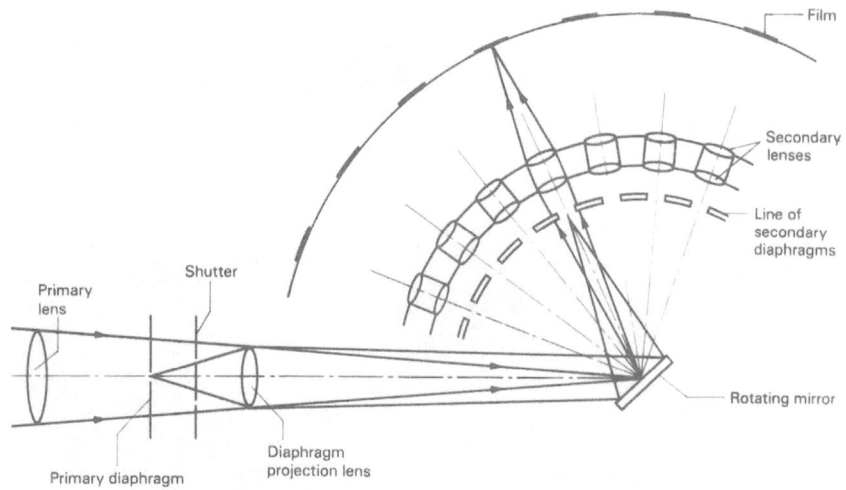

Figure X.4. Skeleton diagram of a high-speed camera with rotating mirror.

sure the image does not suffer any translation on the film, but a small rotation about an axis in its plane, without affecting the definition, thanks to the depth of field of the lenses.

The exposure time is generally limited by a system of diaphragms built in the following way (see also Fig. X.4): the primary lens of the camera is provided with a diaphragm in the form of a lozenge or rectangle, the image of which is formed, by reflection on the rotating mirror, on a crown of diaphragms which are also lozenge-shaped, arranged opposite the secondary return lenses. The scanning of a fixed secondary diaphragm by the moving image of the primary diaphragm constitutes a type of curtain shutter whose opening as a function of time produces a sufficiently sharp maximum. The quality factor of such a camera is generally defined as the ratio of the time separating two successive images to the exposure time. In order to maintain a sufficient camera aperture, this quality factor cannot really exceed 3 in practice.

The number of secondary lenses, and thus of images, is typically of the order of thirty disposed over about a quarter of a circle (certain cameras, called total activity cameras, have more lenses which are disposed on an almost complete circle and are scanned by a prismatic mirror with three reflecting faces). To attain the rate of a million images per second (one image per microsecond) it is thus necessary to obtain a rotational speed of the mirror of the order of 5000 r/s. Such rotational speeds demand air turbines (or helium turbines for very high speeds). The mirror is usually placed in a cavity which has been pumped to primary vacuum in order to avoid air turbulence and to reduce resistance. It is suspended on gas bearings (usually propelling gas) or possibly on oil bearings.

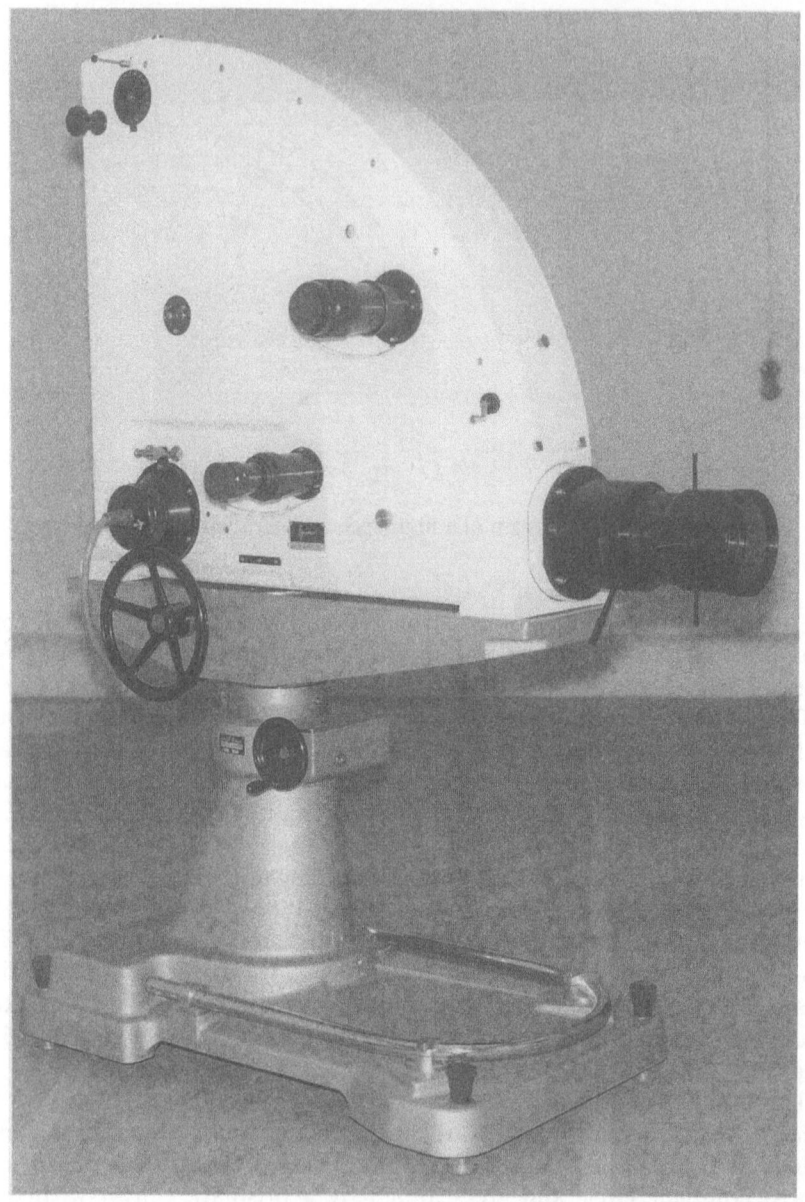

Figure X.5. CIAS camera from the Central Weapons Laboratory.

The size of the mirror is limited by the risk of shattering due to centrifugal force. In certain constructions, beryllium mirrors are used to increase resistance to shattering.

The speed of rotation and the instantaneous position of the mirror are generally known thanks to a simple optical device consisting of a lamp, whose luminous beam is reflected by the rotating mirror, and a photodiode detector, placed at the level of the crown of lenses. This system delivers a "top" which corresponds to the instant of passage of the mirror through a given angular position. An associated frequency meter allows measurements of the rotation speeds which the experimenter can regulate by altering the gas pressure in the turbine. Once the chosen speed is reached, the mechanical shutter of the camera opens. The firing of the experimental device, and possibly of the explosive flash, is thus controlled with a predetermined delay related to a "top" of the passage of the mirror, which allows synchronization of the phenomenon studied and the shooting.

The superimposition of images during successive passages of the mirror is prevented beforehand by the absence of luminosity of the device, and after the phenomenon by a pyrotechnic shutter system (see above). The photograph in Figure X.5 shows a French-designed high-speed camera with rotating mirror.

(d) Particular Recording Techniques (Stereoscopy, Strioscopy)

The high-speed cameras described above are generally used to take classic shots of the phenomena studied. However, these same cameras may be associated with particular recording techniques which allow access to other parameters such as the relief of the object observed and the positions of shock waves in gases, which are not detectable by conventional observation.

The *stereoscopic* technique, well known in photography, allows restitution, either qualitative or quantitative (photogrammetry), of relief by recording two frames of the same object taken from different angles. In the particular case of high-speed phenomena, the two frames must be made in perfect synchronization, which in practice means that they must be recorded with the same camera. For this, a system of return mirrors is used, placed in proximity to the object studied and thus destroyed during the experiment, following the principle shown in Figure X.6. To ensure the equality of the optical paths, the two mirrors (M_1 and M_2) returning the two images corresponding to the two shooting angles are disposed tangentially to an ellipse, which has as foci the object studied Ω and the entry lens of the camera ω. In practice, this optical diagram is complicated by the need to protect the camera against the destructive effects of the experiment, which means that supplementary return mirrors have to be used.

Strioscopy (schlieren technique) is a recording method in which transmission is used through transparent objects (particularly gases). It allows visualization of the refractive index variation zones which are linked to the shock waves. Figure X.7 shows the principle. If one observes the image of a trans-

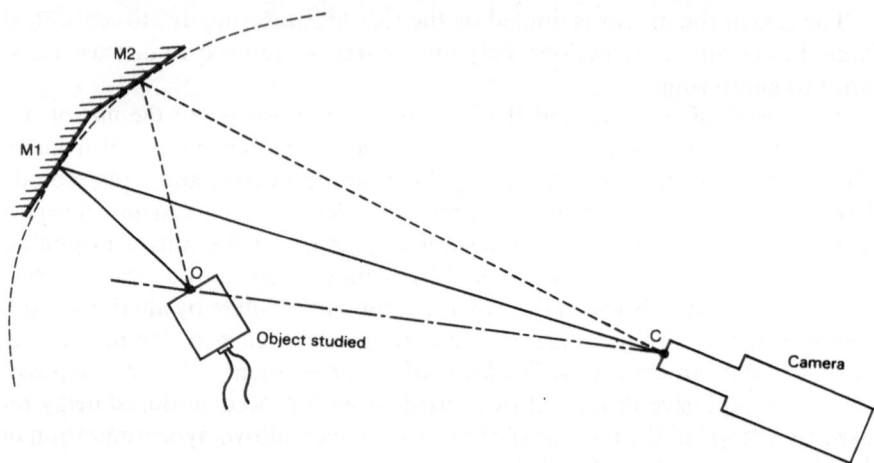

Figure X.6. Optical diagram of stereoscopic observation using a high-speed camera.

parent medium obtained by transmission from a point source of light, the lighting of this image remains uniform so long as the refractive index of the medium remains constant. But if one region of the medium has a refractive index different from that in the rest of the medium having uniform index, then this region will deflect the light rays crossing it and thus modify the illumination of the corresponding image. Such a region is said to be a striation. The shadow projected by this region thus receives stronger illumination on one side than on the other, and the position of the zone of different refractive index can be clearly revealed by a photographic method. A shock wave in a gas forms a striation, since the density of the gas and thus its refractive index is higher in the shock wave than in the rest of the gas. Thus the strioscopic method, linked with a high-speed cinematography technique, allows us to track shock waves in a gas with time.

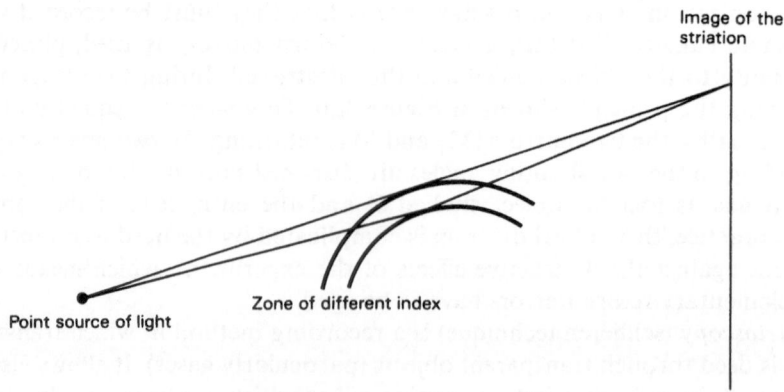

Figure X.7. Principle of strioscopy.

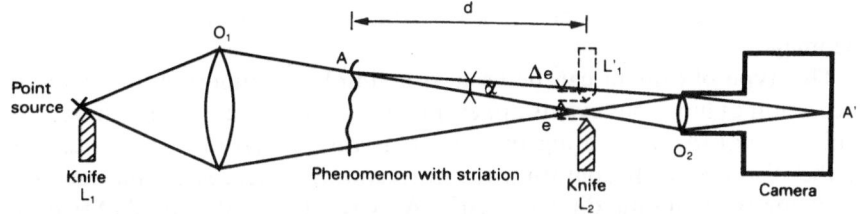

Figure X.8. Strioscopy. Diagram of Doppler–Foucault device.

The device in Figure X.7 is simple, but not highly sensitive. This is why the device in Figure X.8, called the Doppler–Foucault device, is generally preferred.

A point source (He–Ne or other laser) is diaphragmed in one direction by a knife L_1, whose image is formed by a lens O_1 in the plane of a second knife L_2, and forms with it a virtual slit of width e. The camera placed behind captures an image of the phenomenon with a striation, using lens O_2. In the absence of a striation, all the rays entering the lens O_2 pass through the virtual slit. But if, at a point A, the light rays are diverted upwards by an angle α, then the width of the virtual slit will be increased by $\Delta e = \alpha d$, d being the distance between the phenomenon studied and the knife L_2. The illumination of the image point A′ of A is thus increased. Conversely, if the deviation α is made downward, then the illumination of the image of A is reduced. The image of the striation A thus appears as a variation in illumination, whose direction depends upon the direction of the deviation in A. The sensitivity of the method is given by $\Delta e/e = \alpha d/e$ and is therefore limited by diffraction.

1.3. Slit Scanning Cameras

The high-speed cameras—called "integral image" cameras—that we have been describing are mainly reserved, in detonics experimentation, for use in qualitative global observation of the evolution of the phenomena. In fact, the measurements which can be made on the images obtained in this way are generally of low precision, partly because of the difficulty of their chronometric retiming, and partly because of optical errors encountered (distortions, errors due to parallax, luminous parasites due to shock induced ionization of air, etc.). However, by the use of very similar techniques, so-called "slit scanning" or "streak" cameras have been devised which are measuring instruments of high precision.

These cameras form an image of a line of the object studied, limited with the aid of an observation *slit*, and then scan this linear image at a constant velocity, following a direction perpendicular to it. The final image obtained is in two dimensions: the direction of the slit corresponds to spatial information (displacements of the object following this direction), while the perpendicular direction, which is the scanning direction, contains temporal information.

The standard of time is determined by the scanning speed of the slit across the image.

This type of camera can be used as a simple chronograph recording the succession of luminous events appearing on the line of the object viewed. The time interval (Δt) separating two events is then given by the measurement of the distance ($\Delta \ell$) separating the corresponding traces on the recording following the scanning direction: $\Delta t = \Delta \ell / v$ where v is the speed of scanning of the slit across the image. The resolution in time is limited by the geometric resolution of several microns over the recording sensitive surface (grain of film, optical resolution, etc.). To obtain the resolutions needed in detonics (about one nanosecond) it is therefore necessary to attain very high speeds for the scanning of the image, of the order of several millimeters per microsecond. These speeds are of the same order of magnitude as image displacement speeds necessary for sequencing the frames for high-speed cinematography. This explains why the same techniques are used for making slit scanning cameras as for high-speed cameras with integral images: use of a rotating mirror or deflection of an electronic image.

The slit scanning camera can also be used to measure displacement speeds over the object studied. If a point of the object is displaced with velocity V in the direction of the observation slit, its image on the recording will be displaced on a line making an angle α with the scanning direction and we have: $V = v \tan \alpha / g$ where g is the image/object magnification and v is the scanning speed on the recording. Measurement of the angle α, knowing the magnification and the scanning speed, allows us to find the displacement velocity V.

Recording with the aid of a slit scanning camera can be made directly when the phenomenon studied is autoluminous, by the appearance and displacement of luminous zones along the slit. We will see that in the opposite case certain optical devices have been developed to record signals and, in some cases, to obtain parameters which are not directly measurable. Lastly, the slit scanning camera has also found an application in velocimetry by Doppler laser interferometry which will be described later.

(a) Mechanical Cameras

In this type of camera, the image of the slit on the film is scanned by the rapid motion of a rotating mirror of identical design to those used for high-speed cameras with integral images (see §1.2). Figure X.9 shows the optical diagram which is most commonly used.

An entry lens forms an image of the object under investigation on the plane of an observation slit. An image of this slit is formed on the film, by a secondary lens, after reflection from the rotating mirror. The photographic film is plated on a circular crown centered on the axis of rotation of the mirror. If ϕ is the diameter of this crown and Ω is the velocity of angular rotation of the mirror, the scanning velocity v will be given by $v = \phi \Omega$. For a

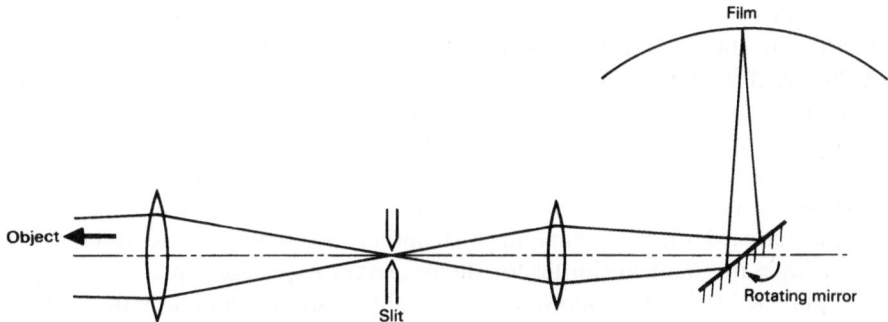

Figure X.9. Optical diagram of a slit scanning camera with rotating mirror.

radius of the order of 1 m and a mirror rotation velocity of 3000 revs/s, the scanning velocity will be about 35 mm/μs. Such scanning velocities allow us to attain time resolutions close to 1 ns. However, to obtain these high resolutions—as well as good spatial resolutions—many precautions must be taken in the construction of these cameras, particularly with regard to the quality of the optics, the flatness of the mirror (subject, as it rotates, to the action of centrifugal force), the positioning, and the plating of the film.

To improve the latter, a variant to the optical diagram shown above (Fig. X.9) has been proposed by Brixner [5] . This arrangement, shown in Figure X.10, allows the use of flat rigid photographic plates. The first lens always forms the image of the object on the slit, while the second lens forms an image of the slit at infinity. After reflection from the rotating mirror, the beam of parallel rays thus obtained is returned by a third lens which reforms the image of the slit on a photographic plate placed in its focal plane. During the rotation of the mirror the image of the slit will thus always remain in the plane of the rigid plate. This type of camera suffers from a limited recording time. Furthermore, the scanning velocity over the film is not uniform and thus needs correction after analysis.

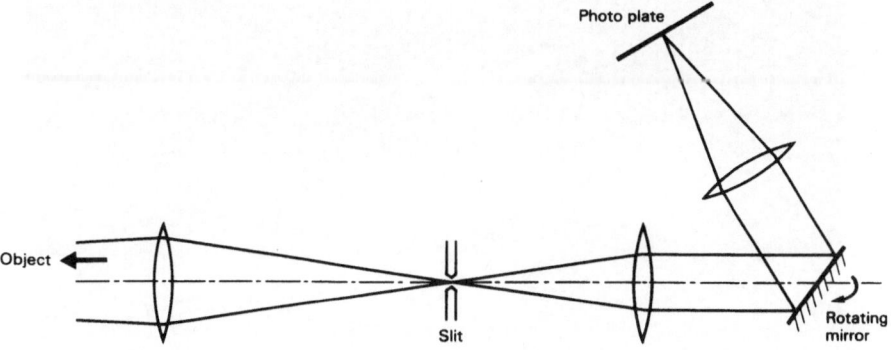

Figure X.10. Optical diagram of a Brixner-type slit scanning camera.

For all such rotating mirror cameras, the speed of rotation, which constitutes the standard of time, is measured very precisely by a device similar to that used for cameras with integral images (see §1.2(c)). Synchronization is also ensured in the same manner; the camera triggers activation of the firing with a predetermined delay in relation to the passage of the mirror through a given angular position. For the spatial standardization of the recording, a static frame (with mirror at rest) is usually first taken of the object, together with a test card, on the same film, with sufficient opening of the observation slit, using a mechanical shutter for the necessary exposure time.

Figure X.11 shows a photograph of a slit scanning camera with rotating

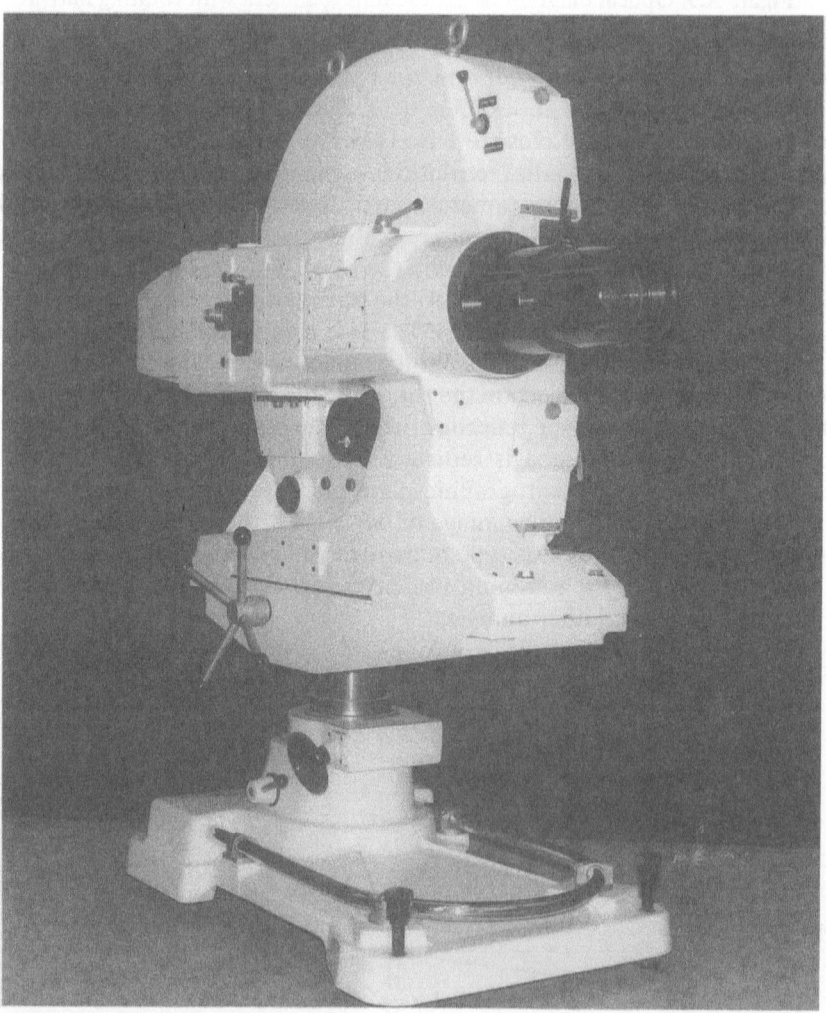

Figure X.11. CFAT camera from Laboratoire Central de l'Armement, France.

mirror, made by Laboratoire Central de l'Armement, France. It should be noted that some manufacturers offer convertible cameras which, with the same rotating mirror, can function with integral images or by slit scanning (when the crown of secondary lenses is withdrawn), or dual use cameras where one part of the rotating mirror is used for integral images and the remainder for slit scanning.

(b) Electronic Cameras

Electronic slit scanning cameras use the same tubes as the integral image cameras described in §1.2(a), with the same camera often providing both functions. A special electronic device applies a voltage pulse to the shutter grid, thus triggering the opening of the camera. For the duration of this pulse, two symmetric saw-toothed signals strike a pair of deflection plates, thus ensuring a linear scanning of the image along a direction perpendicular to these plates (see Fig. X.12).

It should be noted that the scanning affects the entirety of the image formed on the photocathode. To achieve the slit scanning function, lenses are used to form an image of the object studied on an observation slit, and then to return an image of this slit to the photocathode, setting it perpendicular to the direction of scan. The film then records the image of the scanning of this slit over time.

It can be seen that very high scanning velocities can be obtained with

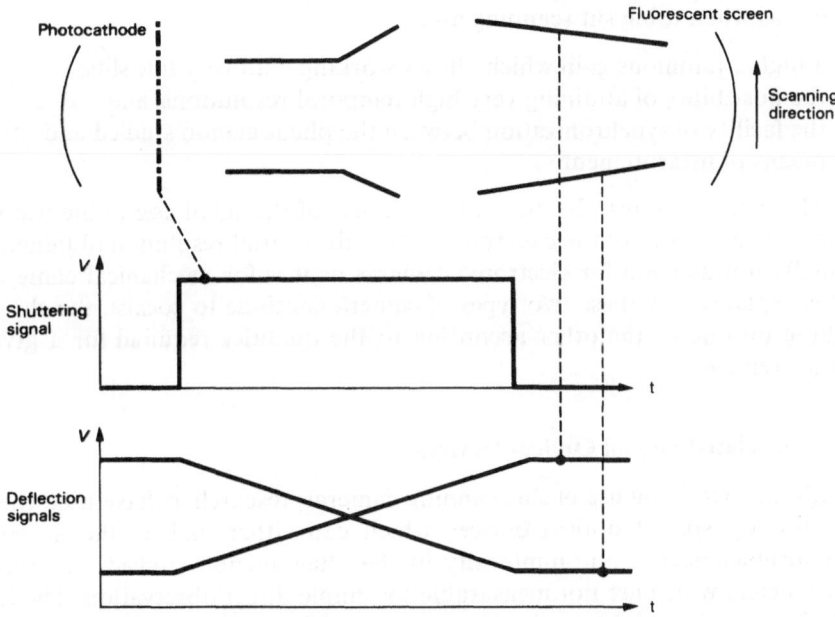

Figure X.12. Functioning of an electronic camera in scanning mode.

such a system, with the highest performance cameras of this type reaching scanning velocities of the order of 100 mm/ns and thus allowing temporal resolutions close to 1 ps. However, the length of recording is usually only about 50 mm, and the total time of measurement at such velocities is only half a nanosecond. For detonic applications scanning velocities ranging from several millimeters per microsecond to some tens of millimeters per microsecond are generally used, and correspond to total measuring times ranging from about ten microseconds down to several hundred nanoseconds. The corresponding temporal resolutions thus vary from several nanoseconds to about one hundred picoseconds.

However, we have to note that the precision of chronometric measurement is not necessarily as good as the temporal resolution inasmuch as for these cameras, unlike rotating mirror cameras, the scan linearity is not perfect. To improve this precision, marking systems may be used by recording the image of a photoemission diode energized by an oscillating circuit on one side of the frame during scanning. This system allows improvements to be made in the scan linearity in order to bring the precision nearer the temporal resolution.

Synchronization is ensured in the same way as in the "integral image" mode (see §1.2(b)). The electronic scanning device usually delivers two impulses which correspond, respectively, to the beginning and the end of the scan, which can be recorded on a chronometer to allow retiming to match other measurements.

The principal advantages of electronic cameras compared to mechanical cameras when using slit scanning are:

- a higher luminous gain which allows working with very fine slits;
- the possibility of attaining very high temporal resolutions; and
- the facility of synchronization between the phenomenon studied and other means of measurement.

However, we must also note that, because of the small size of the frames and the distortions of the electronic optics, the spatial resolution obtained is usually not as good for electronic cameras as it is for mechanical cameras. This explains why these two types of camera continue to coexist, the choice falling on one or the other according to the qualities required for a given measurement.

(c) Associated Special Optical Devices

With the increasing use of slit scanning cameras, researchers have used them to develop special optical devices which can either palliate the absence (or insufficiency) of autoluminosity in the phenomenon studied, or obtain parameters which are not measurable by simple direct observation. The fertile imagination of detonation workers throughout the world has given us many innovations in this field. We will content ourselves here with describing only those devices whose use is, to our knowledge, most general.

The *argon chamber* method uses the light emission from the ionization of gases under shock to detect the emergence of a shock wave on a free surface. As with explosive flashes (see §1.1(a)), argon is generally chosen because of its easy ionization under shock. The argon chamber is formed by a very thin facing (a few hundredths of a millimeter) hollowed in a flat surface of a block of transparent plexiglass. This surface is applied to the equally flat surface where the emergence of a shock is to be determined (the flatness of these two surfaces is particularly crucial). Shortly before the experiment the chamber is filled with argon, usually by continuous sweeping through two orifices. The slit scanning camera views a line of this chamber through the transparent plexiglass.

The shock wave emerging from the free surface is transmitted into the argon and triggers the illumination of the chamber. When this shock reaches the opposite wall of the chamber, it is transmitted to the plexiglass block and makes it opaque, thus blocking the glow of the chamber. The image thus obtained on the recording of the slit camera appears in the form of a luminous line perpendicular to the direction of scan (for a plane frontal shock). By using several argon chambers, placed at different levels, which are aligned along the same slit, the shock velocity can be measured. Figure X.13 shows the principle of such measurement.

This type of chronometric measurement is scarcely used nowadays because of its limited precision. In fact, the appearance of luminosity in the argon suffers a delay which is not always highly reproducible in relation to the outlet of the shock onto the free surface. However, the advantage of the argon chamber technique is that it allows measurement of the flatness of a shock in a continuous manner in one direction, when a large size chamber is used. In fact, the appearance of luminosity in one point of the slit corresponds to the outlet of the shock at that point. The distances between the recorded luminous trace and a straight line perpendicular to the direction of scan thus give the errors in synchronization in the emergence of the shock at different points, and allow calculation of the errors in the flatness of the shock wave (provided the velocity is known) following this direction.

The *reflecting or luminous deposit* method is similar to the argon chamber method: to determine the emergence of shock on a free surface, a fine layer of a substance whose luminous properties vary under shock is deposited on the surface, so that the slit scanning camera can record these variations in luminosity as a function of time, following a line.

Two types of deposit can be distinguished: substances which are luminescent under shock, and reflecting substances. The first, as with argon chambers, generates an intense light when they are submitted to a shock. In this category we find explosive paints (with pentrite–penta-erythritol tetranitrate) and coatings of $Al_2(SiF_6)_3$, which have the advantage of functioning at a level

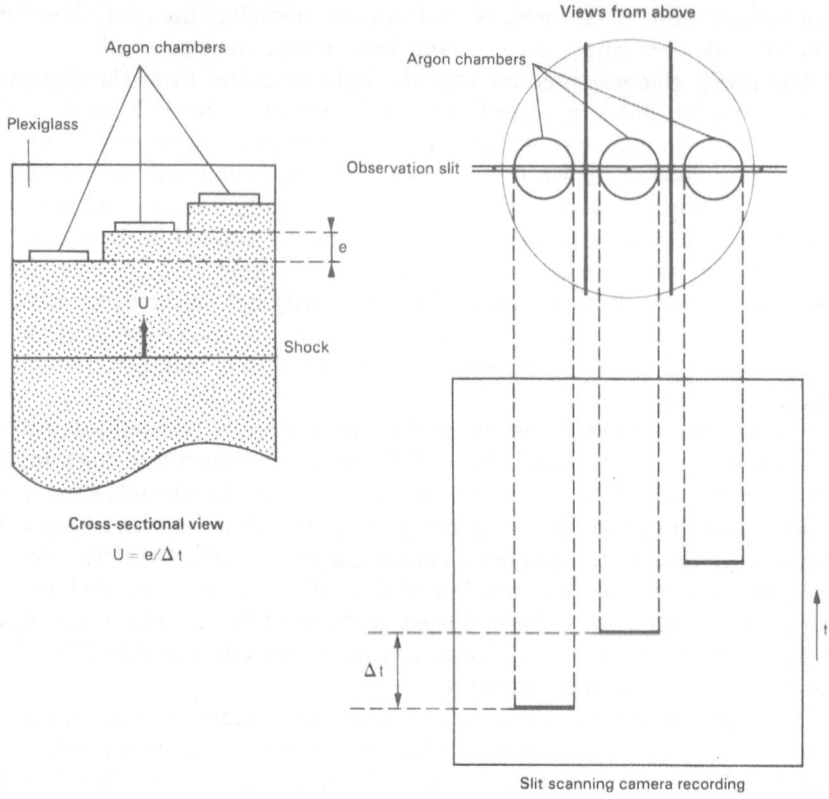

Figure X.13. Simplified diagram for measuring shock velocities using argon chambers.

which allows discrimination between strong and weak shocks. However, those in the second class are not autoluminescent and need an additional light source which is usually an electronic or explosive flash: the intensity of the reflected light declines sharply because of the degradation of the reflectivity under shock and the possible rotation of the free edge itself. The reflecting substances employed appear either in the form of a very fine metallization of the surface (by evaporation under vacuum or cathodic evaporation) or of a metallic paint (with silver), or of a self-adhesive band of microspheres (Scotch-Lite). The first two techniques are those most often used today because of the fineness of the deposit that can be obtained, thus limiting delays in their functioning.

Plate X.1 shows an example combining the use of reflecting deposit with autoluminescence for the study of the generation of detonation by the "double wedge" method.

The *reflected wire* method allows continuous measurement of the velocity of a free surface observed from the front, using a slit camera. A wire (see

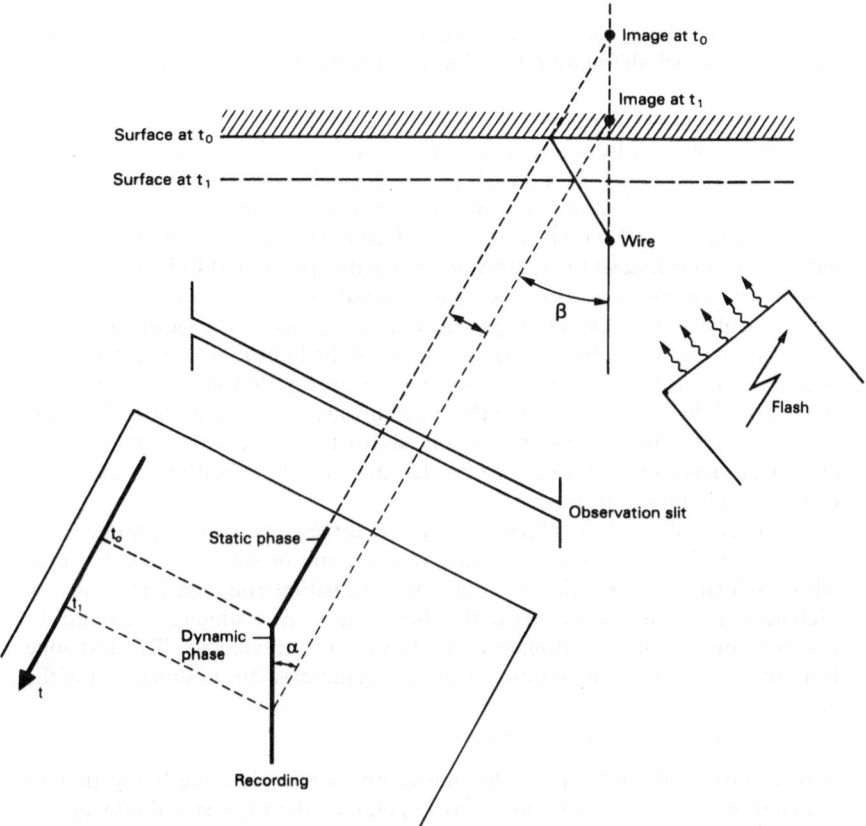

Figure X.14. Principle of the method for measuring the velocity of a surface by reflected wires.

Fig. X.14)—or a series of parallel wires—is held parallel to the free surface studied, which is rendered reflective by polishing or possibly by metallization. This wire is illuminated with the help of an electronic or explosive flash. The slit scanning camera, whose slit is adjusted perpendicular to the wire sees, at an angle β, the virtual image of this wire in the mirror formed by the free surface. When a frontal shock sets the free surface in motion, the virtual image of the stationary wire is displaced by double the distance in the same direction. The trace of this image on the recording of the slit camera will thus follow an oblique line whose angle α relative to the scanning direction is directly related to the instantaneous velocity of the surface by $V = v \tan \alpha / 2g \sin \beta$, v being the scanning speed, and g the magnification. The measurement of the angle α which the track makes at each point with the direction of scanning allows the changes in velocity of the free surface to be followed continuously as a function of time.

Plate X.1. Application of a slit scanning camera used with reflecting deposit for the study of shock-detonation transition by the double wedge method.

The wedge method or "wedge test" has been used for a long time in detonics to study the transition from shock to detonation in a solid explosive. A planar one-dimensional shock is induced by an explosion-driven flying plate (see, §4.1(b))in a target of solid explosive in the form of a wedge or, as here, of a double wedge (see §IX.1). The external surface of the explosive prism is coated with a reflecting deposit (metallization or metallic paint) and lit by two flashes. A slit scanning camera records the light reflected by the surface following a line perpendicular to the edge of the prism. The emergence of the shock at a point of the surface is shown by a sharp reduction of the light reflected at this point, because of the rotation of this surface. The recording thus shows a symmetric reduction of the luminous zone as the shock progresses. Each point of the interface between the light zone and dark zone shows the time of emergence of shock at a given depth of explosive. The displacement of shock within an explosive can thus be followed over time.

If an induced shock is sufficiently intense, three zones can be distinguished in the recording: zone 1 corresponds to a regime of weakly reactive shock whose velocity increases slowly, zone 2 is a transition zone marked by a very high acceleration in the velocity of the shock, and zone 3 which corresponds to propagation of built-up detonation at almost constant velocity. This last zone is marked not by a loss of luminosity but by an increase due to ionization of the air near the surface.

Precise analysis of this recording allows us:

- to determine the velocity of the nonreactive shock (in zone l) and thus to determine a point of the inert shock polar of the explosive (knowing the velocity of the flying plate); and
- to measure the build-up distance for the corresponding shock pressure.

By varying the thickness of the flying plate and thus the impact velocity, a law can be established relating the pressure $\hat{p}$ of the shock induced in the explosive to the build-up distance X^*.

One drawback of this method is that the measurement of velocity is made at a point of the surface which moves with time, and that flatness of the incident shock is assumed. It may also be noted that the intersection between the trace parallel to the scanning direction (corresponding to the static phase) and the oblique trace (corresponding to the dynamic phase of the surface) allows precise determination of the chronometry of emergence of the shock at the point of reflection.

This ingenious measurement technique has, however, lost most of its interest due to the advent of interferometric methods of measuring velocity which are much more precise (§1.4).

The use of *optical fibers*, thanks to their great flexibility, sets us free from

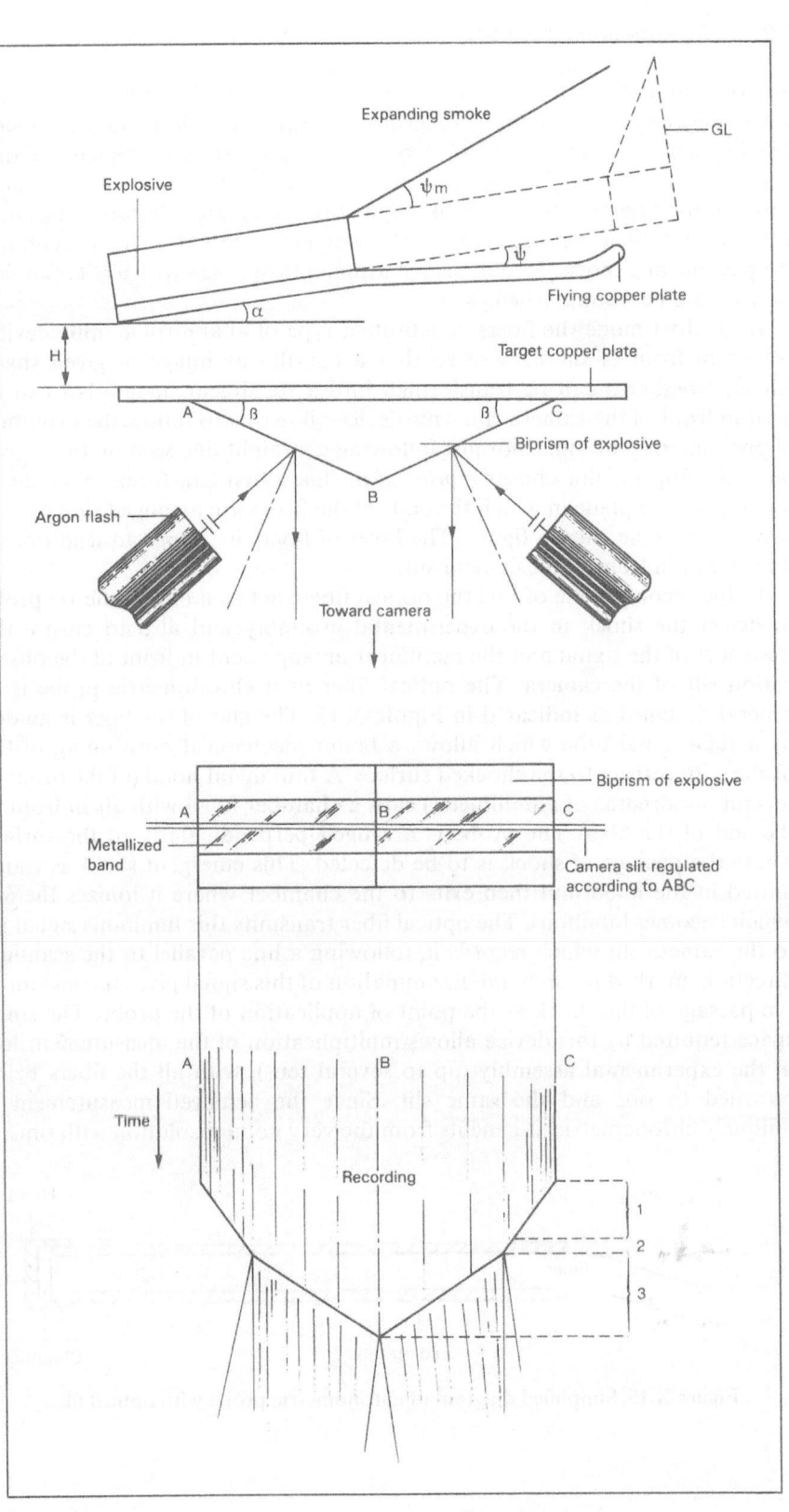

the constraints set by a rectilinear observation slit, and thus allows us to vary the arrangement of the points of the surface seen by the slit scanning camera. The fibers can also serve as a good means of transporting the signal—with a large band-width, small space requirement, and high flexibility of the path— between the experimental assembly in the firing area and the recording shed. The end of a fiber can constitute a chronometric probe for the detection of the passage of a shock. Two modes of using optical fibers with a slit scanning camera can be distinguished.

In the first mode the fibers constitute a type of anamorphic optic device, placed in front of the camera so that a curvilinear image of given shape (circle, cross, etc.) can be transformed into a rectilinear image that can be used in front of the camera slit. This device allows us to follow the evolution of phenomena with time, not just following a straight line seen on the object, but following any line chosen *a priori*. For this, a first lens forms an image of the object on a plane in which the ends of the fibers are arranged side by side according to the chosen figure. The layer of fibers is shaped to lead onto a line placed in front of the camera slit.

In the second mode of use, the optical fibers act as a chronometric probe to detect the shock in the experimental assembly and also to ensure the transport of the signal and the rectilinear arrangement in front of the observation slit of the camera. The optical fiber as a chronometric probe is in general designed as indicated in Figure X.15. The end of the fiber is guided by a rigid metal tube which allows a better mechanical positioning of the probe with respect to the shocked surface. A thin metal hood (of the order of several hundredths of a millimeter) caps a chamber filled with air in front of the end of the fiber. The probe is arranged perpendicularly to the surface where the passage of shock is to be detected. This emergent shock is transmitted in the hood and then exits to the chamber where it ionizes the air, which becomes luminous. The optical fiber transmits this luminous signal up to the camera slit which records it, following a line parallel to the scanning direction, marked beforehand. Examination of this signal gives the instant of the passage of the shock at the point of application of the probe. The small space required by this device allows multiplication of the measurement loci in the experimental assembly (up to several tens), with all the fibers being returned to one and the same slit. Since the achieved measurement is uniquely chronometric, it benefits from the very good resolution with time of

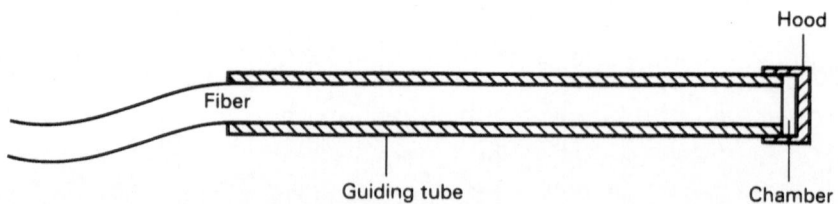

Figure X.15. Simplified diagram of chronometric probe with optical fiber.

electronic cameras. However, the final precision is limited by the reproducibility of the delay caused by crossing the hood of the probe and the delay in ionizing the air in the chamber. In addition, these probes function only when the shocks are intense enough to ionize the air.

1.4. Velocimetry by Doppler Laser Interferometry

(a) General Remarks on the Doppler Effect

The frequency of an electromagnetic wave reflected by a moving surface suffers, by the Doppler effect, a variation directly related to the velocity of displacement of this mirror. Researchers in detonics were quick to utilize this effect for continuous measurement of the changes in velocities of free surfaces and of wave surfaces. The first systems to be developed for this purpose used hyperfrequency transmitters (millimeter waves), which were the only coherent sources practically available at that time.

These waves have the advantage of being able to penetrate matter, so they can be reflected on the strongly ionized surface constituted by a detonation or shock wave front, thus permitting continuous measurement of the wave velocity. However, the time resolution of this method is limited, since the distance of flight needed to measure the velocity is of the same order of magnitude as the wavelength. In practice, millimeter waves allow measurement in only the established regime. Furthermore, because of diffraction, the diameter of the focusing spot that can be obtained is roughly limited to several wavelengths, which is equivalent to a minimum focusing spot of the order of one centimeter for millimeter waves. Thus the velocity measurement is not to be regarded as punctual but as integrated over the reflection domain.

The development of lasers has provided researchers with coherent sources which do not present such problems, so these hyperfrequency velocimetry systems have been steadily abandoned in favor of laser interferometry systems.

In these techniques, a laser beam is projected perpendicular to the moving surface, at the point where one wishes to measure the velocity. The Doppler effect varies the wavelength of the reflected light. As a first approximation

$$\frac{\Delta\lambda}{\lambda_0} = -2\frac{V}{c},$$

where λ_0 is the wavelength of the laser, V is the velocity of the free surface, c is the velocity of light, and $\Delta\lambda$ is the variation in wavelength due to the Doppler effect. As the velocities encountered in detonics are around one kilometer per second (one millimeter per microsecond), the relative variation in wavelength due to the Doppler effect is of the order of 6.7×10^{-6}. In order to measure this variation in wavelength, and thus obtain the velocity of the free surface, we need equipment whose resolution is better than 150,000.

Two interferometric techniques possessing the required resolution have been developed: one using a Michelson interferometer and the other a Fabry–Pérot interferometer.

(b) Michelson Interferometer. VISAR System

The Michelson interferometer is a dual wave interferometer which is shown diagrammatically in Figure X.16. The incident beam is divided in two by a semitransparent strip L. The two beams thus obtained are returned by two mirrors M_1 and M_2 onto this same strip, which reunites them. As the two mirrors are placed at different distances (d_1 and d_2) from the semitransparent plate, a path difference $2(d_1 - d_2) = 2e$ is made between the two resulting beams. In the same way, all the rays entering the interferometer at the same incidence i leave as rays which are all parallel to each other and have the same path difference $2e \cos i$. A coherent point source thus produces an interference figure at infinity. The interposition of a lens allows this figure to be formed in its focal plane. The interference fringes appear in the form of a succession of rings of equal incidence whose intensity is given by $I = I_0 \cos^2(\pi\delta/\lambda)$, where δ is the path difference in the interferometer $2e \cos i$ and λ is the wavelength of the laser source. The first laser interferometry systems developed around a Michelson interferometer used the moving surface studied as one of the interferometer mirrors. Figure X.17 shows the measuring device thus constructed.

A lens L_1 focuses the laser beam on the moving surface studied, after reflection on the semitransparent plate L which divides this beam equally into two. The Michelson interferometer consists of a fixed mirror and the moving surface. There a laser beam of wavelength λ_0 (reflected by the fixed mirror) interferes with a beam of wavelength $\lambda_0 + d\lambda$ which has undergone the Doppler effect (reflected by the moving surface). The interference figure is formed by a lens L_2 in its focal plane. A photomultiplier placed in this plane allows the change in intensity at a point to be followed. This intensity $I = I_0$

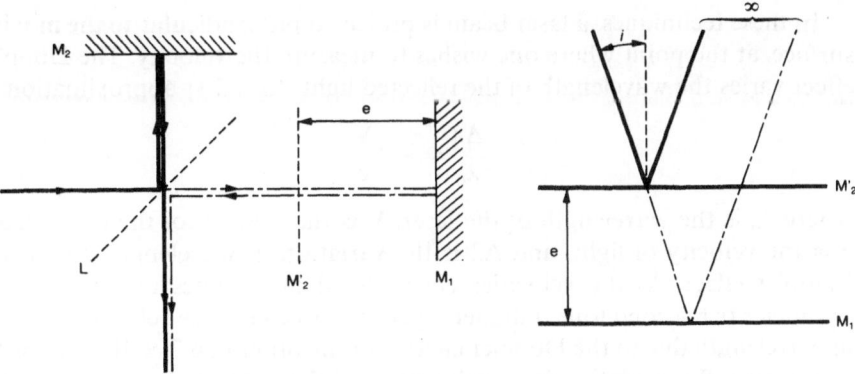

Figure X.16. Michelson interferometer and its equivalent figure.

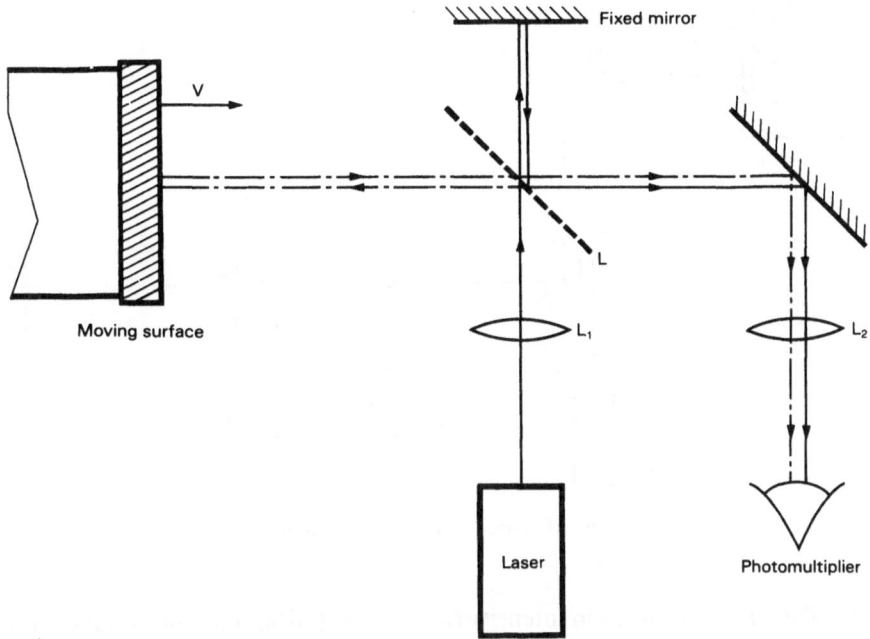

Figure X.17. Diagram showing the principle of a displacement interferometer.

$\cos^2(\pi\delta/\lambda)$ varies with time because of two effects: the changes in wavelength λ due to the Doppler effect, and the variations in the path difference δ due to the displacement of the Michelson mirror formed by the surface. In fact, in detonic applications the first term is negligible compared to the second. The displacement of the surface is written as $\frac{1}{2}\lambda_0 F(t)$, as a function of the number of interference fringes $F(t)$ which have passed in front of the photomultiplier at instant t. We thus have a device which allows tracking of the displacement of a moving surface. Note that, strictly speaking, it is not a Doppler effect velocimeter but a displacement interferometer, the velocity being obtained only indirectly.

The equipment thus conceived has a very high sensitivity since it records the transit of a fringe for a displacement equal to one half of a wavelength, and analysis of the sinusoidal signal measuring the amplitude allows resolution of about $\pm 1/10$ of the fringe. But, conversely, there is the problem of the fringe-scanning frequency for high displacement velocities encountered in detonics: for a He–Ne laser of wavelength 632.8 nm this frequency is 3.16 GHz per mm/μs. The recording systems (photomultiplier, amplifiers, oscilloscopes) do not really allow precise counting of the passage above 1 GHz and thus limit the application of the device to velocities lower than 1 mm/μs, which is usually inadequate.

To reduce this problem of the fringe-scanning frequency, researchers then tried entering only the beam reflected on the moving surface into the Michel-

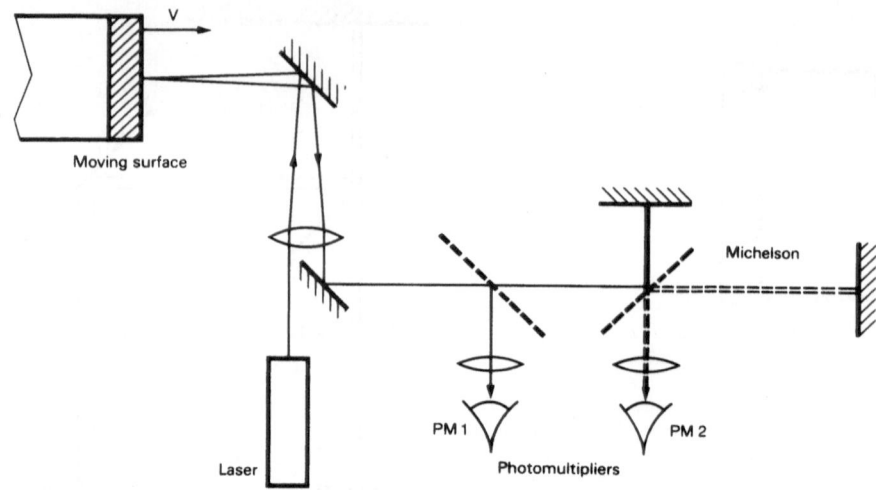

Figure X.18. Principle of velocimetry with a Michelson interferometer.

son. The only variations in intensity ($I = I_0 \cos^2(\pi\delta/\lambda)$) recorded on the interference figure are those due to variations in wavelength caused by the Doppler effect. As long as the velocity of movement remains constant, the wavelength λ does not change and the interference figure remains stationary. Thus the device reacts only to surface accelerations. It is in fact an accelerometer. The velocity is obtained by summing the fringes. Figure X.18 is a diagram showing the principle of the device used.

The velocity V of the moving surface is related to the number of fringes F(t) which have passed in front of the photomultiplier by the relation

$$\frac{dF(t)}{dt} = \frac{2\delta}{c\lambda_0}\frac{dV(t)}{dt}.$$

On integrating this formula, for a surface initially at rest, we obtain $V(t) = V_a F(t)$ where the velocity $V_a = c\lambda_0/2\delta$ is called "equipment velocity." By counting the fringes which pass in front of the photomultiplier PM_2 we can determine the velocity of the moving surface as a function of time. Furthermore, in order to discriminate the variations in intensity linked to the passing of the fringes from those due to possible variations in the reflectivity of the surface (which are common when velocity is caused by reflection of an intense shock), a part of the reflected signal is taken by a plate reflecting to a few percent and then recorded by a second photomultiplier PM_1.

Unlike the displacement interferometer, the range of velocities measurable by this device is not limited. However, the recordable acceleration is limited by the counting speed of the system. This limit depends on the equipment velocity V_a that can be adjusted by playing on the path difference δ in the two arms of the interferometer. However, in the presence of accelerations as great as those linked to the emergence of an intense surface shock, the transit of the fringes is always too fast for the recording system, and some fringes are lost in the counting. By choosing a high equipment velocity and thus reducing δ, it can usually be contrived for only a small number of fringes to be lost so that the ambiguity thus created in the determination of the velocity after shock can be removed by a rough estimation of this velocity. However, we must note that the increase in the time resolution of the system, obtained by reducing the path difference in the interferometer, is paid for by a reduction of the precision in measuring the velocity, since the counting resolution is limited to about one tenth of a fringe. Usually, the time resolution obtained with a Michelson interferometer is at best of the order of one nanosecond, with the measuring velocity precision reaching 2–3%.

The simple device described has, in fact, two serious problems which have led most researchers to abandon it in favor of a modified version proposed by Barker and Hollenbach [1], [2]: the VISAR system (Velocity Interferometer System for Any Reflector).

The first of these faults is in allowing measurements only on perfectly reflecting surfaces to conserve the spatial coherence of the laser beam. In fact, the diffusion of the light by a frosted surface deprives it of much of its spatial coherence and, in spite of the good focusing of the beam on the surface, noticeably reduces the contrast of the interference fringes. To reduce this problem, the VISAR system interposes in the longest arm of the Michelson a device called a field compensator consisting of a glass bar, or possibly a system of afocal lenses. The principle of this device is to ensure that a zero path difference will be restored between the two arms of the Michelson. For this, the glass bar forms an image of mirror M_2 at a distance from the semitransparent plate equal to that which separates it from mirror M_1 (see Fig. X.19). The path difference in the Michelson then depends directly on the

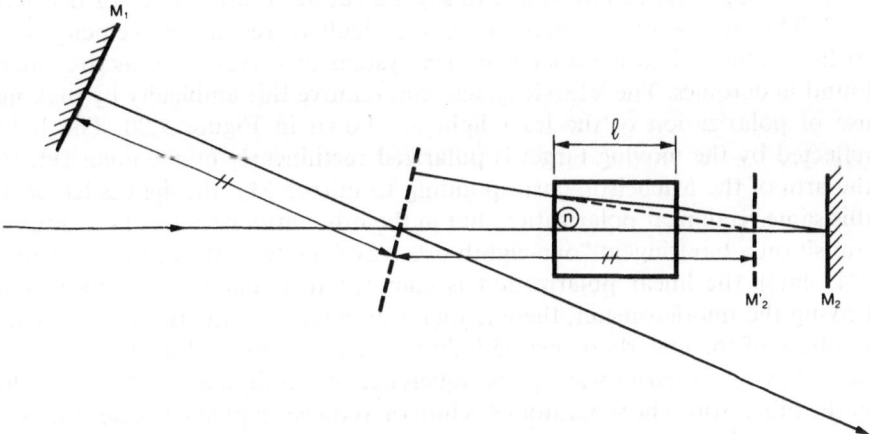

Figure X.19. Device for field compensation by a glass bar.

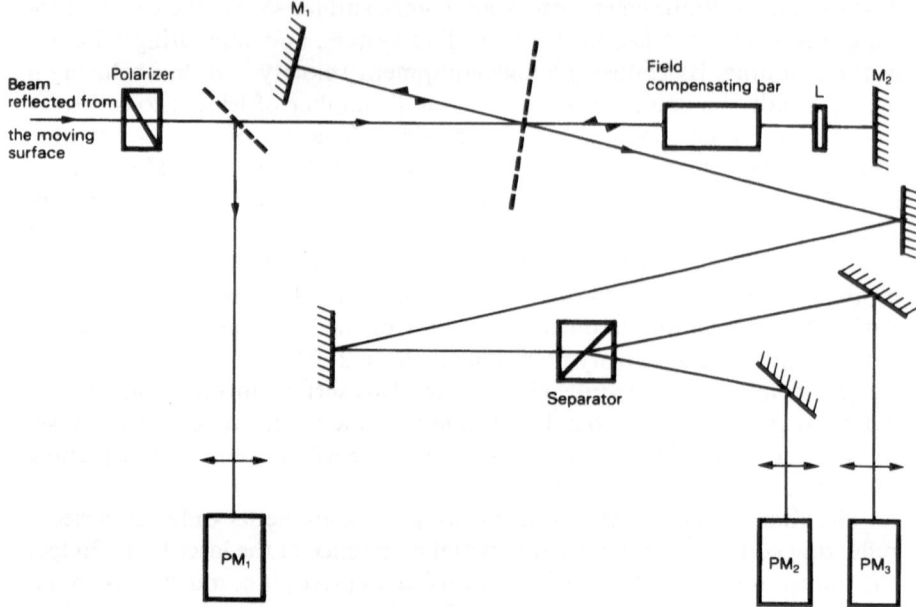

Figure X.20. VISAR interferometry system.

length of the glass bar of index n: $\delta = 2\ell[n - (1/n)]$. It can then be varied if necessary by changing the bar.

The second problem with the velocimeter described here is that it does not allow determination of the direction of passage of the interference rings, since their transit in front of the photomultiplier is detected only by the variation of luminous intensity. Since this transit direction is linked to the sign of the acceleration, it will be impossible to say if a surface is accelerated or decelerated. This ambiguity can make it very difficult to restore the velocity of a surface subjected to a rather complex system of waves, such as are often found in detonics. The VISAR system can remove this ambiguity by making use of polarization of the laser light as shown in Figure X.20. The light reflected by the moving target is polarized rectilinearly by the polarizer. In the arm of the Michelson corresponding to mirror M_1, the light is left with this same rectilinear polarization; but in the other arm, because of the double transit on a birefringent "one-eighth of a wave" plate L (thus equivalent to a $\lambda/4$ plate), the linear polarization is changed to circular polarization. On leaving the interferometer, there is then interference of the two components in phase of the linearly polarized light in the first arm, and the two components having a quarter wave phase difference of the circularly polarized light in the other arm. The separator (Rochon or Wollaston prism) isolates the two systems of rings thus created, and two systems of fringes with a quarter wave phase difference can be observed on the two photomultipliers (PM$_2$ and

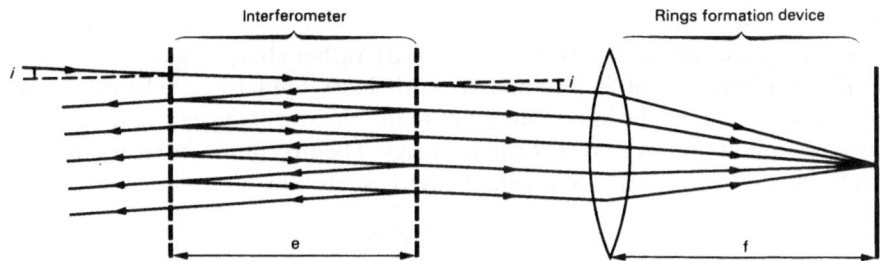

Figure X.21. Principle of the Fabry–Pérot interferometer.

PM_3). The interest in the system resides in the fact that at each change in sign of the acceleration, the signal which was a quarter wave ahead becomes a quarter wave behind. Comparison of the signals in the two photomultipliers thus allows detection of changes in the sign of the acceleration.

(c) Fabry–Pérot Interferometer. DLI System

Unlike the Michelson interferometer, the Fabry–Pérot interferometer is a multiple-beam device. Figure X.21 shows the principle. It consists of two plates with a high reflection coefficient. Because of the interplay of multiple transmissions and reflections, a wave entering the cavity of this interferometer leaves it divided into a multitude of parallel waves with the same incidence. Each wave is phase shifted with respect to the preceding wave by $\phi = 4\pi(e/\lambda)\cos i$. These waves can interfere at infinity. A lens allows formation of the interference figure in its focal plane. The resulting intensity is given by

$$I = \frac{\mathcal{K}}{1 + \frac{4r}{(1-r)^2}\sin^2\frac{\varphi}{2}},$$

where $r < 1$ is the reflection coefficient of the two plates. Figure X.22 is a graph showing intensity as a function of phase. Because of the high value of

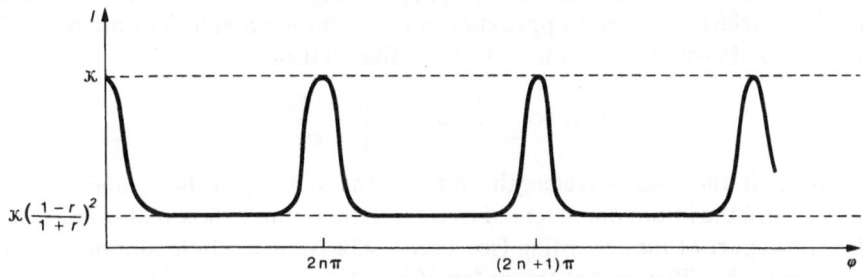

Figure X.22. Intensity of the interference figure in a Fabry–Pérot.

the reflection coefficient r, as distinct from the Michelson, this intensity presents, for phase values $2n\pi$ (n being integral), rather sharp peaks.

If the Fabry–Pérot is illuminated with laser light of wavelength λ_0 at different incidences, the interference figure obtained will appear in the form of a succession of fine concentric brilliant rings. The maximum intensity of each brilliant ring corresponds to incident rays i_n such that

$$4\pi e \cos i_n = 2n\pi\lambda.$$

If f is the focal length of the lens used to form the interference rings, the corresponding ring diameter is $2f\tan i_n$.

The rings thus obtained are finer as the reflection coefficient r of the Fabry–Pérot plates is greater. We can thus see that, if the reflection coefficient is sufficiently great, simple measurement of the diameter of the brilliant ring will enable us to determine i_n with good precision. Now, for a given aperture of the interferometer, the value of i_n corresponding to a given ring depends only on the wavelength λ. Every variation of this wavelength by the Doppler effect is thus translated into a variation in the diameters of the interference rings. Conversely, tracking the changes in ring diameter allows calculation of the variations in wavelength and thus of the changes in velocity of the surface studied. This is the principle on which the DLI (Doppler Laser Interferometer) system of velocimetry is based, as shown in Figure X.23 [12], [19].

The laser beam reflected from the moving surface is returned into the Fabry–Pérot interferometer. Interposition of a lens (or equivalent optical system) makes it possible to obtain the multiple incidences necessary for the formation of the ring system (which is generally limited to three rings). These rings are formed by a lens L_2 in the plane of an observation slit. A lens L_3 recaptures the image of this slit to return it to a slit scanning camera. Since the rings are centered on the observation slit, their recording by the camera allows their diameters to be tracked. As long as the surface remains immobile, the ring diameters remain constant, and the recording is composed of lines which are parallel to the scanning direction. When the surface is set in motion, the wavelength of the reflected light is modified by the Doppler effect, and the diameters of the interference rings vary. The diameter $\phi(t)$ of a given interference ring at instant t depends only on the velocity V(t) of the moving surface. As a first approximation—valid for a sufficient aperture of the Fabry–Pérot (e > 10 mm)—we have the relation

$$V(t) = \frac{\lambda_0 c}{4e}\left(\Delta p - \frac{\phi^2(t) - \phi_0^2}{\phi_1^2 - \phi_0^2}\right),$$

where λ_0 is the laser wavelength and c is the velocity of light, and ϕ_0 and $\phi_1 < \phi_0$ are the initial diameters of two successive rings, chosen at random, in the static part of the recording (see Fig. X.23); Δp is a whole number which represents the difference of the order of interference between the ring whose variation in diameter is being tracked and the ring of initial diameter ϕ_0

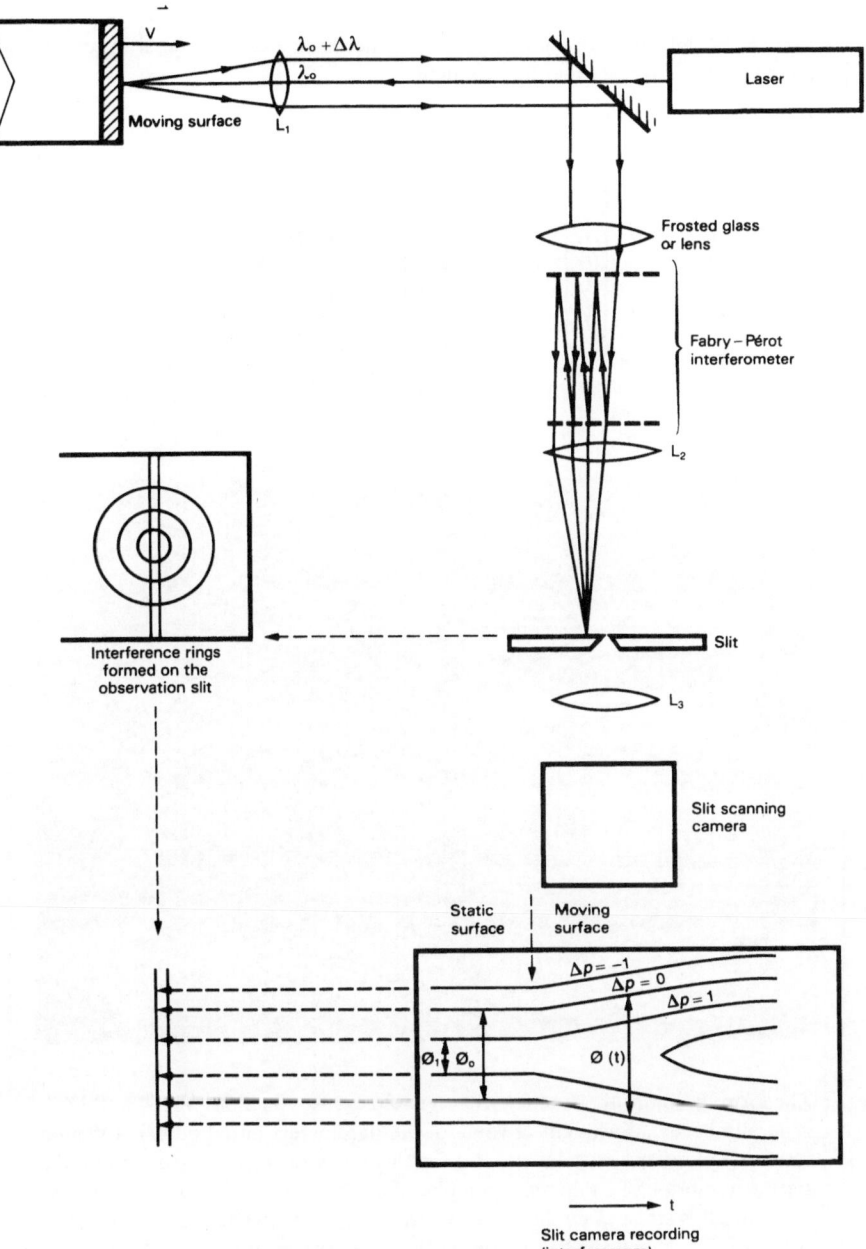

Figure X.23. Principle the DLI system.

Plate X.2 Application of the Doppler laser interferometer to the analysis of the reactive flow of the detonation products.

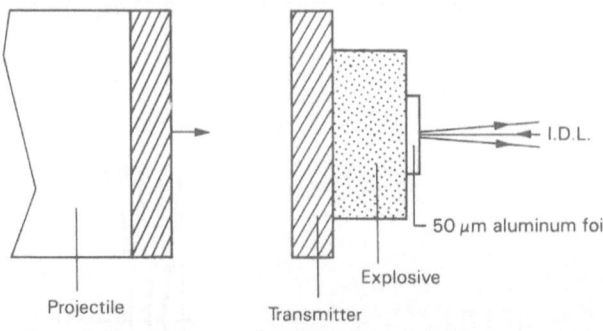

Projectile   Transmitter

50 µm aluminum foil

Explosive

I.D.L.

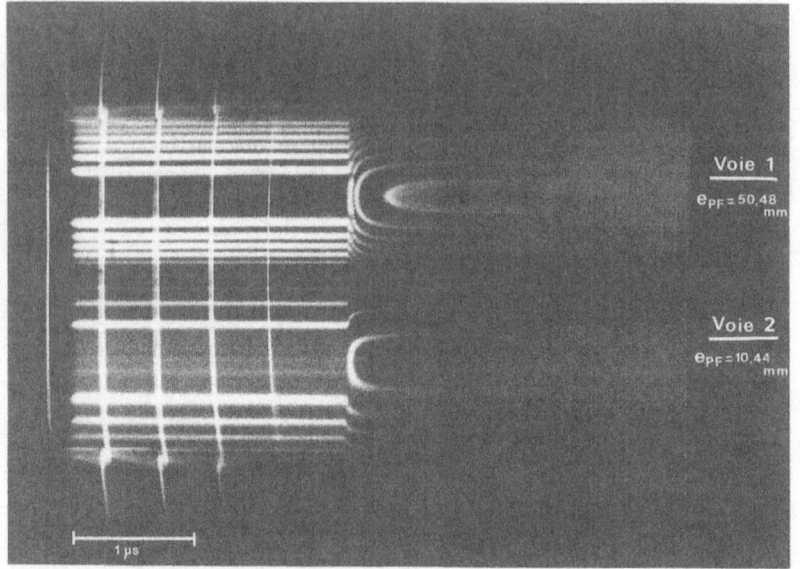

Voie 1

$e_{PF} = 50.48$ mm

Voie 2

$e_{PF} = 10.44$ mm

1 µs

The formulation of the reactivity of an explosive (see Sections VIII.4 and IX.3) generally needs adjustment of some of the parameters based on experimental results. A standard experiment used for this purpose consists of recording the setting in motion of a thin metal foil placed on the surface of a slab of explosive, which itself is stressed by the projectile from a launcher. The absence of disturbance of movement, the continuous character of the recording, good time resolution, and high precision of the DLI velocimetry technique make this an ideal tool for this type of experiment.

($\Delta p = 0$ for this ring itself, $\Delta p = +1$ for the ring of diameter ϕ_1 immediately below, etc.). It should be noted that determination of the velocity from the interferogram needs knowledge only of the laser wavelength and the distance between the plates in the Fabry–Pérot, independently of the optical magnifications introduced by the different elements in the chain.

In the formula establishing the velocity, the terms λ_0, c, and e are usually known with very high precision, so the uncertainty of measurement essentially depends on the spatial resolution of the measured ring diameters. This resolution is itself a function of the fineness of the rings on the one hand and the spatial resolution of the recording system on the other. If sufficient reflected light is available, we can improve the resolution by choosing plates with a very high reflection coefficient, and by recording the interferogram with a mechanical slit camera (see §1.3). Unfortunately, this is not always possible because of degradation phenomena of the surfaces under shock.

For a given spatial resolution of recording, the absolute uncertainty in measuring the velocity is lower as the aperture e of the interferometer is greater. However, this aperture cannot be increased indefinitely since, as well as the loss in time resolution which is involved, as will be seen later, we encounter problems of thermal turbulence in the air in the cavity, and of diffraction due to the finite diameter of the plates, which degrade the fineness of the rings. The maximum aperture which can be used in practice is of the order of one meter. For the very large apertures of the Fabry–Pérot we also have a problem with uncertainty of order p of a dynamic ring faced with a jump in velocity. In fact, the time resolution of the system does not allow the tracking of very high accelerations, such as those found when intense shocks emerge. This sudden increase in velocity will produce more rings in proportion to the size of aperture e. The velocity after shock will then be known only as modulo of a value $\Delta V = \lambda_0 c/4e$, which will be, for example, 77 m/s if $e = 50$ cm. To remove this uncertainty it is thus necessary to know the expected order of magnitude of the velocity with a precision greater than this value. One way of doing this is to take a part of the reflected light and return it over a second interferometric path with a Fabry–Pérot operating with a small aperture.

Under optimum experimental conditions (maximum aperture of the Fabry–Pérot and optimal recording resolution) the absolute uncertainty of velocity measurement is of the order of one meter per second. The DLI technique therefore only allows good relative precision for measuring velocities in excess of a few tens of meters per second. It is particularly well adapted to the range of velocities encountered in detonics of the order of several thousands of metres per second, for which it has a measurement precision of several parts per thousand. The time resolution of this velocity measurement is essentially limited by the response time of the Fabry–Pérot, since the recording system with the slit camera permits resolutions of much less than one nanosecond, particularly with electronic cameras (see §1.3). The response time of the Fabry–Pérot corresponds to the time which the interferometer takes to "empty" the light which has returned at a given instant. As distinct from the Michelson, this time is theoretically infinite, because of the interplay of forward and reverse multiple reflections on the plates. However, in practice, as the intensity of the ray emitted on the nth forward–reverse is equal to that of the ray emitted the (n − 1)th time multiplied by r^2 ($r < 1$, reflection coefficient), only a finite number of forward and reverses take part in the interference, since the sum of the resulting intensities of subsequent forward and reverses is below the detectability threshold of the system. The resulting response time varies as $e/(1 − r^2)$. To reduce this, we can diminish the reflection coefficient of the plates, but, as we have seen above, this damages the fineness of the rings and thus the precision of velocity determination. We can also reduce the aperture of the Fabry–Pérot, but again we have seen that this will increase the absolute uncertainty of velocity measurement. It may be observed that, for the DLI system as for the VISAR system, time resolution and absolute uncertainty of measurement are irreconcilable. However, the situation varies according to the range of velocities to be measured. In fact, for velocities of several thousands of metres per second, we must be

Figure X.24. DLI chain with two measuring paths next to a light gas gun in Centre d'Etudes de Vaujours, Commissariat à l'Energie Atomique, France.

content with an absolute uncertainty of several tens of meters per second—which corresponds to a precision of the order of 1%—and to choose a small Fabry–Pérot aperture of the order of 10 mm; the response time of the system then falls below one nanosecond. For lower velocities one can either keep the time resolution and sacrifice precision, or keep the precision and be content with a time resolution of the order of ten nanoseconds: it is up to the experimenter to define the best compromise according to the application envisaged.

Another particular aspect of the DLI technique must be stressed: as opposed to the VISAR system it needs powerful lasers because of the low luminous gain of the detection system. An argon laser ionized to a power of some 5 W in monomode is generally used. This can pose security problems, both for users and for the explosive. An elegant way of resolving these problems, which simplifies the path and its control, is to ensure the transport of direct and reflected beams by a bundle of optical fibers.

In conclusion, these interferometric techniques which allow continuous measurement of the velocity of a free surface, without any disturbance of movement, constitute, because of their precision and time resolution, by far the most effective means of measuring velocity at present available for detonics, where their applications continue to increase. Figure X.24 shows a two-path chain, installed at Centre d'Etudes de Vaujours, Commissariat à l'Energie Atomique, France. Plate X.2 shows an example of the use of this chain for the study of reactive flows by the "thin metal foil" method.

2. Electronic Techniques

These measurement techniques have benefited from the considerable progress made in recent years in the field of electronic rapid recording. The experimenter now has at his disposal chronometers with a resolution better than one nanosecond, and oscilloscopes with band-width of 500 MHz or even 1000 MHz. Recording techniques are thus in general no longer a limitation to the resolution and precision of measurements. Conversely, the quality of the latter is still noticeably dependent on the constitution and the mechanical positioning of the gauges or probes which generate the signal at the level of the experimental assembly and are destroyed with it, as well as on the care taken with connections and synchronizations.

2.1. Probes and Gauges

It is usual to distinguish three main groups as a function of the parameter one is seeking to measure:

— chronometric probes;
— manometric gauges; and
— velocimetric gauges.

(a) Chronometric Probes

By this term are designated devices which, when placed at a point in an experimental assembly, deliver an electric impulse at the time of arrival of a shock wave or of a free surface, thus allowing recording of the instant of their arrival at this point. These probes are generally used to measure a velocity (of shock or of detonation) or the velocity of a free surface, by determining the transit time over a previously fixed distance. The qualities to be expected are therefore:

- the least possible disturbance of the flow studied (particularly for probes which will be "touched" subsequently);
- a signal of sufficient intensity to stand out from the background noise, with a steep and highly reproducible ascent front; and
- a response time which is very short compared to that of the passage of the shock or of the surface, with the smallest possible scatter of this response time (jitter).

In fact, the ideal chronometric probe, suited to all experimental needs, does not exist (or at least not yet). Each type of probe shown here is more or less suited to one range of incident shock pressures, to one kind of material (solid or liquid, explosive or not, conductor or insulator, etc.), and to one experimental environment.

A *passive probe* functions as a switch, creating a short circuit between two conductors at the instant of the passage of the shock wave or of the free surface. The corresponding electrical impulse is generated by the discharge of a capacitor, as shown in Figure X.25. The capacitor (usually some tens of pF) is charged a voltage of about 100 V delivered by a battery. Closure of contact by the probe creates a discharge front which is propagated in the line connecting the probe to the capacitor and then into the line transporting the signal to the recording system. The damping resistance R (usually 100 Ω) is

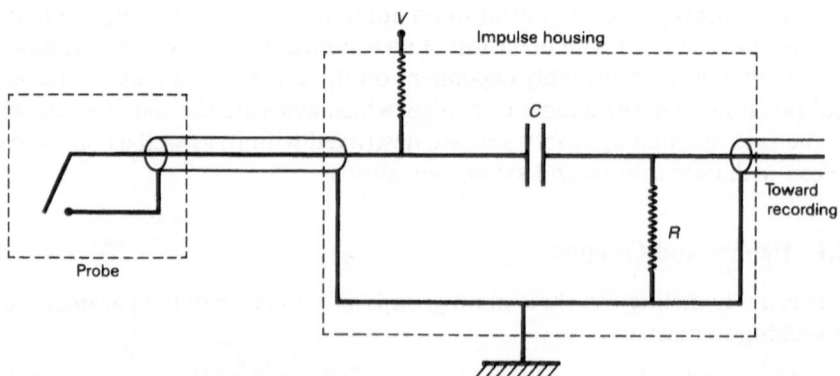

Figure X.25. Simplified diagram of the discharge circuit of a passive chronometric probe, designed to make the discharge aperiodic, thus avoiding subsequent oscillation. To obtain a very steep ascent front (less than one nanosecond), the self-inductance of the discharge circuit is reduced as much as possible by using coaxial structures wherever possible, within the probe itself and within the housing—called the impulse housing—which houses the capacitor and its matching resistance. This impulse housing is also placed as close as possible to the probe to limit degradation of the signal in the first linking cable, which—as it is destroyed during the experiment—is usually of inferior quality to that of the cables which transport the signal to the recording shed. The signal thus produced has a maximum of several tens of volts.

Figure X.26 shows the main types of passive chronometric probes developed. The needle probe was one of the first to be used for this application: displacement of the free surface after reverberation of the shock forming a closed circuit between the metallic needle and the conducting material studied, which is itself earthed. This probe is very crude and does not usually have good response time reproducibility, partly because of the difficulty in

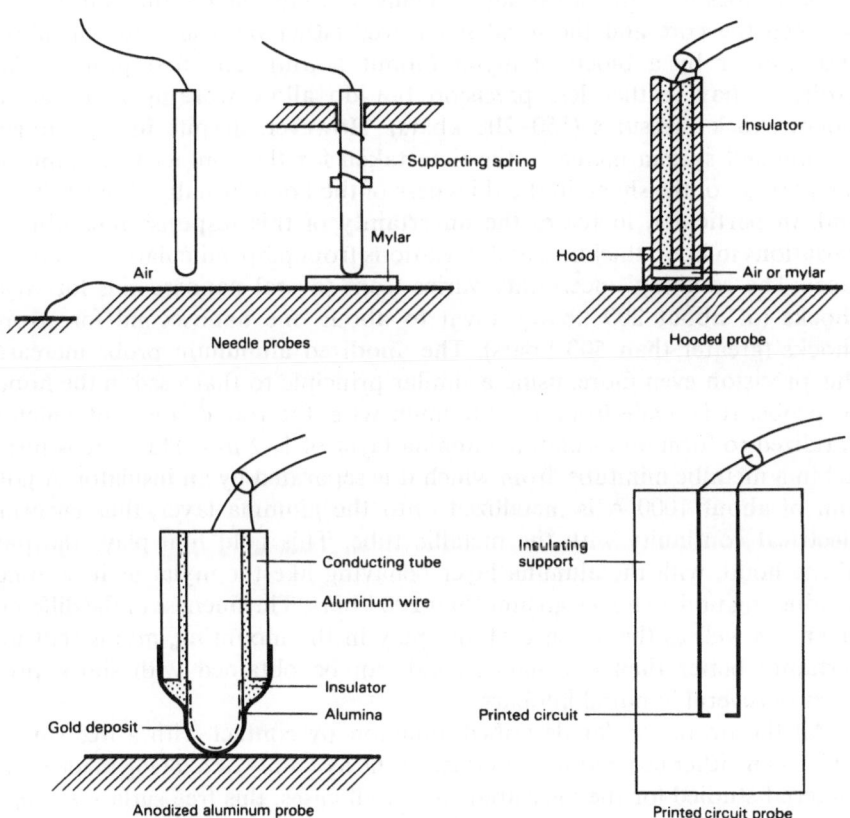

Figure X.26. Main types of passive chronometric probes.

obtaining sufficiently accurate positioning on the surface, partly because of the sensitivity of the break-down mode, between needle and surface, to possible precursors detached by the shock.

If the shocks are sufficiently intense (above 300 kbars), the following modification to the probe will cure these two faults: a thin sheet of an insulating material (usually mylar), made conducting by the passage of an intense shock, is placed between the surface and the needle. This device allows the needle to be positioned very close to the surface (about 5 μm) without risk of breakdown, and with a very high precision because of the rounded shape of the head and a calibrated support spring. In this way uncertainty of the order of one nanosecond is obtained with shocks of several hundred kilobars. One disadvantage of needle probes is that they can only function on conducting materials. Another disadvantage is that they are unsuitable for measuring the velocities of free surfaces. To reduce these two problems, the "hooded probe" has been developed, using the same principles. Contact is established by a shock between a needle and a metallic hood, which is itself fixed to a metallic minitube which surrounds the needle and is separated from it by an insulator. These probes have the advantage of being entirely coaxial. Initial insulation between the core and the hood is ensured either by a layer of air (about 100 μm) or by a block of mylar (about 5 μm). The first probes—"air probes"—have rather less precision but do allow working at lower incident shock pressures (150–200 kbars). However, despite its advantages, the hooded system increases the time taken for the contact to respond to the passage of the shock in the thickness of the hood (usually about 100 μm) and, in particular, increases the uncertainty of this response time (due to variations in hood thickness and deviations from perpendicularity to the surface). The resulting uncertainty varies from several nanoseconds for weak shocks (of about 200 kbars) down to about one nanosecond for strong shocks (greater than 500 kbars). The anodized aluminum probe increases this precision even more, using a similar principle to that used in the hooded probe. It is made from an aluminum wire, the rounded end of which is anodized to form an insulating alumina layer of 1–2 μm. This wire is inserted in a metallic minitube, from which it is separated by an insulator. A gold film of about 1000 Å is metallized onto the alumina layer, thus ensuring electrical continuity with the metallic tube. This gold film plays the part of the hood, with the alumina layer behaving like the mylar as it becomes conducting under shocks greater than 200 kbars. The fineness of the different layers, as well as the absence of any play in the mounting, means that uncertainty better than one nanosecond can be obtained with shock pressures of several hundred kilobars.

All the probes so far described, function by contact with a free surface which can either be a natural interface of the assembly or a facing made in the material studied for the measurement. In all cases, this free surface strongly disturbs the downstream flow, so that there can be no possibility of making several chronometric measurements at different depths along the same "line of flow": each probe must be placed on a stream line which is sufficiently displaced to be outside the disturbance cone caused by any upstream probe. Velocity measurement must therefore take account of the geometric form of the incident wave. However, the printed circuit probe can be buried in the mass of material studied so that it only disturbs the flow slightly, enabling several chronometric measurements to be made along the same flow line. This probe is formed from a very thin insulating support (kapton about 50 μm thick) on which metallic lines (copper of several tens of microns) are deposited, by the same technique which is used for printed circuits. These lines form conducting loops, interrupted at one point for several tenths of one millimeter. Several different loops can be printed on the same support. Such probes are generally used in solid explosive. After the cartridge is cut, in a plane perpendicular to the detonation propagation direction, the probe is inserted in this plane and the cartridge reconstituted by simple mechanical pressure. The contact is closed by the strongly ionized detonation wave. If this probe is used to measure detonation velocities, the transit time measurement must be corrected by allowing for the time of rebuild-up of detonation after travel through the probe. It can also be used to measure velocities of detonation waves passing over a free surface; the probe, with several conducting loops, usually being applied to the surface with mechanical backing. The chronometric uncertainty obtained with these probes is usually just below one nanosecond.

Active probes differ from passive probes. The current type uses the fact that ferroelectric ceramics can be depolarized under shock [11], [5]. Figure X.27 shows a possible means for constructing such a probe. It contains a very small ceramic chip (around one millimeter in diameter and several tenths of a millimeter thick) plated inside a metallic hood by an anvil which is also metallic, the hood being extended by a metallic tube. A metallic wire is connected to the anvil and is insulated from the tube. The front surface of the ceramic must be as flat as possible to reduce looseness in the assembly. It should be noted that the hood may be replaced by simple metallization which decreases the uncertainty of the transit time of the shock wave and

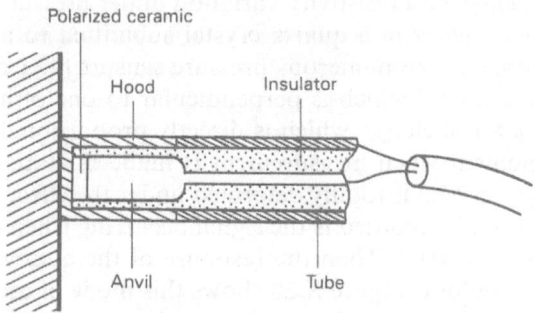

Figure X.27. Simplified diagram of a ferroceramic active probe.

thus improves precision. The ceramic is first polarized by a voltage of several hundred volts applied between its two faces at a temperature above its Curie point.

When this polarized ceramic is traversed by a shock wave, it suffers irreversible transformations, which cause electric charges to be released, thus creating a current in the circuit which connects the two faces of the disc. A signal of several tens of volts may be recorded, which starts at the instant of arrival of the shock on the front face of the ceramic, and whose shape depends on the pressure of the incident shock. The chronometric precision obtained with such a probe is comparable to that of the passive probes described above: it varies from several nanoseconds for low pressures (about 100 kbar) down to lower than one nanosecond for higher pressures (of the order of 500 kbar). It has the advantage of functioning within the range of low pressures for which the functioning mode of most passive probes becomes very uncertain. It should be noted that this type of ceramic probe can also deliver a piezoelectric signal of low amplitude, for shock pressures below one kilobar, which correspond, for example, to an elastic precursor. This does not preclude subsequent functioning by depolarization under a second, more intense, shock. The successive passage of two waves can therefore be recorded by the same probe.

Alumina, which can also be depolarized under shock [8], has also sometimes been used for the construction of active probes. An aluminum probe which has been anodized as described for passive probes can therefore be used as an active probe. However, the amplitude of the depolarization signals is noticeably lower than for ferroceramics.

(b) Manometric Gauges

Some authors [25] have tried to link the amplitude of the depolarization signals delivered by the active probes described above to the pressure of the incident shock. However, the precision of the pressure calibration of these probes remains limited. On the other hand, true manometric gauges have been developed for detonation experiments by using two different principles: the piezoelectric effect and resistivity variation under pressure.

The *piezoelectric effect* in a quartz crystal submitted to a static stress is well known, and is used in numerous pressure sensors [15]: a stress applied to one face of the crystal which is perpendicular to one of its three electric axes creates an electric charge which is directly proportional to the applied stress. At the moment when measurement is made in static or quasi-static gauges, the quartz crystal is totally deformed under the effect of the pressure. But in detonics what is recorded is the signal occurring when the shock wave travels through the crystal. Then the response of the quartz results from a different phenomenology. Figure X.28 shows this mode of use.

When the quartz is traversed by a flat shock wave that is parallel to the electrodes and of constant velocity U, polarization P is proportional to stress

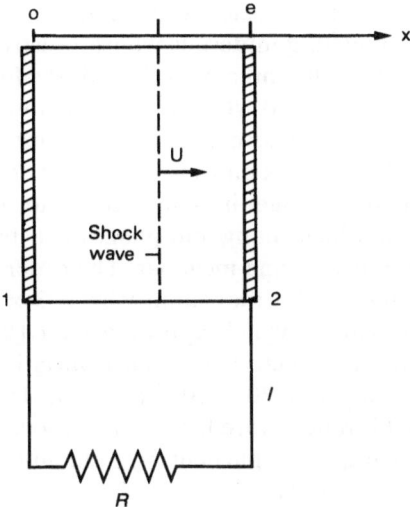

Figure X.28. Functioning of a quartz piezoelectric gauge under shock.

σ: $P = \kappa\sigma$. Furthermore, assuming that resistance R is negligible compared to that of the quartz, a close approximation to the current intensity I is:

$$I \simeq \kappa \frac{S}{e} U(\sigma_1 - \sigma_2),$$

where S is the area of the electrodes, e is the thickness of the quartz, and $\sigma_1(\sigma_2)$ is normal stress on the entry (exit) face. Thus, as long as the shock has not reached the exit face ($\sigma_2 \approx 0$), stress σ_1 is proportional to intensity I of the current, measured over time while recording on a fast oscilloscope the voltage at the terminals of resistance R.

This piezoelectric gauge therefore allows continuous measurement of the changes in pressure on the interface between the material studied and the quartz crystal. The duration of the measurement is a function of the thickness of the quartz. The diameter of the crystal must be large enough for the wave to remain flat in the measurement zone (a guard ring is sometimes used to reduce this diameter). The sensitivity of this device is excellent: it can reach several bars. However, the linear relation between intensity and stress only holds up to about 25 kbar. Above that, the response of the quartz is more complex [17].

A major problem with the piezoelectric gauge described here is that it measures only the equilibrium stress at the interface between the material studied and the quartz crystal. However, use of the *piezoresistive* effect allows the manufacture of gauges which can continuously measure the pressure within an inert or even explosive material. These gauges use the resistivity variation caused by pressure in a conducting wire. Since shock phenomena are always accompanied by considerable temperature increases, experimenters have tested conducting materials whose resistivity varies as little as possible with temperature. The three materials most often used for this purpose are manganin, carbon, and ytterbium. Manganin is an alloy of Cu (86%), Mn (12%), and Ni (2%), whose use as a static piezoresistive manometer for measuring very high pressures dates from the start of this century. Fuller and Price [14] first used a manganin wire manometer to measure dynamic pressures. The range in which manganin gauges can be used is very wide, from several kilobars to nearly one megabar. The order of magnitude of the relative resistance variation of a manganin wire is 3×10^{-3} per kbar. The carbon shock pressure gauges which appeared later [9] show a much improved sensitivity, since their relative resistance variation is about ten times greater than that for manganin: 30×10^{-3} per kbar. However, the range in which they can be used is much more limited: from several kilobars to about 100 kbar. Ytterbium gauges are particularly well suited to measuring low pressures from 1 kbar to 30 kbar.

To limit the flow disturbances due to these pressure gauges, very thin flat foils of piezoresistive material are generally used, inserted between two thin sheets of insulator (kapton, teflon, etc.). The inclusion of flat copper foils between the insulating sheets ensures power supply in the gauge. Such a flat gauge is inserted between two planes in the material studied, perpendicular to the mean flow direction. Variations in resistance are measured by a Wheatstone bridge or by a pulsed source (to avoid preheating of the gauge) of constant intensity, the voltages at the edges of the foil being recorded by an oscilloscope. The response time of the gauge is determined by the time taken for the piezoresistive foil to reach pressure equilibrium with the material studied. This time depends on the thickness of the different elements in the gauge and on their shock impedance as well as that of the material studied. Since theoretical determination of the response of a gauge to pressure is very complex, each type of pressure gauge is calibrated for shock experiments by determining shock pressures with high precision by other means (usually by measuring the shock velocity knowing the shock polar of the material). Gauges sometimes show hysteresis, which can slightly affect measurements made during expansion following a strong shock. The precision in measuring the pressure thus obtained is of the order of several per cent.

(c) Velocimetric Gauges

Several authors have measured the velocity of a free surface by using as gauge a flat air capacitor formed from the surface studied (which must be metallic) and a fixed armature arranged parallel to this surface. Recording of the variations in potential at the edges of this capacitor, at a constant charge, allows calculation of the variation in capacitance, which itself is directly related to the distance between the electrodes. The velocity can then be calculated from the displacement of the surface with time. However, this velocimetric gauge has severe problems: partly because the velocity measurement is not made at a point but integrated over a surface which has to be large enough to limit boundary effects, and partly because the method is sensitive to precursors (metallic particles) which can be detached from the free surface by the initial shock. The precision of measurement is poor, particularly in comparison to that attained by Doppler laser interferometry systems, so this gauge has now been almost completely abandoned.

However, another type of velocimetric gauge, which appeared in 1960 [28], is still of great interest since it is one of the rare ways of directly measuring the material velocity behind a shock or a detonation front. A very thin conducting foil, oriented perpendicularly to the shock propagation direction, is buried in the material studied. At the instant of the experiment, a uniform, constant magnetic field is created perpendicular to this foil and to the shock propagation direction. Since the foil has very low inertia, it is very rapidly brought after the shock to the material velocity attained by the medium behind the shock. Since this conductor is displaced perpendicularly to the direction of the magnetic field, an induced e.m.f. appears at its edges, which is a function of its displacement velocity and can be recorded. Figure X.29 is a simplified diagram of this device.

If ℓ is the length of the foil, B is the magnetic field, and u is the material displacement velocity, Faraday's law gives the induced e.m.f. In the electric circuit which includes the moving foil as $V = B \cdot \ell \cdot u$. B and ℓ are measured before the experiment, and the changes in voltage V, at the edges of an adapted charge resistance, are recorded on an oscilloscope as a function of time. The material velocity u(t) is thus known for each instant. With a magnetic field of several hundredths of one Tesla, for a foil several millimeters long, a signal of the order of one hundred millivolts is obtained for a material

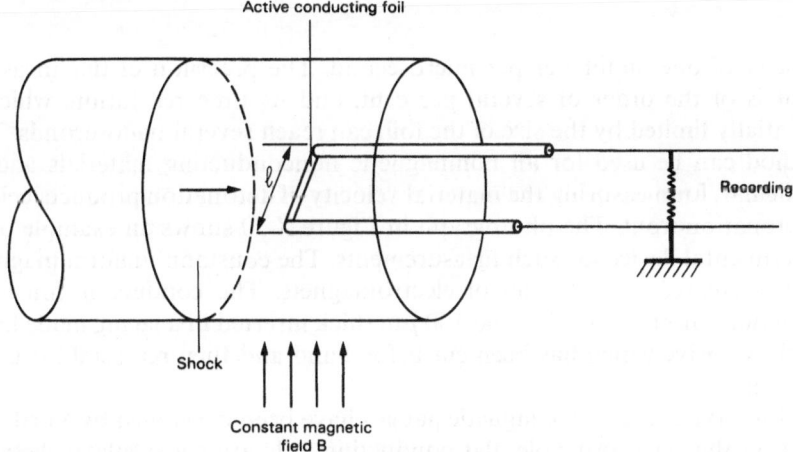

Figure X.29. Simplified diagram of an electromagnetic gauge for measuring material velocity.

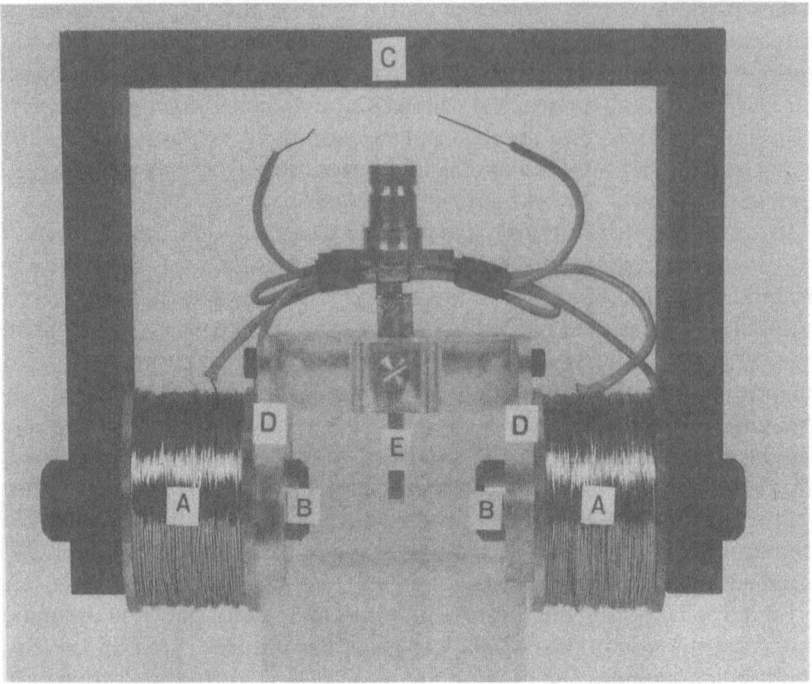

A — Bobbins with 1090 turns of enamelled copper, 0.4 mm diameter.
B — Flux concentration cores.
C — Soft iron stirrup to complete the external magnetic circuit.
D — Plexi supports connecting the electromagnet to the sample housing.
E — Measuring loop.

Figure X.30. Experimental device for measuring the material velocity of detonation products.

velocity of one millimeter per microsecond. The precision of this measurement is of the order of several per cent, and its time resolution, which is essentially limited by the size of the foil, can reach several nanoseconds. This method can be used for all nonmagnetic nonconducting materials and, in particular, for measuring the material velocity of detonation products behind a detonation front. The photograph in Figure X.30 shows an example of an experimental device for such measurements. The constant, uniform magnetic field is created by a system of electromagnets. The conducting foil is an aluminum sheet 2 mm wide and 100 μm thick inserted in a facing made in the solid explosive which has been cut beforehand and then reassembled under pressing.

Other types of electromagnetic gauges have been developed by Vorthman [27] on the same principle: flat conducting foils are encapsulated between two thin insulating layers. Several measuring loops can be incorporated in one and the same gauge. The device is then inserted between the two parts of the explosive slab, cut along a plane at an angle with the detonation propagation direction. This system has the advantage of allowing velocity measurements to be made at several different points during the same experiment, with minimal use of explosive.

2.2. Recording Methods

In the years following the Second World War, it was often the demands of experimenters in detonics which prompted electronic technology and materials for rapid recording. This demand no longer plays such a role in their development, since experimenters can now easily find commercially available equipment with performances broadly corresponding to their precision and time resolution requirements. Since the number of techniques in this area which are genuinely specific to detonics are thus limited, little attention will be paid to descriptions of equipment, but rather to the required performances.

Two main recording techniques are generally used, according to the type of signals collected in the experimental assembly: *oscillography* and *digital chronometry*. Oscillography is particularly well suited to recording analogue signals of voltages from velocimetric or manometric gauges, as well as controlling the shape of the pulse from a chronometric probe. Conversely, digital chronometry is much simpler to operate when measuring velocities of waves or of free surfaces with chronometric probes, when a large number of pulses must be collected in one assembly.

(a) Analogue Oscillography

The recording is made by a single sweep of the oscilloscope, synchronized to the phenomenon studied, the screen being photographed by a camera which has been preset to a long exposure. Several parameters are essential when choosing an oscilloscope for recording high-speed signals. These are now briefly reviewed. First, the band-width characterizes the capacity of an oscilloscope (amplifiers and tube) to transmit transient signals without deforming them. It is conventionally defined by the frequency response to a sinusoidal signal: it corresponds to a maximum attenuation of 3 dB (or 0.707) as defined in Figure X.31.

In fact, for a nonperiodic unique signal, the primary interest is in the degradation of the ascent fronts which is caused by this limitation of the frequency response. In the most damaging case of a perfectly square source signal, the oscilloscope will record a progressive signal (see Fig. X.31). The rise time τ of this signal may be defined as the time required to increase from 10% to 90% of its value. It is generally agreed that τ is related to the band-width B by the simple relation: $\tau \times B = 0.35$.

In theory, if the transfer function of the equipment is known, this recorded signal can be decoded by Fourier series transform to recover the original

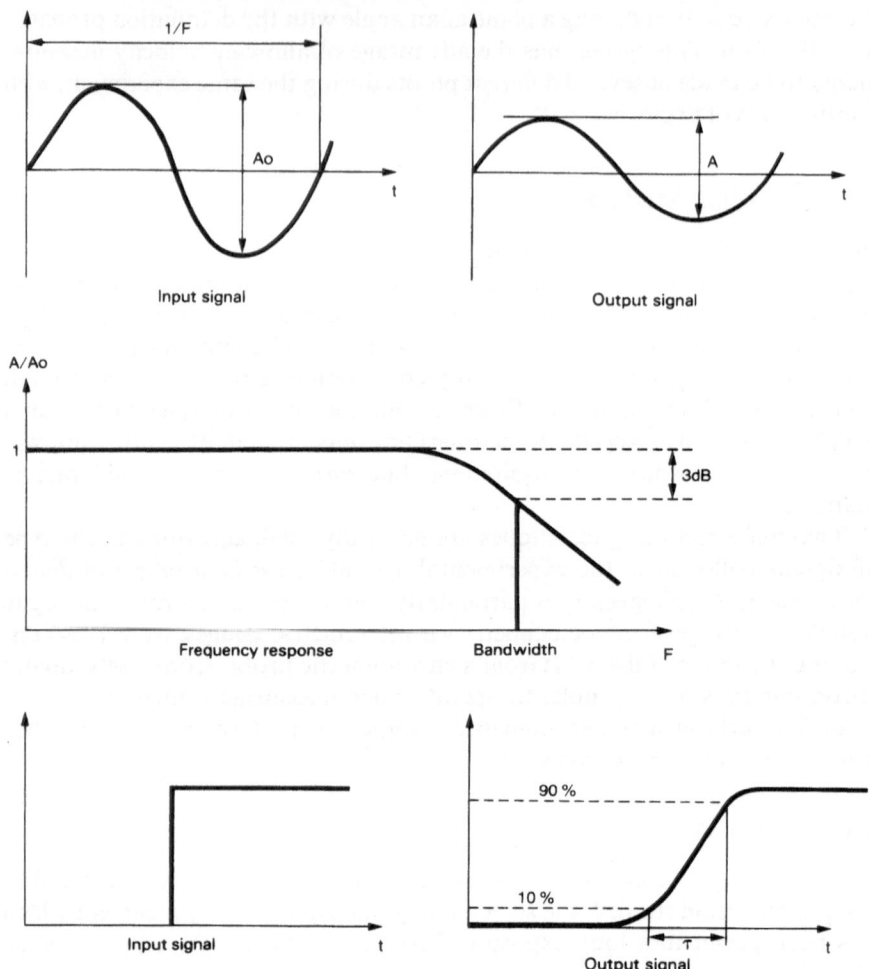

Figure X.31. Response of an oscilloscope to transient signals.

signal. This type of correction is limited in practice, since the attenuation increases very rapidly for high frequencies, and the corresponding information is lost in the background noise. The time resolution of the recording is thus appropriate to signals whose width is at least a fraction of the value of τ. Fortunately, this is only rarely constrictive for detonics measurements, since 1000 MHz (or over) band-width oscilloscopes are available, for which τ is 350 ps.

Yet another factor limits the possibility of recording transient signals by analogue oscillography—the sweeping rate on the film. In fact, for the signal to be recorded, the luminous intensity of the corresponding trace must be sufficient to show on the sensitive surface. Other things being equal, this intensity is weaker when the spot displacement rate is greater. To increase the resulting maximum rate, manufacturers have increased the electron accelerating voltage (up to values of the order of 25 kV) and have improved the electron–photon conversion factors of phosphor screens. In some recent oscilloscopes, plates of microchannels are used, similar to those used for high-speed cameras (see §1.2(b)), placed in front of the screen in order to multiply the electrons by secondary emission. This device makes it possible to move the maximum recording rate of the order of 9 cm/ns for the best conventional equipment, up to maximum rates of nearly 20 cm/ns. The primary conclusion here is that, for certain types of oscilloscopes, the limitation due to recording rate can sometimes be more restrictive than the limitation due to band-width for single sweep recording of highly transitory signals.

The time resolution of measurement with an analogue oscilloscope also depends on the spatial resolution of the recording. This is essentially a function of the spot dimension (usually several hundred microns) and of the sweeping rate. This limitation usually corresponds to several parts per thousand for the duration of sweeping. For the highest sweeping rates, this is often not as restrictive as the band-width limitation, but it becomes so at the lower rates used to increase the measuring range duration.

The chronometric precision is not necessarily as good as the resolution, because of faults in the sweeping linearity guided by the time base. This precision is generally at best about one per cent of the sweeping time. It is possible to improve it by marking time with a stable frequency pulsed signal provided by an oscillating circuit.

In the same way, the precision of the signal amplitude essentially depends on the linearity of the deflection system response. This precision is generally at best about one per cent of the value corresponding to the maximum deflection. Thanks to amplifying systems, the sensitivity range that can be covered by one oscilloscope is very high: a typical range is from less than one millivolt up to several tens of volts. The experimenter can and must first choose the appropriate sensitivity: one which allows all the dynamics of the expected signal to be covered with the greatest precision. He can and must also adapt the sweeping rate chosen to the anticipated duration of recording to benefit from the best chronometric precision. To improve precision, the same signal will often be recorded on several different oscilloscopes (or on two different traces within the same oscilloscope) while using different sensitivities and the highest sweeping rate, these recordings being displaced in time. The resulting synchronization problems will be examined later.

(b) Digital Oscillography: Transient Analyzers

Along with analogue recording techniques for ultrafast signals, digital recording techniques have been developed in recent years, whose performance is now just as high. These have the great advantage of allowing immediate computer processing of the signal information, thus eliminating the subjectivity of a visual examination of the oscillogram or the laborious work of digitizing the frame.

Different types of equipment are commercially available. Their performances are similar, their principal differences lie in the methods of time sampling and of digitizing the signals. Some use an oscilloscope tube whose phosphor screen is replaced by a mosaic of silicon diodes. These diodes are first charged, being discharged when they are touched by an electron beam. The mosaic of diodes thus memorizes the oscilloscope sweep trace. A second tube, placed opposite, reads this information by sweeping the mosaic column by column, thus recharging the discharged diodes. The current thus produced is used to store in a digital memory the number of the column (time information) and the position of the diode within the column (signal amplitude information). Other systems also use an oscilloscope tube, but without time sweeping. The amplitude deflection is recorded in real time by a series of elongated silicon diodes, arranged parallel to the deflection direction. The electron beam is enlarged into a layer to cover the entire series of diodes, while a time marking system allows direct digital coding of the amplitude deflection, over as many bits as there are diodes. A real time sampling circuit periodically takes these digital values and stores them in memory. Lastly, some very recent instruments do not contain cathode ray tubes but use very high performance analogue–digital converters for direct digitization of the amplitude signals which are sampled at very high frequency and stored in memory.

The characteristics which must be taken into account when choosing this type of instrument to record ultrafast transient signals are essentially:

- the band-width, for the same reasons as those invoked for analogue oscillography (the best digital equipment now available reaches 500 MHz);
- sampling frequency which fixes time resolution (at best 500 ps) and the stability of this frequency which limits precision (with a digital oscilloscope this depends on the sweeping linearity, as for an analogue oscilloscope, but with a sampler this can be significantly improved thanks to high stability clocks);
- the number of levels of coding the amplitude signal which give the measuring resolution, and the response linearity which determines its precision (usually comparable to that of analogue oscilloscopes); and
- the number of points which can be stored in the memory, which determines the total recording time at a given frequency (currently a few thousand points in the memory at best).

Because of the immense progress made with this type of instrument, it is reasonable to suppose that in a few years they will steadily supplant analogue oscilloscopes for recording transient signals in detonic experiments.

(c) Digital Chronometry

To determine the velocity of a wave or of a free surface with chronometric probes, it is not usually necessary to record the shape of the pulses delivered, but only to be able to measure the times between the different pulses. Among other considerations, a very small uncertainty (about one nanosecond) is sought for such measurements on time bases which are rather long (up to several microseconds). Now, oscilloscopes and transient analyzers do not often allow such precision, except when significantly shorter time bases are used. The experimenter must then resort to more or less complex synchronization devices where several recorders are displaced in time. The multiplication of measuring points, which is often necessary to take account of the wave form, also makes these recorders difficult to use: the recorders must either be multiplied in parallel, or the signals from different probes must be multiplexed. All these factors show why multipath digital chronometric systems are of interest, since they are very simple to use and offer high precision with very long time bases.

These chronometers generally contain a very highly stable quartz clock, which is used to measure the time interval $\mathscr{T}$ separating a common trigger pulse from the stop pulse received on each of the measuring paths. Since the frequency of the quartz clock is usually insufficient to provide the required precision, each path contains a vernier system—or time expansion circuit— which allows determination of the time interval separating the stop pulse received on this path from the immediately following pulse from the quartz clock. Figure X.32 shows the functioning of this device.

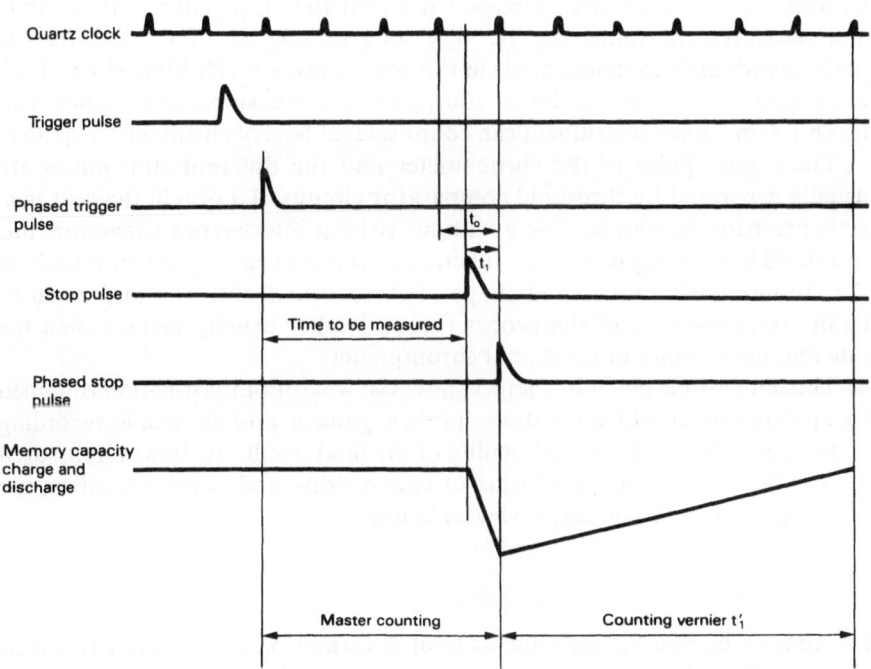

Figure X.32. Principle of a high performance digital chronometer.

The trigger pulse from the chronometer is brought into phase with the immediately following pulse from the clock. The times measured on the different stop paths of the chronometer are related to this phased pulse. The stop pulse received on one path is also phased with the immediately following pulse from the clock. The two phased pulses—trigger and stop—serve, respectively, to start and to stop the master clock pulse counter which thus determines the time $\mathscr{T} + t_1$ ($\mathscr{T}$ is the time to be measured, t_1 is the time interval between the stop pulse and the immediately following pulse from the clock). To determine the time t_1, the time expansion circuit uses a memory capacity: the stop pulse charges this capacity by a current of intensity I up to the arrival of the phased stop signal which discharges it by a current of much weaker intensity I'. The discharge time t_1' will then be much longer than the charge time, so that $t_1'/t_1 = I/I' = $ constant: the ratio κ of the two intensities generally being of the order of 100, the time t_1' can be measured with good precision by the clock vernier pulse counter, started by the phased stop pulse and stopped by a comparator detecting the end of capacity discharge. The time t_1 can thus be determined, and subtracted from the time given by the master counter to obtain the value $\mathscr{T}$ of the time to be measured.

The precision obtained with this device is limited only by the resolution of the vernier system, for the clock stability (which can be as good as 10^{-7}/day) allows excellent precision in the master counter measurements. The precision in measuring t_1 is of the order of t_0/κ_0 (t_0 being the clock period). The chronometric precision between the stop paths is then $2t^0/\kappa$, and measurement times can be very long compared to t_0. The highest performance chronometers made in this way contain a 100 MHz clock. They reach precisions of the order of 100 ps for the measured times which can reach 100 ms. Each instrument can count several tens of chronometric paths.

The trigger pulse of the chronometer and the different stop pulses are usually generated by threshold comparator circuits. To obtain the best possible precision in response time without risking interference triggering, this threshold level is regulated as a function of the expected signal amplitude of the chronometric probes and shape of the ascent front. The dispersion as to the response-time of the probes (see §2.1(a)) is usually greater than the inherent uncertainty of the digital chronometer.

The indications given in Plate 3 show the wealth of information obtained by an experiment which combines probes, gauges, and electronic recording techniques. The existence and quality of the final result are, however, assured only if the less "noble" problems of connections and synchronization are controlled. These are briefly reviewed below.

(d) Connections and Synchronization

In order to be able to transmit without deformation the ultrafast transient signals collected from a firing assembly, the experimenter must pay special attention to the quality of the connecting cables used, particularly when their length is very great, because of the need to protect the recording means against the destructive effects of the explosion. Cable manufacturers usually provide a graph of attenuation per unit of length as a function of the frequency transmitted. For consistency, cables are usually chosen so that the band-width of the connection, defined as an attenuation of 3 dB over its total length, is at least equal to that of the recording system used.

Furthermore, when chronometric precision equal to or less than one nanosecond is sought between the measuring paths, it is essential to take account of the delays introduced by the signal propagation times in the corresponding different cables. The majority of cables have a signal propagation velocity of the order of 0.2 m/ns, which gives a path time of 150 ns. for the connections of about 30 m which are currently encountered in the firing area. As a rule, all the measuring paths are chosen so that they will be of the same "electrical length." If this is not possible, the propagation time in each cable must be measured precisely, and this value used to correct the chronometry of the signals recorded in each path.

To attain high chronometric precision in electronic measurements, there must be careful study of the synchronization of the different recorders with regard to the phenomenon studied, and also with regard to each other. The problem of synchronization is particularly important in the case of oscilloscopes or transient analyzers, since it has been seen that the chronometric precision of such equipment is better when their measurement range is short (for a precision of 1 ns, the measurement range is typically limited to 100 ns). It is therefore generally impossible to synchronize with the pulse for firing the explosive, since the uncertainty of the priming time with a detonator is usually greater than 10 ns. The possibilities of internal triggering of single-sweep oscilloscopes are often used: recording is triggered by the rising front of the first signal received, with an internal delay system allowing recording of the complete signal. When this is not possible, a pulse delivered by a chronometric probe can be used, placed slightly ahead of the measuring gauges, to activate the recording. To synchronize several recording devices with each other, activating make-and-breaks are used which, on receiving a single pulse, deliver several pulses which are strictly synchronized. When displacement of the time ranges is needed for different recordings, the activating pulses thus obtained can be transferred by delay generators (which are sometimes integrated with the recorder). These delay generators can be made of a simple delay line for the shortest times: the pulse is delayed by a path along a length of cable determined beforehand. For the longer times, monostable circuits are generally used, where the time constant can be altered in discrete steps, which, after a trigger pulse, deliver a delayed new pulse. All these electronic systems now offer high precision operation. However, when different recordings need retiming with regard to each other, one and the same pulse is often recorded on two recorders with overlapping time ranges.

3. Radiographic Techniques

Radiography is practically the only method which allows observation of

Plate X.3. Application of electronic techniques to the experimental determination of the surface of state of a solid explosive.

The experimental device opposite associates chronometric probes with piezo-resistive gauges which may be of carbon or manganin according to the range of pressures to be measured. The shock generator used is a powder launcher; the projectile, whose front face is of copper, impacts a transmitter which is also of copper. The hooded chronometric probes, linked to a digital chronometer, allow measurement of the velocity V of the projectile, by measuring the flight time over a previously determined distance. Knowing the copper shock polar, the pressure and the material velocity behind the incident shock in the transmitter can then be determined. This shock is transmitted to the solid explosive sample studied, which is inserted between the transmitter and a plexiglass anvil. The two piezoresistive gauges, where the resistance variations are recorded by oscilloscopes in Wheatstone bridge assemblies, allow the changes in pressure to be tracked both at the transmitter–sample interface and at the sample–anvil interface. Gauge 1 successively shows the increase in pressure corresponding to the shock transmitted into the sample, and the decrease due to the arrival of the release wave resulting from the transmission of the shock from the sample into the anvil whose shock impedance is weaker. Gauge 2 shows the increase in pressure corresponding to the shock transmitted from the sample into the anvil and then the decrease due to the arrival of the release wave linked to the reflection of this shock from the free surface of the anvil. These different episodes are clearly shown by the oscillograms which record the voltages at the terminals of these two gauges and by examining the pressure values obtained from the calibration curves of the gauges. Measurement of the time interval $t_1 - t_2$ separating the pressure rises in the two gauges allows, knowing the sample thickness e_0, determination of the velocity U of the traveling shock. Knowing the initial sample density ρ_0, the state of initial shock in the transmitter, and the release isentrope originating from this point, a state (p_1, u_1) of the explosive sample shock polar can be determined in the plane (p, u). Direct measurement of the pressure corresponding to the first level recorded on gauge 1 can confirm the validity of this determination of p_1. Furthermore, knowing the shock polarity of the material forming the anvil, measurement of the second level of pressure recorded on gauge 2 allows determination of a state (p_2, u_2) of the isentrope originating from state (p_1, u_1).

Furthermore, the existence of pressure levels after the arrival of the first shock on gauges 1 and 2 can confirm that this shock is weak enough to leave the explosive inert. In fact, when the shock is intense enough to induce chemical reactions, there is a progressive increase in pressure on the two interfaces after the passage of the first shock.

flows within a material without causing disturbance. This technique is therefore of great interest in detonics, for observing positions and shapes of waves and interfaces. It will also be seen that it is the only way for direct determination of the instantaneous local density of materials in a flow (e.g., behind a shock front or a detonation front). We shall now discuss in turn the three main aspects of this technique:

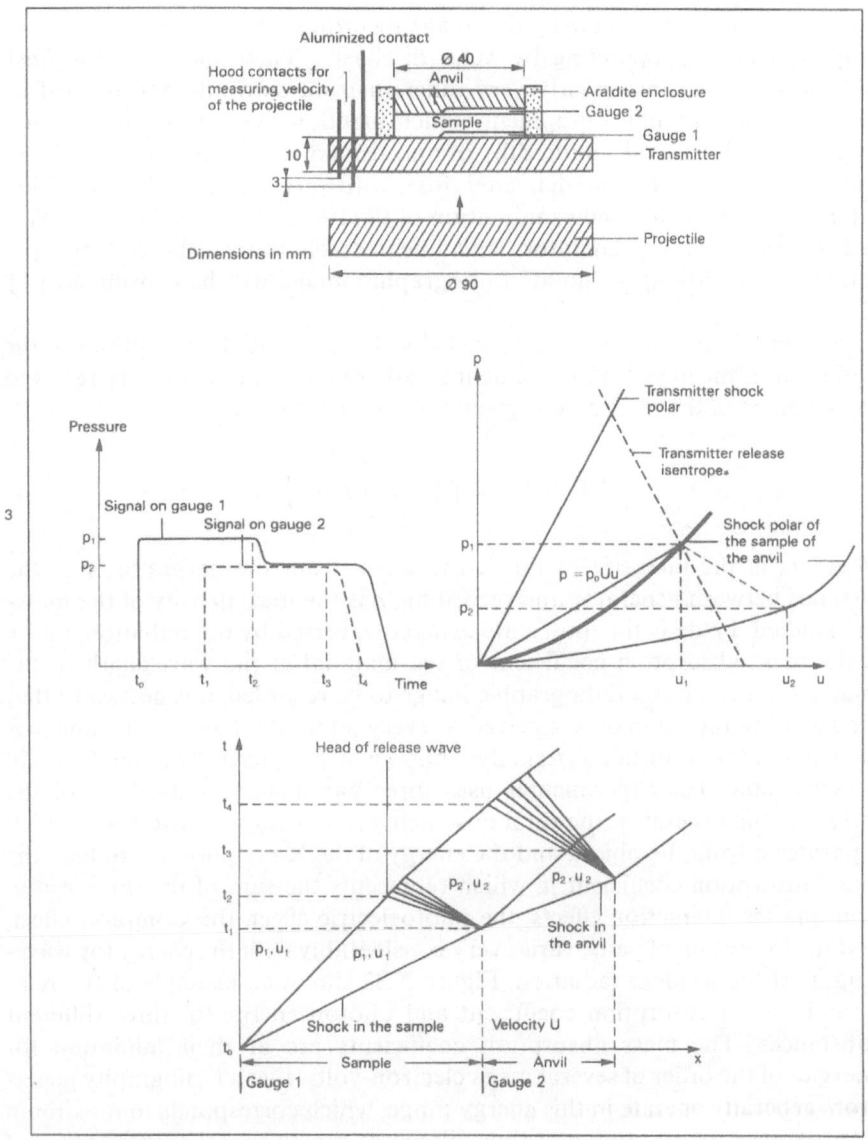

- radiographic generators;
- recording techniques; and
- plate processing techniques.

3.1. Flash Radiographic Generators

To limit the effects of kinetic flow on a radiographic plate in detonic experiments, it is necessary to use exposure times considerably below one microsecond, as for photographic techniques. However, these very brief exposure times are not obtained here by the image recording system but by the radio-graphic generator operating by pulse discharge. These machines for *flash radiography* are consequently very different from conventional medical or industrial radiographic equipment, which function over much longer times (typically from several hundredths of one second to several hours) and can thus operate, for the same delivered dose, with very much weaker radiation intensities. Before detailed examination of the two technologies used in flash radiography—Marx generators and linear accelerators—the essential pa-rameters for forming a "good" radiographic image will be considered [4], [18].

A radiographic image is achieved due to differential absorption of the radiation in the object. For monochromatic radiation, the intensity received at one point of the recording is given by the relation

$$I/I_0 = r^{-2} \exp\left(-\mu \int \rho \, dx\right),$$

where I_0 is the intensity of the X-ray beam from the generator, r is the distance between generator and recording, ρ is the local density of the mate-rial studied, $\int \rho \, dx$ is the total surface mass traversed by the radiation, and μ is the mass absorption coefficient of the material at the wavelength of the incident X-rays. For a radiographic image to be recorded, it is necessary that the resulting radiation dose received at every point of interest in the image is included in the recording system dynamics, which is generally limited, as will be seen later. The experimenter uses three parameters at the level of the radiographic generator: the total dose delivered $\int I_0 \, dt$, the distance r which separates it from the object, and the energy of the X-ray photons. In fact, the mass absorption coefficient μ, which represents the sum of the three radia-tion–matter interaction effects, the photoelectric effect, the Compton effect, and the formation of pairs, varies very considerably with the energy (or wave-length) of the incident radiation. Figure X.33 shows an example of the rela-tions between absorption coefficient and photon energy for three different substances. The mass absorption coefficients are at their minimum for energies of the order of several mega electron-volts. Flash radiography gener-ators generally operate in this energy range, which corresponds to maximum transparence to radiation and thus allows, at the given dose, penetration of the maximum surface mass of a given material. However, it is also for this energy range that there is minimum contrast between surface masses pene-trated in the same material on the one hand, and between two materials for the same surface mass on the other, since the values of the absorption coeffi-cients of the different materials are close to one another in this range. If a sufficient incident dose is available the experimenter can optimize the con-trast on the plate by altering the wavelength so that the doses issuing from the different surface masses penetrated, in the zone of interest in the object, use the dynamics of the recording system to the best possible advantage.

Mass absorption coefficient (cm²/g)

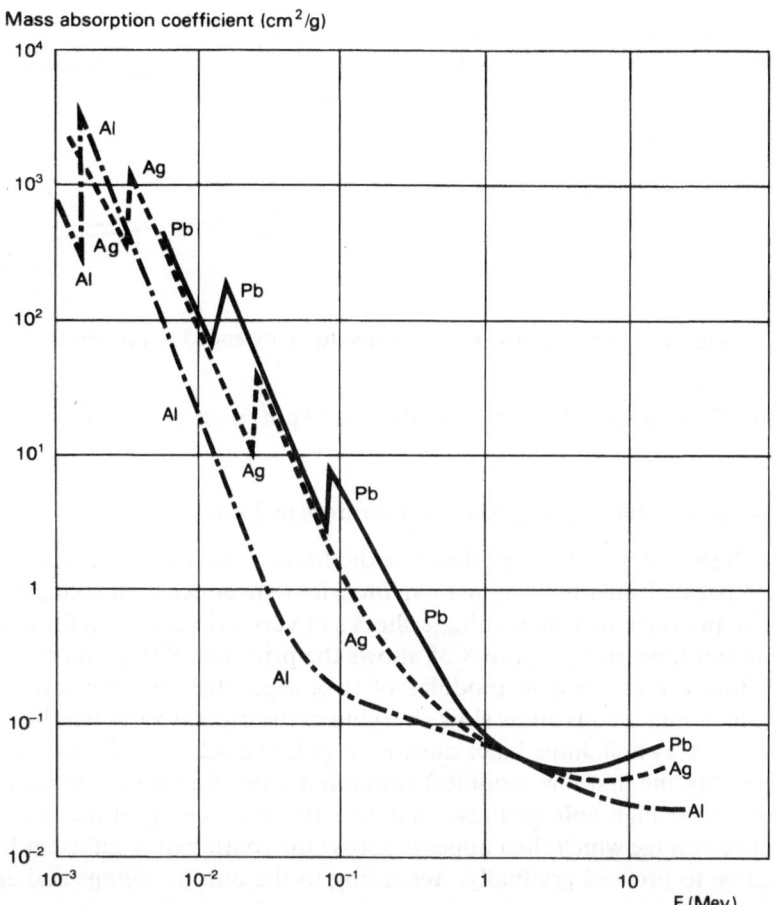

Figure X.33. Mass absorption coefficients of X-ray photons in Al, Ag, and Pb.

Since the X-ray beam from the generators is produced by bremsstrahlung, it is never monochromatic, but shows a continuous spectrum whose maximum energy is a function of the energy of the electrons striking the target. By altering the voltage that is accelerating these electrons, the maximum energy of the X-ray photons emitted can be varied in order to achieve this optimization.

One last important parameter for the quality of radiographic plates, and thus for the precision of measurement, is the size of the emissive source or "focal spot." In fact, since the radiation source is never strictly a point, every contour of the object suffers a penumbral or geometric fuzziness on the recording as shown in Figure X.34. Since the object can neither be placed very close to the recording device (because of the destructive effects of the experiment) nor too far from the generator (because the intensity decreases as the

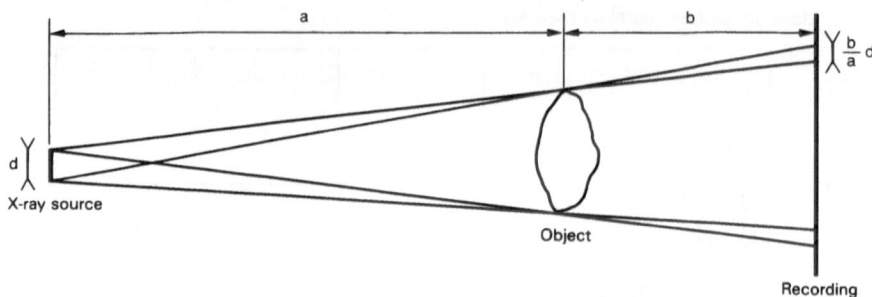

Figure X.34. Geometric fuzziness due to an extended X-ray source.

square of the distance), the area of the focal spot must always be as small as possible.

(a) X-Ray Machines Using Marx Generators as Power Source

These flash radiographic machines use the Marx principle of charging capacitors in parallel and discharging them in series to produce high voltages. Such a device provides very high voltage shocks of very brief duration for entering the radiographic tube. Figure X.35 shows the principle of this generator. The capacitors are grouped in modules of two, separated by inductance coils. These inductance coils allow slow charging of the capacitors in parallel, while showing a very high impedance during the pulsed discharge. The first module (or possibly the first few modules) contain(s) a priming electrode which, on release of the high voltage pulse, activates the functioning of the first spark gap. The voltage which then appears across the spark gap is sufficient for the discharge to proceed gradually. According to the output voltage and energy required, the generator can contain up to several tens of capacitor modules. These modules are generally mounted in a metallic tank filled with an insulating fluid. This device allows insulating distances to be reduced, and thus improves compactness, while also limiting electromagnetic interference which can be troublesome when electronic recording systems are adjacent. Such a generator delivers, on an adapted circuit, a very brief voltage pulse whose magnitude is close to the product of the charging voltage of the capac-

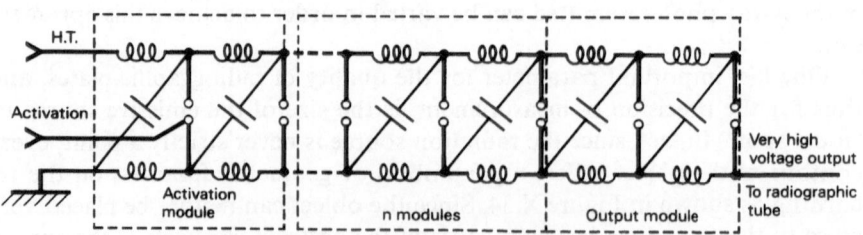

Figure X.35. Principle of a Marx generator.

itors and the number of modules. Since the variation of intensity with time is very pronounced, the self-inductance of the discharge circuit must be very low to avoid losses. The connection between generator and tube must therefore be coaxial, and as short as possible. For this reason, the tube is often integrated with the Marx generator tank.

Flash radiographic machines cannot use hot-cathode X-ray tubes (with emission of electrons by a thermoelectronic effect) like those conventionally used in industrial and medical radiography, because of the very high current densities needed (of the order of 1,000,000 A/cm^2). Two sorts of tubes are in general use: field-emission tubes and vacuum discharge tubes.

Field-emission tubes contain a conical tungsten anode placed in the axis of a cylindrical cathode with a metallic comb. When a voltage pulse is applied to the anode by the Marx generator, the very fine points of this comb generate very intense fields (3×10^9 V/m), which allow electrons from the metal to cross the potential barrier of the surface by the tunnel effect (field emission). The conical shape of the anode allows the surface for electron impact to be increased, to avoid vaporization of the tungsten, while keeping a very small apparent emissive surface vis-à-vis the object. The anode, which can thus serve for very many X-ray flashes, need not be collapsible, but is placed in a sealed glass ampoule under high vacuum. Figure X.36 shows the principle of this device.

Vacuum discharge tubes use the phenomenon of the breakdown which is produced, if the voltage is sufficiently high, between two electrodes in a high vacuum (better than 10^5 torrs). The most up-to-date tubes of this type use an annular cathode and a pointed tungsten anode, which allows the discharge to be focused on a tiny surface. Figure X.37 shows the principle of such a tube. For the lowest voltages (< 300 kV) the discharge of the tube must be primed by an activating electrode guided by a high voltage pulse

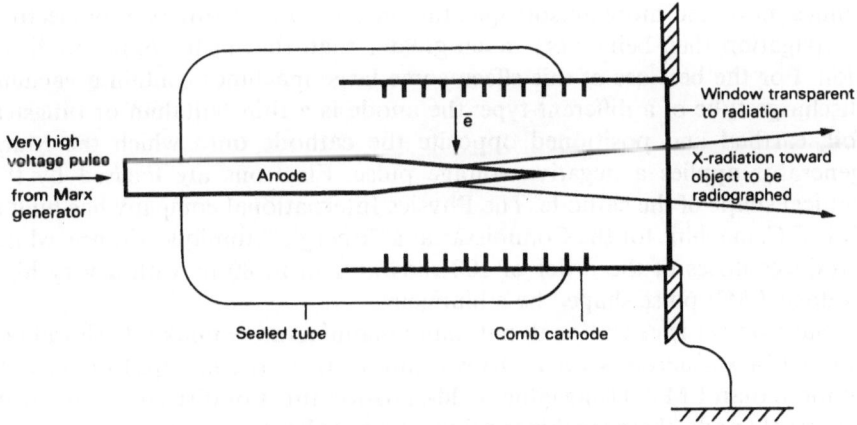

Figure X.36. Diagram of a field-emission X-ray tube.

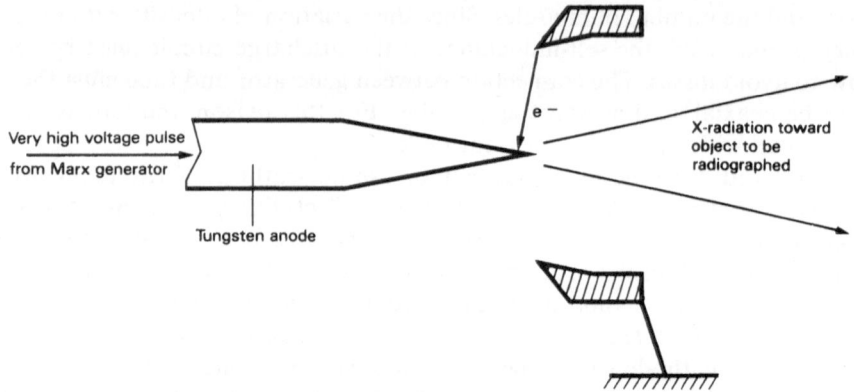

Figure X.37. Diagram of a vacuum discharge X-ray tube.

taken from an intermediate stage of the Marx generator (triode tube). Because of the very strong electron bombardment suffered by the tip of the anode, it is rapidly blunted, and thus the size of the focal spot of the tube increases steadily. The anode must therefore be renewed after a certain number of X-ray flashes: some manufacturers offer tubes which can be dismantled but require maintaining of the vacuum, while others offer sealed interchangeable tubes which avoid any need for pumping.

The photograph in Figure X.38 shows a 600 kV flash radiographic machine with maintained vacuum discharge tube manufactured by the SCANDITRONIX company.

The vacuum discharge tube which is shown, like the field-emission tube, works by reflection, since the X-rays used for radiography are those produced in a direction practically opposite to that of the electron bombardment. Now, for high energies, (> 500 kV), the production of radiation becomes more and more anisotropic: the intensity in the direction of electron propagation then being very much greater than that in the opposite direction. For the best use of this effect, some large machines contain a vacuum discharge tube of a different type: the anode is a thin tantalum or tungsten foil, earthed and positioned opposite the cathode onto which the Marx generator applies a negative voltage pulse. Electrons are focused by the conical shape of the cathode. The Physics International company has built a G.R.E.C. machine for the Commissariat à l'Energie Atomique, France, which produces doses of the order of 100 rads at 1 m in 80 ns with a very high voltage 7 MV pulse shaped by a blumlein.

Such power is exceptional, but many manufacturers market flash radiography Marx generators where the working voltages rise in steps from 100 kV to more than 1 MV. Using either field-emission tubes or discharge tubes with a conical anode, these machines deliver doses of between several tens to more

Figure X.38. SCANDITRONIX 600 kV generator and tube with maintained vacuum.

than 100 mrads at 1 m, for times around several tens of nanoseconds and on focal spots of about 1–5 mm.

(b) Linear Accelerators

The linear accelerator, which allows high energy electrons to be focused on a metallic target, can operate by pulse discharge and thus form a flash radiographic machine. Figure X.39 shows the block diagram of such a machine. It contains a diode or triode type of electron gun with a tungsten hot-cathode. The burst of electrons is obtained by applying a very high voltage pulse of very brief duration (several tens of nanoseconds), which can be provided by a Marx generator, to this gun. The electron bunch thus formed is accelerated in resonant cavities placed in series (usually three or four according to the machine). These cavities are fed by hyperfrequency sources which usually

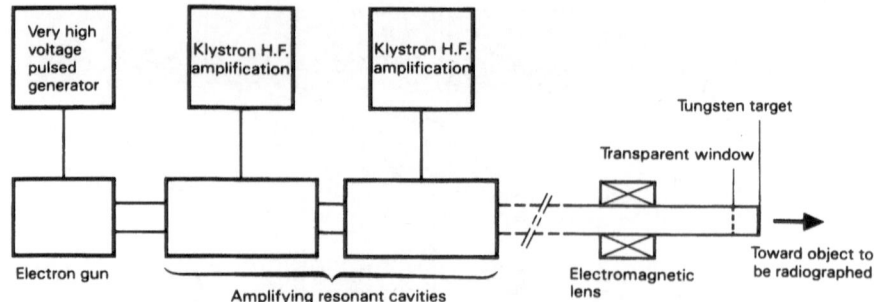

Figure X.39. Block diagram of a linear accelerator for flash radiography.

contain a klystron amplifying the H.F. signal from a pilot source. The component of the electric field produced is displaced parallel to the axis of the cavity and "hitches onto" the electron bunch, accelerating it while transferring part of its energy. These very high energy accelerator systems also function by pulse discharge but for durations much longer than that of the electron burst. On leaving the last acceleration cavity, the electron beam is focused by an electromagnetic lens on a tungsten target to produce X-rays by bremsstrahlung. A window which is transparent to electrons (e.g., of beryllium), placed slightly ahead of the target, serves to protect the accelerating sections, which function under very high vacuum, from tungsten vaporized by electron impact.

One great advantage of the accelerator, compared to Marx generator X-ray machines, is that it is capable, by means of a special development of the electron gun activation system, of delivering many X-flashes for the same target during one experiment, and can thus provide cineradiography. It can also be made to function continuously for durations of several tens of microseconds, to provide slit scanning recordings.

High performance radiographic machines have been devised using this linear accelerator technology, such as the PHERMEX at the Los Alamos National Laboratory and ARTEMIS at the Commissariat à l'Energie Atomique, France. Such machines can deliver three X-ray flashes in about 50 ns, carrying doses of several tens of rads at 1m, and the interval between flashes can be reduced to several hundred nanoseconds. The X-ray spectrum resulting from accelerating electrons at about 50 MeV is centered round several mega electron-volts, and the focal spot is reduced owing to electromagnetic focusing of the electrons (of the order of 1 mm).

3.2. Recording Techniques

In radiography, recording involves converting information about the "radiation dose" transmitted through the object at each point into information about "optical density" [7]. As to recording, flash radiography poses two specific problems when used in detonics:

- the X-photons are usually of distinctly higher energy than those used in industrial and medical radiography and interact much less with the detectors; and
- the recording must be protected from the destructive effects of the experiment while remaining as close as possible to the object to limit geometric fuzziness (see §3.1).

There are two different techniques which can be used to resolve these problems: the use of films together with screens, assembled in cassettes which are protected from the effects of the explosion, or the use of scintillators which are destroyed in the experiment and whose image is taken at a distance by an electronic camera. The possibilities of cineradiography, both with integral images and with slit scanning, will also be examined. After that, several devices will be presented which, when interposed between the generator and the recording system, allow plate quality to be improved and the plates calibrated.

(a) Film–Screens

Photosensitive film, because of its ease of use and particularly because of the quality of the images obtained, is still the first choice for recording radiographic images. However, the probability of interaction of X-photons with the layer of active grains in the emulsion varies inversely with their energy, and it is therefore practically impossible to record a direct image with flash radiography generators using only one film. It is therefore necessary to devise a system of screens which convert part of the incident X-radiation into a radiation to which the film is more sensitive. These screens are placed in contact with the film, either in front (on the side of the radiation source), or behind, or on both sides. The recording performances that must be considered are therefore those for the screen(s)–film conjunction, their modes of response to radiation being closely interlinked. Two main groups of screens may be used: metallic and fluorescent.

Metallic screens produce electrons by the photoelectric effect and by the Compton effect, which show up on the sensitive film. Lead screens are most often used. Use of a relatively thick screen in front allows part of the radiation diffused on the object to be filtered, this being of lower energy than directly transmitted radiation. This allows improved image definition. The rear screen also emits photoelectrons toward the film and retrodiffuses some of the photons, and also electrons liberated by the front screen. The efficiency of these metallic screens is still low, and their use in flash radiography is possible only when radiation doses are sufficiently high, which is quite rare with present-day machines.

Fluorescent screens contain substances which become luminescent under the action of X-photons. The substance most often used at present is calcium tungstate, where fluorescence is centered in the blue (about 430 nm) and is thus very actinic. Screens which have more recently appeared on the market, seemingly very interesting, use rare earths (particularly yttrium). The luminescent substance is finely ground and suspended in a conditioned plastic binder, then inserted as a thin layer between protective supports. To improve the efficiency of the system, the photosensitive film generally has an emulsion on each side and is sandwiched between two fluorescent screens. On the outer face, these screens are often fitted with a thin aluminum foil which is transparent to X-ray photons but reflects the photons from the luminescence towards the film. The nature of the sensitive emulsion must be adapted to the luminous spectrum emitted by the screen.

The response in optical density of the screen–film assembly to radiation is generally given by a characteristic curve similar to that shown in Figure X.40. This reception system is used as far as possible in the range where the optical density D—defined as the logarithm to the base 10 of the ratio of incident light intensity to the intensity transmitted through the developed film— varies linearly as a function of the dose of radiation received ($I \cdot t$). For a given variation in the dose received, the contrast obtained is better the higher the proportionality coefficient (contrast factor). However, when this factor increases, the recording dynamics generally decrease. Apart from the two factors mentioned (contrast and dynamics), the "speed" of the system must also be considered, i.e., the reciprocal of the minimum exposure time neces-

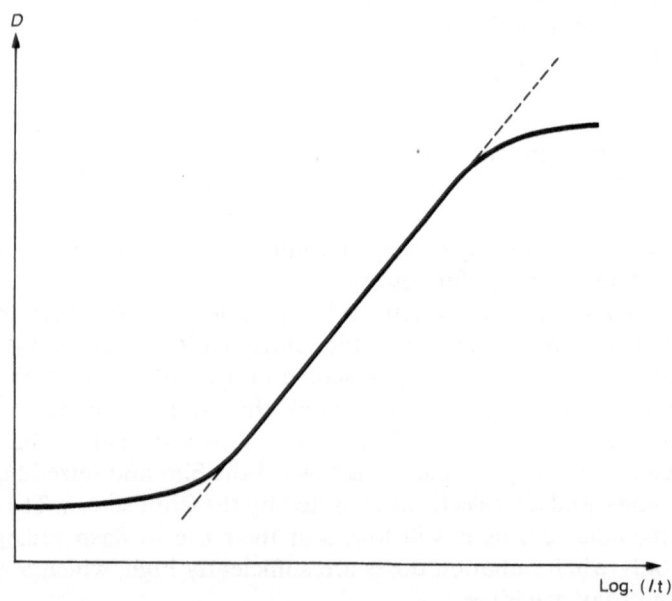

Figure X.40. Characteristic response curve of a screen–film assembly to X-radiation.

sary to achieve a plate with a given radiation intensity: it is higher as the sensitivity of the film is higher.

In fact, the response curve of a detection system—and thus its "speed" and contrast factor—depends on the energy of the incident X-ray photons. The screen—film assembly must be adapted to the object radiographed, to the available dose of radiation, and also to the X-ray spectrum delivered by the generator.

Another factor which affects the quality of the plate obtained is the intrinsic fuzziness of the recording. This fuzziness is due to several different factors: the grain of the films, the grain of the screens, and the diffusion of photons in the thickness of the screens. As a very general rule, this fuzziness is generally more marked when the system "speed" is high. The user must seek a compromise between minimum sensitivity needed for recording and maximum permissible fuzziness. Although several authors have made systematic comparisons between different assemblies of screens and films [7], the choice for the experimenter is purely empirical among the continually increasing diversity of products commercially available. Preliminary trials on static models can be made in order to define the system giving the best results with a given radiographic generator for the class of objects studied.

(b) Cassettes and Shielding

It has been shown above that, to reduce geometric fuzziness, the detection system must be placed as close as possible to the object radiographed, but without damage to the developed plate from secondary effects of the detonic experiments: blast and splinters.

One approach to this problem is to interpose between the object and the radiographic cassette—which contains and presses together screens and film—layers of protective material to deflect or arrest the splinters, as well as a large metallic shield (pierced with a window the size of the plate) to reduce the blast effects. Figure X.41 shows such a device. Protection against splinters, generally achieved by an assembly of metallic plates or cones, must be optimized to minimize the absorbed radiation dose: use of light metals or plastics with high perforation resistance and shape modifications.

Another possibility for protecting the plate is to use a very light cassette which will move with the explosive blast. This "flying cassette" technique usually permits the use of a smaller shield, since this is no longer an immovable wall receiving the splinters, but moves with them. The main difficulty with the cassette is the contradictory requirements of lightness and solidity (to ensure the necessary protection of the exposed film against shocks).

As a general rule, protection of the plate against these destructive effects is one of the most serious practical problems encountered when using flash radiography for detonic experiments. The practical knowledge of the experimenter in this field is a major factor in obtaining high quality results.

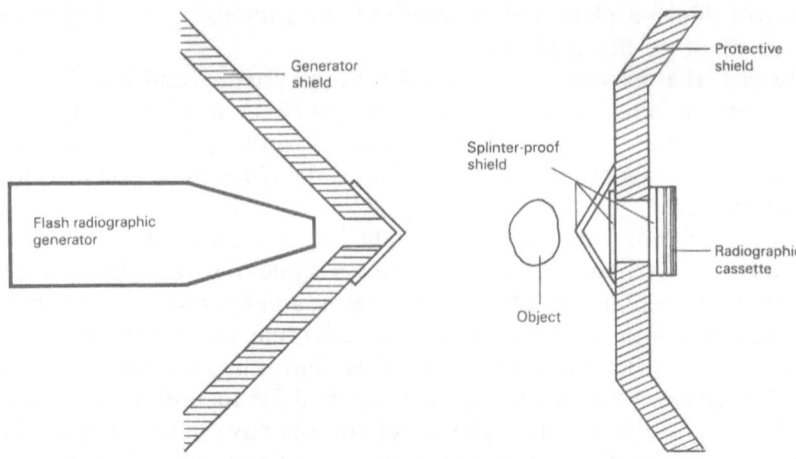

Figure X.41. Protection of a radiographic cassette by massive shield.

(c) Scintillators

Another recording technique, linking a scintillator with an electronic camera, has a priori the great advantage of overcoming this protection problem. In fact, only the X-ray photon to optical photon converter remains close to the object radiographed, being destroyed in the experiment. The electronic camera records from a distance the optical image created by the X-ray flash on the scintillator. Figure X.42 shows this device. The luminous efficiency of the

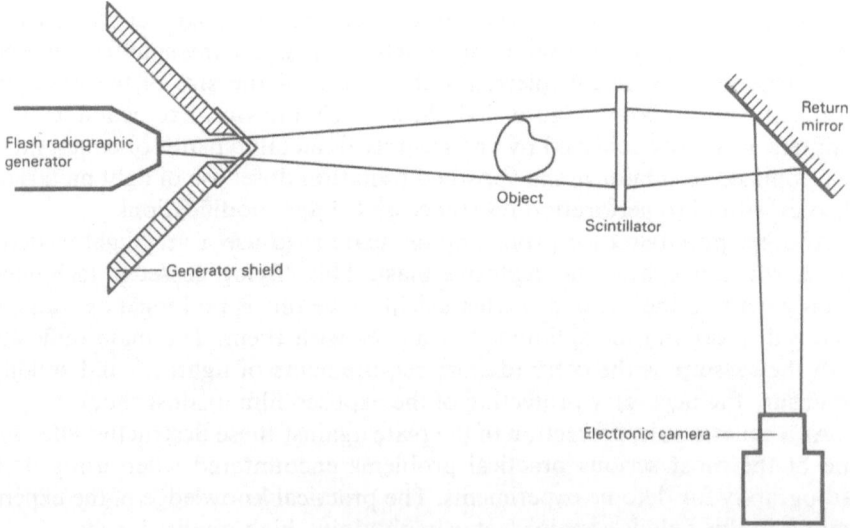

Figure X.42. Recording of radiographic image with scintillator and electronic camera.

system (mirror and camera) which recaptures the image is very low compared to that of a film plated against a screen. To record a sufficiently bright image, it is essential to use an electronic camera fitted with a brilliance amplifier (see §1.3(b)). Unfortunately, the image definition obtainable with this type of camera is limited compared to that obtainable with a screen—film assembly. Use of this technique is thus restricted to some specialized needs, such as cineradiography.

(d) Cineradiography (Images and Slit)

As with optical observations, it is often valuable in radiography to be able to track phenomena. For this purpose it is possible to carry out several identical experiments by taking radiographic plates at different instants. This method, which requires multiple experiments, has limitations because of problems of reproducibility of the phenomena and of the framing. For this reason it is usually preferable to be able to track these changes in a single experiment by use of true cineradiography, either with integral images or by slit scanning. When the object studied has an axis of revolution, several flash radiography tubes can be used, directed at different incidences perpendicular to the axis of revolution. These tubes are then activated at successive instants, reception being ensured by a different radiographic cassette for each tube. Collimation of different beams avoids interference between the different emission recep-tion assemblies. Figure X.43 shows such a device.

The major limitation of this method, apart from its restriction to objects with an axis of revolution, is that it requires as many flash radiographic generators as framing instants required per experiment. It is thus almost impossible to operate with powerful radiographic machines. Another method which does not have these limitations uses linear accelerators (see §3.1(b))

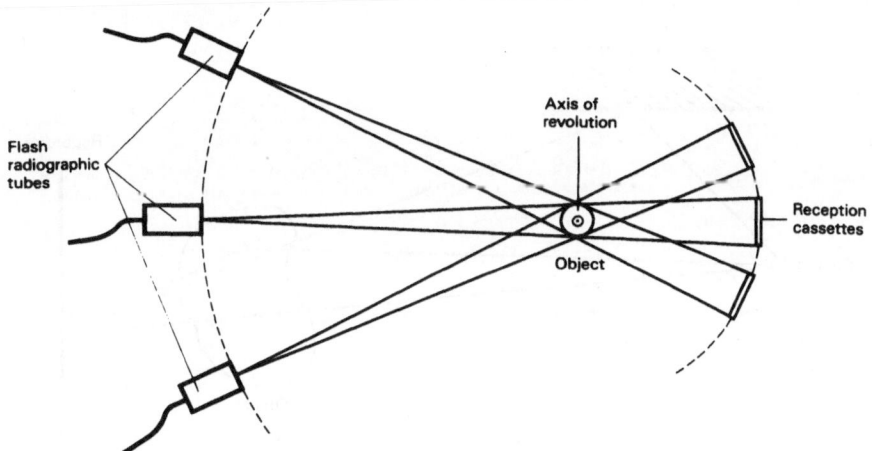

Figure X.43. Cineradiography system by multiple X-ray tubes.

which can deliver several powerful X-ray flashes at very short time intervals (from several hundred nanoseconds to a few microseconds). In this case, reception cannot be ensured by a screen–film cassette, on which the different images are superimposed. The reception system with scintillator and electronic camera shown above (see Fig. X.42) is therefore used. Images are separated by very brief exposure times of the electronic camera and very weak remanence (after-imagery) of the chosen scintillator. The definition can be improved by using one electronic camera for each recorded image.

Another option offered by linear accelerators, because of their long pulse duration (several microseconds), is slit scanning cineradiography. This technique is similar to the slit scanning cinematography discussed in §1.3. The object studied is radiographed along an analysis slit, while an electronic camera scans the radiographic image of this slit as a function of time. Then the whole object is usually irradiated using a long-duration X-ray flash produced by the accelerator. The observation strip is provided by a linear scintillator strip placed behind the object in the direction to be observed, the slit scanning electronic camera (see §1.3(b)) recapturing the scintillator image.

(e) Collimators, Filters, and Penetrameters

Apart from geometric fuzziness and intrinsic recording fuzziness already noted, there is a third type of fuzziness affecting the definition of radiographic images: diffusion fuzziness. Figure X.44 shows how this effect damages the image. Every point I of the recording simultaneously receives X-radiation which is directly transmitted from the source through the object, which forms the radiographic image, and X-radiation diffused onto different points of the object or onto the walls of the firing chamber (basically by the Compton effect).

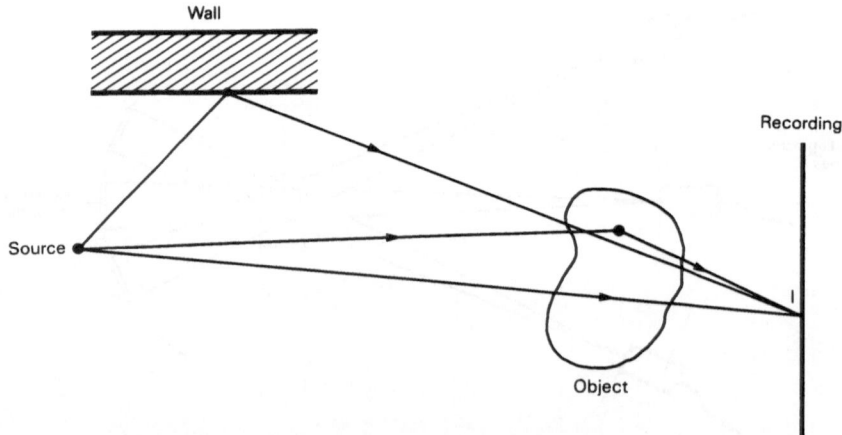

Figure X.44. Diffusion fuzziness in radiography.

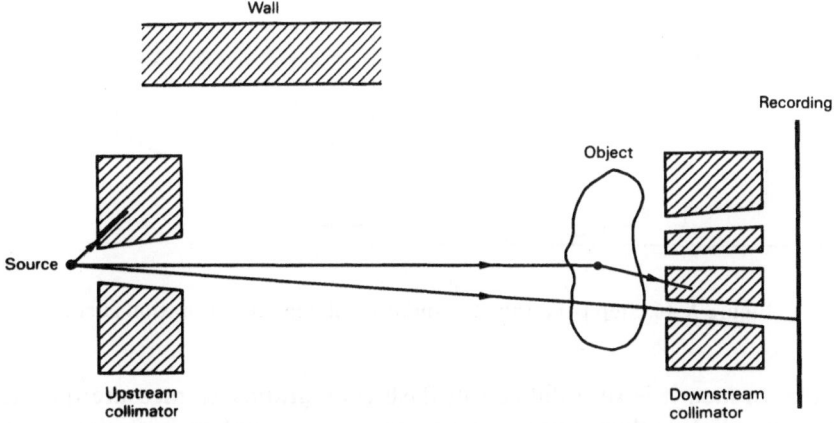

Figure X.45. Reduction of diffusion fuzziness by collimation.

The use of *collimators* is the first way of reducing the importance of this fault. A collimator is usually formed from a massive block of a material with high atomic number (which is very absorbent of radiation) pierced with one or more holes of modified shape. Collimation can be achieved at two levels, as shown in Figure X.45:

- between the source and the object, it limits divergence of the beam and can eliminate rays diffused by the walls or by objects other than the one under investigation; and
- between the object and the recording, it permits, in certain zones of interest in the image, suppression of the diffused radiation generated in the object outside a limited solid angle around the direct incident direction.

This last mode of collimation is usually the most efficient, but does have the disadvantage of reducing the extent of the observable range on the object

Another way of controlling diffused radiation is by the use of filters to eliminate the lowest energy components of the spectrum. Such a filter is made from a thin foil of a high atomic number metal (usually lead). This filter can be placed upstream or downstream of the object. If placed upstream, it increases the ratio between the energy of the directly transmitted photons and that of the diffused photons, since Compton diffusion absorbs a larger proportion of the energy of the incident photons if they are of higher energy. If placed downstream the filter acts differently, by preferential absorption of the diffused photons which are of lower energy than the transmitted photons. However, these filters have the major disadvantage of reducing the radiation dose available for recording. They are rarely used in flash radiography, where it is very unusual to have an excessive incident dose.

Finally, a third way to improve experiments is to use a *penetrameter* which permits "individual" calibration of the recording in the surface mass pene-

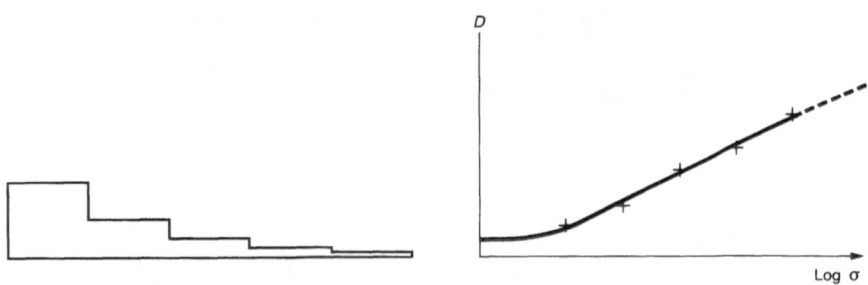

Figure X.46. Step penetrameter and optical density calibration curve.

trated. In fact, it is very difficult in flash radiography to measure the exact emitted radiation dose and the spectrum of this radiation, since functioning of the generator is not perfectly reproducible from one X-ray flash to another. Furthermore, calibration of the detection system response as a function of the dose and of the spectrum received is very difficult, and can fluctuate according to the plate development conditions. It is therefore impossible in practice to make a theoretical link between optical density on the film and surface mass of the material penetrated by the radiation. For practical calibration, a penetrameter made of different thicknesses of the material of which the object is made, and covering the range of surface masses penetrated, is radiographed alongside the object studied. The device most often used is the step penetrameter as shown in Figure X.46.

Step thicknesses are generally arranged in geometric progression, in accordance with the shape of the detection system response. This penetrameter must be placed as close as possible to the object so that the radiation it receives is approximately the same, in both dose and spectrum. Examination with a densitometer can then determine the calibration curve for optical density of the film as a function of the surface masses of the material penetrated (Fig. X.46).

3.3. Plate Processing Techniques

Flash radiography can be used, as high-speed cinematography is frequently used, to track the global evolution of internal phenomena in the experimental assembly, inaccessible by other methods. But being limited to this type of use would mean failing to recognize the wealth of information contained in a radiographic plate, which includes: spatial information on the instantaneous shapes and positions of interfaces and shock waves (which are detectable by the marked variations in density which accompany them), and also information on the surface masses penetrated at each point from the local optical density values of the plate by a suitable calibration. The complexity in extracting and exploiting such information has led to increasing use of digital image processing, in which there have been considerable developments in recent years. Description of the numerous and often complex mathematical techniques used to analyze two-dimensional image signals is largely outside the scope of this work. Only the principal experimental tools they offer to experimenters in connection with flash radiography plates will be noted here. It will also be seen that, as well as possibilities for improving and restoring images and for automated extraction of contours and shapes, these techniques even allow, under certain conditions, the use of two-dimensional information in radiographic plates to obtain three-dimensional information on the instantaneous density at every point of the object, which is particularly valuable in detonic experiments.

(a) Data Acquisition

Depending on the ultimate purpose, a wide range of equipment can be used to extract relevant data from radiographic plates.

If simple exploitation of spatial data is all that is sought, optical magnification techniques of the profile projector type, which project without distortion, are often used. Such machines generally allow close positioning (to several microns) of the points chosen visually by the operator. The digital coordinates of these points are recorded for subsequent data processing.

However, to exploit the local optical densities of the plates, a microdensitometer must be used. This equipment has a very stable light source, diaphragmed to a fine pencil of light, linked to a precise and sensitive photodetector (photomultiplier or photodiode). The light intensity transmitted through the plate is therefore measured through a very small observation window, allowing the optical density to be found. A table which can be moved in two orthogonal directions allows the plate to be moved under the reading head for very precise determination of the coordinates of the point measured.

If digital processing of the image is required, equipment for digitizing the plates is used. Such a machine functions as a microdensitometer, but the plate is moved under the reading head mechanically for line-by-line scanning. A sampling device, at a certain number of points per line at constant advance intervals, obtains information about the optical density, which is digitally coded. Sampling steps in the lines are usually equal to the steps separating the lines, the window for measuring optical density being a square of the same dimension. The average value of the optical density on the surface is thus allocated to an elementary square of the image sample or pixel (picture cell). A matrix is formed by the digitization process, with line number n and column number j, when the size of the digitization step is known; this gives the position of the pixel on the plate, while element D_n^j represents the average optical density of this pixel. Subsequent digital processing is carried out on this matrix. Current high-performance equipment can digitize a plate for 1024×1024 samples with a step of the order of ten microns and coding of the optical density on eleven bits (2048 levels). The huge amount of data to be handled by the image processing programs, the memory capacities, and the calculations required, can be imagined.

(b) Preliminary Processing of Numerical Images

Starting with a digitized image, a certain amount of preliminary processing can be performed [16], [23], both to improve legibility and to correct faults inherent in the shooting technique. The first process, which is simple to operate, is to reinforce the contrast in a plate when it is too faint. For this purpose, it is sufficient to modify the optical density value at each pixel by applying one-to-one mapping between values before and after processing, to increase the dynamics for the plate assembly or for a chosen range of values. The image thus processed is then viewed on a video screen. In this way the operator can improve the legibility of certain details and thus facilitate later visual examination of the points. It is possible, for the same purpose, to change a monochromatic image into a false-colored image by allocating a color to each range of optical density values. Each pixel is thus colored according to the corresponding optical density, and the image thus created viewed on a color video monitor.

A second process aims at reducing background noise in the image due to the recording system (grain of films and of screens). Appropriate methods depend on the observation that noise shows a spectrum of spatial frequencies that are higher than the other components of the image because it is not spatially correlated. One method of reducing the effect of these high frequency components is to weight the optical density value for each pixel according to the values for its nearest neighbors. For this purpose, the optical density value for each pixel is replaced either by the arithmetic mean of the optical densities of the pixel itself and the eight pixels around it, or by a barycentric mean, calculated from these same values by weighting each of them with a position coefficient relative to the central pixel. Another method, which is rather more laborious, is to perform a two-dimensional Fourier transform of the image to decompose it into its spatial frequencies spectrum. This Fourier transform is then treated with a low-pass numerical filter to eliminate high spatial frequency components. The inverse Fourier transform then restores an image with improved signal-to-noise ratio, and consequently, as well as improving the visual quality of the image, makes it possible to increase the precision of subsequent restoration or reconstruction of density, which are usually sensitive to noise.

A third type of preliminary process, of value for some flash radiographic plates, is to reduce the geometric fuzziness linked to the extent of the focal spot of the generator (see §3.1). For this purpose, it is necessary to model the emission of the source considered as a plane image. By working with illumination instead of optical density—then it is necessary to know the recording system response—it can be shown that the real image of the object is the product of the convolution of the "clean" image of the object by an image called "source-image."

$$\text{Real image} = \text{"clean" image} \times \text{source-image.}$$

If a model of the source-image is defined, the experimental image can be deconvoluted to obtain an expression close to the clean image of the object. This deconvolution can be made by the Fourier transform (designated here as FT):

$$\text{FT (real image)} = \text{FT ("clean" image)} \times \text{FT (source-image),}$$

$$\text{"Clean" image} = \text{FT}^{-1}\{\text{FT (real image)/FT (source-image)}\}.$$

It is also possible to use iteration. The simplest method is to presume a "clean" image which is then convoluted by the source-image [26]. If this last image is subtracted from the experimental image, an error image is obtained. The presumed "clean" image is then corrected by this error image to give a new presumed "clean" image, and this is then used for further iteration. For low spatial frequencies, the method converges towards the "clean" image. The information corresponding to the highest frequencies cannot be restored, as it has been swamped by the fuzziness in the background noise of the real image. The fuzziness reduction thus obtained is greater in proportion to source modeling precision and low background noise of the real image.

Finally, processing of digitized images offers a wide range of practical options which can facilitate subsequent examination processes: recentering, enlargement or reduction, symmetrization about an axis, subtraction of the image from a superimposed object (protection, flange, etc.), trimming of the useful zone, etc.

(c) Extraction of Contours

To find the position of a contour corresponding to the projection on the film plane of a surface of discontinuity (material interface or shock wave or detonation wave), the zone having a high optical density gradient, which is linked to this discontinuity, is sought. This examination may be purely visual, using the sensitivity of the eye to different shades of grey to find the points directly from the original plate. It can also be completely automated with the use of a program to handle the digitized image which searches for the maximum density gradient points by differentiation (first or second derivative), and then interpolates a contour between these points. However, experience shows that this type of program is very difficult to use on an image affected by the imperfections that are to be found on most flash radiographic plates. The solution which gives the best results is generally an interactive examination technique in which, with the help of programs for reinforcing contrast or for false-color restitution, an operator defines the general form of the contour by a few plottings. The program is used to find maximum gradients between points and to smooth the resulting contour. The operator then inspects this contour visually to ensure that the program has not followed steep gradient zones which are not linked to the discontinuity sought (using expected contour shape and regularity as criteria). The precision of this type of examination is essentially limited by the residual fuzziness of the image (possibly after preliminary processing).

(d) Density Measurements, Tomography

It has been shown above (see §3.2(d))that calibration of a radiographic plate, with a penetrameter made of the same material as the object and suitably arranged, will allow correspondence of each optical density value to the value of the surface mass $\int \rho \, dx$ penetrated by the radiation between the source and the image point under consideration. By applying the transfer function resulting from this calibration to the digital image of the optical densities, it can be transformed into the image of the surface masses penetrated on the object.

When density is uniform along a radiation propagation direction—which is, for example, the case for one-dimensional plane flow radiographed perpendicular to the flow direction—then if the thickness of the object penetrated by the radiation is known, either *a priori* or by another observation at $\pi/2$ from the first, the value of the density in this direction can be calculated from the surface mass penetrated. Unfortunately, such a case is very rare in detonic experiments and usually the measured surface mass gives information on the density of the object, *integrated* along the projection direction. To extract the essential information, which is the local value of the density in the object, it is necessary to use a *derivative* method: *tomography*.

A first instance which is particularly simple, and fortunately also very common, is when the object has a symmetry axis. By taking the radiograph perpendicular to this axis, an image is obtained which has a symmetry axis. Cutting this image by a straight line perpendicular to this axis then corresponds to the projection of a section of the object following a plane perpendicular to its axis of rotation as shown in Figure X.47. In this section of the object, because of the rotational symmetry, the local density ρ depends only on the distance r from the axis. These density values can be made discrete as a function of r by a step ℓ, identical to that used for sampling the image, in the corresponding section (see Fig. X.48). Every pixel of the image section is associated with a surface mass σ_n linked to the densities ρ_j in the different samples of the section by the linear relation

$$\sigma_n = \sum_{j \geq n} \ell_n^j \rho_j,$$

where the coefficient ℓ_n^j, corresponding to the layer thickness j of the object penetrated by the radiation arriving on pixel n, can be easily determined, if the layer radius r and the distance of the pixel from the symmetry axis ℓ are known. It is then sufficient to invert the triangular matrix $[\ell_n^j]$ to determine the local density ρ as a function of the surface mass penetrated σ. By per-

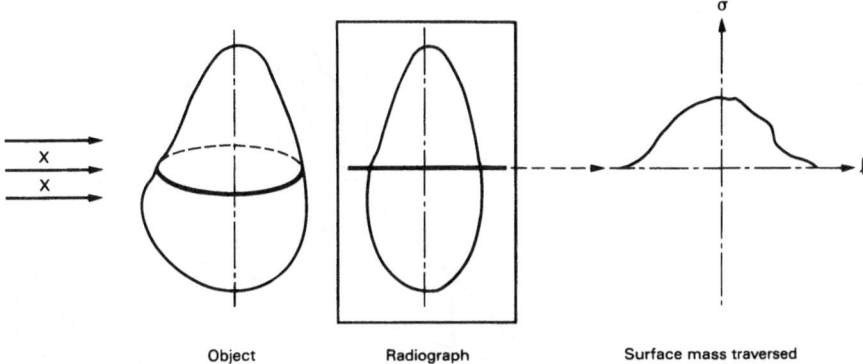

Object   Radiograph   Surface mass traversed

Figure X.47. Radiograph of an axisymmetric object in a direction perpendicular to its axis.

forming this operation for all the sections perpendicular to the symmetry axis, a complete map of the object densities can be constructed.

In the general case where the object does not possess rotational symmetry, a single radiograph will not be enough for tomographic reconstruction since an infinity of different objects can have the same radiographic image. Several radiographs of the same object must therefore be taken at the same instant, but from different angles. In fact, in flash radiography, identical experimental assemblies are generally used, which are fired successively in front of the same generator activated at the same instant, but oriented along different axes. These different plates are generally taken perpendicular to one and the same axis of the object, which is chosen arbitrarily and called the tomographic axis. The three-dimensional object can then be considered as a stack of two-dimensional layers, with projection being made in different directions. Figure X.49 shows such an arrangement. This technique is the same as that used

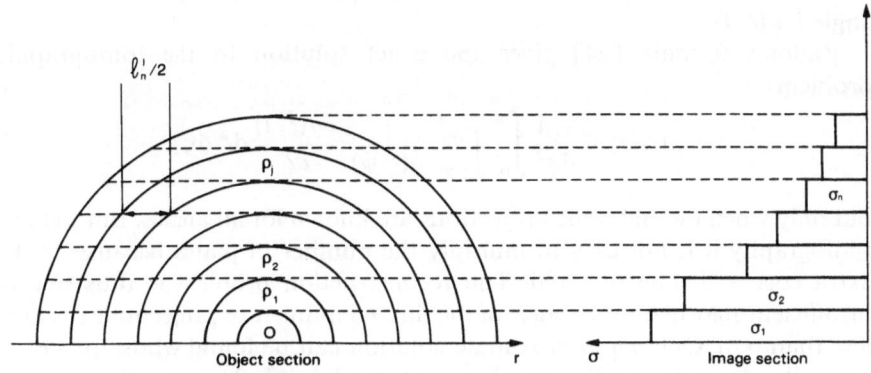

Object section   r   σ   Image section

Figure X.48. Density discretization in the object section.

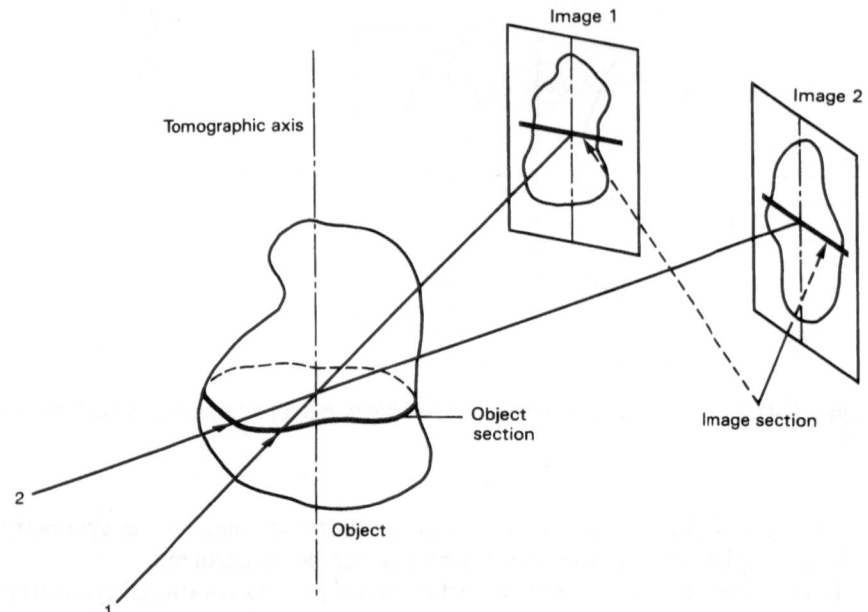

Figure X.49. Radiographs of the same object taken at different angles perpendicular to the same axis.

in industrial and medical tomography. It allows the problem of three-dimensional reconstruction to be simplified to a two-dimensional problem. Radon [24] has shown that, starting from a knowledge of all the projections taken at different angles, it is possible to go back to the two-dimensional function—integrated by the projection operation—as shown in Figure X.50. In the present case, the two-dimensional function is the density $\rho(r, \psi)$ in a tomographic section of the object, and the projection is the surface mass penetrated at a distance ℓ from the tomographic axis for a plate oriented at angle i: $\sigma(\ell, i)$.

Radon's formula [24] gives the exact solution to the tomographic problem

$$\rho(r, \psi) = \frac{-1}{4\pi^2} \int_0^\pi \int_{-\infty}^{+\infty} \frac{1}{\delta(r, \psi)} \cdot \frac{\partial \sigma(\ell, i)}{\partial \ell} \, d\ell \, di,$$

but only when radiographic projections are known for all angles. But in flash radiography it is not easy to multiply the number of plates because of the extra cost and time required. The reconstruction problem is thus one of insufficient information because of the limited number of projections (usually less than ten). Only an approximate solution can be found whose precision basically depends on the number and quality of the recordings made. Different methods of tomographic reconstruction are available, the two most

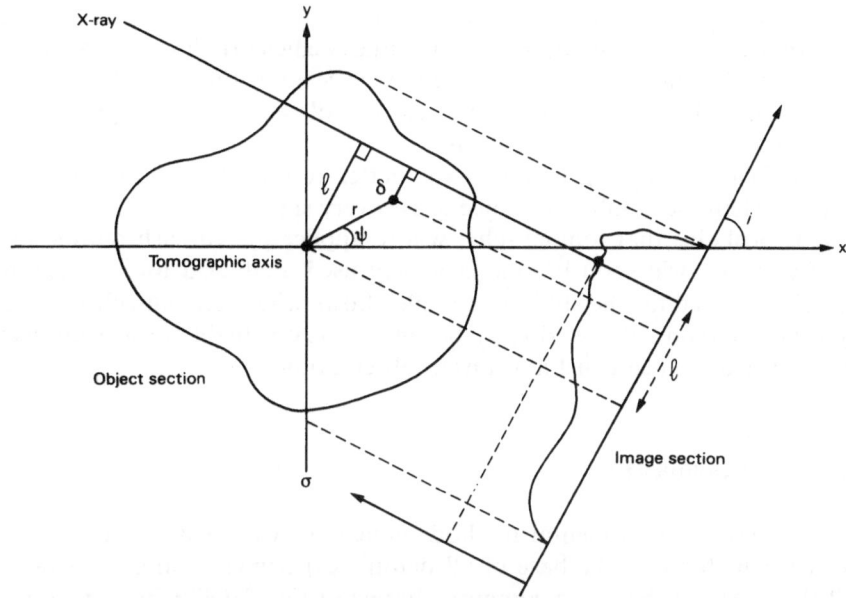

Figure X.50. Correspondence between object section and image section.

widely used being the convolution method and the Algebraic Reconstruction Technique (A.R.T.).

The *convolution method* is an approximate resolution technique based on Radon's formula in which each projection is treated by a filter—which is the inverse Fourier transform of the module in the spatial frequencies plane—followed by integration of the convolution products obtained at different projection angles i. It can be shown that the result is equivalent to Radon's formula for continuous integration over i. The approximation obtained when only a limited number of projections are taken results from the integration error arising from the discrete values i_n. It may be noted that this method, which is that generally used in medical tomography, can be applied in the case of one radiograph for an object which has rotational symmetry. In fact, the projections of such an object perpendicular to its symmetry axis are all identical, and as many projection angles i_n can be inserted into the resolution method as are needed for the integration precision required.

The *A.R.T. (Algebraic Reconstruction Technique) method* does not make use of Radon's formula. It is an iterative solution technique in which an arbitrary density distribution is made on a grid covering the section of the object studied, its projection in terms of surface mass at the angle of the first shot then being calculated. This calculated value is compared at each point with the experimental value, and the field density then corrected so that the mass penetrated by each ray corresponds to the measured penetrated mass. This correction may be additive (by adding or subtracting the same mass to the cells of the grid traversed by each ray), or multiplicative (when the densities of these cells are multiplied by the same coefficient). This new calculated density field is then used for the second projection, then for all the other projections. This iterative method is used until there is convergence to an approximate solution of the problem.

This last technique generally gives better results than the convolution method when the number of projections is very small.

It is probable that tomography, whose application to flash radiography is very recent, will soon become a widely used analytical tool for detonic experiments and for the study of complex flows which are difficult to access by other methods. It has already become a very valuable measuring technique, instead of being just a means of observation.

4. Stress Generators

This review of experimental methods concludes with a description of the devices which are at the base of all detonic experiments: stress generators. All these systems have the common characteristic of delivering a very high energy in a very short time onto the sample of the explosive studied. The principal differences between the different generators are in the primary energy source (electrical, mechanical, chemical, etc.), in the mode of transmitting the energy to the sample (projectile impact, shock wave transmission, etc.), and finally in the time and space distribution of this deposit of energy. Details follow of two categories currently used: explosive generators and launchers, with other stress generators deliberately being left to one side, such as those with lasers, ion beams, and electron beams, since their use in detonics is infrequent.

4.1. Explosive Generators

These various generators all use energy released from detonation of a solid explosive. They include simple detonators, which provoke quasi-punctual priming for detonation in the sample studied, and also wave conformators, which exploit the geometric properties of detonation propagation, wave transmission between adjacent media or projection of a "liner" in order to transform this quasi-punctual priming into priming distributed over a surface.

(a) Detonators

Among the "pyrotechnic artifices" used to prime the detonation of solid explosives—called *detonators*—only electric devices prove to be practically synchronizable with high-speed measuring equipment.

In the *hot wire* device, passage of an electric current through a filament

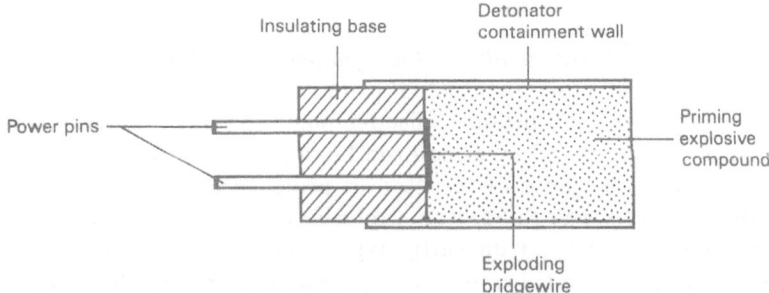

Figure X.51. Cross section of exploding bridgewire detonator.

embedded in a "primary" explosive compound, releases heat in it by the Joule effect, thus priming its decomposition; while a second explosive substance, which is less sensitive but more powerful, acts as a relay to transmit the stress to the explosive sample studied. According to the mode of heating, functioning times can vary from several milliseconds to several tens of microseconds, so dispersion is considerable and safety in use low (a few milliJoules of static electricity can trigger them). The *exploding bridgewire* device aims to remedy some of the problems encountered with the previous method. It is composed (see Fig. X.51) of an insulating base supporting two power pins connected on the inner side by a very fine metallic wire (with a diameter of about one hundred microns and a length from one to several millimeters). The containment cylinder—usually metallic—fitted to this base is filled with a "secondary" priming explosive compound, usually fine granular PETN, which will not be detonated by simple conductance heating. The detonation regime is then achieved as follows. As a result of the rapid discharge of a capacitive circuit between the pins, the wire is adiabatically heated by the Joule effect, which leads to its almost instantaneous vaporization; the consequent release of this metallic vapor at very high initial density creates a divergent shock wave which induces the detonation of the priming explosive compound; the insulating base and the containment wall promote the transfer of the detonation to the main charge. This mode of operation needs development in the wire of a current of more than several hundred amperes. A high operating speed is obtained by a high voltage at the terminals of the firing unit capacity (usually several kilovolts) and by a low self-inductance in the power supply circuit (discharge pulse transferred by coaxial cable); the total operating time of the detonator is less than a few microseconds, thus reducing the dispersion time to several tens of nanoseconds. The priming surface with such a detonator is usually of the order of one square centimeter, which, considering the scale of the explosive assemblies studied, can usually be considered as point firing. Operational security with an exploding bridgewire detonator requires the exclusive use of "secondary" explosives and a high pulsed voltage for nominal functioning.

(b) Plane Wave Generators

The majority of experiments for the dynamic characterization of explosives are carried out in a one-dimensional plane configuration. It will be seen later that launchers usually produce effective plane shock waves. However, their caliber is often low and operation can be problematic, particularly with explosive charges, because of the need to protect the equipment against the destructive effects. For this reason, explosive devices called *plane wave generators* were developed at an early stage, which, upon point priming by a detonator, generate a shock with nice planarity over a significant area.

The most up-to-date device is the "plane wave cone" (see Fig. X.52) which uses the difference between the detonation velocities of two explosives, one explosive with a lower value D_1 forming a conical core and another explosive with a higher value D_2 forming a shell around this cone. If the half-angle α at the peak of the cone is such that $\sin \alpha = D_1/D_2$, then priming of the fast exploding shell at its tip by a detonator will induce a plane detonation wave in the slow explosive core. At the base of the cone, this plane wave is transmitted to a cylinder of "reinforcing" fast explosive. The surface covered by the plane wave only depends on the cone height and can thus be very large. The error in synchronization for the wave emerging on the downstream face of the reinforcing cylinder is usually several tens of nanoseconds, and shows axial symmetry. This generator is very simple to operate but has two drawbacks: the shock which it delivers is not sustained—because of the expansion of detonation products—and its intensity cannot be controlled. One way to reduce these problems is to use the plane wave cone as a projectile launcher, by using the shock accumulation mechanism: a metal plate (or stack of plates) is placed in contact with the reinforcing cylinder, leaving a sufficient space between this plate and the sample studied so that the flying plate can

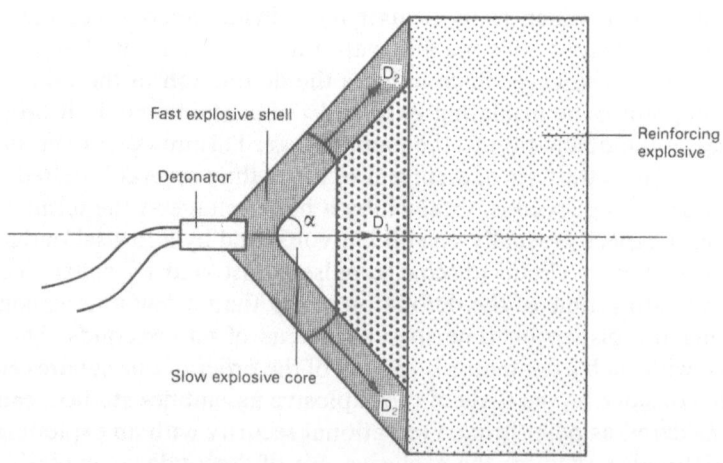

Figure X.52. Cross section of "plane wave cone."

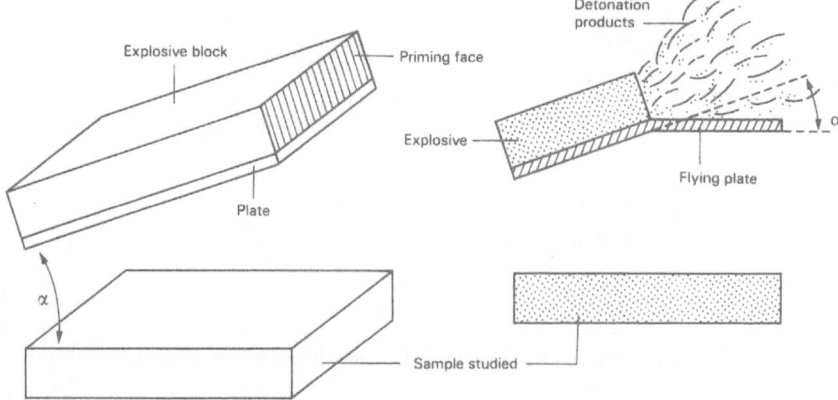

Figure X.53. Principle of a plane wave generator with projected plate; cross section of the device when functioning.

attain its maximum velocity. By altering the thicknesses of the plate and the reinforcing cylinder, or by altering the shock impedance of the material, one may vary the intensity of the sustained plane shock which is induced in the sample by the flyer.

Another plane wave generator device, very often used in explosive dynamic characterization experiments, uses projection of a plate by an explosive: a thin metallic plate is projected by an explosive block which is primed in such a way that a plane detonation wave is propagated within the block perpendicular to this plate. Figure X.53 shows the principle of such a generator. Priming of the explosive prism is generally provided by a linear wave generator (see §4.1(c)). After an acceleration phase, the plate becomes effectively flat again, while taking a direction which makes a constant angle α with its initial position. The sample being studied is set so as to form, with this plate when at rest, a dihedron of the same angle α. Impact of the plate thus creates a sustained plane shock therein. The projection speeds obtained in this way are appreciably lower than those obtained by direct projection, and this generator produces a range of shock pressures which are definitely less than those obtainable with the plane wave cone.

(c) Other Wave Shapers

In order to study the influence of the wave front shape on detonation, several different wave shapers have been used, which induce different detonation wave shapes after point priming. Only the more common types will be described here.

To obtain priming along a line, the same mechanism as that used for the plane wave cone (§4.1(b)) can be used, by taking advantage of the difference between the detonation velocities of two explosives. A linear generator

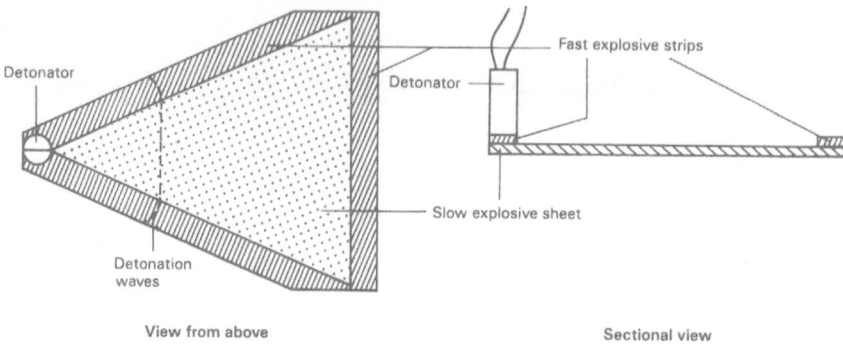

Detonator

Fast explosive strips

Detonator

Slow explosive sheet

Detonation
waves

View from above

Sectional view

Figure X.54. Linear wave generator with explosive sheet.

can thus be constructed with a thin section of a plane wave cone cut along its plane of symmetry. The same principle can also be applied to flexible explosive sheets as shown in Figure X.54; the fast explosive is cut into strips which are stuck onto the slow explosive sheet. (N.B. Because of the flexibility of the explosive leaf, this device also allows priming along a nonrectilinear arc.) Another way of obtaining linear priming is to alter the distances covered by the detonation wave, starting at the priming point, in a thin explosive sheet, using the configuration given to this assembly. The simplest way to achieve this is to use a conic sector next to a plate as shown in Figure X.55. By inclining the plane in relation to the cone's axis at an angle equal to the half-angle at the top, all the paths OAB, OA'B', etc., taken by the detonation wave between the detonator and the priming line, can be equalized.

To generate a conical wave, the plate projection mechanism is used, as for

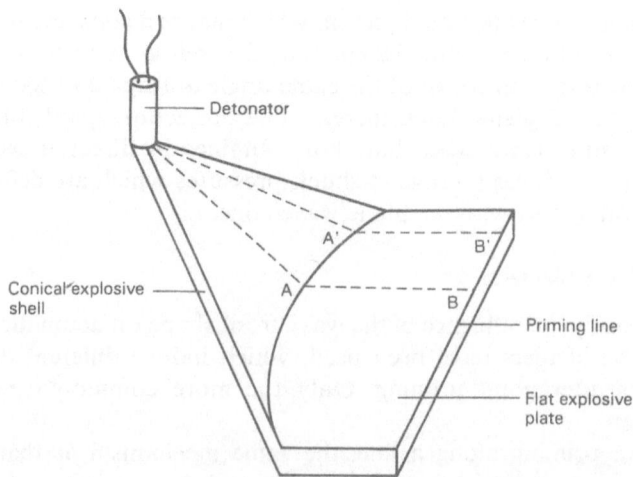

Detonator

A'
B'

Conical explosive
shell

A
B

Priming line

Flat explosive
plate

Figure X.55. Linear wave generator by cone–plane intersection.

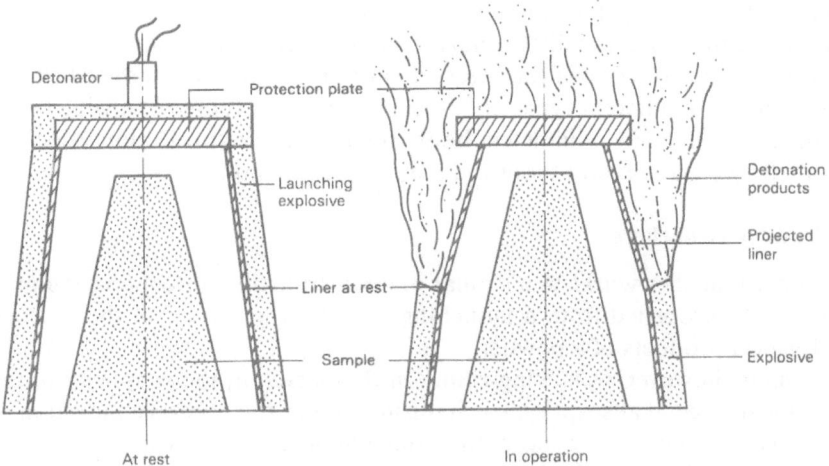

Figure X.56. Principle of a conical wave generator.

the plane wave generator shown in §4.1(b). Figure X.56 shows the operating principle. A thin conical plate of explosive projects a metallic interior liner on to the sample, which is also conical. The device is constructed so that, after projection, the liner forms a cone with the same angle at the top as the sample, and thus synchronizes an impact which creates a conical detonation wave. The priming of the explosive which causes projection is achieved on a crown by an explosive disc which itself is primed at the center. A thick plate placed under this disc protects the tip of the sample from premature priming. The angle at the top of the conical wave induced by this type of generator can be altered by altering the angle of the explosive cone which causes projection, its thickness and that of the metallic liner. One of the advantages of this type of generator is that it can lead to strong detonation configurations, because of the convergence effect [6].

In principle, it is very simple to obtain a divergent spherical wave, since it is sufficient to prime a sphere of explosive at its center point. However, in practice, since detonators are not of infinitely small size and because the priming caused by them is strongly directive, a detonation wave generated by central priming will not be truly spherical until the radius is sufficiently large (several centimeters). An ingenious technique [10] allows a smaller radius: two detonators are arranged face to face, at opposite ends of an axis going through the center of the sphere, at a distance of the order of twice the depth necessary for detonator priming.

4.2. Launchers

Launchers are used to send a projectile, usually metallic, at high velocity onto the explosive sample, which, on impact, induces a shock (usually planar). Their principal advantages, compared to plane wave explosive generators, are that they usually produce more effective plane shock waves and it is much easier to control the pressure and duration of the induced shock. The existence of different types of launchers allows coverage of a wide range of calibers (from a few millimeters bout 100 mm) and of projection velocities (from a few tens to several thousand meters per second).

(a) Powder Launchers

A powder launcher works in a similar manner to artillery weapons, the projectile being propelled from a launching tube by the release of the gaseous combustion products of a powder.

It differs, however, in lacking rifling in the barrel, in the simplicity of the charging and recoil absorption mechanisms, and in the general conception of the projectile. Figure X.57 shows the principle of such a launcher.

The projectile is usually in the form of a hollow shoe of plastic material (nylon, teflon, rilsan, etc.) to reduce the weight projected. A truncated cone-shaped flange at the base of the shoe ensures gas tightness by friction with the launching barrel. The front face of the projectile is a flat metallic disc hooped to the shoe. The nature and thickness of the disc can be varied to modify the pressure and duration of the induced shock for a given launching velocity. In turn, the amount of powder allows the projection velocity to be adjusted for the given mass of projectile. The powder, encased in a cartridge, is ignited by an electrical igniter set on the screwed breech which closes the combustion chamber. The length of the launching barrel must be sufficient for the projectile to reach its maximum velocity at the level of the vents which, by removing the gas pressure, permit constant velocity over the last few centimeters before impact. The target, equipped with instruments and containing the sample, is usually fixed directly at the mouth of the launcher. Before the experiment, the launching barrel is evacuated by a pumping system to primary vacuum to reduce the precursors of compression on the target. Shock flatness is limited by the deformations in the front face of the projectile due to the very high accelerations in the initial launching phase.

The caliber of such launchers is usually between 50 mm and 100 mm. The

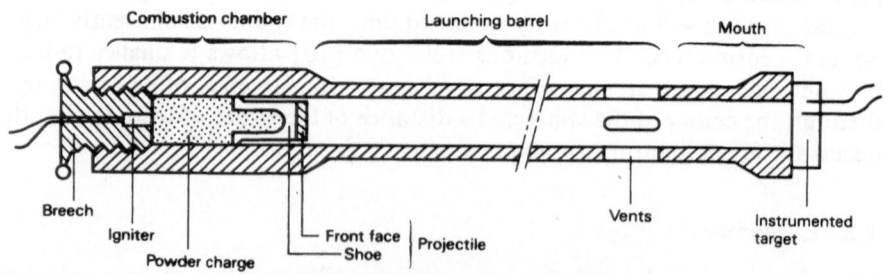

Figure X.57. Principle of a powder launcher.

projection velocities attained vary from several tens of meters per second to about 2000 m/s. This last value is, in practice, the physical limit for the performance of such a launcher, whatever the mass of powder used or the length of the barrel, because of the low value of the maximum velocity of release into the vacuum of the combustion gases from the powder. If higher velocities are required, the projectile must be propelled by the release of a gas with much lower molecular weight (hydrogen, helium, air, argon).

(b) Compressed Gas Launchers

For these launchers the projectile, the launching tube and the target are almost identical to those described above; but the propelling agent is not provided by the combustion of a powder but by a reservoir of compressed gas under high pressure (see Fig. X.58) filled in a quasi-static way either by direct decanting from cylinders or by a pump. The propelling gas may be argon, air, helium, or hydrogen, according to the performance required. To keep the projectile in place during the slow filling phase, the high pressure enclosure of the launching tube is insulated by a mechanical device. The rapid opening of this device, once the filling pressure is reached, triggers release and thus launching. This insulating/triggering operation is ensured either by a fast opening valve or, more usually, by a prenotched diaphragm which bursts at a given filling pressure.

The range of calibers and launching velocities with these launchers is very similar to that for powder launchers. The compressed gas system has the advantage of reducing the initial accelerations communicated to the projectile and thus the deformations of its front face, and therefore allowing better impact flatness. Although hydrogen or helium can be used, the performance of this type of launcher is limited by the technical feasibility of compression and quasi-static storage of the gas propellant.

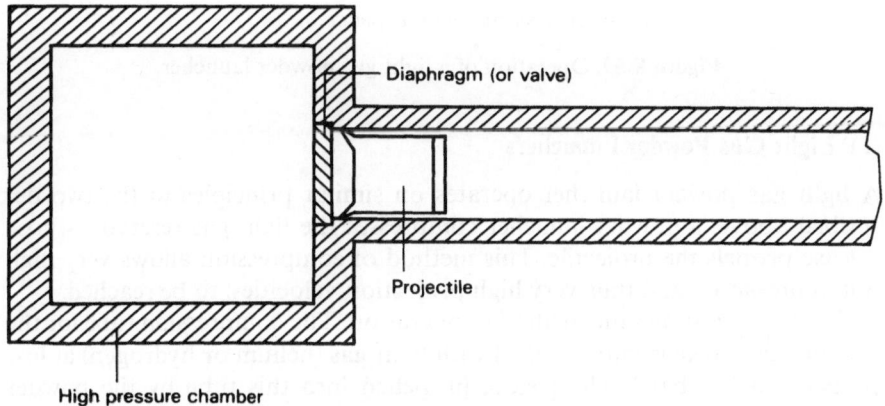

Figure X.58. Principle of a compressed gas launcher.

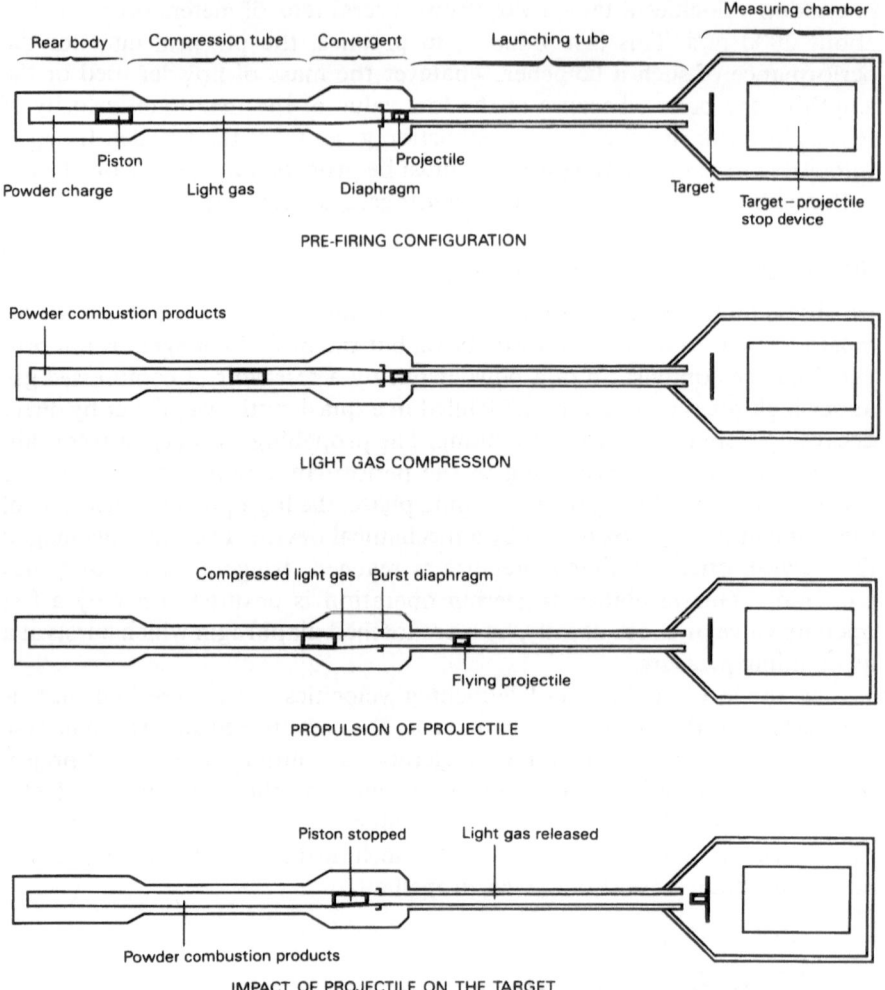

Figure X.59. Operation of a light gas powder launcher.

(c) Light Gas Powder Launchers

A light gas powder launcher operates on similar principles to the two described above: a powder launcher compresses the light gas reservoir whose
release propels the projectile. This method of compression allows very high
initial pressures, and thus very high projection velocities, to be reached.

Figure X.59 shows the method of operation. The compression tube for the
first powder stage is initially filled with light gas (helium or hydrogen) at low
pressure (a few bars). The piston, propelled into this tube by the powder
charge, compresses this gas in front of it, the high-pressure reservoir thus
formed being restrained by a prenotched diaphragm. When the pressure of

Figure X.60. Light gas powder launcher at Centre d'Etudes de Vaujours, Commissariat à l'Energie Atomique, France.

the light gas reaches the chosen rupture value of the diaphragm, the light gas is released into the launching tube while propelling a projectile similar to those in the launchers described above. During this time the piston is stopped by plastic deformation in the extrusion die formed by the convergent. Because of the elastic stresses transmitted by the tube assembly before the arrival of the projectile, the target may not be placed on the launching tube mouth, but is fixed on an independent support some distance away, inside a chamber pumped to primary vacuum at the same time as the launching tube.

This type of launcher can project with very high velocities, 8,000–9,000 m/s. The useful caliber is often rather limited by the size and weight of the mechanical parts needed to resist the enormous stresses developed, especially in the compression stage. The caliber is usually about 30 mm, but the launcher constructed at the Centre d'Etudes de Vaujours, Commissariat à

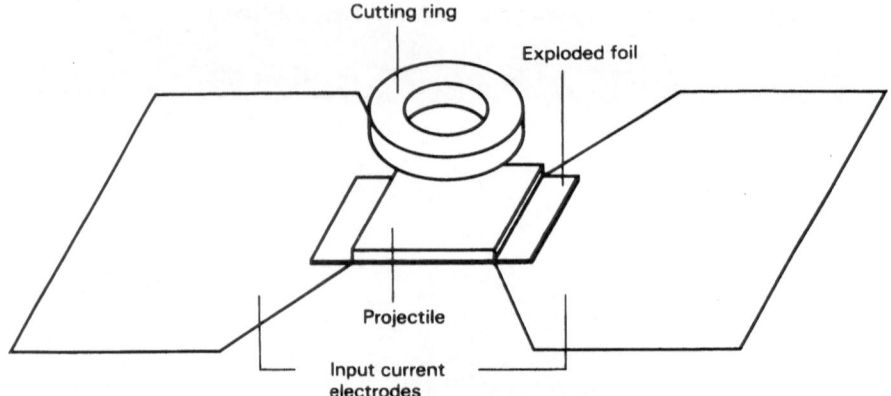

Figure X.61. Principle of an electric launcher with exploded foil.

l'Energie Atomique, France, has a caliber of 80 mm for an overall length of 57 m and a weight, not counting supports, of 52 metric tonnes. A photograph of this launcher is shown in Figure X.60.

(d) Exploded Foil Launchers

Exploded foil launchers are designed on a different principle from those above: the discharge of a capacitor bank causes the explosion of a thin metal foil which in turn projects another thin foil with which it is initially in contact. Figure X.61 shows this device. The exploded foil is usually a small copper or aluminum rectangle a few millimetres wide and a few tens of microns thick. The discharge current of the capacitor bank, storing several tens of kiloJoules at a voltage of a few tens of kilovolts, vaporizes the foil by the Joule effect. The release which follows this sudden vaporization accelerates the insulating projectile (mylar or kapton) placed on the foil. The cutting ring above and in contact with this foil removes the edges of the projectile, accelerating only a quasi-planar chip into the vent. The useful diameter of the projectile is only a few millimeters. A metal foil, a few tens of microns thick, may be placed in some cases on the front face of the insulating material to increase the induced shock pressure.

The projection velocities obtained with this device vary from a few hundred meters per second to nearly 16 km/s. Because of its low cost and ease of operation, it is often used in current research into the priming of explosives.

References

[1] BARKER, L.M. Laser interferometry in shock wave research. *J. Exp. Mech.*, **12** (1987), p. 209.

[2] BARKER, L.M., HOLLENBACH, R.E. Laser interferometer for measuring high velocities of any reflecting surface. *J. Appl. Phys.*, **43**, 11 (1972), p. 4669.

[3] BEAMS, J.W., KUHLTHAN, A.R. *et al.* Spark light source of short duration. *J. Opt. Soc. Amer.*, **37** (1947), p. 868.

[4] BEAUDOIN, L. *La radiographie éclair.* Presses Universitaires de France, Publications INSTN, Paris/France (1968).

[5] BRIXNER, B. A high-speed rotating-mirror frame camera. *J. SMPTE*, **59** (1952), p. 503.

[6] BRUN, L., CHÉRET, R., VACELLIER, J. Considérations sur les détonations fortes. *Proc. Symposium H.D.P.*, Paris/France (1978), p. 269.

[7] BRYANT, L. F. Flash radiographic techniques: exposure, recording, triggering and film protection. LANL Report, LA-UR-76-1328 (1976).

[8] CHAMPION, A.R. Electrical response of anodized aluminum layers to shock wave compression. *J. Appl. Phys.*, **40** (1969), p. 9.

[9] CHAREST, J.A. Development of a carbon shock pressure gage. DNA 3101 F, Defence Nuclear Agency, Washington, July 1973.

[10] CHERET, R. Contribution à l'étude des détonations sphériques divergentes dans les explosifs solides. Thèse de Doctorat ès Sciences, Poitiers/France (1971). Rapport CEA no. 4243.

[11] CUTCHEN, J.T. Polarity effects and charge liberation in lead zirconate–titanate ceramics under high dynamic stress. *J. Appl. Phys.*, **37** (1966), p. 4745.

[12] DURAND, M. *et al.* Interferometric laser technique for accurate velocity measurements in shock wave physics. *Rev. Sci. Inst.*, **48**, 3 (1977), p. 275.

[13] ESCHARD, G., POLAERT, R. Tube obturateur pour photographie ultra-rapide. *L'Onde Électrique*, **48** (1968), p. 494.

[14] FULLER, P.J.A., PRICE, J.H. Dynamic pressure measurements to 300 kbars with a resistance transducer. *British J. Appl. Phys.*, **15** (1964), p. 751.

[15] GRAHAM, R.A., NEILSON, F.N., BENEDICK, B.W. Piezo-electric current from shock-loaded quartz; a submicrosecond stress gauge. *J. Appl. Phys.*, **36** (1965), p. 1775.

[16] HUANG, T.S. *Picture Processing and Digital Filtering.* Springer-Verlag, New York (1975).

[17] INGRAM, G.E., GRAHAM, R.A. Quartz gauge technique for impact experiments. *Proc. 5th Symposium on Detonation*, Pasadena/CA (1970), p. 369.

[18] JAMET, F. La radiographie éclair. *Revue de la Défense Nationale*, **35**, 11 (1979), p. 113.

[19] LAHARRAGUE, P., DURAND, M., LE BIHAN, A. Systèmes de mesure de la vitesse d'un projectile utilisant un laser et un interféromètre de Perot–Fabry. *L'Onde Électrique*, **50**, 9 (1970), p. 804.

[20] LAURENT, B., LOICHOT, R. Réalisation d'un flash électronique et son application à la prise de vue par caméra ultra-rapide. *Proc. 16th International Congress on High-Speed Photography*, Strasbourg/France (1984).

[21] MILLER, C.D. Half-million stationary images per second with refocused revolving beams. *J. SMPJE*, **43** (1940), p. 479.

[22] MURAOUR, H. Sur l'utilisation des rencontres d'onde de choc dans l'argon comme source lumineuse brève et puissante. *Mémorial de l'Artillerie Française*, **1**, 23 (1949), p. 105.

[23] PRATT, W.K. *Digital Image Processing* Wiley–Interscience, New York (1978).

[24] RADON, J. Uber die Bestimmung von Funktionnen durch ihre Integralwerte

Langs gewisser Maunigfaltigkeiten. *Berichte Sächsische Akademie der Wissenschaften*, **69** (1917), p. 262.

[25] REYNOLDS, C.E., SEAY, G.E. Two waves shock structures. *J. Appl. Phys.*, **33** (1962), p. 2234.

[26] VIBERT, P., AMALRIC, Y. X-ray flash radiography images restauration. *Proc. "Image Detection and Quality,"* Paris/France (1986), p. 337; SPIE Publications.

[27] VORTHMAN, J.E., ANDREWS, G., WACKERLE, J. Reaction rate from electromagnetic gauge data. *Proc. 8th Symposium on Detonation*, Albuquerque/NM (1985), p. 99.

[28] ZAITSEV, V.M., POKHIL, P.F., SHVEDOV, K.K. The electromagnetic method for the measurement of detonation products velocity. *Dokl. Akad. Sci. USSR*, **132**, 6 (1960), p. 1339.

CHAPTER XI

Elementary Configurations of Simple Detonation

1. Tradition: Good Points and Bad Points

In all laboratories engaged in the characterization of explosives and the study of detonation phenomena, there is a long-standing tradition which prefers simple detonation "regimes" where waves are propagated by mere translation with a permanent velocity D parallel to the direction i related to the location of the firing station. (N.B. The superscript p shows only, as in §II.3.3, that such propagation is endowed with a privileged plane direction which is normal to i; however, it must be borne in mind that referring to this regime as "plane" is only a widely accepted misuse.) But it is worth paying attention to such a persistent, widespread tradition, to find its good points ... and underline its bad points.

The first reason is that such a regime is "naturally" established in a tube of liquid or gas explosive or in a stick of solid explosive, provided only that the tube or stick is sufficiently long and wide and that the stress on the priming section is sufficiently great and intense (see Chap. IX). It is therefore not by chance that this regime is the one which, having been observed for about one hundred years, has produced the concepts of explosive and detonation.

But the tradition could not have developed unless, to the experimental convenience of generating the regime, another factor was added which is just as important, that of justifying a velocity measurement whose precision depends only on those factors available for measuring the length and time of propagation (usually a precision of 0.2%, as noted in §II.3.3).

These two reasons amply justify the attention given to these regimes in experimental programmes, but they do not justify exclusive attention while neglecting all others, under the influence of simplistic ideas which are clearly false, but still widespread, that can be caricatured as follows:

- the wave is plane;
- the relative downstream flow of the wave is permanent;

- the transverse component of the velocity-vector is stationary on the axis; and
- D^p is the Chapman–Jouguet (C–J) velocity.

The expositions in Chapters III and IX have treated these ideas fairly. However, to exorcize and discredit them completely in the minds of readers who may be still undecided, this chapter is designed to show—for axisymmetric (Section 2) and spherical (Section 3) configurations—one regime where the established detonation is quasi C–J and another regime where it is strong, according to the priming conditions. The fourth and last section contains current considerations on the break of a built-up detonation in the medium adjacent to the explosive on a free boundary. The reader will thus be introduced to regimes which are geometrically and kinematically more complex.

2. Axisymmetric Configurations

2.1. Traditional Regime

In principle, if not in practical details, the experimental device resulting in the traditional regime discussed above in Section XI.1 hardly varies from one laboratory to another:

- a priming system is embedded in the end of the tube or stick of diameter ϕ;
- observations are made for propagation distances X greater than the run to detonation X* (as a general rule, X* is estimated to be less than 4ϕ);
- in the simplest version, the velocity D^p is measured by a slit scanning optical technique along a generatrix (see Section X.1) and/or by an electronic technique (see Section X.2);
- in a better equipped version, as well as the preceding measurement, the shape of the detonation wave is measured by a slit scanning optical technique for a diameter of the face of the tube (or stick) opposite the priming face; and
- in the fully equipped version, the shock propagation is measured in a known medium, placed in contact with the detonation output face.

Figures XI.1 and XI.2, referring to the works of Aveillé et al. [1], illustrate the last case, with regard to the solid explosives comp. X/X_1 and comp. T/T_1, referred to in Tables XI.1 and XI.2. The information given below indicates the layouts used and the values measured.

Samples

The samples are obtained by isostatic moulding and machining. The quality of the stick faces (0.01 mm) allows assembly and centering without the use of an adhesive matrix. All the sticks in a stack have the same density, within 0.002 g/cm^3.

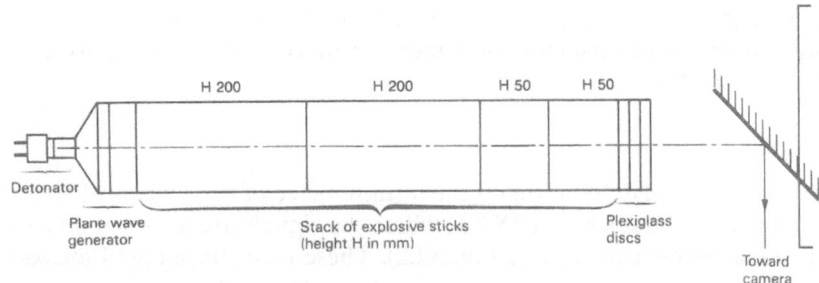

Figure XI.1. Diagram showing solid explosive assembly for measuring the velocity and shape of detonation waves in the traditional regime ($\phi = 50$ mm). The stack of plexiglass discs at the end of the structure is used to measure the induced shock velocity.

Electronic Measurements

The travel of the wave along the axis is tracked by passive probes (see Fig. XI.1), containing printed circuits, arranged at 120° on a flexible mylar insulated support 0.08 mm thick, operating by ionization. They are inserted at the stick interfaces and each causes closure of a rapid discharge circuit. The pulses formed are led coaxially to a digital chronometer with 100 ps resolution.

Wave travel times are thus recorded at each interface. The difference in length between the measuring bases (second and third sticks with lengths of 200 and 50 mm, respectively) allows for the disturbance caused by the probes: thus correcting the "joint effect" which causes a systematic error of about 10 ns in the transit time.

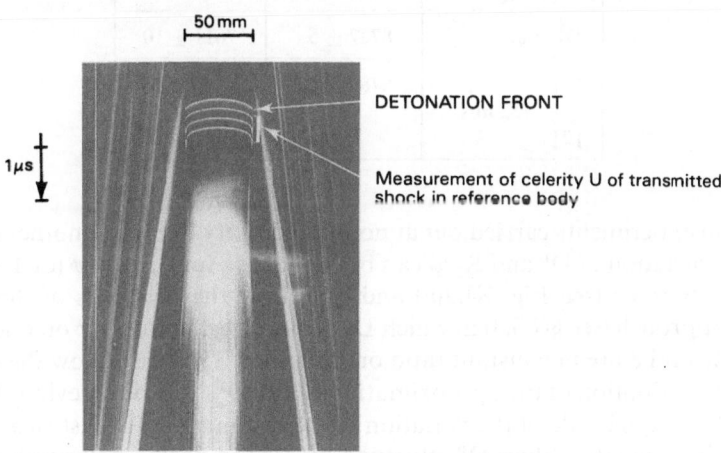

Figure XI.2. Recording on the film of a slit scanning camera during the detonation of the assembly shown in Figure XI.1.

For high values of ϕ (100 mm and 200 mm) the output face is fitted with optical fibers and coaxial microcontacts arranged in order along two perpendicular diameters.

Optical measurements

The emergence of detonation on the terminal face of the stack is observed by air ionization chambers (see §X.2.1) where the signals are recorded on the film of a slit scanning camera (see Fig. XI.2). These recordings are of interest for two reasons:

- they show that the wave is concave towards the detonation products, and is thus quasi C–J by nature (see §III.4.2); and
- they allow the wave axial curvature $1/R_\Sigma$ to be estimated (curvature of the best circular approximation of the trace portion between $-\phi/4$ and $+\phi/4$.)

Results

For experiments made with 50 mm diameter, linear smoothing of the velocity measurements as a function of density gives a value denoted as D_{50}^p and a value of $\partial D_{50}^p / \partial \rho_0$ for a nominal value of the density ρ_0 (see Table XI.1).

Table XI.1. Typical results of the traditional characterization of a solid explosive.

		comp. X/X_1	comp. T/T_1
ρ_0 g/cm^3		1.822	1.880
$\partial D_{50}^p / \partial \rho_0$ m·s^{-1}/g·cm^{-3}		4.6 ± 0.1	2.8 ± 0.3
D_{50}^p m/s		8787 ± 5	7611 ± 10
[1]	D_∞^p m/s	8789 ± 10	7715
[2]			7737 ± 15

From experiments carried out at nominal density but with another diameter, the variation of D^p and R_Σ/ϕ can be found as a function of ϕ (or $1/\phi$). The results obtained (see Fig. XI.3(a) and (b)) show the weakness of the traditional approach (see §II.3.3) in which D^p varies linearly with $1/\phi$ on one hand, while R_Σ and ϕ are in constant ratio on the other. They also show the danger in a hasty adoption of the approximation $D_* = D_{\phi_0}^p$ without previous knowledge of the amplitude of the variation of D^p beyond ϕ^0. The last two lines of Table XI.1 give the values D_∞^p obtained by second degree polynomial extrapolation for X/X_1 and by second and third degrees for T/T_1. When using D_∞^p (see §XII.1.6) it is necessary to note the uncertainty attached to the actual extrapolation.

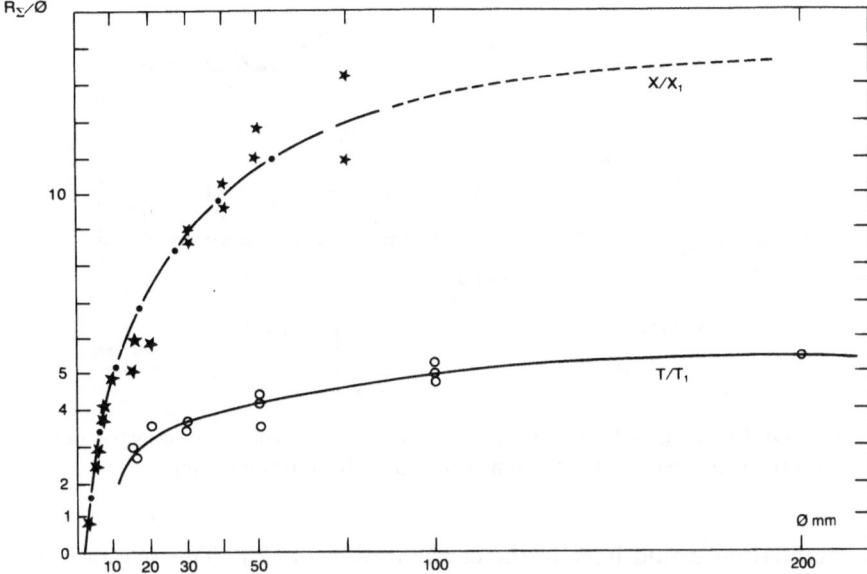

Figure XI.3(a). Variation of R_Σ/ϕ with ϕ (after [1]).

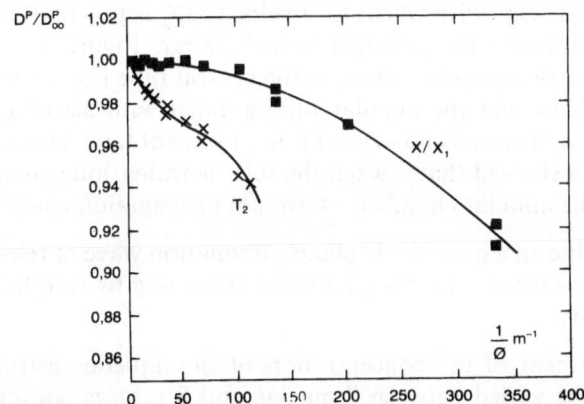

Figure XI.3(b). Variation of D^p/D^p_∞ with $1/\phi$ (after [2]).

2.2. Lateral Priming Regime

In terms of the boundary conditions, it can be said that the traditional experiment, recalled and discussed in §2.1, is characterized by a priming boundary occupying all or part of the entry face (of the tube or stick) while the lateral surface and the output face have free boundaries.

What happens when this distribution between priming boundary and free boundaries is different? In particular, what happens if a stress progresses on the lateral surface with a constant velocity higher than the velocity D^p_ϕ which

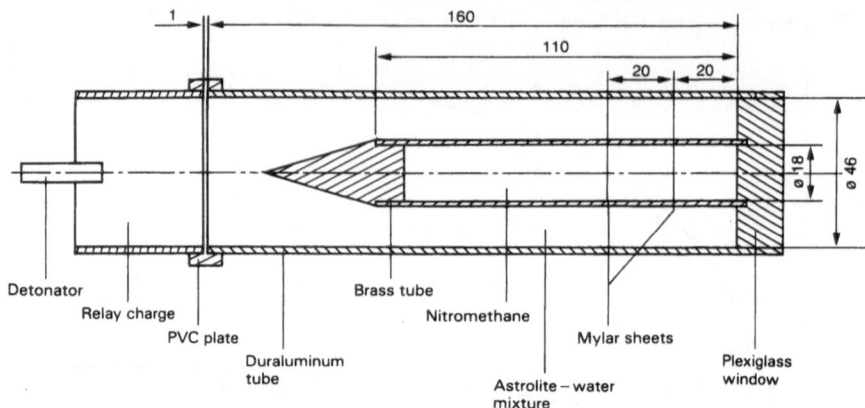

Figure XI.4. Experimental device for the generation of a strong axisymmetric detonation in nitromethane (after [21]). Dimensions are given in millimeters.

the detonation would have in the configuration of §2.1? This question was raised by Chéret [5], and partly answered by two series of experiments reported by Krishnan *et al.* [17], Scllam *et al.* [21], and Hamada *et al.* [15].

The principle in these experiments is to create the required stress by careful use of a "fast" explosive whose C–J velocity D'_* is greater than that D_* of the explosive forming the principal "core" charge. Figure XI.4, taken from [21], shows the device obtained when the central tube is filled with commercial nitromethane and the annular tube is filled with astrolite in aqueous solution. The dimensions and materials used were obtained from preliminary research which showed that—when the tube is rather long compared to the thickness of the annular chamber—terminal propagation consists:

- in the astrolite, of a quasi C–J "plane" detonation wave of velocity $D'i$; and
- in the nitromethane, of a complex wave Ω moving by translation with the same velocity.

Simple adjustment of the concentration of the aqueous astrolite solution allows D' to be varied between 7 mm/µs and 8 mm/µs, an interval which compares with the value 6.29 mm/µs of the standard C–J velocity of nitromethane.

The following subsections discuss the analysis of the wave Ω.

Optical Measurements

On its end section, the nitromethane core is in contact with a plexiglass "window" whose shock impedance is less than that of the nitromethane detonation products. Interaction of Ω with the window reduces the emissivity of the detonation products together with the plexiglass transparency. On the film of a slit scanning camera, whose optical axis coincides with the assembly axis and whose entry lens sees a diameter of the output face, as a total result, there appear significant traces of arrival times on the window of the frontal parts of Ω. Starting with these traces, and using the precautions given above to ensure that Ω is stationary, it is possible to "reconstruct" Ω in the neighborhood of the window.

This "end" observation can be completed—using the same camera—by that of the interaction of Ω with two mylar sheets (0.1 mm thick, 10 mm wide) placed normally to i, respectively, at 20 ± 0.2 mm and 40 ± 0.02 mm from the window. This interaction causes a momentary extinction of detonation and thus an interruption in the transmission of light towards the camera. The dark traces recorded on the film thus show the time of arrival on the mylar of emissive parts of Ω. These "intermediate" traces can be reconstructed in the same way as the "end" traces.

Each recording shows three zones (①, ②, and ③ in Figure XI.5(b)) clearly differentiated by their optical density, which decreases from the axis to the periphery. They are limited in the scanning direction by two arcs AJ and JI' (see Fig. XI.5(b)):

- AJ is completely situated in the most luminous zone; it is concave toward the window; it is normal in A at i and tangential in J to the line separating zones ① and ②; its curvature in A increases as D' increases (see Table XI.2); and
- JI' extends from the interface between zones ① and ② to the edge of the tube; it is concave towards the detonation products; in J it is inclined on i by an angle ψ whose sine is little different from D_*/D' (see Table XI.2); when crossing the interface between zones ② and ③ an arc J'I is formed (revealed by interaction with the plexiglass window) which makes in I with i an angle close to that made by the shock wave induced in the nitromethane across the tube by the detonation of the astrolite.

Interpretation

The rules in §III.4.2 and the quasi nullity of $[(D'/D_*) \sin \psi - 1]$ show that the arc AJ is the meridian of a strong lenticular detonation, while the arc JI' is that of an annular quasi C–J detonation. These conclusions allow interpretation of the discontinuity of illumination between zones ① and ② as the manifestation of a sharp evolution of the velocity and temperature profiles (see §III.3.2). They also allow interpretation of the difference in illumination between zones ② and ③ and the significance of the arc J'I:

- the shock induced (across the tube) by the detonation of the astrolite curves and intensifies from I to J', but is always too weak to supply the ignition entropy s^i directly to the nitromethane; and
- ignition is obtained only at the cost of subsequent compressions due to "the accumulation of shocks" in the tube itself, in such a way that the different points of I'J' correspond to quasi C–J detonation of thin streams

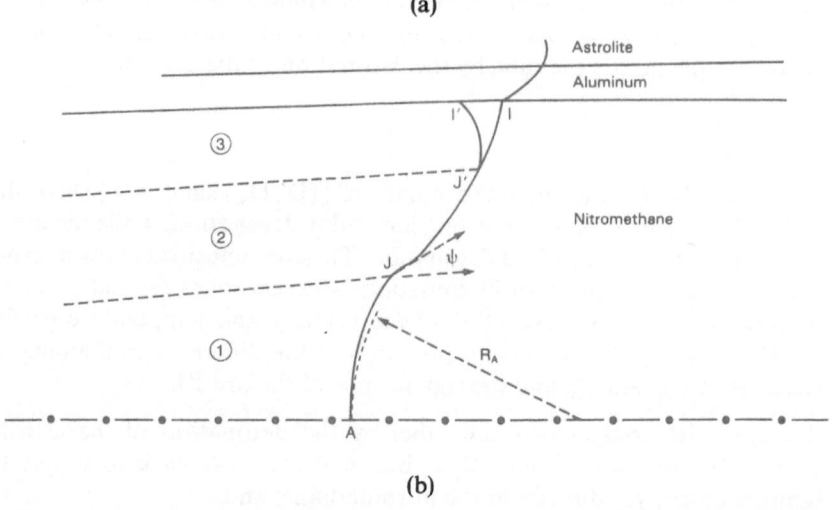

(a)

(b)

Figure XI.5. Detonation of the structure shown in Figure XI.4: (a) result of the densitometric analysis of the recording on the film of a slit scanning camera: time increases from left to right; and (b) wave system Ω reconstructed from the recording.

Table XI.2. Values of angle ψ and radius of curvature R_A defined in Figure XI.5(b) as a function of velocity D' (after Sellam, doctoral thesis, Poitiers, 1986).

D' mm/μs	7.3	7.51	7.72	7.87	8
R_A mm	44.3	31.2	25.5	18.9	19
$\dfrac{D'}{D_*}\sin\psi - 1$		$9\cdot10^{-3}$	$17\cdot10^{-3}$	$9\cdot10^{-3}$	$12\cdot10^{-3}$

of precompressed, preheated nitromethane and not to nitromethane at rest; the corresponding C–J temperatures are found to be lower as is also the emissivity of the detonation products.

Conclusion

The configuration described and interpreted in this section is much less "natural" than the traditional configuration. It is nevertheless of great interest insofar as it provides:

- an example of strong stationary detonation, the velocity of which can be chosen over a continuum; and
- an example of the coexistence of strong and quasi C–J detonations and of the absence of continuous transition from one to the other (see §III.3.2 and §III.5.2).

3. Spherical Configurations

3.1. Explosion Regime

Among the rules established in §III.4.2 can be read: "an autonomous spherically diverging detonation is quasi C–J" from which it follows at once that the velocity of an autonomous spherically diverging detonation wave tends toward D_* when the radius of the wave increases indefinitely (Figure II.6 shows an early illustration of this asymptotic type of behavior). In other words, since autonomy is always achieved, the explosion regime together with the traditional "plane" regime provides another way of determining the C–J velocity, the value of which is necessary for the dynamic characterization of explosives (see Chap. XII).

But it must be recognized that this way has hardly been explored for condensed explosives [2], [3], [7], [11]. The reasons for this "neglect" are not only the theoretical research difficulties (see §III.5.2), but also the experimental constraints found with this configuration. Although obvious, they are worth recalling here since they largely explain the layouts used in designing the explosive structure and the measuring system:

- first, there is a need for the design of a priming device which, (i) occupies little space compared to the total volume of explosive, (ii) causes little disturbance of the spherical symmetry, (iii) is powerful enough to ensure ignition, and (iv) is reproducible enough to permit reliable analysis of a limited number of experiments;
- an explosive structure is needed of such dimensions that, after ignition, the build-up time of a detonation (see §XI.1.3) is short enough for most of the propagation to be that of built-up detonation;
- observation and/or measurement methods must be designed which do not disturb the spherical symmetry and which are appropriate to the non-permanent character of the flow: in particular, when the velocity is obtained by measuring the time taken for the wave to cross a known distance, a compromise must be found between local measurement and precise measurement; and
- finally, when suitable dimensions have been defined for the structure as a function of the first two points, it must be constructed in a well-defined and uniform state, so that the nonpermanent character of the measurements can be unambiguously attributed to the very nature of the flow.

The first condition is more easily met with an inhomogeneous explosive (e.g., an aggregate explosive) than with a homogenous explosive (e.g., a liquid explosive) as can readily be seen from the results in Chapter IX. This difficulty explains the device used by Chéret and Verdès [11] to prime the explosion of nitromethane (see Fig. XI.6). However, the last three conditions, which require only elementary precautions for a liquid explosive, are very restrictive for solid explosives: they limit the research area to a few moulded aggregate compositions based on cyclonite (RDX), octogen (HMX), or triaminotrinitrobenzene (TATB), which must take the shape of "spheres" [3], [7] or "logospheres" [2] (see Fig. XI.7(a–c)). The following sections indicate the layouts used, the measurements made, and the results obtained.

Experiments with Nitromethane

The experiment carried out by Chéret and Verdès [11] is a generalization of that devised by Campbell [6]. A tank (see Fig. XI.6(a)) filled with pure ("spectroscopic") nitromethane is set between an argon flash (see §X.1.1) and a slit scanning camera (see §X.1.3); the first lens in the camera forms on the slit the image of a straight line $\omega_1 \omega_2$ which starts from the center O of the tank and is parallel to one of the tank faces. When at rest, the nitromethane is transparent and transmits all the beam; but at a time τ subsequent to the firing time τ_0 when the detonation (or shock) wave has reached radius X, the beam is blocked over a diameter 2X, and a segment of width 2gX will not be recorded on the film (see Fig. XI.6(b)), g being the recording magnification. Analysis of the variation of 2gX with time (by the procedure detailed in [11]) gives a "continuous" estimation of the wave velocity $D^s(X)$.

The explosive structure itself is a compromise between the different experi-

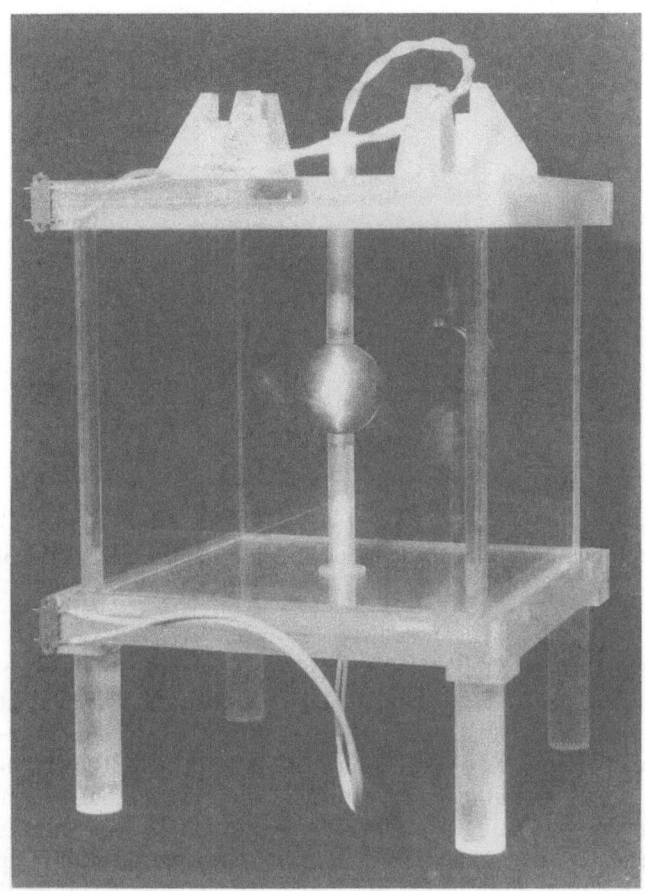

Figure XI.6(a). Experimental set-up for measuring the divergent spherical detonation velocity of nitromethane (after [11]).

mental constraints. The base and walls of the tank are 10 mm thick glass slabs, glued together. The walls perpendicular to the optic axis, through which measurement is made, have "optical polish" and are carefully mounted to ensure parallelism. The tank is closed by an "adjusted" plexiglass covering cap, whose multiple roles will appear later. A sphere of powerful solid explosive serves as priming relay: it is formed from two hemispheres joined along a vertical plane and pierced with two diametrically opposed vertical holes of 10.2 mm diameter which will receive two exploding bridgewire detonators of 10 mm diameter. It is kept in a central position by two plexiglass tubes (external diameter 12 mm, internal diameter 10.4 mm), one being fixed to the bottom of the tank and the other to the covering cap, extending the holes for the insertion of the detonators; the assembly (sphere and tubes) is made rigid by bonding with adhesive, then made tight and chemically inert by spraying

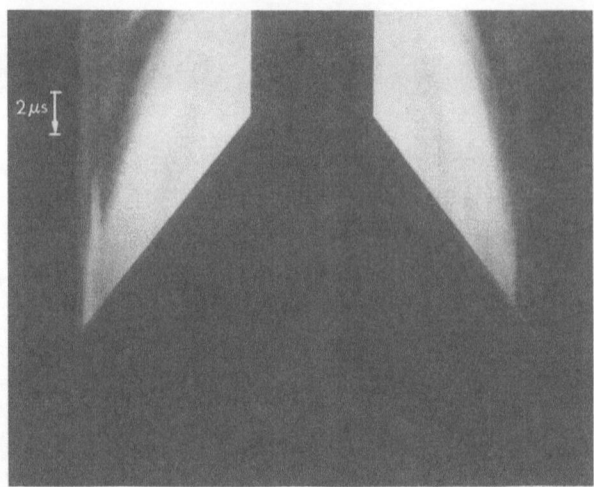

Figure XI.6(b). Recording on the film of a slit scanning camera during the spherically diverging detonation of nitromethane primed by a sphere of high explosive (after [11]).

with aerosol varnish. The covering cap is pierced with three more holes: one to allow filling of the tank, while the other two allow introduction of the needles of a sighting-mark and their integration with the whole set-up, during the preliminary focusing of the camera and measuring of the magnification g.

The variation in initial temperature of the nitromethane is measured in the interval $(\tau_0 - 10 \text{ min}, \tau_0 - 5 \text{ min})$; it is about $0.1°$ C. Its value is measured at the instant $\tau_0 - 5 \text{ min}$: $13.5°$ C. For the estimated initial temperature of $13.4°$ C, the law established by Davis *et al.* (see [21] in Chap. XII) gives a corresponding velocity $D_\infty^p = 6.336 \pm 8$ m/s.

Solid Explosive Structures

The experiments made in 1969/1970 by Chéret and Verdès, reported in [7] and [4], were carried out on "spheres" of F209 explosive, made from a mixture whose essential components are cyclonite (RDX) and TNT: hot-setting uniaxial compression followed by slow cooling produces blocks from which pieces of the required shape and size can be obtained by machining. Since a general study of the procedure showed that the blocks are most nearly homogeneous when the mould (whose axis of symmetry is parallel to the compression axis) has similar radial and axial dimensions, an attempt was made to make the best possible use of the largest mould: a hemispherical mould which can yield a 300 mm diameter hemisphere, and then a 150 mm diameter sphere. Because preliminary velocity measurements had shown that a study limited to waves with maximum radius 75 mm was insufficient, preference was given to the manufacture of spheres by joining two hemispheres, necessary to allow study of explosions up to 150 mm radius. (N.B. The experiments reported by Bahl [3] are limited to 50 mm radius.)

Each hemisphere is machined to size by grade 10 machining: its radius being 50 ± 0.05 mm, 75 ± 0.08 mm, 100 ± 0.092 mm, 150 ± 0.105 mm at temperature $20° \pm 2°$ C. After machining to shape, the density ρ_0 of each piece is determined hydrostatically at $20° \pm 1°$ C with a precision of 0.2%: for all the pieces used the values of ρ_0 were found to be regularly distributed between 1.712 g/cm^3 and 1.728 g/cm^3, with a mean value of 1.720 g/cm^3, subsequently called the nominal value. The only modification after weighing is the drilling of housings for the detonators, which is strictly made on a pair of hemispheres fitted together: the metal device which holds this assembly on the machine tool is such that the distance between the centers of the two hemispheres is less than the tolerance of machining to radius. The two matched hemispheres are then separated for transport from the workshop to the firing area where the detonators are introduced before their assembly.

The above procedure has been carried out everywhere for "logosphere" type structures, as shown in Figure XI.7(c). Their major advantage over spheres is their greater observation distance (up to 230 mm) particularly appreciable for TATB compounds, but at the price of very complex instrumentation, which will be described in the next two sections.

Optical Measurements on Spheres

The velocity of an explosion is found by marking its travel between a sequence of known points aligned along Oω (see Fig. XI.7(b)). The following technique is used to produce these points in the sphere and use them to detect wave travel.

Polished plexiglass sticks (4 mm diameter, length $\Lambda = 9.80$ or 14.80 ± 0.01 mm), separated from each other by piano wire rings (200 ± 5 μm thick) are placed in a hollow frosted plexiglass cylinder (external diameter 6 mm, internal diameter 4 mm). After assembly in the laboratory, some air remains trapped in a "chamber" between two sticks because of the insertion of the ring. The rod thus made (see Fig. XI.8(a)) is introduced into a cylindrical cavity (6 mm diameter) drilled along Oω up to 7.5 mm from O. When detonation reaches a "chamber" the air ionizes and glows. The corresponding flash is recorded on the film of a slit camera, whose optical axis has been made to coincide with the rod axis. If the "inscription rate v" on the camera film is known and the distance λ between two consecutive flashes measured, the average detonation velocity D^s over the length Λ of the corresponding base line is given by D$^s = v\Lambda/\lambda$. The film also records the arrival of the detonation on the surface of the sphere (see Fig. XI.8(b)) and allows the wave sphericity to be controlled during each shot: such a control has shown that no faults can be seen near the rod, while disturbances to wave front propagation by the rod can be considered as negligible.

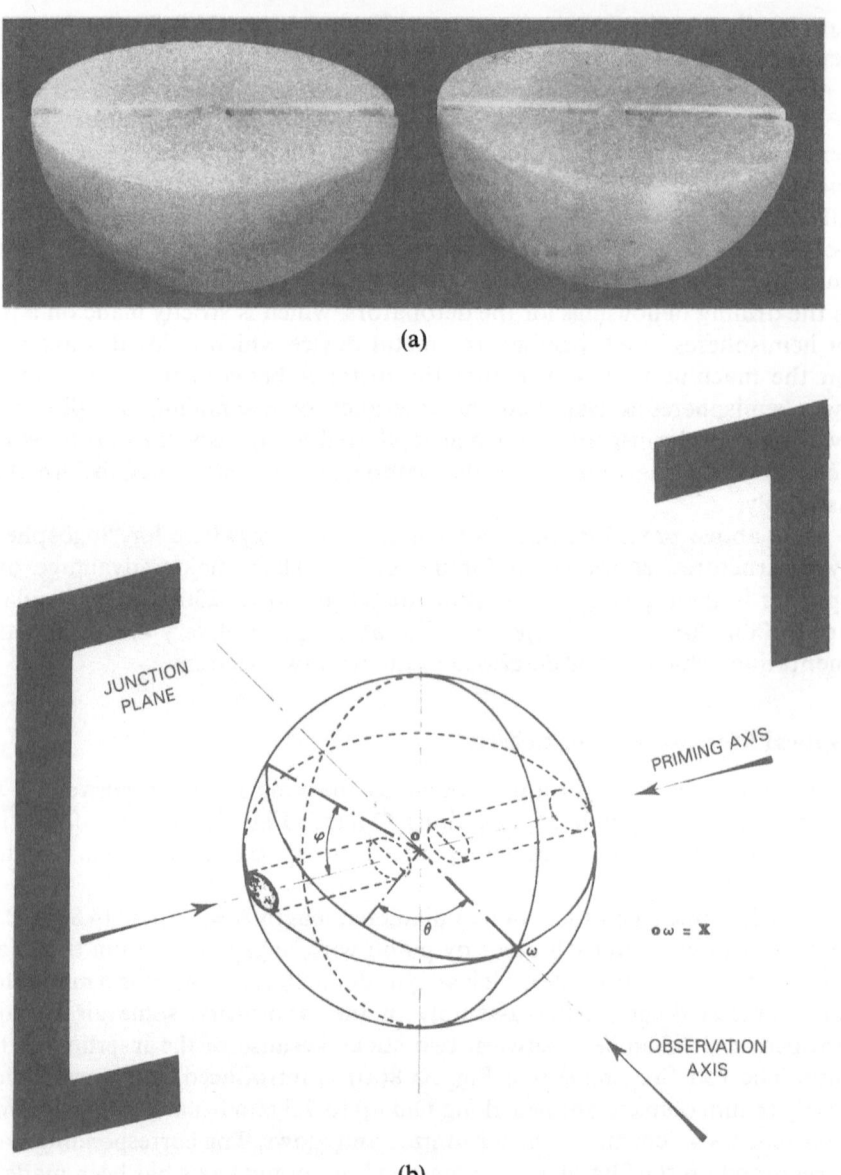

(a)

(b)

Figure XI.7. (a) Two hemispheres before assembly (after [7]); (b) diagram of an assembled sphere (after [7]); and (c) meridian section of a logosphere structure (after [2]).

(c)

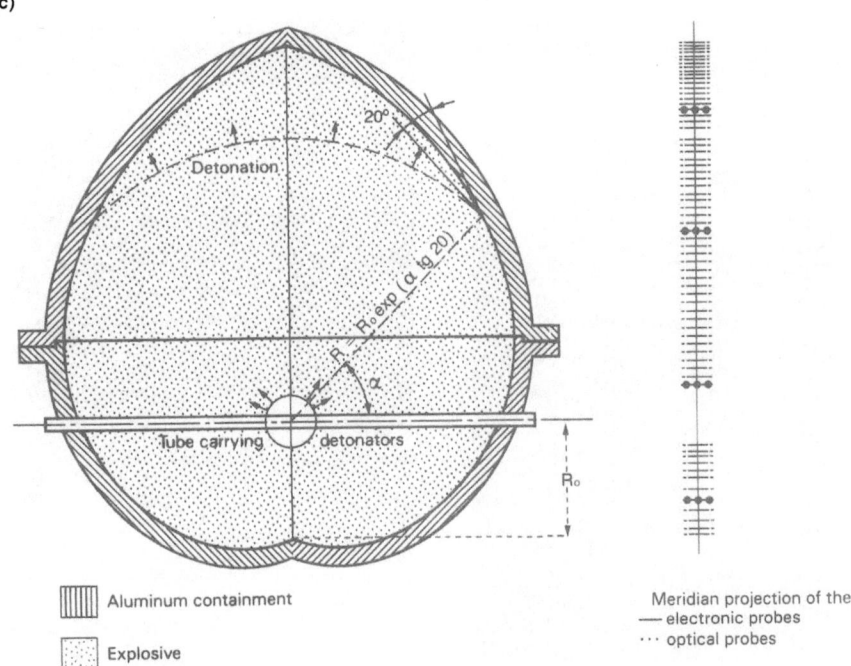

Aluminum containment

Explosive

Meridian projection of the
—— electronic probes
··· optical probes

Figure XI.7(c) (*Cont.*)

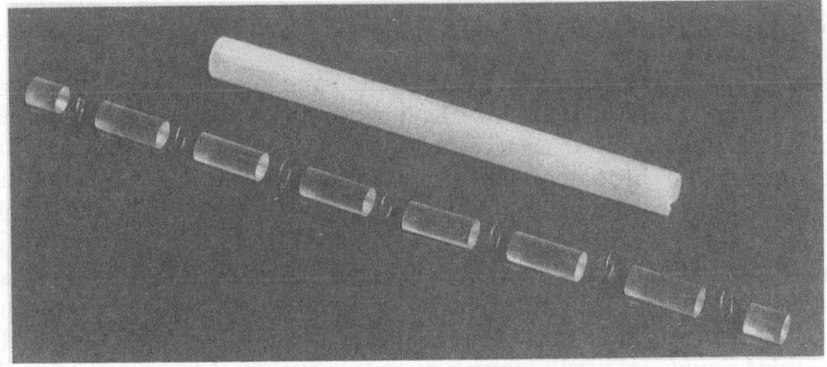

Figure XI.8(a). Exploded view of a rod.

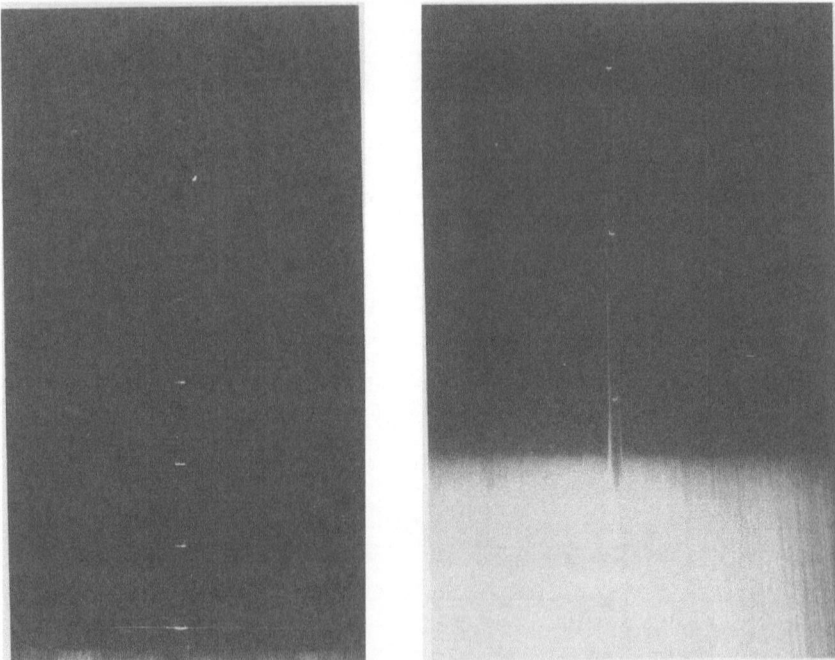

Figure XI.8(b). Recordings on a slit camera film of flashes emitted by air chambers (scale 1:1): (1) $\Lambda \cong 15$ mm, $v = 12.6$ mm/μs, sphere of 100 mm radius; and (2) $\Lambda \cong 10$ mm, $v = 25.2$ mm/μs, sphere of 150 mm radius.

Electronic Measurements on Logospheres

Passage of the explosion along the free boundary of the logosphere is detected by two series of probes. The first are short-circuit type probes: each causes closure of a fast discharge circuit, and the pulses formed are led coaxially to a digital chronometer with 1 ns resolution. The second are ionization chamber-type probes with optical fiber head: they deliver a light signal which is recorded on an electronic camera at about ± 2 ns. Probes of both types are inserted in an aluminum probe support plate placed normally to the axis of the detonators, in positions X^j preset at about ± 0.01 mm. Each structure thus has up to 80 values of X^j distributed from 70 mm to 230 mm. The recorded signals are used to form a discrete sequence of passage instants t^j of the wave at radii X^j. To obtain continuous representation from this sequence, use is made of the experimental observation noted in §III.1.6 and often repeated, that the normal local detonation velocity is all the less variable as the average local curvature is smaller. By taking into account the range of variation of X (from 70 mm to 230 mm) and the small size of the variation of D (less than 1%) it is easy to verify that a linear representation of D in 1/X is sufficiently accurate. By integrating the simple expression of D in one-dimensional geometry

$$D = dX/dt;$$

we obtain from this an expression for t as a function of X of the form

$$t - t^0 = \frac{1}{D_\infty^s}\left[X - X^0 + a \operatorname{Log} \frac{X - a}{X - X^0}\right].$$

Calculating the best root mean square approximation for all the pairs (t^j, X^j) leads to a value of D_∞^s.

Results

Interpretation of the raw data—which have been obtained as described in the last three subsections—differs according to whether they are for:

- a liquid explosive with continuous recording of velocity during a single experiment at temperature T_0; and
- a solid explosive with recording of discrete values that are usually obtained during distinct experiments which must be restored by linear smoothing, to reference loading density ρ_0.

However, it is possible in both cases to define, when radius X increases indefinitely, a limit D_∞^s to which the explosion velocity tends at temperature T_0 in the first case, at density ρ_0 in the second. Provided that allowance is made for these initial conditions, D_∞^s can be compared to the limit D_∞^p of the traditional regime (see Table XI.3).

Table XI.3. Comparison of extrapolated velocities D_∞^p and D_∞^s for four explosive substances.

		D_∞^p m/s	D_∞^s m/s	References
Pure nitromethane	$T_0 = 13.4°$ C	6336 ± 8	6342 ± 24	[11]
F209	$\rho_0 = 1.720$ g/cm^3	8193	8227	[7]
Comp. X/X$_1$	$\rho_0 = 1.822$ g/cm^3	8789 ± 10	8802 ± 20	[2]
Comp. T/T$_1$	$\rho_0 = 1.880$ g/cm^3	7715	7737 ± 20	[1]
		7737		[2]

The difference $D_\infty^s - D_\infty^p$ appears here as positive or zero according to the explosive considered. This diversity is surprising in view of the similarity of detonation phenomena established for all condensed explosives. For want of a better hypothesis, we conjecture that

$$D_\infty^s = D_\infty^p,$$

and attribute the differences shown in Table XI.3 to the uncertainty of D_∞^p because of the polynomial approximation used, as is shown by the two values given for comp. T/T_1.

Independently of this conjecture, the sections above show the superiority of the explosion regime to the traditional regime for the determination of D_*: a single experiment on a logosphere or on a large radius sphere rather than a large number of experiments whose results are blurred by uncertainty because of linear smoothing of density and extrapolation to infinite diameter.

3.2. Implosion Regime

Despite the experimental difficulties which arise when researching perfect spherical symmetry in a large solid angle, the explosion regime is very much easier to obtain than its false twin, the implosion regime. Nevertheless, the latter is by no means purely hypothetical since suitable priming devices have been developed; these are known as "spherical wave generators," abbreviated to S.W.G. The results below are partly due to [10].

Explosive Structures

Experiments carried out by Chéret, Chaissé, and Zoé [10] were made on HMX-based F626 explosive spheres, for which traditional characterization (see [13]) leads to

$$\begin{cases} \rho_0 = 1.835 \text{ g/cm}^3, \\ D_* = 8765 \text{ m/s}, \\ \Gamma_* = 3.005, \\ G_* = 0.582. \end{cases}$$

These are intended to determine the path diagram (X, t) of detonation following priming of a sphere of diameter ϕ. Because the S.W.G. does not have exact spherical symmetry, but only an invariance property in relation to Δ_α axis rotations and Π_β plane symmetries, experimental conditions are selected which ensure slight loss of sphericity. Therefore, only two experiments (ϕ 300 mm and ϕ 280 mm) have been designed and considered satisfactory: at radius $[(\phi/2) - 20]$ mm the loss of sphericity is less than 0.2 μs.

Electronic Measurements

Measurement is made by recording the time t^j at which the wave reaches a location X^j predetermined to about ± 15 μm. Passage of the wave is detected by short-circuit type probes (25 μm kapton support, 35 μm copper electrodes, 25 μm screen), which operate by closing a fast discharge circuit; the pulses formed are transferred by coaxial lead to a digital chronometer with

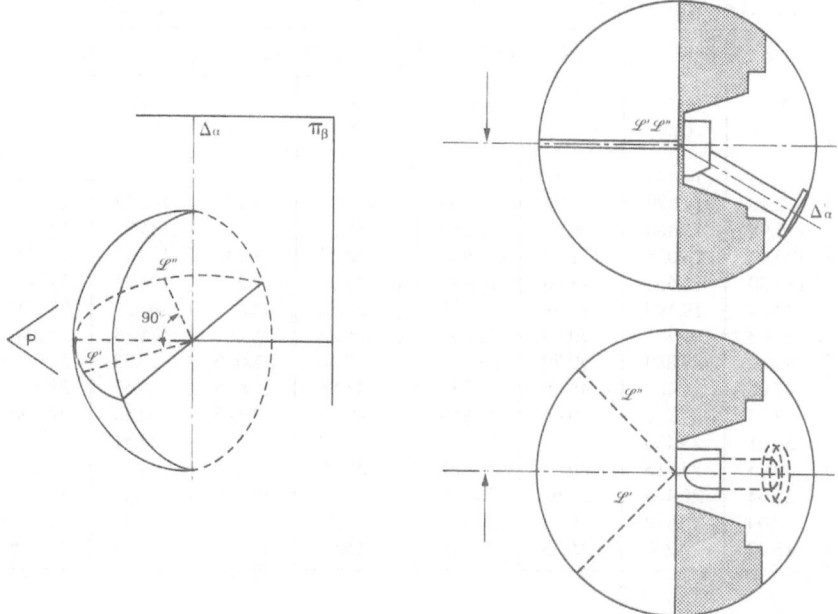

Figure XI.9. Diagram of explosive structure for measurement of convergent spherical detonation velocity (after [10]).

5 ns resolution. The probes are located on two orthogonal radii $\mathscr{L}'$ and $\mathscr{L}''$ belonging to a plane P normal to an axis Δ_α; furthermore, $\mathscr{L}'$ and $\mathscr{L}''$ on P are chosen so as to be symmetrical about a plane Π_β which includes the preceding axis Δ_α (see Fig. XI.9). Measurements are thus made in the junction plane of two quarters of a sphere whose common diameter is the external bisectrix of the angle $(\mathscr{L}', \mathscr{L}'')$.

To avoid premature destruction of the coaxial wiring, an inert medium and suitable shields are substituted for the explosive, as shown in Figure XI.9. The substitution is made in such a way as to preserve the symmetry of the truncated hemisphere with respect to Π_β and the symmetry about Δ'_α of the priming device. The results are shown in Table XI.4.

Interpretation

The measurements were interpreted from the angle of a test of the law proposed in §III.5.3 (*in fine*).

$$\frac{D}{D_*} = 1 + \frac{C_*}{4m^2}\left[\left|\frac{r^*}{r}\right|^{Nm/2} - 1\right]^2, \tag{XI.1}$$

where C_* and $m(\bar{\gamma})$ are given, respectively, by the equations (III.52) or (III.52′)

Table XI.4. Moments t^j of passage to radii X'^j and X''^j in implosion experiments.

ϕ 280 mm		No. 814F79		ϕ 300 mm		No. 815F79	
X'^j (mm)	t^j (μs)	X''^j (mm)	t^j (μs)	X'^j (mm)	t^j (μs)	X''^j (mm)	t^j (μs)
134.83	14.425	135.45		145.32	17.410	145.17	17.410
130.20	14.970	130.93	14.830	140.24	18.105	140.23	18.055
125.98	15.460	126.40	15.350	135.32	18.675	135.32	18.605
120.92	16.025	121.50	15.900	130.37	19.245	130.35	19.175
116.30	16.555	117.00	16.420	125.37	19.805	125.35	19.740
92.84	19.190	93.59	19.065	100.32	22.655	100.34	22.570
70.03	21.725	70.60	21.655	75.52	25.380	75.35	25.340
50.10	23.880	50.70	23.805	50.48	28.095	50.35	28.050
45.06	24.400	45.76	24.330	45.56	28.625	45.53	28.575
39.98	24.925	40.65	24.850	40.60	29.145	40.50	29.100
35.01	25.430	35.73	25.345	35.69	29.645	35.51	—
30.05	25.915	30.63	25.855	30.68	30.145	30.51	30.110
25.04	26.400	25.62	26.320	25.71	30.610	25.47	30.590
20.04	26.840	20.73	26.760	20.71	—	20.44	31.035
15.02	27.200	15.75	27.110	15.68	31.400	15.44	31.355

(*in fine*) and (III.54) (*in fine*) from which the following expression is derived:

$$C_* = \frac{\Gamma_*^2}{2(\Gamma_* + 1)(2\Gamma_* - G_*)},$$ (XI.2)

$$m(\tilde{\gamma}) = \left[1 + \frac{2}{\tilde{\gamma}} + \sqrt{\frac{2\tilde{\gamma}}{\tilde{\gamma} - 1}}\right]^{-1}.$$ (XI.3)

The test consists of ensuring the existence of:

- realistic initial conditions t^*, $|r|^*$ for each sphere; and
- a realistic value of m for the whole of the two spheres.

First consider the equation to which (XI.1) reduces in the neighborhood of $|r| = |r|^*$ under the hypotheses which lead to the simplified expression (III.52') given above in (XI.2)

$$\frac{D}{D_*} = 1 + C_*\left(\frac{N}{4}\right)^2\left(\left|\frac{r}{r^*}\right| - 1\right)^2.$$ (XI.4)

If D is replaced by $d|r|/dt$, a differential equation is obtained whose exact integral is

$$\left|\frac{r}{r^*}\right| = 1 - \frac{3}{\sqrt{C_*}}\tan\frac{D_*(t - t^*)}{2|r^*|}\sqrt{C_*}.$$

This integral gives the best (root mean square) representation of the first ten pairs in the table when t^* and $|r|^*$ have the following values:

$\phi/2$ mm	140	150
t^* μs	13.188	17.077
$\|r\|^*$ mm	145.5	148.85

Let us go back now to the numerical integration of (XI.1) for the initial conditions thus determined and let us seek the value of m which, for each sphere, gives the best root mean square representation for all the pairs (X^j, t^j). A unique value of m = 0.275 is obtained, for which $\tilde{\gamma} = 2.5$. The identity of the two different estimates of m is the anticipated argument in favor of the validity of the law (XI.1).

4. Break of a Quasi C–J Autonomous Detonation

4.1. Introduction

The attraction and the grasp of the C–J condition are such that detonation theories are most often used to establish, for the wave and the associated flow, properties independent of the conditions imposed on the priming boundary $\mathscr{B}^a$ and on the free boundaries $\mathscr{B}^\ell$. This point of view has its limits, as has been seen in §III.5.2, with regard to divergent spherical detonation and the role of the velocity $\dot{x}_B$ in the determination of the wave nature, and also in §III.6.3 with regard to traditional quasi C–J detonation and the role which the medium next to the explosive plays in the determination of the eigenvalue $D(\phi)$ and critical values D_c and D'_c. With regard to the latter, the possibility has been admitted, but not proved, of unambiguously defining the co-incidence η_1 of the wave on the flow relative to the interface, using only the conditions that relative pressure and velocity will be compatible. In fact, this definition is far from obvious; moreover, the phenomenon of detonation-break on a free boundary is of paramount importance, both for knowledge of the explosive by measurements made at such a boundary and for the estimation of the effects of the explosive on its immediate environment. This is why it is necessary, in this last section, to clarify the debate by using recent results from [8], [9], although all the developments are far from being available.

After posing the problem in the most general manner (§4.2), we consider the geometrically simple case of a divergent cylindrical detonation breaking on a plane, which lends itself well to the introduction of the local concept of *automorphy* (§4.3). We then return to the traditional "plane" configuration and propose a way out of the paradox of the "classic" theory where no solution exists (§4.4). This proposal may explain the existence of configurations which are noteworthy in that they are, in the neighborhood of a free boundary, invariably linked to the principal reference frame of the free boundary at rest $\mathscr{B}'_0$ (§4.5).

4.2. From Frontal Break to Oblique Break

To clarify the geometry of the problem, the following notation is used:

- I is the current point on the intersection $\Sigma_* \cap \mathcal{B}^\ell$ of the free edge $\mathcal{B}^\ell$ and the quasi C–J wave Σ_*;
- $\Pi(I)$ is the plane in I normal to $\Sigma_* \cap \mathcal{B}^\ell$;
- $\mathcal{B}_0^\ell(I)$ is the upstream arc of $\mathcal{B}^\ell \cap \Pi(I)$;
- $\mathcal{B}^\ell(I)$ is the downstream arc of $\mathcal{B}^\ell \cap \Pi(I)$;
- $t(I)$ is the unit vector of $\mathcal{B}_0^\ell(I)$ in I oriented from upstream to downstream of Σ_*; and
- $\eta = (\pi/2) - [N(I), t(I)]$ is the co-incidence of $N(I)$ on $t(I)$.

There are two quite different situations, according to whether $\eta = 0$ (i.e., $N(I)$ normal to $t(I)$, Σ_* and $\mathcal{B}^\ell$ tangential in I, "frontal" break) or $\eta \neq 0$ (oblique break).

In the first situation ($\eta = 0$), the theory depends on the ideas proposed by Jouguet [16], developed in numerous articles cited and classified by Courant and Friedrichs [12], then clearly reviewed by Pack [19]: that the appearance of a shock Σ_m in the adjacent medium is accompanied by reverberation in the detonation products of a shock Σ or of a Riemann centered expansion wave (see the equations given in §III.5.3) according to whether the shock pressure in "m" which would cause a velocity jump u_* therein is greater or less than the pressure p_*.

In the second situation ($\eta \neq 0$), theoretical treatment is based on the aerodynamic researches of Mach [18] and the authors cited in [12], and their transfer to two compressible media by Drummond [14] and Sternberg and Piacesi [22], who raise and discuss the paradox of "oblique break without hydrodynamic solution." Building on recent experimental results, a new theoretical analysis is proposed here, based on continuous variation of the flow around I when passing from frontal break to oblique break. To simplify the

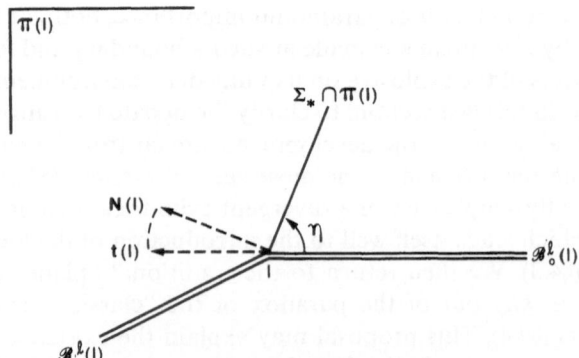

Figure XI.10. Diagram showing break in the plane $\Pi(I)$ normal in I to the surfaces Σ_* of the detonation wave and $\mathcal{B}^\ell$ of the free boundary.

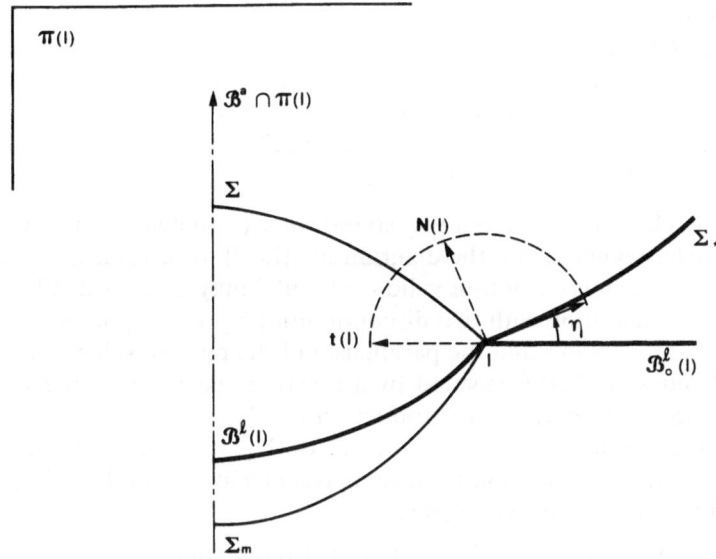

Figure XI.11. Diagram showing break of a divergent cylindrical detonation on a free boundary parallel to the priming axis.

strictly geometric aspects of this problem, we only consider the case (see Fig. XI.11) where the initial free boundary $\mathscr{B}_0^{\ell}$ is a plane and the priming boundary is a straight line $\mathscr{B}^{a}$ parallel to $\mathscr{B}_0^{\ell}$ and inside the explosive structure $\mathscr{E}_0$: the divergent cylindrical detonation is then quasi C–J according to the propagation rules in §III.4.2. Let H be the distance from $\mathscr{B}^{a}$ to $\mathscr{B}_0^{\ell}$, and let h_I be the distance from I to the plane projecting $\mathscr{B}^{a}$ on $\mathscr{B}_0^{\ell}$.

Consider a couple $\mathscr{E}_0/m$ such that, in a frontal break situation, a shock Σ is reverberated in the detonation products. For such a couple, the geometric hypotheses on $\mathscr{B}^{a}$ and $\mathscr{B}_0^{\ell}$ predict frontal break followed by oblique break with η being as small as required. In the relative reference frame related to I, $\mathscr{B}^{\ell} \cap \Pi(I)$ is the stream line both for the explosive and for the adjacent medium (see Fig. XI.12), and the relative vector velocity common to the upstream of Σ_* and Σ_m is

$$\mathbf{W}_0 = \mathbf{W}_m = \frac{\mathbf{D}_*}{\sin \eta}\, \mathbf{t}(I).$$

The spreading out of the jump relations across Σ_*, Σ, and Σ_m in the neighborhood of $\eta = +0$ shows the existence and uniqueness of one state downstream of Σ and one state downstream of Σ_m which in I obey the equality of the pressures and the parallelism of the relative velocity vectors on both sides of $\mathscr{B}^{\ell}(I)$. Furthermore, given the dependence of $\mathbf{W}_0$ and $\mathbf{W}_m$ on $1/\sin \eta$ and the jump relations, each of these downstream states corresponds, if η is small enough, to supersonic downstream flow; and finally $d\eta/dt > 0$. The configuration $\Sigma_* \Sigma \Sigma_m$ is then stable, in accordance with the reasoning of Walsh *et al.* [23].

When I is far from the symmetry axis, angle η increases as arctan (h_I/H): a unique, stable configuration with three discontinuities exists, while changing to suit η. This situation holds until $\eta = (\pi/2) - 0$ provided that neither of the following circumstances occurs:

(i) the shock Σ in I is reduced to a sound wave (zero deflection and pressure jump) (see point $K°$ on the diagram in §II.4.3) for a value $\eta^\omega < \pi/2$ of η; then the reference solution valid on $[0, \eta^\omega]$ may be extended beyond η^ω by a configuration with two discontinuities Σ_* and Σ_m, where the equality of the pressures and the parallelism of the relative velocity vectors on both sides of $\mathscr{B}'(I)$ is ensured by a reverberated Prandtl–Meyer expansion within the detonation products; nor

(ii) the relative flow downstream of Σ or Σ_m is sonic in I for a value $\eta^-(\mathscr{E}_0, m) < \pi/2$ of η; and then, some recent experimental results (see §4.3) lead to the following conjecture:

$$\left.\begin{array}{l}\text{When } h_I \text{ increases from initial } H \tan \eta^-, \text{ flow} \\ \text{results from a continuous extension of the ``classic} \\ \text{solution with reverberated shock'': detonation } \Sigma_* \\ \text{inclined on } \mathscr{B}_0' \text{ of the angle arctan}(h_I/H), \text{ existence} \\ \text{downstream of } \Sigma_* \text{ of a shock } \Sigma \text{ followed by a } sonic \\ \text{relative flow, shock } \Sigma_m \text{ followed by a Prandtl–} \\ \text{Meyer expansion.}\end{array}\right\} \quad \text{(XI.5)}$$

Figure XI.12(a–c), drawn in the real plane perpendicular to the priming line $\mathscr{B}^a$, and (XI.12a', b', c') drawn in the plane of the variables deflection θ and pressure p, show the pattern of flow in the neighborhood of I, and, in particular, the way in which the flow of detonation products and the flow of the adjacent medium adjust to one another:

• in the reference solution, the relative flow is supersonic everywhere, and is formed by the juxtaposition of uniform zones: the state at the interface $\mathscr{B}'(I)$ results from the mutual adjustment of the stream lines on both sides;
• in extension (i) of the reference solution, flow remains supersonic everywhere, but comprises a nonuniform zone, i.e., an expansion of detonation products which continuously changes their state from $[p_*, \theta_*(\eta)]$ to $[p(\eta), \theta(\eta) > \theta_*(\eta)]$; and
• in extension (ii) of the reference solution, the relative flow remains supersonic except downstream of Σ where it is sonic; in addition, it comprises a nonuniform zone, i.e., an isentropic expansion of the adjacent medium which causes it to change continuously from an intermediate state $[p'(\eta), \theta'(\eta) < \theta^S(\eta)]$ to the state $[p^S(\eta), \theta^S(\eta)]$.

Some analogy between the two extensions should not conceal a striking difference: in extension (i) (just as in the reference solution) the final deflection

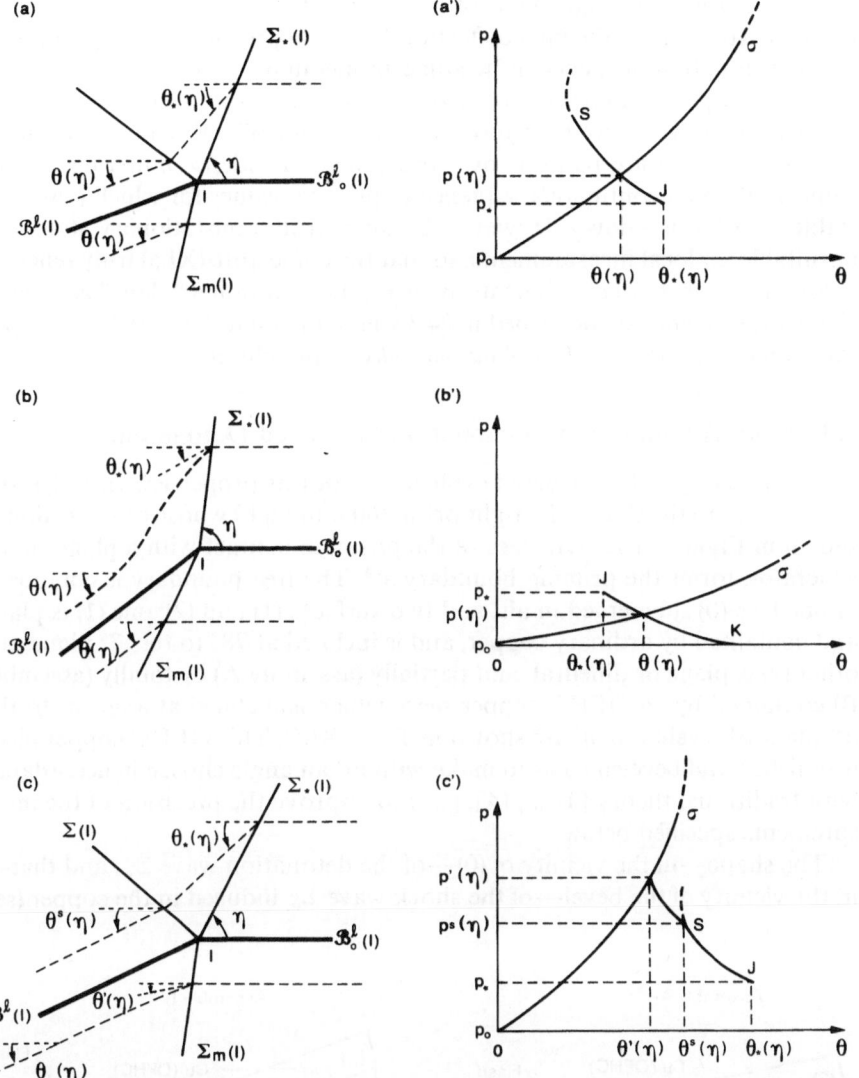

Figure XI.12. Break of a divergent cylindrical detonation on a plane parallel to the priming axis; determination of the pressure and deflection conditions on both sides of the free boundary. Parts (a) and (a') are relative to the reference flow (weak coincidence); parts (b) and (b') of the one part, (c) and (c') of the other part are relative to the two possible extensions of the reference solution.

results from mutual adjustment of the stream lines on both sides of $\mathscr{B}'(\mathrm{I})$, while in extension (ii) the final deflection $\theta^S(\eta)$ and pressure $p^S(\eta)$ are imposed by the detonation products in the adjacent medium.

The divergent cylindrical geometry envisaged in this section lends itself particularly well—as we have just seen—to the introduction of extensions of the reference solution because, over time, the co-incidence η first has a zero value and then a continuous sequence of positive values for which a unique, stable solution is known. However, because of its nonstationary state it is unsuitable for local measurements, so that the conjecture (XI.5) truly relies on a discrete sequence of experiments made on the traditional "plane" geometry. These experiments are described in §4.3 where the material in [9] is incorporated and where the concept of *automorphy* is introduced.

4.3. From Automorphous Detonation to Guided Detonation

A rough casting of TATB-based explosive T_2 (for its properties, see [2], [20]) is machined to the shape of a right prism (60 mm high) whose cross section is shown in Figure XI.13. One face of the prism, in contact with a plane shock generator, forms the priming boundary $\mathscr{B}^a$. The free boundary is composed of one face (0) submerged in air, and two surfaces (1) and (2); one (1) is plane and contained by ordinary copper, and is inclined at 78° to the $\mathscr{B}^a$ plane; the other (2) is plane or dihedral, and partially (assembly A) or totally (assembly B) contained by an OFHC copper piece which is inclined at angle η^a to the $\mathscr{B}^a$ plane (the values of η^a are shown in Table XI.5). This OFHC copper piece is polished and beveled so as to make with $\mathscr{B}^a$ an angle chosen in accordance with traditional theory [18], [14], [22] to improve the precision of the measurements specified below.

The shape—in the vicinity of (0)—of the detonation wave Σ_*, and that—in the vicinity of the bevel—of the shock wave Σ_m induced in the copper (see

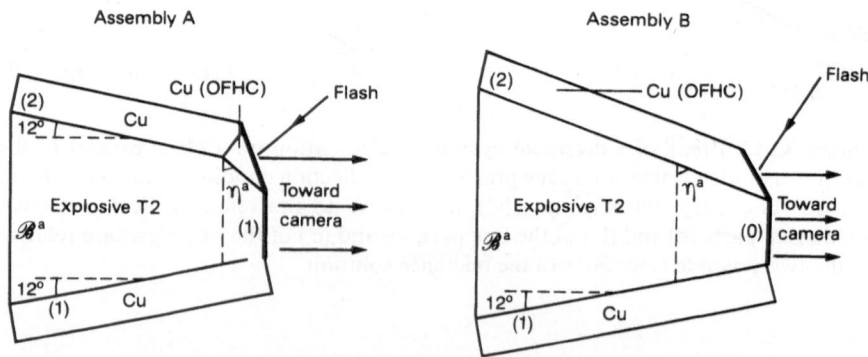

Figure XI.13. Cross-sectional diagram of experimental set-up.

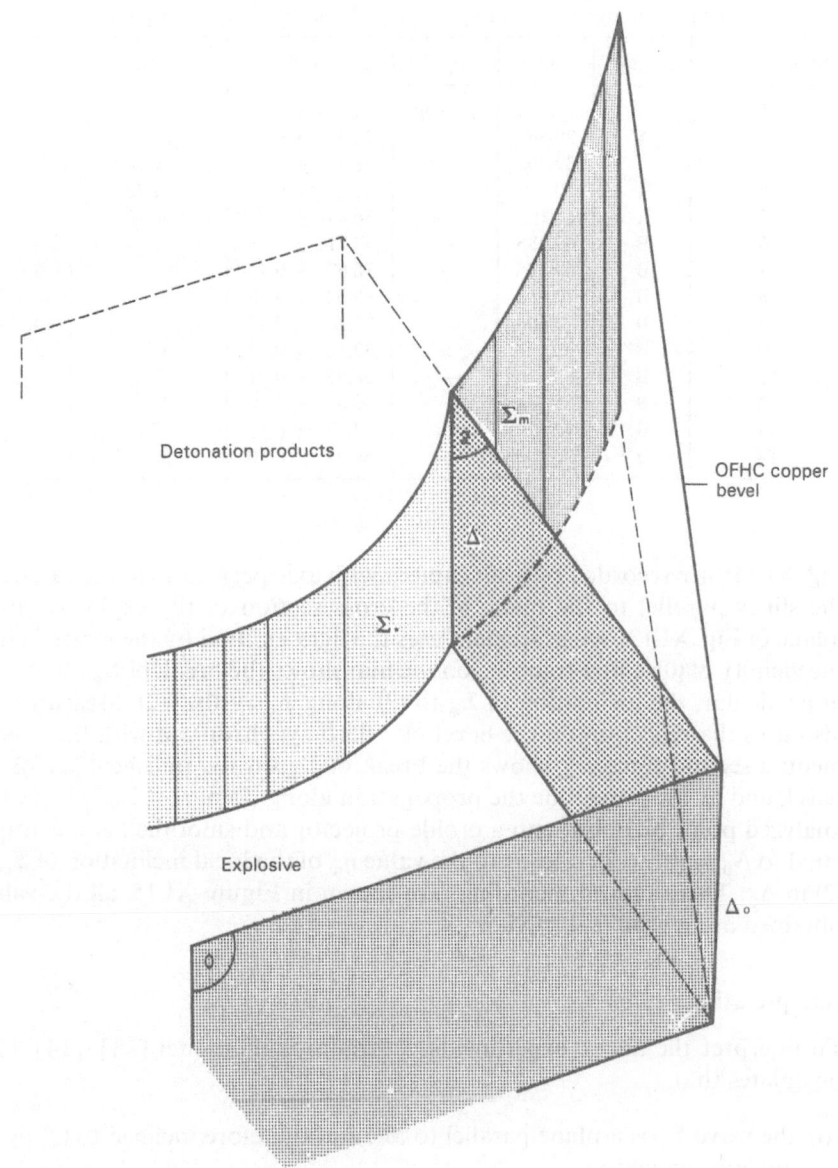

Figure XI.14. Isometric projection of waves Σ_* and Σ_m in the vicinity of free boundaries (0) and (2).

Table XI.5. Experimental values of η_* and η_m; theoretical values η'_m (after [9]).

Shot number	Set-up	η^a	η_*	η_m	η'_m [14], [18]	η'_m [22]
1	A	20.224	$\eta_* \simeq \eta^a$	14.153 ± 0.05	13.77	
2	A	29.866	—	20.35 ± 0.31	20.29	
3	A	39.908	—	26.46 ± 0.2	26.34	
4	A	50	—	32.29 ± 0.78	31.72	
5	A	60.103	—	36.44 ± 0.77	36.22	
6	B	67.285	—	37.71 ± 0.23		40.28
7	B	69.72	—	38.67 ± 0.41		40.27
8	B	70.082	—	38.52 ± 0.38		40.27
9	B	72.606	—	39.38 ± 0.42		40.14
10	B	80	$\eta_* > \eta^a$	39.18 ± 0.59	38.39	
11	B	85	—	38.85 ± 0.21	37.75	
12	B	88	—	39.82 ± 0.10	37.31	
13	B	90	—	39.73 ± 0.16	37.15	
14	B	90	—	39.92 ± 0.10	37.15	

Fig. XI.14), are recorded by a slit camera with axis perpendicular to $\mathscr{B}^a$, while the slit is parallel to the plane of the cross section of the explosive prism (plane of Fig. XI.13). Measurement uses the light emitted by the ionized air in the vicinity of (0); a first trace $\mathscr{L}_*$ on the film shows the break of Σ_* on (0) and, in particular, the inclination of Σ_* to (2) along $\Delta_0 = (0) \cap (2)$. Measurement also uses the reflection on the bevel of a flash synchronized with the experiment; a second trace $\mathscr{L}_m$ shows the break of Σ_m on the polished face of the bevel, and at the same time the propagation along (2) of $\Delta = \Sigma_* \cap \Sigma_m$; when analyzed point by point with a profile projector and smoothed and extrapolated to $\Delta_0 = (0) \cap (2)$, it leads to the value η_m of the local inclination of Σ_m to (2) in Δ_0. Three typical recordings are shown in Figure XI.15; all the values obtained are give in Table XI.5.

Interpretation

To interpret the above experiments, a "traditional" model [18], [14], [22] postulates that:

(i) the wave Σ_* is a plane parallel to $\mathscr{B}^a$, and therefore inclined to (2) by an angle η_* equal to η_a;

(ii) the relative flow in the frame linked to Σ_* is two-dimensional stationary; and

(iii) the shock Σ_m is sustained, at least locally in the vicinity of (2).

From these hypotheses, three particular values

$$\eta^\alpha \simeq 64°, \qquad \eta^\beta \simeq 75°, \qquad \eta^\gamma \simeq 78°,$$

appear on determining the pressure $p(\eta)$ and deflection $\theta(\eta)$ which ensure

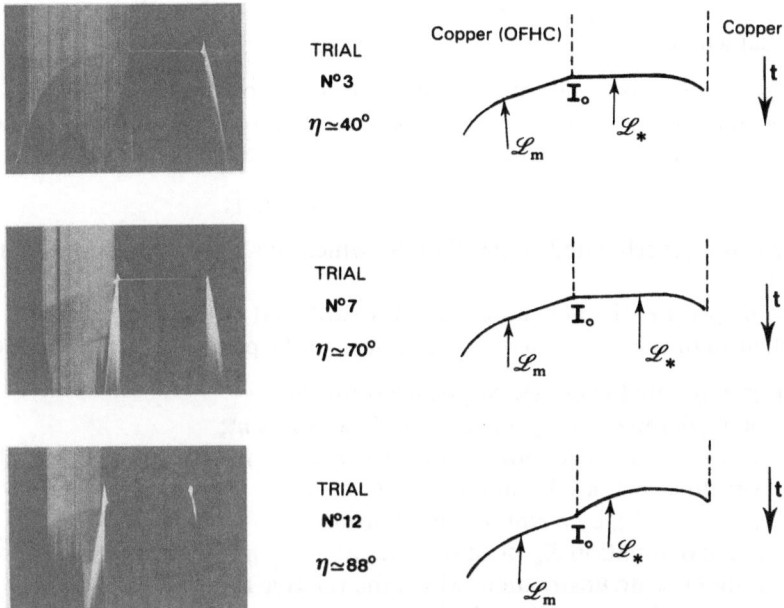

Figure XI.15. Slit camera recordings: light trace $\mathscr{L}_m$ shows shock Σ breaking at polished face of copper bevel, light trace $\mathscr{L}_*$ shows detonation Σ_* break in air.

equality of pressures and parallelism of velocity vectors on both sides of partition (2) [see (I.29) with $M = 0$]:

- for $\eta^a \leq \eta^\alpha$ and $\eta^\beta \leq \eta^a \leq \eta^\gamma$, there is a unique solution where a shock Σ rectifies the flow of detonation products downstream of Σ_*;
- for $\eta^a > \eta^\gamma$, there is a unique solution where Prandtl–Meyer expansion prolongs the deflection of detonation products downstream of Σ_*; and
- for $\eta^\alpha < \eta^a < \eta^\beta$, there is no solution, unless it is assumed [22] that a Mach configuration is set up in the detonation products.

The values η'_m obtained from this "traditional" theory are given in the last two columns of Table XI.5. Comparison with experimental values shows that this model ceases to be valid when η^a exceeds a value close to η^α not only in the interval $(\eta^\alpha, \eta^\beta)$, where recourse to the assumption in [22] is necessary, but also—contrary to all expectations—beyond η^β up to 90°. Furthermore, independently of any reference to the "traditional" model, examination of experimental values confirms two laws:

- $\eta_m(\eta^a)$ remains continuous when η^a exceeds the value η^α and may be considered, within the limits of measurement, as being constant beyond a value close to η^β; and
- $\eta_m(\eta^a)$ only differs from η^a beyond a value close to η^β and η^γ.

These laws are valuable since, combined with assumption (XI.5), they suggest a model where:

- hypotheses (i), (ii), and (iii) are called into question; and
- development of the flow, as η^a increases from 0° to 90°, is continuous, being controlled not by the values η^α, η^β, η^γ, but by the particular values

$$\eta^- < \eta^\alpha; \qquad \eta^+ \in \,]\eta^\beta, \eta^\gamma[;$$

for which a reverberated shock Σ exists which has a sonic downstream relative flow.

To be precise, the new model sets forward that equality of pressures and parallelism of velocity vectors on both sides of the partition (2) are obtained:

- for $\eta^a < \eta^-$, by the classic *stationary* solution:
  — plane detonation Σ_*, inclined to (2) at angle η^a;
  — a shock Σ downstream of which the relative flow is *supersonic*; and
  — a *sustained* shock Σ_m in the vicinity of Δ;
- for $\eta^a \in [\eta^-, \eta^+]$, by a *stationary* extension:
  — plane detonation Σ_* inclined to (2) at angle η^a;
  — a shock Σ downstream of which the relative flow is *sonic*; and
  — a shock Σ_m *followed by the Prandtl–Meyer expansion*;
- for $\eta^a > \eta^+$ by a *locally stationary* extension:
  — a detonation Σ_* which is not plane but locally inclined to (2) in Δ at angle η^+, whatever the value of η^a;
  — a shock Σ downstream of which the relative flow is *sonic*; and
  — a shock Σ_m locally sustained on Δ and inclined to (2) at angle $\eta_m(\eta^+)$, whatever the value of η^a.

This model takes full account of all the data given in Table XI.5. Further confirmation is found from photometric information contained in the recordings: the light trace $\mathscr{L}_m$ which shows the break of shock Σ_m on the polished surface of the bevel is definitely less intense in shots 6–9, precisely those where, in the above model, shock Σ_m is followed by the Prandtl–Meyer expansion which limits its effects.

Suppose now that a divergent cylindrical detonation experiment, as described in §4.2, is made on the T2/copper couple. Then assumption (XI.5) leads to the belief that, when h_1 increases from $H \tan \eta^+$:

- η keeps the value η^+ so that the configuraton $\Sigma_* \Sigma \Sigma_m$ remains locally stationary in I; and
- outside point I, the configuration develops in such a way that the entirety of the flow corresponds to all the boundary conditions at $\mathscr{B}^\ell$ and $\mathscr{B}^a$.

In other words, conjecture (XI.5) predicts the existence of two successive propagation phases along the free edge: one $h_1 \leq H \tan \eta^+$ where detonation is *automorphous* in the sense that Σ_* is not altered by the presence of $\mathscr{B}^\ell$, the other $h_1 > H \tan \eta^+$ where $\mathscr{B}^\ell$ disturbs the shape which Σ_* would take, solely because of the priming conditions along $\mathscr{B}^a$, and leads to progressive reduction of the automorphous part of Σ_*. To avoid any confusion, it should be emphasized that automorphy, like autonomy, is not an intrinsic property of detonation, since its existence depends on the adjacent medium: in a less formal, but still instructive manner, we can say that automorphy is to the free edge what autonomy is to the priming boundary. It should also be noted that the automorphous → nonautomorphous transition may exceptionally not appear in this simple form if an elastic wave can detach itself from the shock Σ_m, i.e., if there is a value $\eta^e(\mathscr{E}_0, m) < \eta^+(\mathscr{E}_0, m)$ linked to the velocity c_m^L of the longitudinal elastic waves in "m" by

$$c_m^L = D_*/\sin \eta^e.$$

Then the configuraton $\Sigma_* \Sigma \Sigma_m$ is disturbed in Δ as soon as its stability is no longer ensured, i.e., as soon as η attains the value η^e. The transformation

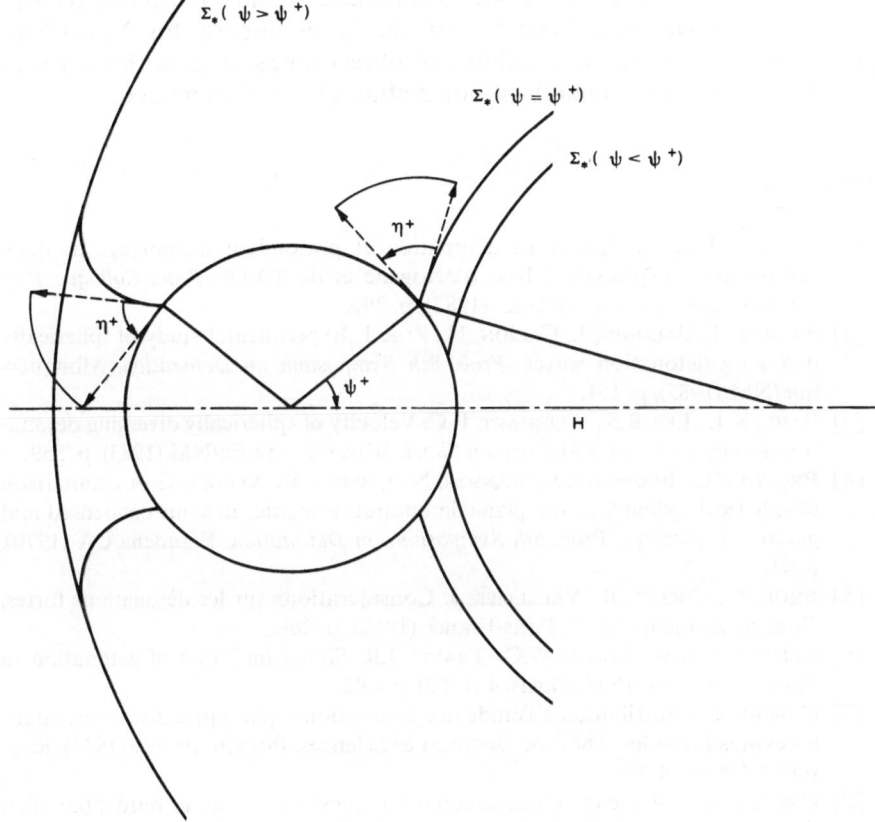

Figure XI.16. Diagram of divergent cylindrical detonation break on a cylindrical free boundary parallel to the priming axis. Automorphous break ($\psi < \psi^+$), limit break ($\psi = \psi^+$), and guided break ($\psi > \psi^+$).

which changes the adjacent medium from rest to a state identical in stresses and deflection to that of the detonation products downstream of Σ belongs to the field of elastoplastic behavior and is outside the scope of this work.

Forgetting this exceptional circumstance, a consequence of the automorphous $\rightarrow$ nonautomorphous transition will now be considered: *guided* propagation along a free boundary.

Returning to the case of a divergent cylindrical detonation, we now consider a free boundary in the shape of a hollow cylinder, of circular section (diameter ϕ) with generatrix parallel to the priming boundary. Figure XI.16 shows a cross section of the explosive structure. As in the preceding sections, detonation remains automorphous as long as the angle ψ under which the break lines are seen remains less than a value ψ^+ defined by

$$\psi^+ + \text{Arctan} \frac{\sin \psi^+}{1 - \cos \psi^+ + (2H/\phi)} = \eta^+.$$

As soon as ψ exceeds ψ^+, co-incidence η maintains value η^+ so that the free boundary seems to "guide" and "retard" the detonation (see Fig. XI.16). This phenomenon is the key to a number of observations made with respect to regimes more complex than those noted above, but still of interest.

References

[1] AVEILLÉ, J. *et al.* Célérité de détonation et profondeur d'amorçage de deux compositions explosives à base d'octogène et de TATB. *Proc. Colloque Pyr. Fond et App.*, Arcachon/France (1982), p. 396.

[2] AVEILLÉ, J., BACONIN, J., CARION, N., ZOÉ, J. Experimental study of spherically diverging detonation waves. *Proc. 8th Symposium on Detonation*, Albuquerque/NM (1985), p. 151.

[3] BAHL, K.L., LEE, R.S., WEINGART, R.C. Velocity of spherically diverging detonation waves. *Proc. APS Meeting on Shock Waves*, Santa Fe/NM (1983), p. 559.

[4] BROCHET, C., BROSSARD, J., MANSON, N., CHERET, R., VERDÈS, G. A comparison of spherical, cylindrical and plane detonation velocities in some condensed and gaseous explosives. *Proc. 5th Symposium on Detonation*, Pasadena/CA (1970), p. 41.

[5] BRUN, L., CHERET, R., VACELLIER, J. Considérations sur les détonations fortes. *Proc. Symposium H.D.P.*, Paris/France (1978), p. 269.

[6] CAMPBELL, A.W., DAVIS, W.C., TRAVIS, J.R. Shock initiation of detonation in liquid explosives. *Phys. Fluids*, **4** (1960), p. 498.

[7] CHERET, R. Contribution à l'étude des détonations sphériques divergentes dans les explosifs solides. Thèse de Doctorat ès Sciences, Poitiers/France (1971). Rapport CEA no. 4283.

[8] CHERET, R. Émergence d'une détonation quasi C.–J. sur le bord libre d'un domaine explosif. *C. R. Acad. Sci. Paris*, **301**, Series II (1985), p. 657.

[9] CHERET, R., AVEILLÉ, J., CARION, N. Émergence d'une détonation quasi C.–J. sur le bord libre d'un domaine explosif. *C. R. Acad. Sci. Paris*, **303**, Series II (1986), p. 1.

[10] CHERET, R., CHAISSÉ, F., ZOE, J. Some results on the converging spherical deto-nation in a solid explosive. *Proc. 7th Symposium on Detonation*, Annapolis/MD (1981), p. 602.

[11] CHERET, R., VERDÈS, G. Détonation sphérique divergente du nitrométhane. *Mémorial de l'Artillerie Francaise*, **48**, 3 (1974), p. 687.

[12] COURANT, R., FRIEDRICHS, K.O. *Supersonic Flow and Shock Waves.* Interscience, New York (1948).

[13] DROUX, R., MOUCHEL, C. Étude du comportement sous choc d'explosifs hétérogènes. *Proc. Symposium H.D.P.*, Paris/France (1978), p. 103.

[14] DRUMMOND, W.E. Explosive induced shock waves. Part II. Oblique shock waves. *J. Appl. Phys.*, **29**, 2 (1958), p. 167.

[15] HAMADA, L., PRESLES, H.N., BROCHET, C. BOURIANNES, R., CHERET, R. Charac-terization of an overdriven detonation state in nitromethane. *Prog. Astronaut. Aeronaut.*, **94** (1985), p. 343.

[16] JOUGUET, E. *Mécanique des Explosifs.* Octave Doin, Paris/France (1917).

[17] KRISHNAN, S., BROCHET, C., CHERET, R. Mach reflexion in condensed explosives. *Propellants and Explosives*, **6** (1981), p. 170.

[18] MACH, E. *Sitzungsberichte Akad. Sci.*, **78**, Wien/Österreich (1878), p. 819.

[19] PACK, D.D. Reflection and transmission of shock waves. *Phil. Mag.*, **2**, 14 (1957), p. 182.

[20] PINÈGRE, M. *et al.* Expansion isentropes of TATB compositions released into argon. *Proc. 8th Symposium on Detonation*, Albuquerque/NM (1985), p. 815.

[21] SELLAM, M., PRESLES, H.N., BROCHET, C., CHERET, R. Characterization of strong detonation waves in nitromethane. *Proc. 8th Symposium on Detonation*, Albu-querque/NM (1985), p. 425.

[22] STERNBERG, H.M., PIACESI, D. Interaction of oblique detonation waves with iron. *Phys. Fluids*, **9**, 7 (1966), p. 1307.

[23] WALSH, J.M., SHREFFLER, R.G., WILLIG, I.J. Limiting conditions for jet forma-tion in high velocity collisions. *J. Appl. Phys.*, **24**, 3 (1953), p. 349.

CHAPTER XII

Numerical Predictions

1. Purpose and Value of Prediction

1.1. Prediction Levels and Elementary Problem

It must be stated at the outset (so as to warn the reader) that numerical predictions for explosives are, at first sight, somewhat disconcerting because of the diversity of the phenomena invoked, the quantities measured, the approximations made. This disorder is not an effect of Art but a consequence of the ambivalence of the concerns which dominate the use of an explosive structure:

(i) to control the risks encountered during manufacturing, storage, and dismantling phases,

(ii) to control, in time and intensity, performance during the normal and unique sequence of use.

The first concern is the definition and implementation of trials and tests, analyzing incidents and accidents which have been duly listed. The reader interested in this aspect can usefully turn to the irreplaceable work of Médard [44].

The second concern is not only the writing of numerical codes but also carrying out experiments to validate them. The subject of this twelfth and last chapter is to approach the question very generally and to present an updated picture of its implications for different disciplines and techniques; it is left to the reader to fit his prediction effort to the economic value of the explosive structure concerned or of the project of which it forms a part.

To introduce the "elementary" problem which dominates numerical predictions, it is convenient to consider the circumstances which most often accompany the "birth" and "death" of an explosive structure:

- manufacturing conditions generally allow the initial state to be considered as a uniform state of rest;
- conditions at the priming boundary generally mean that propagation is ensured simply because of the release of chemical energy, to the exclu-

sion of any compression wave driving the flow of detonation products up to the ignitor shock; and

- conditions on the free edge and/or priming boundary generally lead to flow with detonation of positive curvature;

so that, because of the propagation rules established in §III.4.2, detonation is usually quasi C–J in the sense of §III.1.6. We may be satisfied with this information and determine only C–J detonation state(s) of an explosive substance. (N.B. The uniqueness subject to inequalities (II.16) has been shown in §II.3.1.) But we may also wish to attain a complete description of the flow with detonation, knowing (Chap. VIII) that this objective implies knowledge of the surface of state ($\mathscr{D}$) of the "final substance," i.e., of the detonation products in thermochemical equilibrium.

These two approaches clearly show the scale of the objectives: in one a state (the C–J state) is to be determined, in the other the whole of a surface of state (or at least a piece, in the hypothesis that it would be possible to set limits to the thermodynamic path followed by the final substance).

But it happens that the difficulty of the problem is virtually independent of the objective of the prediction. In both approaches, the common, unavoidable problem is the determination of the current point of the surface of state ($\mathscr{D}$). In fact, once this elementary problem is solved, it is a simple matter to determine any line on ($\mathscr{D}$) and, in particular:

- the arc (H_+) and the point J, target of the first approach; and
- a network of curves which gives a reply to the second.

The following section is devoted to analysis of preliminary questions which arise from this elementary problem.

1.2. Preliminary Questions

The substance qualified as "final" in the detonation process has only ephemeral existence: the velocities which create it (a few millimeters per microsecond), the pressures achieved in it (some $10^1 - 10^2$ kbar), the gradients which cross it neither allow leisurely observation nor isolation of a state correlated in a simple manner to a state of ($\mathscr{D}$). Indirect paths must therefore be followed: by relying on global observation of the flows of detonation products, arguments are found in favor of hypotheses upon which thermodynamic knowledge of ($\mathscr{D}$) can be built, and finally the validity is checked by comparison with appropriate measurements.

What elementary detonation experiments (see Chap. XI) primarily tell us is that detonation products have all the appearances of a fluid. Furthermore, when the explosive is of organic type (i.e., composed of atoms of carbon, oxygen, hydrogen, nitrogen), they also tell us that free carbon may figure in the detonation products. This is why, from the first attempts at numerical characterization of condensed explosives (see Kistiakowsky and Wilson

[32a, b]), three hypotheses have been advanced: the detonation products

($\mathcal{H}_1$) form a macroscopically homogeneous mixture of one gaseous phase (composed of I chemical species, each marked with superscript i) and J condensed phases (marked with superscript j) compared to which the internal energy and volume are additive;

($\mathcal{H}_2$) behave as a single fluid which, at pressure p and at temperature T possesses a specific volume v and specific internal energy e; and

($\mathcal{H}_3$) are in chemical equilibrium.

In order to determine this equilibrium, such a model requires, as well as the atomic composition $A^1_{\alpha_1} \ldots A^\ell_{\alpha_\ell} \ldots A^L_{\alpha_L}$ of the explosive substance, the introduction of variables inherent in each of the species and/or phases:

ε^λ_ℓ number of atoms (A') in the chemical individual (λ); and

N^λ $(\lambda = i$ or $j)$ number of moles of (λ) in the unit mass of the mixture.

$N' = \sum_i N^i$

v^j specific volume of the phase (j); and

v' volume of the gaseous species resulting from the detonation of the unit mass of the mixture

$$v' = v - \sum_j N^j M^j v^j. \tag{XII.1}$$

In order to express the fluid behavior, it is also necessary to introduce functions $\varphi(v', T, \ldots, N^i, \ldots)$ and $\varphi^j(v^j, T)$ which give the respective pressures in the gaseous phase and the condensed phase (j). The first "preliminary question" now appears: separate and *a priori* knowledge of the surface of state of each phase.

Other difficulties arise with regard to listing the species potentially present in the mixture and the phase(s) in which each species occurs therein. In the example of an organic explosive, there can be hardly any doubt that such gases as CO, CO_2, H_2O, NH_3, CH_4, etc., must be taken into consideration, but it is not obvious that atomic O, H, N species can be eliminated, nor that the free radical OH should be considered; among other considerations we lack simple and/or decisive arguments to affirm that free carbon is present in the form of graphite, diamond, etc., or even a transient quasi-crystalline phase which is absent from the inventory drawn from static studies. Hence the second "preliminary question" which bears on the formulation of the problem: Has the mixture been represented in a realistic manner? This question is all the more important when the only possible answer is a clearly laborious procedure: introducing the largest possible number of species and phases (compatible with the available means for calculation and with theoretical knowledge of the functions φ and φ^j), even if it entails stating *a posteriori* that, for some group of explosives, such species or such phase could be absent in the list.

Before returning to the first preliminary question and discussing the possible answers to it (see Section 2), we will deal with the nature of the system of equations to be solved (§§1.3–1.5) and the validation criteria of the numerical results (§1.6).

1.3. Equations of the Current Point of ($\mathscr{D}$)

In each state of the detonation products of an explosive substance of composition $A_{\alpha_1}^1 \dots A_{\alpha_\ell}^\ell \dots A_{\alpha_L}^L$ per unit mass, the following have been defined:

- $I + J$ composition variables N^λ; and
- $4 + J$ state variables p, T, v, v', v^j.

The surface ($\mathscr{D}$) is generated when these $4 + I + 2J$ variables evolve in accordance with $2 + I + 2J$ linking equations:

- the homogeneity equation (XII.1);
- L linear equations of conservation of atoms

$$\sum_\lambda N^\lambda \varepsilon_\ell^\lambda = \alpha_\ell, \qquad \ell = 1, \dots, L; \tag{XII.2}$$

- $J + 1$ equations of state

$$p = \varphi(v', T, \dots, N^i, \dots) = \varphi^1(v^1, T) = \cdots = \varphi^j(v^j, T); \text{ and} \tag{XII.3}$$

- $I + J - L$ equations of chemical equilibrium.

These latter are obtained by seeking, at constant p and T, either directly or by iteration, the *constrained minimum*—by the L equations (XII.2)—of the specific Gibbs free energy g of the molar system ($N^i, i = 1, \dots, I; N^j, j = 1, \dots, J$). The crux of the problem now resides in the clarification of g.

Let $\mu_G^i(p, T)$ be the molar chemical potential of the gas (i) in the gaseous mixture ($N^i, i = 1, \dots, I$) at pressure p and temperature T. Let $\mu^j(p, T)$ be the proper molar chemical potential of the condensed phase (j). Taking hypotheses ($\mathscr{H}_1$) and ($\mathscr{H}_2$) into account (see §1.2), we can write

$$g = \sum_i N^i \mu_G^i(p, T) + \sum_j N^j \mu^j(p, T). \tag{XII.4}$$

Each of the above terms must therefore be evaluated in turn: for this purpose, it is convenient to introduce:

- a pressure p^0 and a temperature T^0 called "standard";
- the proper molar chemical potential $\tilde{\mu}^i(p, T)$ of the gas (i) if it were ideal at pressure p and temperature T; and
- the function $\mathscr{F}$ associated with the equation of state $p = \varphi(v', T, . N^i.)$ of the gaseous phase by

$$\mathscr{F}(v', T, . N^i.) = pv'/N'\Re T, \tag{XII.5}$$

where $\Re$ is the product of Boltzmann constant k and Avogadro number $\mathscr{N}$.

Chemical thermodynamic calculations, detailed in Appendix E, then lead to the following expressions:

$$\frac{1}{\Re T}\mu_G^i(p, T) = \frac{1}{\Re T}\tilde{\mu}^i(p^0, T) + \text{Log}\,\frac{N^i\Re T}{p^0 v'}$$

$$+ \int_{v'}^{\infty}\left(\mathscr{F} - 1 + \frac{\partial\mathscr{F}}{\partial N^i}\right)\frac{dv'}{v'}, \tag{XII.6}$$

$$\mu^j(p, T) = \mu^j(p^0, T) + M^j\int_{v^j(p^0, T)}^{v^j} v^j(\partial\varphi^j/\partial v^j)\,dv^j. \tag{XII.7}$$

It is now sufficient to rely on the thermodynamic identity $(g.3)$ of Appendix A expressed in the form which uses the chemical potential and molar enthalpy H

$$\frac{\partial(\mu/T)}{\partial T}(p, T) = -\frac{H}{T^2} \tag{XII.8}$$

to express $\tilde{\mu}^i(p^0, T)$ and $\mu^j(p^0, T)$ in terms of the quantities tabulated in specialized works (e.g., Janaf, Thermochemical tables). In fact, simple integration of (XII.8) first gives

$$\left.\begin{aligned}
\frac{1}{T}\tilde{\mu}^i(p^0, T) &= \frac{1}{T^0}\tilde{\mu}^i(p^0, T^0) - \int_{T^0}^{T}\tilde{H}^i(T)\frac{dT}{T^2}, \\
\frac{1}{T}\mu^j(p^0, T) &= \frac{1}{T^0}\mu^j(p^0, T^0) - \int_{T^0}^{T}H^j(p^0, T)\frac{dT}{T^2}.
\end{aligned}\right\} \tag{XII.9}$$

Returning to the definition of chemical potential and introducing the molar enthalpies $\tilde{S}^i$ and S^j we finally obtain

$$\left.\begin{aligned}
\frac{1}{T}\tilde{\mu}^i(p^0, T) &= \frac{1}{T}\tilde{H}^i(T^0) - \tilde{S}^i(p^0, T^0) - \int_{T^0}^{T}\Delta_{T^0}^{T}\tilde{H}^i(T)\frac{dT}{T^2}, \\
\frac{1}{T}\mu^j(p^0, T) &= \frac{1}{T}H^j(p^0, T^0) - S^j(p^0, T^0) - \int_{T^0}^{T}\Delta_{T^0}^{T}H^j(p^0, T)\frac{dT}{T^2}.
\end{aligned}\right\} \tag{XII.10}$$

Such detailed formulation deserves comment. It is usual to say that the enthalpy of a species is defined within a constant, but less usual to underline that this indeterminacy is only true for each species taken in isolation. Thus in a mixture the indeterminacy holds for the enthalpy of the simple species (often by agreement taken as equal to zero in the standard state), while those of compound species are perfectly determined compared with this reference by the standard enthalpies of formation.

By combining any supplementary equation (E) with equations (XII.1), (XII.2), (XII.3), and $I + J - L$ equations of chemical equilibrium, one line of $(\mathscr{D})$ can be determined. This is an isobar if equation (E) is p = constant, an isochore if (E) is $v = $ constant. It is the detonation arc (H_+) of the Crussard curve if (E) is the equation

$$e - e_0 = \tfrac{1}{2}(p + p_0)(v_0 - v), \qquad v \leq v_0.$$

It is an isentropic if (E) is the differential equation $(i.3)$ established in Appendix A:

$$\frac{\partial e}{\partial T}(v, T)\, dT + T\frac{\partial p}{\partial T}(v, T)\, dv = 0. \qquad (XII.11)$$

It can readily be seen that the first two families cited arise immediatey from the determination of the current point of $(\mathscr{D})$. However, it can also be seen that the determination of (H_+) and of the isentropes follows only on condition that, on the one hand, the difference $e - e_0$ and, on the other hand, the thermodynamic coefficients $(\partial e/\partial T)(v, T)$ and $(\partial p/\partial T)(v, T)$ have been successfully expressed relative to the $4 + I + 2J$ independent variables $(p, T, v, v', v, .N^\lambda.)$. The following two sections are devoted to establishing these so-called "normal forms."

1.4. Normal Form of the Internal Energy Variation $e - e_0$

Let $e_G(v', T, .N^i.)$ be the specific internal energy of the gaseous phase of composition $(.N^i.)$ when the volume of the N' moles is v' and the temperature is T. Let $e^j(p, T)$ be the specific internal energy of the condensed phase (j) at pressure p and temperature T. Taking account of the hypothesis $(\mathscr{H}_1)$ (see §1.2) we can write

$$e - e_0 = \left(\sum_i N^i M^i\right)\left[e_G(v, T, .N^i.) - e_0\right] + \sum_j N^j M^j\left[e^j(p, T) - e_0\right].$$

$$(XII.12)$$

To calculate $e - e_0$, as in calculating the Gibbs free energy g, it is necessary to revert to a "standard" state with temperature T^0 and pressure p^0. It is also necessary to introduce other quantities as listed and defined below:

- e^0, standard specific internal energy of the explosive substance;
- e_f^0, standard specific energy of formation of the explosive substance on the basis of the elements taken in the standard state;
- $(H_f^0)^\lambda$, standard molar enthalpy of formation of the species (λ) considered in the phase which it possesses within the detonation products; and
- v_f^λ, variation in the number of gaseous moles in the reaction of formation of species (λ) under the condition defined above.

Chemical thermodynamic calculations detailed in Appendix E now lead, on the sole condition that p^0 is sufficiently low, to the expression

$$
\left.
\begin{aligned}
e - e_0 &\simeq -e_f^0 + \sum_\lambda N^\lambda [(H_f^0)^\lambda - v_f^\lambda \Re T^0] + e^0 - e_0 \\
&+ \sum_i N^i \Delta_{T^0}^T \tilde{H}(T) + \sum_j N^j \Delta_{T^0}^T H^j(p^0, T) - N' \Re(T - T^0) \\
&+ N' \Re T^2 \int_\infty^{v'} \frac{\partial \mathscr{F}}{\partial T} \frac{dv'}{v'} \\
&+ T^2 \sum_j N^j M^j \int_{v^j(p^0, T)}^{v^j} \frac{\partial}{\partial T}\left(\frac{\varphi^j}{T}\right) dv^j \\
&- p^0 \sum_j [v^j(p^0, T) - v^j(p^0, T^0)].
\end{aligned}
\right\}
\qquad \text{(XII.13)}
$$

In (XII.13), all the quantities in the right-hand member are expressed as a function of the state functions $\mathscr{F}$ and φ^j or of tabulated values, with the sole exception of e_f^0 which cannot be found directly by experiment. One stage therefore has to be surmounted, that which allows evaluation of e_f^0 as a function of the standard specific heat of total combustion q_{comb}^0 measured with a calorimetric chamber; this relation does not have a general form since it depends on the nature of the chemical species formed during total combustion; an example of evaluation will be given in §3.2 for the usual case of organic explosives.

1.5. Normal Form of the Thermodynamic Coefficients

Equation (XII.11) of an isentrope in $(\mathscr{D})$ shows that for its effective determination it is necessary to know how to calculate the derivatives $\partial e/\partial T(v, T)$ and $\partial p/\partial T(v, T)$ at point (v, T) of $(\mathscr{D})$. This calculation can vary greatly according to the mathematical technique used to find the chemical composition N^λ which leads to $g(p, T, .N^\lambda.)$ being a minimum at a given p and T.

A first method, which is the most usual today, applies an algorithm suitably derived from the simplex algorithm [55], without explicitly establishing the $I + J - L$ chemical equilibrium equations. The differentials shown above may then result only from numerical differentiation, i.e., may be approximated only as the ratio of the difference Δe of the internal energy (Δp of the pressure) to difference ΔT of temperature at constant specific volume v. Then, strictly, the quality of the estimate can appear only as the result of a convergence study on all the domain of $(\mathscr{D})$ explored.

A second method, now abandoned because it is less suited to modern computers than the first, consists in establishing an explicit expression of the equilibrium equations by considering $I + J - L$ independent equilibria of the form

$$
0 \rightleftharpoons \sum_\lambda v_q^\lambda \mathscr{E}_q^\lambda, \qquad q = 1, \ldots, I + J - L, \qquad \text{(XII.14)}
$$

to which the $I + J - L$ chemical equilibrium equations correspond

$$0 = \sum_i v_q^i \mu_G^i(p, T) + \sum_j v_q^j \mu^j(p, T), \qquad q = 1, \ldots, I + J - L. \quad \text{(XII.15)}$$

Then the above derivatives can result explicitly from suitable eliminations within the linear system obtained by analytical differentiation of (XII.1), (XII.2), (XII.3), (XII.13), and (XII.15) (in the last the expressions of μ_G^i and μ^j to be taken into account are (XII.6) and (XII.7) established above). In fact, once differentiation has been accomplished, any thermodynamic coefficient becomes accessible with precision: not only $\partial e/\partial T(v, T)$ and $\partial p/\partial T(v, T)$ but also $\partial e/\partial v(v, T)$ and $\partial p/\partial v(v, T)$, and then all the others by using the thermodynamic identities in Appendix A. These especially include the quantities which play a major role in the theory of the Crussard curve, the Grüneisen coefficient G, and the square of the velocity of sound a^2 from the identities (i.9) and (j.4).

These calculations do not present any theoretical difficulty; for this reason, it is sufficient to present the results in a form suited to numerical solving, such as is found, except for the notations, in Chéret [12]. The following symbols are used in Table XII.1(a):

δ the determinant of order $2 + I + 2J$ delimited by double solid lines;

δ_T the determinant deduced from δ by substituting for its first column that on its left subtitled (dT);

δ_v the determinant deduced from δ by substituting for its first column that on its left subtitled (dv);

Δ_T the determinant of order $3 + I + 2J$ obtained by bordering δ on its left by the complete column subtitled (dT) then by the line written above δ; and

Δ_v the determinant of order $3 + I + 2J$ obtained by bordering δ on its left by the complete column subtitled (dv) then by the line written above δ.

With the help of these five determinants, which can be calculated from the expressions in Table XII.1(b), the six thermodynamic coefficients cited above may be generally expressed

$$\frac{\partial p}{\partial T}(v, T) = \frac{\delta_T}{\delta}, \qquad \frac{\partial e}{\partial T}(v, T) = -\frac{\Delta_T}{\delta},$$

$$\frac{\partial p}{\partial v}(v, T) = \frac{\delta_v}{\delta}, \qquad \frac{\partial e}{\partial v}(v, T) = -\frac{\Delta_v}{\delta}, \qquad \text{(XII.16)}$$

$$G = -\frac{\delta_T}{\Delta_T}, \qquad a^2 = -\frac{v^2}{\delta}\left(\delta_v + T\frac{\delta_T^2}{\Delta_T}\right).$$

Only one exception remains, for the states of ($\mathscr{D}$) where the concentration of the chemical species j_0 is zero, in which case the determinant to be considered is that which is derived from the "current" determinant defined above by canceling column $2 + I + J + j_0$ and line $2 + I + J + j_0$.

Table XII.1(a). Matrix underlying the explicit numerical calculation of the normal form of the thermodynamic coefficients. Quantities A, B^i, B_q, C_q^j, D_q^j, E, F_q, G, H^j, J^i, J^j, are defined in Table XII.1(b).

0	G	0	$p \cdot \dfrac{\partial \log \mathscr{F}}{\partial \log T}$	J^i	J^j	H^j	
	0	0		ε_i^λ		0	L
0	F_q	0	$\dfrac{B_q}{v'}$	D_q^i	0	C_q^j	
1	0	0	1	O	$M^k v^k$	$M^k N^k$	
	$\dfrac{E}{T}$	$\dfrac{v'}{N'\Re T}$	$\dfrac{A}{v'}$	$-\dfrac{B^i}{N'}$	O		
0	$\dfrac{\partial \varphi^j}{\partial T}$	1	O			$\dfrac{\partial \varphi^j}{\partial v^j}$	J
dv	dT	dp	dv'	dN^i	dN^j	dv^j	

$$\longleftarrow I \longrightarrow \longleftarrow J \longrightarrow \longleftarrow J \longrightarrow$$

"Singular" results will be given in §4.3, whose nature has been understood only because of the intrinsic precision of the calculations (XII.16).

1.6. Validation Criteria

Before discussing in greater detail the problems set by the equation of state of the gas phase of the detonation products and effective numerical solving of the equations of the current point of $(\mathscr{D})$, it can already be seen, simply from the preceding subsections, that the quality of the evaluation of a point of $(\mathscr{D})$ depends on factors of widely differing types:

- replies given to the preliminary questions (*a priori* list of species and phases, equation of state of each phase present);
- tabulated thermochemical data; and

Table XII.1(b). Definition of the submatrix elements defined by dotted lines in Table XII.1(a).

A	$\mathscr{F} - v'\dfrac{\partial\mathscr{F}}{\partial v'}$
B^i	$\mathscr{F} + N'\dfrac{\partial\mathscr{F}}{\partial N^i}$
B_q	$\sum_i v_q^i B^i$
C_q^j	$-v_q^j \dfrac{M^j v^j}{\Re T}\cdot\dfrac{\partial\varphi^j}{\partial v^j}$
D_q^i	$-\dfrac{v_q^i}{N^i} + \displaystyle\int_\infty^{v'}\left(\sum_i v_q^i\dfrac{\partial\mathscr{F}}{\partial N^i} + v_q\dfrac{\partial\mathscr{F}}{\partial N^i} + N'\dfrac{\partial}{\partial N^i}\sum_i v_q^i\dfrac{\partial\mathscr{F}}{\partial N^i}\right)\dfrac{dv'}{v'}$
E	$\mathscr{F} + T\dfrac{\partial\mathscr{F}}{\partial T}$
F_q	$\dfrac{v_q}{T} - \dfrac{1}{\Re T^2}\sum_\lambda v_q^\lambda[(H_f^0)^\lambda + \Delta_{T^\circ}^T\tilde{H}^\lambda(T)] - \displaystyle\int_\infty^{v'}\left(v_q\dfrac{\partial\mathscr{F}}{\partial T} + N'\sum_i v_q^i\dfrac{\partial^2\mathscr{F}}{\partial N^i\partial T}\right)\dfrac{dv'}{v'}$ $\quad - \sum_j\dfrac{v_q^i M^j}{\Re T^2}\left[\displaystyle\int_{\tilde{v}^j}^{v^j}v^j\left(\dfrac{\partial\varphi^j}{\partial v^j} - T\dfrac{\partial^2\varphi^j}{\partial v^j\partial T}\right)dv^j + \dfrac{d\tilde{v}^j}{dT}\cdot\dfrac{\partial\varphi^j}{\partial\,\mathrm{Log}\,v^j}\right]$
G	$N'\Re - \sum_\lambda N^\lambda C_p^\lambda + N'\Re\dfrac{\partial}{\partial T}\displaystyle\int_\infty^{v'}T^2\dfrac{\partial\mathscr{F}}{\partial T}\cdot\dfrac{dv'}{v'}$ $\quad + \sum_j N^j M^j\left(\dfrac{d\varepsilon^j}{dT} + \dfrac{\partial}{\partial T}\displaystyle\int_{\tilde{v}^j}^{v^j}T^2\dfrac{\partial}{\partial T}\left(\dfrac{\varphi^j}{T}\right)dv^j\right)$
H^j	$N^j M^j T^2\dfrac{\partial}{\partial T}\left(\dfrac{\varphi^j}{T}\right)$
J^i	$(H_f^0)^i - v_f^i\Re T^\circ + \Delta_{T^\circ}^T\tilde{H}^i(T) - \Re(T - T^\circ)$ $\quad + \Re T^2\displaystyle\int_\infty^{v'}\left(\dfrac{\partial\mathscr{F}}{\partial T} + N'\dfrac{\partial^2\mathscr{F}}{\partial N^i\partial T}\right)\dfrac{dv'}{v'}$
J^j	$(H_f^0)^j - v_f^j\Re T^\circ + \Delta_{T^\circ}^T\tilde{H}^j(T) + M^j T^2\displaystyle\int_{\tilde{v}^j}^{v^j}\dfrac{\partial}{\partial T}\left(\dfrac{\varphi^j}{T}\right)dv^j$
$v_q = \sum_i v_q^i;\quad \tilde{H}^j(T) = H^j(p^\circ, T);\quad \tilde{v}^j(T) = v^j(p^\circ, T);$ $\varepsilon^j(T) = p^\circ[v^j(p^\circ, T) - v^j(p^\circ, T^\circ)]$	

- adopted method of resolution and numerical precision decided upon.

Then, which yardstick is to be used to measure this quality? Which criteria are to be used?

During recent decades, replies in the literature have been ambiguous and inconsistent, which simply reflects the uncertainty in detonation propagation theories.

Starting from the asymptotic expansion method and its results (see Chap. III) it is now possible and necessary to adopt a definite position:

To assess the validity of the calculations for $(\mathscr{D})$ it is necessary to locate point J on $(\mathscr{D})$, and calculate the detonation velocity D_*, compare D_* with the spherically diverging detonation velocity extrapolated to infinite radius D_∞^s; a *necessary condition* for the evaluation of $(\mathscr{D})$ to be satisfactory is that $D_* - D_\infty^s$ will be of the order of experimental error for D_∞^s (generally measured with a precision of a few parts per thousand).

This criterion demands the following comments:

- measurement of D_∞^s is more difficult, and thus less often undertaken, than that of D_∞^p, thus forcing substitution of the less reliable criterion $D_* = D_\infty^p$ for $D_* = D_\infty^s$ (see §XI.3.1);
- the necessary condition is far from being sufficient since—as can easily be understood—this criterion only ensures a local quality in the vicinity of J, but by no means in the entire domain of interest which covers two orders of magnitude in pressure.

To compensate for the local imperfection of the criterion $D_* = D_\infty^p$, most authors have usually "completed" it by comparison of the calculated pressure p_* (or the calculated velocity $|u_*|$) to *one* pressure $\bar{p}$ (or one velocity $|\bar{u}|$) measured in an impedance mismatch experiment at the end of a tube or cartridge (see §XI.2.1). While these comparisons are of semiquantitative interest, since it cannot be denied that there is some correlation between p_* and $\bar{p}$, $|u_*|$ and $\bar{u}$, at the present state of knowledge it is difficult to model this correlation so that validation based on criterion $p_* = \bar{p}$ (or $|u_*| = \bar{u}$) remains very risky. To understand the possible danger it is sufficient to recall that, not very long ago, a too sketchy interpretation of impedance mismatch experiments led Davis [21] to announce, wrongly, "the failure of the Chapman–Jouguet theory."

To increase the scope of the validation it seems preferable to use the "*inverse* method" proposed by Manson [40]. This method consists of:

- introducing the variation of the velocity D_* under one of the forms

$$
\left.
\begin{aligned}
d \operatorname{Log} D_* &= \mathscr{T}_0 d \operatorname{Log} T_0 + \mathscr{V}_0 d \operatorname{Log} v_0, \quad \text{(a)} \\
d \operatorname{Log} D_* &= \mathscr{T}_0' d \operatorname{Log} T_0 + \mathscr{P}_0' d \operatorname{Log} p_0, \quad \text{(b)}
\end{aligned}
\right\} \quad \text{(XII.17)}
$$

according to whether we are dealing with a solid explosive or a liquid explosive; and

- linking coefficients $\mathscr{T}_0$ and $\mathscr{V}_0$ (or $\mathscr{T}_0'$ and $\mathscr{P}_0'$) to the initial state (v_0, T_0, p_0)

and to the C–J state (v_*, T_*, p_*) by careful use of the jump relations and the Chapman–Jouguet condition.

The relations established in [40], rewritten with our notation after correcting factual errors with the permission of the author, are

$$
\left.\begin{array}{l}
\dfrac{1}{A_T}\left[\dfrac{2D_*^2}{T_0}\mathscr{T}_0 + B_T\right] = \dfrac{1}{A_v}\left[\dfrac{2D_*^2}{v_0}\mathscr{V}_0 + B_v\right], \quad \text{(a)} \\[3mm]
\dfrac{1}{A_T'}\left[\dfrac{2D_*^2}{T_0}\mathscr{T}_0' + B_T'\right] = \dfrac{1}{A_p'}\left[\dfrac{2D_*^2}{p_0}\mathscr{P}_0' + B_p'\right], \quad \text{(b)}
\end{array}\right\} \quad \text{(XII.18)}
$$

respectively, for the variations (XII.17a) and (XII.17b) where the eight quantities $(A_T, A_v, B_T, B_v;\ A_T', A_p', B_T', B_p')$ are defined by

$$
\left.\begin{array}{ll}
A_T = (C_v)_0\,\dfrac{2v_0^2 + G_0 v_0(v_0 - v_*)}{\gamma_0(v_0 - v_*)^2}, & B_T = (C_v)_0\,\dfrac{v_0}{v_0 - v_*}, \\[3mm]
A_v = (TC_v)_0\,\dfrac{2G_0 v_0}{(v_0 - v_*)^2} + \dfrac{\gamma_0 D_*^2 - a_0^2}{\gamma_0(v_0 - v_*)}, & B_v = \dfrac{\gamma_0(2v_* - v_0)D_*^2 - v_0 a_0^2}{\gamma_0 v_0(v_0 - v_*)}, \\[3mm]
A_T' = A_T + \left[\dfrac{\partial v}{\partial T}(p, T)\right]_0 A_v, & B_T' = B_T + \left[\dfrac{\partial v}{\partial T}(p, T)\right]_0 B_v, \\[3mm]
A_p' = \left[\dfrac{\partial v}{\partial p}(p, T)\right]_0 A_v, & B_p' = \left[\dfrac{\partial v}{\partial p}(p, T)\right]_0 B_v.
\end{array}\right\}
$$

$$ \text{(XII.19)} $$

The remarkable property of the relations (XII.19) is the fact that—apart from the initial state (v_0, p_0, T_0), thermodynamic coefficients $[G_0, \gamma_0, a_0, (C_v)_0]$ associated with this state, and velocity D_*—they contain only one C–J value, the specific volume v_*; each equation in (XII.18) is thus capable of transformation into an equation which allows v_* to be calculated (and thus p_* and $|u_*|$) from measurements on the initial substance, measurements of D_*, and the coefficients $(\mathscr{T}_0, \mathscr{V}_0)$ or $(\mathscr{T}_0', \mathscr{P}_0')$. It must be stressed, however (see Bauer et al. [4] for compressed gas explosives), that the inverse method is practical and fruitful only if D_∞^p is a close approximation to D_*. Once again, spherically diverging detonation experiments become of interest.

2. Functions of State

2.1. Introduction to Estimation of $\mathscr{F}$

Returning to equation (XII.5) which defines the function $\mathscr{F}$, it can be seen that the problem of its determination can also be posed as: What is the pressure within a mixture of N^i moles $(i = 1, \ldots, I)$ if they occupy a volume v' at temperature T without reacting? This is an unusual problem for the physicist, inasmuch as the existence in the mixture of moles of different species cannot be ignored, but it is not a problem for the chemist inasmuch as the existence of chemical equilibrium within the mixture is excluded.

Why is it so difficult to improve our knowledge of this function $\mathscr{F}$? The answer is twofold. On the one hand, experiments can provide information only on the restrictions of $\mathscr{F}$ for each chemical species taken in isolation. On the other hand, the theoretical problem is by no means easy, since the volumes and temperatures considered are such that the repulsions between molecules of different species are not negligible.

In the early 1960s, the more sophisticated theoretical evaluations of Fickett [25] for conventional explosive substances showed differences in $D_* - D_\infty^p$ of up to 500 m/s, i.e., more than ten times the measurement error on D_∞^p. This failure explains why semiempirical estimations of $\mathscr{F}$ were then developed, which are still of current interest (see §2.2) in spite of the giant strides made in theoretical estimations since 1980 (see §2.4). Both have one or more adjustable parameters whose "method of use" differs from one author to another according to the objective sought.

2.2. Semiempirical Estimations of $\mathscr{F}$

To link up with the following subsections, the quantity which dominates analysis of the problem by statistical mechanics, the partition function $\mathfrak{Z}(v', T, . N^i .)$, will now be introduced.

Statistical thermodynamics (see, e.g., Landau and Lifschitz) shows that the specific Helmholtz free energy of the mixture can be written

$$f' = -N'\mathfrak{R}T \, \mathrm{Log} \, \mathfrak{Z}.$$

The corresponding pressure is given by the identity $(f.2)$ in Appendix A, rewritten here as

$$p = -\frac{\partial f'}{\partial v'}(v', T, . N^i .).$$

These two equations lead immediately to

$$p = N'\mathfrak{R}T\frac{\partial \, \mathrm{Log} \, \mathfrak{Z}}{\partial v'},$$

and then to

$$\mathscr{F} = \frac{\partial \, \mathrm{Log} \, \mathfrak{Z}}{\partial \, \mathrm{Log} \, v'}. \tag{XII.20}$$

It can now readily be seen that development of the calculations basically depends on the choice of function $\mathfrak{Z}$.

The literature is not lacking in examples for this choice. We first note, as did Heuzé [27], those which postulate that the function $\mathfrak{Z}$ is:

Table XII.2. The four possible choices for the pair $(\mathscr{F}, K^i)$.

$\mathscr{F}(x)$	$K^i(T)$	Adjustable parameters
$1 + x + 0.625x^2 + 0.278x^3 + 0.193x^4$	B^i	None
$1 + xe^{\beta x}$	$\kappa \dfrac{k^i}{(T + 400)^{1/2}}$	κ, β
$1 + x + 0.625x^2 + 0.287x^3 - 0.093x^4 + 0.014x^5$	$\dfrac{k^i}{T^\chi}$	$\chi = \tfrac{1}{3}$ or $\tfrac{1}{4}$
$(1 + x + x^2)(1 - x)^{-3}$	$\Lambda \dfrac{\pi}{\sigma} \sum_{p,q} N^p N^q \left(\dfrac{r^p + r^q}{2}\right)^3$	Λ

(i) a function of a single variable

$$x = \frac{1}{v'} \sum_i N^i K^i(T);$$

(ii) such that $\text{Log } 3 + \text{Log } x \underset{x \to 0}{\to} 0$.

With these assumptions (XII.20) becomes

$$\mathscr{F} = -\frac{\partial \text{Log } 3}{\partial \text{Log } x}.$$

Table XII.2 specifies the four possible choices for the pair $(\mathscr{F}, K^i)$ and specifies the parameters where adjustment allows approximation to the criterion $D_* = D_\infty^p$.

The first, called the van der Waals approximation, is based on the study of a gas containing hard spheres; B^i is the covolume of species (i).

The second, called the Becker–Kistiakowski–Wilson approximation, is based on the work of Becker [6a, b, c] and the successive improvements by the latter two [32a, b]. The publications of Brinkley and Wilson [8a, b], Cowan and Fickett [19], and Mader [38a, b, c, d] may be used to follow the history of the parameter k^i which has the dimension of a covolume and which is determined, each time that this is possible, to account for the Hugoniot curve of species (i) taken in an initial condensed state.

The third was recently proposed by Heuzé in a contribution to the calculation of gaseous explosives under high initial pressures [27]: for each of the values of χ ($\tfrac{1}{3}$ or $\tfrac{1}{4}$) the author proposes a set of coefficients k^i.

The fourth results from the reports of Percus and Yevick [46], Wertheim [54], Thiele[50], and Edwards and Chaiken [23]. To the extent that it is based on considerations concerning the interaction between molecules of radii $\underline{r}^p$ and $\underline{r}^q$, this last so-called Percus–Yevick estimation is an excellent transition towards the "*ab initio*" estimations which will be seen in §2.4. Before this, the original "JCZ" estimations—after Jacobs [30], Cowperthwaite and Zwisler [20a, b]—will be introduced and explained.

2.3. JCZ Estimations of $\mathscr{F}$

The authors, in attempting to reduce to known situations, assume that the pressure p within the N' moles results from the combination of:

- the pressure in an FCC (face-centered cubic) crystal lattice of energy e'_s in volume v' and at absolute zero temperature; and
- the pressure resulting from the thermal contribution

They therefore set

$$p = -\frac{de'_s}{dv'} + \frac{N'\Re T}{v'}\left[1 + \frac{\partial \operatorname{Log}(3_s + 3_G)}{\partial \operatorname{Log} v'}\right], \qquad \text{(XII.21)}$$

which is equivalent to

$$\mathscr{F} = 1 - \frac{v'}{N'\Re T}\frac{de'_s}{dv'} + \frac{\partial \operatorname{Log}(3_s + 3_G)}{\partial \operatorname{Log} v'}, \qquad \text{(XII.22)}$$

where 3_s and 3_G have the dimension of partition functions. The there basic stages in the calculation will thus be the expression of e'_s and the careful choice of 3_s and 3_G.

The energy e'_s is calculated using a cellular model based on the work of Lennard-Jones and Devonshire [35]. Its explicit expression depends on the form chosen for the molecular interaction potential $\mathscr{V}(r)$, where r designates the distance between two fictitious molecules. In the JCZ 1 and 2 versions the chosen potential is of the type "Mie 12-6" (also called Lennard-Jones and Devonshire)

$$\mathscr{V}(r) = \varepsilon\frac{\ell m}{\ell - m}\left[\frac{1}{\ell}\left(\frac{r}{r}\right)^{\ell} - \frac{1}{m}\left(\frac{r}{r}\right)^{m}\right], \qquad \ell = 12; m = 6; \quad \text{(XII.23a)}$$

while in the JCZ 3 version, the potential is of the "modified 13,5-6 Buckingham" type (also called Mason–Rice).

$$\mathscr{V}(r) = \varepsilon\frac{\ell m}{\ell - m}\left[\frac{1}{\ell}\exp \ell\left(1 - \frac{r}{r}\right) - \frac{1}{m}\left(\frac{r}{r}\right)^{-m}\right], \qquad \ell = 13.5; m = 6;$$
$$\text{(XII.23b)}$$

Since the exponents "repulsive" ℓ and "attractive" m are fixed, there are only two parameters: equilibrium distance r and depth ε of the potential well at equilibrium. Since the authors finally preferred the latter version when using the validation criteria shown above (see §1.6), only the expression obtained for e'_S in the potential case (XII.23b) is given

$$\left.\begin{array}{l} \dfrac{e'_S}{\mathcal{N} N' \underline{\varepsilon}} = \dfrac{m\ell}{2(\ell - m)} \left[\dfrac{B^{(\ell)}}{\ell} \exp \ell(1 - \bar{v}^{1/3}) - \dfrac{B^{(m)}}{m} \dfrac{1}{\bar{v}^m} \right], \\[4mm] \bar{v} = \dfrac{\sqrt{2}}{\mathcal{N}} \dfrac{v'}{N' \underline{r}^3}, \end{array}\right\} \tag{XII.24}$$

where $B^{(\ell)}$ ($B^{(m)}$) is the ratio of the repulsive (attractive) potential created in a node by the entire lattice to that created by the nearest neighboring "molecule." Using the assumption made by Jacobs for a FCC lattice, $B^{(m)}$ and $B^{(\ell)}$ can be calculated exactly

$$\left.\begin{array}{l} B^{(m)} = 14.45, \\[2mm] B^{(\ell)}(v') \approx 12 \text{ in the range of interest in } v'. \end{array}\right\} \tag{XII.25}$$

The term $\mathcal{Z}_S$ is calculated from the fact that, *at high densities*, it predominates over $\mathcal{Z}_G$ and must therefore differ little from the value determined in the FCC solid lattice using the repulsive part of the potential. For the JCZ 3 potential (XII.23b) and within the framework of Einstein's harmonic approximation (see [20b] corrected by G. Pittion-Rossillon), we find

$$(\mathcal{Z}_S)^{-2/3} = \frac{\mathcal{N}\underline{\varepsilon}}{\mathfrak{R}T} \frac{\ell m}{\ell - m} \frac{B^{(\ell)}}{12\pi} \bar{v}^{1/3}(\ell \bar{v}^{1/3} - 2) \cdot \exp \ell(1 - \bar{v}^{1/3}). \tag{XII.26}$$

The term $\mathcal{Z}_G$ is calculated from the fact that, *at average densities*, it predominates over $\mathcal{Z}_S$ and that the corresponding pressure must differ little from the development of the virial calculated from the repulsive part of the potential

$$\frac{\partial \text{Log } \mathcal{Z}_G}{\partial \text{Log } v'} = b_{(1)} \frac{N'}{v'} B + b_{(2)} \left(\frac{N'}{v'} B\right)^2 + \cdots,$$

where the values of $b_{(i)}$ are well known up to the sixth order (see [28])

$$b_{(1)} = 1; \qquad b_{(2)} = 0.625; \qquad b_{(3)} = 0.286875;$$

$$b_{(4)} = 0.110156; \qquad b_{(5)} = 0.038672; \qquad b_{(6)} = 0.0138;$$

but where B is the "covolume" associated with the intermolecular potential within the fictitious fluid. For a potential of *hard spheres* of diameter d, it is well known (see, e.g., [28]) that B is $2\pi \mathcal{N} d^3/3$. For the JCZ 3 potential (XII.23b), and within the framework of the Barker and Henderson [3a, b, c] approximation, it follows that

$$B(T) = \frac{2\pi}{3} \mathcal{N} \underline{r}^3 \left[1 + \frac{\mathscr{C}}{\ell} + \frac{1}{\ell} \text{Log} \frac{m}{\ell - m} \frac{\underline{\varepsilon}}{kT} \right]^3 \equiv \frac{2\pi}{3} \mathcal{N} d_{BH}^3,$$

where $\mathscr{C}$ is Euler's constant (0.577215665).

A final series of assumptions is necessary, which links the "molecular" interaction parameters $\underline{\varepsilon}$ and $\underline{r}$ within the fictitious fluid, to the molecular interaction parameters $\underline{\varepsilon}^i$ and $\underline{r}^i$ within each chemical species present in the gas phase (the only cases calling on either theory or experiment). Cowperthwaite and Zwisler have opted for simple algebraic relations inspired by the "Lorentz" rule for the equilibrium distance, and the "Berthelot" rule for the equilibrium potential. More precisely, they assume

$$
\begin{aligned}
\underline{r}^3 &= \frac{1}{N'^2} \sum_{p,q} N^p N^q \left(\frac{\underline{r}^p + \underline{r}^q}{2} \right)^3, \quad \text{(a)} \\
\underline{\varepsilon} &= \frac{1}{N'^2} \sum_{p,q} N^p N^q \sqrt{\underline{\varepsilon}^p \underline{\varepsilon}^q}, \quad \text{(b)}
\end{aligned}
\right\}
\qquad \text{(XII.27)}
$$

The practical interest of JCZ 3 will be considered in Section 4, but two indisputable merits must be stressed here and now: the fact that it is the first estimation adjusting to the level of the $2I$ molecular interaction parameters $\underline{\varepsilon}^i$ and $\underline{r}^i$, and the fact that it has resulted in an explosion of studies on *ab initio* estimations (§2.4), if only by the rethinking generated by controversial advances and hypotheses.

2.4. *Ab initio* Estimations of $\mathscr{F}$

This section is based on a preliminary documentary work by G. Pittion-Rossillon.

In view of the above presentation of the JCZ estimations and, in particular, of the rather brief character of the transformation (XII.21), it may be seen that there is room for real progress in *ab initio* formulation of $\mathscr{F}$, i.e., starting from elementary molecular interactions. While some progress has been made by a very small number of authors, consensus between them is far from being established, if only for the reason that not one of them has yet reached the ultimate goal: strictly *ab initio* achievement of the criterion $D_\infty^S = D_*$ for all explosives, without notable exception. In the absence of such consensus, all controversy is avoided here, but directions opened up and sensitive points targeted.

Two "doctrines" have steadily developed:

(a) the first, more innovatory but still capable of substantial development has developed from:
  (i) a hard-spheres mixture model by Mansoori, Carnahan, Starling, Leland [42];
  (ii) a pure fluid model by Weeks, Chandler, Andersen [2], [53], simplified by Verlet and Weiss [51] and extended to mixtures by Lee and Levesque [34]; it is illustrated by the articles of Chirat, Pittion-Rossillon, Baute [5], [15], [16], [17];
(b) the second, more conventional but better established, is found in extensions of the studies by Mansoori *et al.* [41, 42] and Rasaiah and Stell [47] improved by Ross [49]; it is illustrated by the articles of Ree [48a, b, c].

(a) "Doctrine of the Effective Free Energy"

In [53] Weeks *et al.* suggested that it is the repulsive part of the intermolecular potential which determines the major properties of the dense fluid. In the case of a spherical potential $\mathscr{V}(r)$ this working hypothesis leads to the equation

$$\mathscr{V}(r) = \mathscr{V}_{(0)}(r) + \mathscr{V}_{(1)}(r),$$

where $\mathscr{V}_{(0)}$ is a principal repulsion term while $\mathscr{V}_{(1)}$ is an attraction perturbation term. In a general way, a distance a may be introduced such that

$$\begin{aligned}
\mathscr{V}_{(0)}(r) &= \mathscr{V}(r) - \mathscr{V}(a), & \mathscr{V}_{(1)}(r) &= \mathscr{V}(a) & \text{if } r \le a, \\
\mathscr{V}_{(0)}(r) &= 0, & \mathscr{V}_{(1)}(r) &= \mathscr{V}(r) & \text{if } r > a.
\end{aligned} \right\} \quad \text{(XII.28)}$$

However, the first idea—which is often returned to—is to choose the equilibrium distance $\underline{r}$ as the value of a. The transformation (XII.28) leads to seeking the partition function $\mathfrak{Z}$ as a factorized form $\mathfrak{Z}_{(0)} \times \mathfrak{Z}_{(1)}$, with the specific Helmholtz free energy f' in the form

$$f' = f'_{(0)} + f'_{(1)}.$$

In [42] Mansoori *et al.* established a procedure to calculate the Helmholtz free energy of a mixture of hard spheres with diameters d^i ($i = 1, \ldots, I$). This procedure introduces the volumetric fractions η^i and their sum η by

$$\eta^i = \frac{\pi}{6} \frac{N^i(d^i)^3}{v'}, \qquad \eta = \sum_i \eta^i;$$

a geometric coupling coefficient between species

$$\Delta^{pq} = \frac{N^p N^q (d^p - d^q)^2 (d^p d^q)^{1/2}}{N' \sum_i N^i(d^i)^3},$$

two means d' and d"

$$d' = \sum_{p>q} \Delta^{pq}(d^p + d^q)(d^p d^q)^{-1/2},$$

$$d'' = \sum_{p>q} \Delta^{pq}(d^p d^q)^{1/2} \cdot \sum_i N^i(d^i)^2 \bigg/ \sum_i N^i(d^i)^3,$$

and a weighted mean $\bar{\eta}$ of the volumetric fractions occupied

$$\bar{\eta} = \left[\sum_i (\eta^i/\eta)^{2/3}(N^i/N')^{1/3} \right]^3.$$

Finally, it provides the part of $f'_{(0)}$ due to nonideality in the form of an expansion in $1 - \eta$

$$-\tfrac{3}{2}(1 - d' + d'' + \bar{\eta}) + \frac{3d'' + 2\bar{\eta}}{1 - \eta} + \frac{3}{2} \frac{1 - d' - d'' - \bar{\eta}/3}{(1 - \eta)^3}$$

$$+ (\bar{\eta} - 1) \, \mathrm{Log}(1 - \eta).$$

It is necessary to stress an essential feature in this approach, that of avoiding the introduction of a mean hard sphere, whose physical reality seems disputable to the authors when the individual diameters d^i are not very close. To end the preliminaries, it is still necessary to define the choice of d^i: that is when the options start. The general method adopted by Chirat, Pitton-Rossillon, and Baute is to start with the diameter d^i_{BH} (see §2.3) calculated for the principal part of the potential proper to the species (i)

$$d^i_{BH} = \int_0^\infty \left(1 - \exp\left\{ \frac{\mathscr{V}^i_{(0)}(r)}{kt} \right\} \right) dr,$$

and then to correct it to allow for the environment with the formula

$$\frac{d^i}{d^i_{BH}} = 1 + \psi(\eta) \int_0^\infty \left(\frac{r}{d^i_{BH}} - 1 \right)^2 d \exp\left\{ -\frac{\mathscr{V}^i_{(0)}(r)}{kT} \right\},$$

where the function $\psi(\eta)$ is chosen in the WCAα versions ($\alpha = 1, \ldots, 4$) according to the original criterion of Weeks *et al.* [53], and in WCASC according to a self-consistent (S.C.) criterion proposed by Verlet and Weiss [51].

In [34] Lee and Levesque calculate $f'_{(1)}$ for a mixture. Having defined for each pair (p, q) of the mixture

$$\text{an "effective" diameter} \quad d^{pq} = \tfrac{1}{2}(d^p + d^q),$$

$$\text{an equilibrium distance} \quad \underline{r}^{pq} = \tfrac{1}{2}(\underline{r}^p + \underline{r}^q),$$

$$\text{an equilibrium potential} \quad \underline{\varepsilon}^{pq} = (\underline{\varepsilon}^p \underline{\varepsilon}^q)^{1/2},$$

$$\text{a repulsion exponent} \quad \ell^{pq} = (\ell^p \ell^q)^{1/2},$$

they show that the attractive contribution $f'_{(1)}$ can be approximated by the sum

$$\frac{2\pi}{v'} \sum_{p,q} N^p N^q \int_{d^{pq}}^\infty g^{pq}(r) \cdot \mathscr{V}^{pq}_{(1)}(r) \cdot r^2 \, dr,$$

where $g^{pq}(r)$ is the radial distribution function of hard spheres with diameter d^{pq}. Options (WCA1, 2, or 3 for one part, WCA4 and WCASC for the other) now start, according to the method chosen to calculate the integral.

(b) "Doctrine of the Effective Potential"

In comparison with this first "doctrine," for which only a theoretical basis has been given, it is instructive to consider the second whose strong points, made by Ree in [48a, b, c], are as follows:

(i) The properties of the true potential proper to each species cannot be considered as having only a negligible effect on the properties of the mixture. This is why it is necessary to give each species an "effective potential" of the exp. 6 type ($\underline{r}^i, \underline{\varepsilon}^i, \ell^i$) subject to distinguishing:

- those which obey the law of corresponding states (e.g., O_2, N_2, CH_4, CO_2, CO), when argon is used as reference and one chooses

$$\underline{r}^i = \underline{r}^{Ar}\frac{T^i_{crit}}{T^{Ar}_{crit}}, \qquad \underline{\varepsilon}^i = \underline{\varepsilon}^{Ar}\frac{v^i_{crit}}{v^{Ar}_{crit}}, \qquad \ell^i = 13;$$

- those where the molecules show a strong electrostatic interaction (e.g., H_2O, NH_3), in which cases a potential corrected by a factor $(1 + \theta^i/T)$ is used; and
- those where the molecules suffer a large number of three-body collisions (e.g., H_2), when a special adjustment procedure is used and $\ell^i = 11.1$.

(ii) It may be sufficient to consider each phase of the mixture as an average fluid endowed with an effective potential of type exp. 6 where the parameters $\underline{r}, \underline{\varepsilon}, \ell$ would depend on the composition $(.\,N^i.\,)$ by

$$\underline{r}^3 = \sum_{p,q} \frac{N^p}{N'} \cdot \frac{N^q}{N'}(\underline{r}^{pq})^3,$$

$$\underline{\varepsilon} = \sum_{p,q} \frac{N^p}{N'} \cdot \frac{N^q}{N'}\left(\frac{\underline{r}^{pq}}{\underline{r}}\right)^3 \underline{\varepsilon}^{pq},$$

$$\ell = \sum_{p,q} \frac{N^p}{N'} \frac{N^q}{N'}\left(\frac{\underline{r}^{pq}}{\underline{r}}\right)^3 \left(\frac{\underline{\varepsilon}^{pq}}{\underline{\varepsilon}}\right)\ell^{pq},$$

where the parameters $\underline{r}^{pq}$, $\underline{\varepsilon}^{pq}$, ℓ^{pq} are given by the equations already presented for the first "doctrine."

(iii) The Helmholtz free energy of the mean fluid f' may be calculated by the so-called MCR variational procedure given in Mansoori and Canfield [41] and Ross [49].

2.5. Introduction to the Estimation of φ^j

From the example of solid carbon which is often present in the detonation products of an organic explosive, the necessity has been seen, in §1.2, of considering condensed species (j) within the "final substance" resulting from the detonation of an explosive, and in §1.3, of giving each of them a law of state $p = \varphi^j(v^j, T)$.

One of the first ways to handle the function φ^j is, as it were, to avoid it, while being satisfied with the incompressibility hypothesis. Density ρ^0 which it possesses under the standard conditions (p^0, T^0) is therefore assigned to species (j). The procedure is rather crude, but it is convenient, and above all it is justified in practice when the concentration of the species (j) is low, e.g., for alumina in an explosive doped with aluminum. However, it is quite unsuitable for an organic explosive where the concentration of carbon in the detonation products is high. There is a final problem, not noticed for some considerable time, that it is only applicable when there is no ambiguity in the choice of ρ^0. Such an ambiguity does occur in the case of carbon, since one may *a priori* opt for the density of graphite (2.25 g/cm^3) or of diamond (3.51 g/cm^3) or even for intermediate values based on differing arguments! We will return to this very topical question in §4.3, only noting here that the question was not even raised until about 1980, the option being taken in favor of graphite.

Independently of this latest, recent objection, the reason put forward concerning high-carbon explosives has resulted in great efforts to estimate the φ^j function of graphite. The Cowan and Fickett formulation [19] is very widely used: pressure p is given as a function of compression $\eta = \rho/\rho^0$ and temperature T by

$$p = a(\eta) + b(\eta)\cdot\frac{T}{10^3} + c(\eta)\left(\frac{T}{10^3}\right)^2,$$

$$a(\eta) = -2467 + 6769\eta - 6956\eta^2 + 3040\eta^3 - 386\eta^4,$$

$$b(\eta) = -19.54 + 23.37\eta,$$

$$c(\eta) = 0.61758 - 0.57995\eta^{-1} + 0.2278\eta^{-2}.$$

(XII.29)

The coefficients are obtained by adjustment of the Hugoniot curve for graphite and of the coefficients of compressibility and dilatation in the standard state; the range of validity shown is $0.95 < \eta < 2.5$, $T < 2$ eV.

The formulation (XII.29) has not always been unanimously accepted: Ree [48a, b, c] prefers an equation from the work of Murnaghan [45] by taking account of the low values of $\eta - 1$.

However, it now appears that the choice of φ^j is less critical than the *a priori* choice on the condensed species present in the final substance.

3. Numerical Codes

3.1. Algorithms

As was shown at the beginning of §1.3, the system to be solved is composed of $L + 1$ linear equations (homogeneity equation (XII.1) and the L equations of conservation of atoms (XII.2)) and by $I + 2J + 1 - L$ nonlinear equations ($J + 1$ equations of state and $I + J - L$ equations of chemical equilibrium). Two remarks must be made in this connection, which illuminate all the studies generated by the system:

(i) its nature is hybrid in the sense that it does not use any solution method which is proved and clearly marked as to convergence and precision; and

(ii) its mathematical complexity depends largely on the chemical complexity of the "final substance" and on the physical complexity of the state functions $\mathscr{F}$ and φ^j, so that enormous progress has been made since the introduction and development of computer calculations.

Table XII.3. Simplified history of calculation codes in the United States and France.

	Method	L	Λ	J	J'	Estimation of ℱ	Reference
RUBY	Minimum of the Helmholtz free energy	4	16	2	1	BKW	[36], 1962
FORTRAN BKW		5 10	15 25	1 5	1 1		[38b], 1962 [38d], 1962
TIGER		10	40	10	10	JCZ 3	[20a], 1973
CHEQ		not published				MCR	[37], 1982
	Chemical equilibrium equations	4 4	7 13	1 1	1 1	"modified" van der Waals	[7], 1960 [52], 1960
LA MINEUR		4	11	1	1	BKW	[12], 1971
ARPEGE	Minimum of the Helmholtz free energy	10	35	>2	>2	BKW	[14], 1974
ETARC		not published				any estimation	not published

In fact, the purely mathematical difficulty is not the least of all those which have to be overcome by authors of codes for calculating explosives, so that the solving algorithm has been, and is, considered valuable information that needs to be protected with the same secrecy that surrounds the manufacture of certain explosives. We therefore limit ourselves here to some generalities, referring unsatisfied readers to the information given in the references cited in Table XII.3.

For the sake of clarity, it is convenient to distinguish between two periods. The first extends over the twenty years starting in 1941, when automatic calculation methods were few and limited; chemical equilibrium equations were written exactly (see Brinkley [9]), but, in seeking to solve the complete system, an attempt was made to reduce the number of species taken into account and the complexity of the functions of state. The second period started shortly after 1960: the first IBM 7030 computer appeared; methods for minimizing the Helmholtz free energy became widespread, based on the articles by White, Johnson, Dantzig [55] and Dorn [22]; the CDC and CRAY scientific computers brought gains of several orders of magnitude both in computing speed and in memory capacity.

Having already referred indirectly to the first period by noting the history of the BKW estimate of ℱ, we will now limit ourselves to analyzing the second, stressing the driving forces in development. There are two:

(i) to increase the numerical capacity of the code to take into account progressively larger numbers of atoms, species, and phases; and

(ii) to increase the logic capacity of the code to take into account the possible disappearance of J' condensed phases ($1 \leq J' \leq J$). The history of codes in the United States and France is given in Table XII.3 with reference to these quality criteria.

However, since 1980 and the appearance of the first *ab initio* estimations of $\mathscr{F}$, a new trend is appearing. The performances noted above have reached such a level that the corresponding criteria lose their dominance compared to other logic capacities: that of treating the equilibrium between two solid phases (specially for graphite/diamond), or that of allowing segregation of the gas phase [37]. We will return to these new aspects in §4.3.

3.2. Thermochemical Data

We return again to §1.3 where the equations were established which control the calculation of the current point of ($\mathscr{D}$). The specific Gibbs free energy g was considered there, and transformed in order to obtain an expression which, apart from the functions $\mathscr{F}$ and φ^j, only contained quantities listed in specialized studies. It is thus shown that it is necessary to know:

- the variation $\Delta_{T^0}^T \tilde{H}^i(T)$ of molar enthalpy between T^0 and T for the gas species (i) supposed ideal;
- the variation $\Delta_{T^0}^T H^j(p^0, T)$ of molar enthalpy between T^0 and T for the condensed species (j) at standard pressure p^0;
- the standard molar entropy $\tilde{S}^i(p^0, T^0)$ for the gas species (i) supposed ideal; and
- the standard molar entropy $S^j(p^0, T^0)$ for the condensed species (j).

In general, enthalpies are set out in tabular form in various publications, of which the most systematic and most complete is the *JANAF Thermochemical Tables* published by the U.S. Department of Commerce. But these tables do not lend themselves conveniently to a numerical solving of the equations. It is, therefore, almost obligatory to choose a simple algebraic formula—but one which conforms to elementary theoretical considerations—and to adjust its coefficients to find the values in the tables with a maximum error fixed in advance. Entropy values are also available in published tables but, even more than for enthalpies, it must be remembered that these tables lag behind specialist publications, so that reference must still be made to values cited in monographs.

We now return to §1.4 where the normal form of the variation $e - e_0$ in the specific internal energy was established. In (XII.13) we see:

- the standard specific energy e_f^0 of formation of the explosive derived from the elements taken in their standard state;
- the variation in the specific internal energy $e^0 - e_0$ between the initial state and the standard state of the explosive;
- the standard molar enthalpy $(H_f^0)^\lambda$ of formation derived from the elements of the chemical species (λ) in the phase in which it appears within the "final substance"; and

• the variation v_f^λ in the number of gas moles in the formation reaction considered above.

By way of illustrating the task to be accomplished, we cite the choices made in [12] with regard to organic explosives of composition $C_x H_y O_z N_w$. The standard state adopted is

$$p^0 = 1 \text{ atm} = 1.013249 \text{ bar},$$

$$T^0 = 298 \text{ K} \approx 25° \text{ C}.$$

The variations in enthalpy are represented by the Kelley function

$$B_0 + B_1 \left(\frac{T}{1000}\right) + B_2 \left(\frac{T}{1000}\right)^2 + B_3 \left(\frac{T}{1000}\right)^3 + B_4 \left(\frac{1000}{T}\right),$$

where the coefficients $B_0, \ldots, B_4$ are adjusted in such a manner that the representation does not differ from the tabulated values (in the most recent or most reliable study) by more than 150 cal/mol-deg in the interval (T^0, T_m^λ). Table XII.4 collects the values $B_0^\lambda \ldots B_4^\lambda$, T_m^λ, $(S^0)^\lambda$, $(H_f^0)^\lambda$, M^λ as well as the references which have served as a basis for the chosen determination; it also recalls, to avoid any ambiguity, the values of v_f^λ; it must be completed by the value taken for the ideal gas constant $\Re = k \mathcal{N}$: 1.98726 cal/mol-deg.

The quantities $e^0 - e_0$ and e_f^0 have still to be calculated. For the first, only some thermodynamic coefficients of the initial explosive substance are required, as may be seen by referring to the identities $(f.5)$, $(j.2)$, and $(j.3)$ in Appendix A. The second is usually evaluated from the standard specific heat q_{comb}^0 of total combustion measured in a calorimetric chamber, in line with conventional thermochemical reasoning (see, e.g., Médard [43])

$$e_f^0 = q_{comb}^0 + x(H_f^0 - \Re T^0 v_f)^{CO_2} + \frac{y}{2}(H_f^0 - \Re T^0 v_f)^{H_2O \text{ liq}}.$$

The value used for $(H_f^0)^{H_2O \text{ liq}}$ is that cited by reference (1) in Table XII.4

$$(H_f^0)^{H_2O \text{ liq}} = -68,317 \text{ calories/mole}.$$

3.3. *Ab initio* Parameters and Floating Parameters

The review of estimations of $\mathcal{F}$ given in §§2.1–2.4 has shown—at widely varying levels—parameters which are expected to be fixed either by physical shock experiments and/or by molecular dynamics numerical experiments ("*ab initio*" parameters) or by means of an adjustment to satisfy the validity criterion $D_* = D_\infty^S$ of §1.6 ("floating" parameters).

In the first category there are I covolumes k^i of the semiempirical estimations, the two I parameters $\underline{r}^i$ and $\underline{\varepsilon}^i$ of the JCZ 3 estimation, the three I parameters $\underline{r}^i$, $\underline{\varepsilon}^i$, ℓ^i of the *ab initio* estimations.

The second category includes the parameters κ and β of the BKW estimation, Λ of the Percus–Yevick estimation, etc., but also, less explicitly, the factor $\psi(\eta)$ of the WCA estimations, and even the θ^i coefficients introduced

Table XII.4. Thermochemical data used for the calculation of organic explosives by the LA MINEUR [12] code.

Molecule	CO_2	CO	H_2	H_2O_s	N_2	O_2	OH	NO	NH_3	CH_4	C graphite
i	1	2	3	4	5	6	7	8	9	10	11
$M(i)$ g	44.011	28.011	2.016	18.016	28.016	32	17.008	30.008	17.032	16.043	12.011
ν_r^i	0	0.5	0	−0.5	0	0	0	0	−1	−1	0
$(S°)^i$ cal/mol deg	51.072 (1)	47.2167 (4)	31.208 (1)	45.0917 (3)	45.7708 (2)	49.0065 (2)	43.8927 (5)	50.347 (1)	46.0451 (6)	44.5206 (7)	1.359 (1)
$(H_f°)^i$ (1) cal/mol	−94,057	−26,417	0	−57,798	0	0	0	21,652	−11,040	−17,895	0
T_m^i °K	2,700	2,700	5,300	5,300	5,300	5,300	5,300	2,700	5,300	2,100	2,300
B_0^i cal/mol	−3.452739	−5.066827	−1.7528	−2.44499	−2.32486	−2.556259	−1.76867	−1.902582	−4.64018	−1.787050	−1.273643
B_1^i cal mol^{-1} deg^{-1}	9.176837	8.897902	6.166396	7.24428	6.984570	7.479087	6.24306	6.235555	9.750312	3.253907	0.234401
B_2^i cal mol^{-1} deg^{-2}	2.483248	27.34317	0.590306	1.566302	0.517637	0.478689	0629352	1.173571	2.45929	9.033199	1.995252
B_3^i cal mol^{-1} deg^{-3}	−0.377975	−0.000201	−0.031679	−0.120067	−0.043507	−0.028547	−0.045640	−0.176923	−0.764039	−1.3311	−0.373582
B_4^i cal mol^{-1} deg	0.143598	1.468743	−0.035360	−0.036404	0.050451	0.076246	0.042213	−0.018248	0.488752	0.016100	0.115404

(1) *JANAF Thermochemical data.* U.S. Department of commerce.
(2) Yungman, V.S. and Gurvich, L.C., *Rus. J. Phys. Chem.* (1961), **35**, 1073.
(3) Durand, J.L. and Brandmaier, H.E., WSS/01, paper 62-13, 1962.
(4) Gurvich, L.V. et al., *Tr. Gos. Inst. Prikl. Khim* (1962), **49**, 61.
(5) Friedman, A.S. and Haar, L., *J. Chem. Phys.* (1955), **23**, 869.
(6) Mac Bride, B. et al., NASA SP 3001, Lewis Research Center, Cleveland, Ohio, 1963.
(7) Mac Dowell, R., *J. Chem. Phys.* (1963), **39**, 526.

Table XII.5. BKW covolumes and exp. 6 potential parameters: comparison of values used by different authors.

	Covolume BKW k^i (cm^3/mole)		$\frac{r}{(\text{Å})}$	ε/k (K)	ℓ	$\frac{r}{(\text{Å})}$	ε/k (K)	ℓ
				Exp. 6 potential parameters				
	Mader [38b]	Finger [26]		Ree [48b]			Chirat and Pittion, R. [16]	
H$_2$	180	180	3.43	36.4	11.1	3	30	13.5
N$_2$	380	380	4.09	101.9	13	4.11	117	13.5
O$_2$	350	325	3.84	125.0	13	3.73	132	15
CO	390	390	4.12	108.3	13	4.09	108	13
CO$_2$	600	670	4.17	245.6	13	4.17	245	13
CH$_4$	528	528	4.22	154.1	13	4.198	155.8	13.5
NO	386	386	3.97	112.9	13	3.9	105	12
H$_2$O	250	360	3.06	356.0	13	3.37	135	13.5
NH$_3$	476	476	3.44	474.0	13	3.5	138	17

by Ree in the intermolecular potential of H$_2$O on the one hand, and NH$_3$ on the other hand.

To attempt to size up the "*ab initio*" character of the parameters qualified as such, Table XII.5 compares the values published by a few of the more prolix authors. It is clear from this that the covolumes k^i are nearly constant, except possibly for those associated with the most asymmetric molecules CO$_2$ and H$_2$O. But *a contrario* it also appears that consensus is far from being established for the exp. 6 potential parameters, even for a molecule as simple as N$_2$. This observation calls for prudence both in classifying parameters and in establishing a hierarchy of codes with respect to the quality of the prediction. The following section, in which results are compared, is instructive in this regard.

4. Results

4.1. Points of Comparison

As has been shown in §§XII.1–XII.3, the constituents of a numerical code for calculating explosives are multiple. It is therefore not surprising to find in specialized publications a great number of variants where it is often difficult to specify the precise differences, whether they stem from the thermochemical data or from the parameters.

Furthermore, few explosives have been the object of measurements of D^p which are precise "enough" and of large "enough" diameter, "enough" being understood to mean that they provide a good approximation of the quantity D^p_∞ to be used in the validity criterion in place of D^s_∞.

For these two reasons, rather than an undifferentiated comparison spread over a large number of codes and a large number of explosives, we prefer a comparison limited to:

- a few codes which have been fully described: FORTRAN BKW, LA MINEUR, TIGER BKW, TIGER JCZ 3, ETARC WCA4, CHEQ; and
- two well-defined explosives which have been intensively studied for more than forty years: PETN (solid) and nitromethane (liquid).

Several observations can be made when reading Tables XII.6 and XII.7, where results and references for PETN and nitromethane, respectively, are collected:

(1) the agreement of D_* and D^p_∞ to within 1 m/s given by LA MINEUR results not from a prediction in the strict sense, but from an *ad hoc* adjustment of parameters κ and β in the BKW estimation of $\mathscr{F}$: it is therefore not a proof of the *a priori* validity of the code, but the consequence of the procedure of using the code;

(2) without ad hoc adjustment of floating parameters the relative deviation

Table XII.6. Penta-erythritol tetranitrate (PETN) at different densities ρ_0.
Comparison between D^p ($\phi \geq 1$ in.) and D_*. Comparison of evaluations of p_* and T_*.

Experimental results			Results obtained from experimental codes				
ρ_0 (g/cm³)	D^p (mm/μs) $\phi \geq 1$ inch	Ref.	Code	D_* (mm/μs)	p_* (kbar)	T_* (K)	Ref.
0.95	5.33	[†]	CHEQ	5.15	67		[48b]
1	5.55	[18]	LA MINEUR	5.549	88	3984	[12]
			ETARC WCA4	5.494	80	4899	[16]
1.67	7.98	[≠]	FORTRAN BKW	8.056	280	3018	[39]
			LA MINEUR	7.979	264	2958	[12]
1.77 (crystal)	8.30	[†]	FORTRAN BKW	8.421	318	2833	[39]
			LA MINEUR	8.424	301	2757	[12]
			TIGER BKW	8.46	328	2602	[26]
			TIGER JCZ3	8.21	288	4237	[26]
			ETARC WCA4	8.416	315	4349	[16]
			CHEQ	8.33	289	4428	[48b]

[†] cited in [48b] in referring to [29].
[≠] cited in [39].

Table XII.7. Nitromethane at standard conditions ($p_0 = 1$ atm, $T_0 = 25°$ C). Comparison of values D_∞^p and D_*. Comparison of evaluations of p_* and T_*.

Experimental results at $T_0 = 25°$ C				Results obtained from calculation codes				
Parity	ρ_0 (g/cm^3)	D_∞^p (mm/μs)	Ref.	Code	D_* (mm/μs)	p_* (kbar)	T_* (K)	Ref.
>99%	1.131	6.29	[21]	FORTRAN BKW	6.390	130	3167	[39]
				LA MINEUR	6.290	124	3136	[12]
				TIGER BKW	6.84	144	2600	[26]
				TIGER JCZ3	6.11	119	3467	[26]
				ETARC WCA4	6.66	125	3388	[17]
				inverse method	—	126.5 ± 5.4	—	[21]
97%	1.127	6.25	[10]	Inverse method	—	118 à 3% près	—	[10]

between D^p (or D_∞^p) and D_* can reach 1.5%, while the relative error in measuring D^p is six times less;

(3) If the TIGER BKW evaluation (suspect in more than one way) is excepted, it may be seen that the different evaluations of p_* differ by ±5% from the LA MINEUR value, both for crystalline PETN and for 99% nitromethane; this relative difference reflects both the distance between D^p (or D_∞^p) and D_* and the high sensitivity of p* to a variation of D_* obtained by varying the floating parameters of the code;

(4) in the case of 99% nitromethane, there is excellent agreement between the value of p* obtained from LA MINEUR and that obtained from the inverse method; and

(5) the BKW evaluations of temperature T_* are systematically lower than *ab initio* evaluations by more than 1500 degrees for PETN, but little more than 150 degrees for nitromethane (again excepting the TIGER BKW value) without any apparent explanation for this discrepancy.

To summarize, it must be said that, in spite of considerable theoretical progress and almost unlimited digital processing capacities, the codes diverge considerably, even in the simplest cases. It is probable that unison will eventually be achieved, but the least that can be said is that the way in which it will be achieved has not yet been found. In any case, observations (1)–(5) above show the great advantage—*for the numerical evaluation of C–J state quantities*—of the careful use of floating code parameters so that the most directly and easily accessible experimental quantity, i.e., D^p or D_∞^p, may be "restored" by the code. They also explain why users of explosives, lacking a perfected tool, are forced to determine $(\mathscr{D})$ not point by point but by means of an analytical representation of $(\mathscr{D})$ suitable for numerical solving of equation (VIII.5).

4.2. Analytical Representation of $(\mathscr{D})$

The very structure of system (VIII.5) invites a search for a representation of $(\mathscr{D})$ in which the independent variables are specific volume v^b and pressure p^b, while the dependent variables are specific internal energy e^b and temperature T^b, in the *usual domain of thermodynamic changes in the detonation products*. According to the introductory statement in §1.1, this domain is a narrow interval of $(\mathscr{D})$ bounded by the isentrope $s = s_*$, framing the Chapman–Jouguet pressure p_*

$$s_* \leq s \ll 2s_*,$$

$$\frac{v_*}{2} \ll v < \infty. \qquad (XII.30)$$

This reference to a sufficiently well-defined domain explains the method generally used to construct the representation of $e^b(v^b, p^b)$ that will be discussed first, leaving examination of T^b for a later state. (N.B. For reasons of clarity, the superscipt b is omitted in the remainder of §4.2.)

The specific internal energy, the pressure, and the Grüneisen coefficient along the isentrope $s = s_*$ are designated $e_i(v)$, $p_i(v)$, and $G_i(v)$, respectively. In the vicinity of this line in $(\mathscr{D})$ and for a fixed v, the following expansion can be written

$$e(v, p) = e_i(v) + \frac{v}{G_i(v)}(p - p_i) + O(p - p_i)^2 \qquad (XII.31)$$

and it should also be noted that

$$e_i(v) - e_i^\infty = \int_v^\infty p_i(v)\, dv. \qquad (XII.32)$$

Equations (XII.31) and (XII.32) can be used to construct, for the interval (XII.30), a representation of the form

$$e(v, p) - e_i^\infty = \frac{v}{G_i(v)}(p - p_i) + \int_v^\infty p_i(v)\, dv, \qquad (XII.33a)$$

which can be conveniently transformed to

$$p = p_i(v) - \frac{G_i(v)}{v}\int_v^\infty p_i(v)\, dv + \frac{G_i(v)}{v}(e - e_i^\infty). \qquad (XII.33b)$$

Having to choose the function $G_i(v)$, Fickett and Wood [24], and a number of authors following them, chose to represent

$$G_i(v) \equiv \mathscr{G} = \text{constant},$$

which ignores the variation of G between its value G_* in the C–J state (0.5–0.7 according to the explosives) and its value $\gamma - 1$ at ordinary pressures.

However, for the function $p_i(v)$, the literature is characterized by a profusion of options and variants. It is left to the reader to study the numerous expressions; Table XII.8 shows only those which are demonstrably instructive:

- the term Av^{-Q} shows the affiliation to the polytropic model (see §II.4.1);
- the term $A'v^{-(1+\mathscr{G})}$ shows the affiliation to an ideal gas model with $\gamma = 1 + \mathscr{G}$;
- the exponential terms show the care taken to ensure agreement in conserving the computable character of the integral in (XII.33b); and
- the parameters are adapted in number to the available experimental data.

We will explain the last point and at the same time throw light on the last column "adjustment variance" in Table XII.8.

In a general way determining the parameters can result only from the local knowledge considered to be available as to the isentrope $s = s_*$. In this respect the C–J state itself is of first rank importance. In fact, Γ_* designating the local polytropic coefficient of the C–J state ($\Gamma_* = a_*^2/p_* v_*$), the jump relations (I.26b), and the Chapman–Jouguet relation lead to

$$
\left.
\begin{aligned}
\Gamma_* &= \frac{\rho_0 D_*^2}{p_* - p_0} - 1, \quad \text{(a)} \\[2mm]
v_* &= v_0 \frac{\Gamma_*}{\Gamma_* + 1}. \quad \text{(b)}
\end{aligned}
\right\}
\tag{XII.34}
$$

which may be used to determine Γ_* then v_* derived from D_* (estimated by D_∞^P) and p_* (estimated by the inverse method or by a numerical code). Then the identities

$$p_i(v_*) = p_*,$$
$$\dot{p}_i(v_*) = \Gamma_* p_* / v_*,$$

constitute two relations between the parameters. However, they may be completely determined only by an adjustment of states $s = s_*$ which are:

- close to or far from the C–J state according to the weight given to the pressures close to p_* and to the pressures close to the ambient pressure; and
- usually estimated by the method following from the experiment noted in §XI.2.1.

The representation of Fickett and Wood [24] constitutes an interesting limiting case: according to whether one chooses to use for $\mathscr{G}$ the value G_* (calculated by a numerical code) or the value $\gamma - 1$ of the ideal gas (with $\frac{7}{5} < \gamma < \frac{9}{7}$) it is obviously adapted to the study of the vicinity of $p = p_*$ or of that of the low one pressures (1 bar to a few tens of bars). But, *a contrario*, this example shows clearly that only the representations with an adjustment variance at least equal to two may be admitted in the entire domain defined by

Table XII.8. Principal expressions proposed for $p_i(v)$; corresponding expressions of $p - \mathscr{G}/v(e - e_i^\infty)$ according to (XII.33b).

$p_i(v)$	$p - \dfrac{\mathscr{G}}{v}(e - e_i^\infty)$ according to (XII.33b)	Reference	Parameters	Adjustment variance
Av^{-Q}	$A\left(1 - \dfrac{\mathscr{G}}{Q-1}\right)v^{-Q}$	[24]	A, Q, $\mathscr{G}$	1
$Av^{-Q} + A'v^{-(1+\mathscr{G})}$	$A\left(1 - \dfrac{\mathscr{G}}{Q-1}\right)v^{-Q}$	cited in [56]	A, A', Q, $\mathscr{G}$	2
$Av^{-Q} + A'v^{-(1+\mathscr{G})} + B\exp(-Rv)$	$A\left(1 - \dfrac{\mathscr{G}}{Q-1}\right)v^{-Q} + B\left(1 - \dfrac{\mathscr{G}}{Rv}\right)\exp(-Rv)$	[56]	A, A', B, Q, $\mathscr{G}$, R	4
$A'v^{-(1+\mathscr{G})} + \displaystyle\sum_{q=1,2} B_q\exp(-R_qv)$	$\displaystyle\sum_{q=1,2} B_q\left(1 - \dfrac{\mathscr{G}}{R_qv}\right)\exp(-R_qv)$	[33]	A', B_1, B_2, $\mathscr{G}$, R_1, R_2	4

(XII.30). The practical interest of the formulations of Wilkins, or Jones–Wilkins–Lee [33] is thus explained, although their defects cannot be concealed. To cite one example among others: elementary examination of the JWL expression for $p_i(v)$ shows that $\Gamma_i(v)$ tends to $1 + \mathcal{G}$ as much when v tends to zero as when v tends to infinity, and thus presents an extremum whose physical reality is, to say the least, doubtful. In any case, introduction of the exponential term by Wilkins et al. [56] is a response to a deliberate decision to introduce a maximum for $\Gamma(v)$ in the interval (v_*, ∞), based on experimental results which now appear very disputable. However that may be, with its two exponential terms, the JWL expression for $p_i(v)$ doubles the "risk" of a maximum. Effectively, for hexanitrobenzene, Ree [48a] reports the existence of two maxima and one minimum in the interval (v_*, ∞) of practical interest. It must therefore be noted that although the JWL expression of $p_i(v)$ has been adjusted to take account of the cylinder test, although the results of this test depend essentially on work produced by the expansion of detonation products, and although the function $\Gamma_i(v)$ is directly linked to this expansion, it cannot be expected that the JWL expression will provide an adequate physical variation of Γ_i which, it must be remembered, is a second derivative of the internal energy. In conclusion, it seems rather too weak to say that the problem of the representation $e(v, p)$ of $(\mathcal{D})$ in the domain (XII.30) is still open: a new way is shown by Chéret [11] and summarized in the following subsections.

We know three things about the variation of $\Gamma_i(v)$:

(i) at high dilutions, its value $1 + \mathcal{G}$ is that of an ideal gas composed of diatomic and triatomic molecules, and thus between $\frac{7}{5}$ and $\frac{9}{7}$;

(ii) at high compression, its value is that of an ideal gas composed of nucleons and electrons, and thus close to $5/3$; and

(iii) in the Chapman–Jouguet state, its value Γ_* does not differ from 3 by more than 20% [see, e.g., values in Tables XII.6 and XII.7 derived from (XII.34a)].

The simplest variation of Γ is thus of the shape indicated in Figure XII.1: monotonic increasing for $v \le \tilde{v}$, monotonic decreasing for $\tilde{v} \le v < +\infty$. Moreover, the very definition of Γ_i

$$\frac{dp_i}{dv} = -\Gamma_i \frac{p_i}{v},$$

and its immediate consequence

$$\frac{d^2 p_i}{dv^2} = -\frac{\Gamma_i}{v} \frac{dp_i}{dv} - \frac{p_i}{v} \frac{d\Gamma_i}{dv} + \Gamma_i \frac{p_i}{v^2},$$

show that the condition (II.16c) of positive curvature of the isentrope $p_i(v)$ is certainly satisfied in any interval where the function $\Gamma_i(v)$ is itself monotonic decreasing ($d\Gamma_i/dv < 0$). From this last observation and from Figure XII.2 the idea follows naturally of trying to find a representation of $(\mathcal{D})$ in the domain $(\tilde{v} < v < \infty, s_* \le 1 \le 2s_*)$ derived from a function $\Gamma_i(v)$ which will be

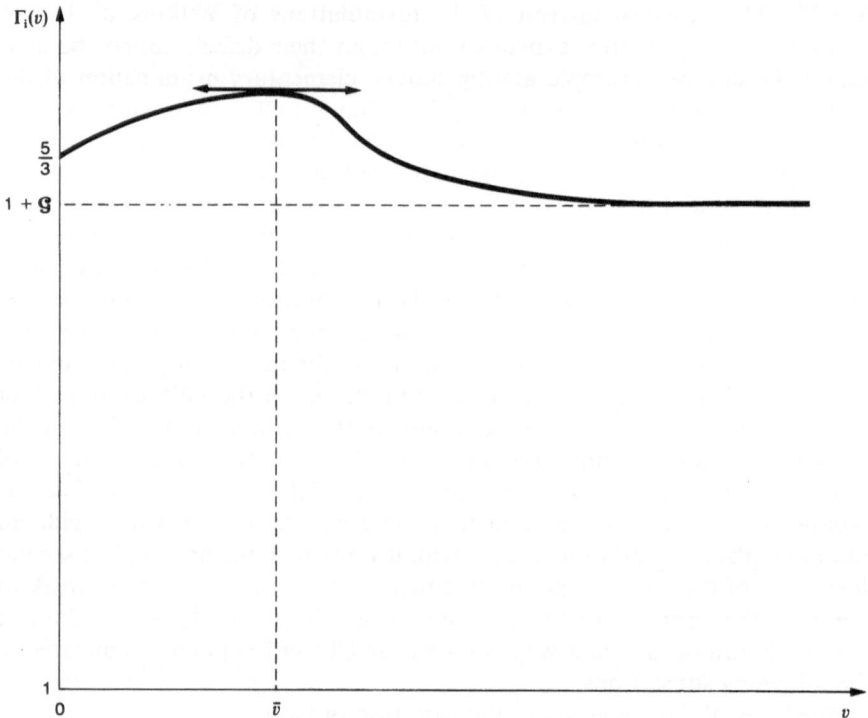

Figure XII.1. Diagram showing the simplest variation of the function $\Gamma_i(v)$.

a maximum in $v = \tilde{v} < v_*$ decreasing toward $\Gamma_\infty = 1 + \mathscr{G}$ as v increases from $\tilde{v}$ to infinity.

Before going further into the choice of $\Gamma_i(v)$, it is useful to note two consequences of such a monotonic variation of $\Gamma_i(v)$. In the first place, two elementary integrations show that the variation of specific internal energy can be put in the form

$$e_i(v) - e_i(\infty) = \frac{vp_i(v)}{\Gamma_i(v) - 1} + O(\tilde{\Gamma} - \Gamma_\infty), \qquad \text{(XII.35)}$$

formally identical to that of a polytropic fluid except for the term $O(\tilde{\Gamma} - \Gamma_\infty)$. Moreover the Riemann integral of the isentropic flow $s = s_*$

$$\sigma_i = -\int_\infty^v a_i \, d \operatorname{Log} v \equiv \int_\infty^v (\Gamma_i p_i v)^{1/2} \, d \operatorname{Log} v \qquad \text{(XII.36)}$$

can be put in the form

$$\sigma_i = \frac{2a_i}{\Gamma_i(v) - 1} + O(\tilde{\Gamma} - \Gamma_\infty),$$

formally identical to that of a polytropic fluid except for the term $0(\tilde{\Gamma} - \Gamma_\infty)$.

It remains to define the function $\Gamma_i(v)$. The inspiration for this is the analogy which the representation of Γ_i presents on the half-line $v \geq \tilde{v}$ with that of an intermolecular potential from the equilibrium distance to infinity. Subsection 2.3 showed two very widely distributed formulations: Mie type and the modified Buckingham type. For reasons for symmetry, the first type is taken here, in the form,

$$\Gamma_i(v) = 1 + \mathcal{G} + R \left[\frac{\ell + 1}{\left(1 + \text{Log}\,\frac{v}{\tilde{v}}\right)^{m+1}} - \frac{m + 1}{\left(1 + \text{Log}\,\frac{v}{\tilde{v}}\right)^{\ell+1}} \right],$$

$$R > 0; \quad \ell > m \geq 0.$$

(XII.37)

Supposing that $y = v/\tilde{v}$ and $\eta = 1 + \text{Log}\,v/\tilde{v}$, the following is deduced by differentiation

$$\frac{1}{(\ell + 1)(m + 1)R} \frac{d\Gamma_i}{dy} = \left(\frac{1}{\eta^{\ell+2}} - \frac{1}{\eta^{m+2}} \right) \frac{1}{y},$$

$$\frac{1}{(\ell + 1)(m + 1)R} \frac{d^2\Gamma_i}{dy^2} = \left[\left(\frac{m + 2}{\eta^{m+3}} - \frac{\ell + 2}{\eta^{\ell+3}} \right) + \left(\frac{1}{\eta^{m+2}} - \frac{1}{\eta^{\ell+2}} \right) \right] \frac{1}{y^2},$$

which show that $d\Gamma_i/dv$ is zero at the two ends of the interval $(\tilde{v}, +\infty)$, is negative everywhere else, and reaches a minimum for a unique value $\bar{\bar{v}}(\ell, m)$ such that

$$\frac{m + 2}{\bar{\bar{\eta}}^{m+1}} - \frac{\ell + 2}{\bar{\bar{\eta}}^{\ell+1}} + \frac{1}{\bar{\bar{\eta}}^m} - \frac{1}{\bar{\bar{\eta}}^\ell} = 0.$$

Using the same notation, it follows by integration

$$\frac{p_i}{\tilde{p}} = \begin{cases} y^{-(1+\mathcal{G})}\eta^{-R(\ell+1)} \exp\left\{ \frac{R}{\ell}(1 - \eta^{-\ell}) \right\} & \text{if } m = 0, \\[3mm] y^{-(1+\mathcal{G})} \exp\left\{ -R \left[\frac{\ell + 1}{m}\left(1 - \frac{1}{\eta^m}\right) - \frac{m + 1}{\ell}\left(1 - \frac{1}{\eta^\ell}\right) \right] \right\}, & \text{if } m > 0 \end{cases}$$

(XII.38)

where $\tilde{p}$ is the value which p_i takes in $v = \tilde{v}$. It is of little value to use (XII.37) and (XII.38) to explicit (XII.35). However, it is useful to note that its practical form is:

$$e_i - e_0 = e_i^\infty - e_0 + \frac{v p_i(v)}{\Gamma_i(v) - 1} + O(\tilde{\Gamma} - \Gamma_\infty),$$

where the variation $e_i^\infty - e_0$ of specific internal energy between the final substance relaxed infinitely at constant entropy $(s = s_*)$ and the initial substance is computable as a function of the C–J state

$$e_i^\infty - e_0 = e_* - e_0 - \frac{v_* P_*}{\Gamma_* - 1} + O(\tilde{\Gamma} - \Gamma_\infty).$$

This calculation is remarkably simple if p_0 is disregarded compared to p_*; in which case (see §III.5.3) the jump relations (I.26b) and the Chapman–Jouguet condition (II.20) can be written

$$\begin{cases} \dfrac{a_*}{D_*} = \dfrac{v_*}{v_0} = \dfrac{\Gamma_*}{\Gamma_* + 1}, \\[2ex] p_* = \rho_0 \dfrac{D_*^2}{\Gamma_* + 1}, \\[2ex] e_* - e_0 = \dfrac{p_* v_*}{2\Gamma_*}, \end{cases}$$

so that within $0(\tilde{\Gamma} - \Gamma_\infty)$

$$e_i^\infty - e_0 = -\frac{\Gamma_* + 1}{2\Gamma_*(\Gamma_* - 1)} p_* v_* = -\frac{p_* v_0}{2(\Gamma_* - 1)} = -\frac{D_*^2}{2(\Gamma_*^2 - 1)},$$

and finally

$$e_i(v) - e_0 = \frac{v p_i(v)}{\Gamma_i(v) - 1} - \frac{D_*^2}{2(\Gamma_*^2 - 1)} + 0(\tilde{\Gamma} - \Gamma_\infty). \qquad \text{(XII.39)}$$

Provided that, on the basis of (XII.38)

$$\lim_{v \to \infty} \frac{v p_i(v)}{\Gamma_i(v) - 1} = 0$$

a simple yet expressive result can be drawn from (XII.39): if the detonation products are expanded completely starting from the C–J state, then the internal energy of each unit of explosive mass will be reduced, compared to the initial state, by

$$\Delta e = \frac{D_*^2}{2(\Gamma_*^2 - 1)} + 0(\tilde{\Gamma} - \Gamma_\infty).$$

This result is of great practical significance: the quantity Δe gives, in some manner, a measure of the maximum specific energy to be expected from an explosive. In this sense, Δe or a conventional fraction of Δe can be chosen as specific energy e_X of the explosive in the so-called "with energy release in the mesh" variant of the computational algorithm for flow with detonation (see §VIII.3.2). Taking account of the orders of magnitude ($D_* \approx 8$ mm/μs, $\Gamma_* \approx 3$), the origin of the "tonne of TNT" as unit of energy that replaces the gigacalorie to measure the energy of a chemical or nuclear explosion is also explained in this way.

At this stage there are six parameters in the model; $\tilde{v}$, $\tilde{p}$, R, $\mathcal{G}$, ℓ, m. However, as for the other representations, two "a priori" relations generally exist among them which are provided by our knowledge of the C–J state: D_* estimated by D_∞^p, p_* estimated by the inverse method or by a numerical code.

Then the identities $p_i(v_*) = p_*$ and $\Gamma_i(v_*) = \Gamma_* \equiv (\rho_0 D_*^2/p_*) - 1$ allow R and $p_*/\tilde{p}$ to be expressed as functions of $\mathscr{G}, y_*, \ell, m$:

$$R = (\Gamma_* - 1 - \mathscr{G})\left(\frac{\ell + 1}{\eta_*^{m+1}} - \frac{m + 1}{\eta_*^{\ell+1}}\right)^{-1},$$

$$\frac{p_*}{\tilde{p}} = \begin{cases} y_*^{-(1+\mathscr{G})} \cdot \eta_*^{-R(\ell+1)} \exp\left\{\frac{R}{\ell}(1 - \eta_*^{-\ell})\right\} & \text{if } m = 0, \\[3mm] y_*^{-(1+\mathscr{G})} \exp\left\{-R\left[\frac{\ell + 1}{m}\left(1 - \frac{1}{\eta_*^m}\right) - \frac{m + 1}{\ell}\left(1 - \frac{1}{\eta_*^\ell}\right)\right]\right\} & \text{if } m > 0, \end{cases}$$

$$\text{(XII.40)}$$

so that the adjustment variance is reduced to four if $m \neq 0$, to three if $m = 0$.

Study of the variation of $\Gamma_i(v)$ shows that choosing $m = 0$ has the advantage of simplicity and of separating the floating parameters:

- $\mathscr{G}$ controls the representation of $(\mathscr{D})$ outside the isentrope;
- ℓ controls the variation of $\Gamma_i(v)$ in the vicinity of $\tilde{v}$ and at infinity; and
- y_* controls the position of the C–J state with respect to $\bar{v}/\tilde{v}$, i.e., with respect to the minimum of $d\Gamma_i/dv$;

which supports a reasoned adjustment procedure for experiments (see §XI.2.1) where a "plane" detonation breaks frontally in different media. For the first of these, the experimental pairs (material velocity u, pressure p) are to be compared with the "model" pairs defined by the expressions (XII.36) for σ_i, (XII.37) for Γ_i, (XII.38) for p_i, and the Riemann equation

$$u - u_* + \sigma_i - \sigma_* = 0. \tag{XII.41}$$

And what about representation $p(v^b, T^b)$, also required for the complete solution of equation (VIII.5)? This question is seen to be very different from the preceding inasmuch as it has already been clearly grasped when discussing the estimation of the function $\mathscr{F}$, defined in §1.3 by the equation (XII.5) rewritten below in the form

$$p = \frac{N'\mathfrak{R}T}{v'}\mathscr{F}(v', T, .N^i.).$$

However, it does not amount to this, insofar as an attempt is now being made to find a representation of $(\mathscr{D})$ and therefore an expression in which neither the species composition nor the phase composition appears. Trials in this area are few. That of F. Chaissé et al. [to be published] is distinguished by its coherence: it relies on the formalism of the virial and seeks p in the form

$$\frac{pv}{cT} = 1 + \frac{b_{(1)}(\overline{T})}{\bar{v}} + \frac{b_{(2)}(\overline{T})}{\bar{v}^2} + \frac{b_{(3)}(\overline{T})}{\bar{v}^3} + \frac{b_{(4)}(\overline{T})}{\bar{v}^4},$$

in which appear:

- $b_{(i)}$ functions (see, e.g., Hirschfelder [28]) which are known if an effective intermolecular potential for the equivalent monofluid is defined; and
- reduced variables

$$\bar{v} = kv \bigg/ \sqrt{\frac{2\pi}{3} c \underline{r}^3},$$

$$\bar{T} = kT/\underline{\varepsilon},$$

where $\underline{r}$ and $\underline{\varepsilon}$ are the parameters of the L.-J. and D. potential (XII.23a), and c is a parameter whose dimension is that of a specific heat.

4.3. Singular Lines on ($\mathscr{D}$)

Among the causes of imperfection in the numerical predictions, there is one not yet mentioned, whose nature is of a theoretical order and whose occurrence is linked to the existence on ($\mathscr{D}$) of singular lines on crossing which the number of phases "jumps" by one unit. To approach this difficulty, a thermodynamic identity is necessary, which will be established in the following paragraph before two examples are considered.

Consider a fluid (specific volume, v; specific entropy, s; specific internal energy, e) in the vicinity of a line $\mathscr{L}$ of the surface of state separating a phase α from a homogeneous mixture of two phases α and β in homobaric–homothermal equilibrium (pressure, p; temperature, T). Denote by ε and $1 - \varepsilon$ the specific proportions of β and α and choose v and T as normal variables: the hypotheses of homogeneity and homobaric–homothermal equilibrium imply

$$\left.\begin{aligned}
v &= \varepsilon v^\beta(v, \text{T}) + (1 - \varepsilon)v^\alpha, && \text{(a)} \\
e(v, \text{T}) &= \varepsilon E^\beta(v^\beta, \text{T}) + (1 - \varepsilon)E^\alpha(v^\alpha, \text{T}), && \text{(b)} \\
p(v, \text{T}) &= p^\alpha(v^\alpha, \text{T}). && \text{(c)}
\end{aligned}\right\} \qquad \text{(XII.42)}$$

We wish to find the sign of the jump in the slope of an isentrope at the junction of $\mathscr{L}$, i.e., the difference

$$\frac{\partial p}{\partial v}(v, s) - \frac{\partial p^\alpha}{\partial v^\alpha}(v^\alpha, s^\alpha).$$

For this we evaluate each term of the difference by making use of identities $(j.4)$ and $(i.3)$ of Appendix A

$$\frac{\partial p}{\partial v}(v, s) = \frac{\partial p}{\partial v}(v, \text{T}) - \frac{\text{T}}{c_v}\left[\frac{\partial p}{\partial \text{T}}(v, \text{T})\right]^2,$$

$$\frac{\partial p^\alpha}{\partial v^\alpha}(v^\alpha, s^\alpha) = \frac{\partial p^\alpha}{\partial v^\alpha}(v^\alpha, \text{T}) - \frac{\text{T}}{c_v^\alpha}\left[\frac{\partial p^\alpha}{\partial \text{T}}(v^\alpha, \text{T})\right]^2,$$

then, like Bethe (see Ref. [4] in Part One) we form

$$c_v \left[\frac{\partial p}{\partial v}(v, s) - \frac{\partial p^\alpha}{\partial v^\alpha}(v^\alpha, s^\alpha) \right] = c_v \frac{\partial p}{\partial T}(v, T) - T \left[\frac{\partial p}{\partial T}(v, T) \right]^2$$

$$- c_v \left[\frac{\partial p^\alpha}{\partial v^\alpha} - \frac{T}{c_v^\alpha} \left(\frac{\partial p^\alpha}{\partial T} \right)^2 \right] \qquad \text{(XII.43)}$$

and finally eliminate the quantities referring to the mixture in the right-hand side of the equation, i.e., $\partial p/\partial T$ and c_v. The first, because of (XII.42c), can be written

$$\frac{\partial p}{\partial T}(v, T) = \frac{\partial p^\alpha}{\partial T}(v^\alpha, T) + \frac{\partial p^\alpha}{\partial v^\alpha}(v^\alpha, T) \cdot \frac{\partial v^\alpha}{\partial T}(v, T). \qquad \text{(XII.44)}$$

The second, by making use of (XII.42a and b) and of $\varepsilon = 0$, can be written

$$c_v = c_v^\alpha + \left[\frac{\partial e^\alpha}{\partial v^\alpha} - \frac{e^\beta - e^\alpha}{v^\beta - v^\alpha} \right] \frac{\partial v^\alpha}{\partial T},$$

which, from the Clapeyron formula

$$\frac{e^\beta - e^\alpha}{v^\beta - v^\alpha} = T \frac{\partial p}{\partial T}(v, T) - p,$$

and the identity $(f.4)$ in Appendix A, becomes

$$c_v = c_v^\alpha + T \left[\frac{\partial p^\alpha}{\partial T} - \frac{\partial p}{\partial T} \right] \frac{\partial v^\alpha}{\partial T};$$

that is, by using (XII.44),

$$c_v = c_v^\alpha - T \frac{\partial p^\alpha}{\partial v^\alpha} \left(\frac{\partial v^\alpha}{\partial T} \right)^2.$$

Finally, the right-hand side of equation (XII.43) is written as the sum of three terms

$$c_v \frac{\partial p}{\partial T} - T \left[\frac{\partial p^\alpha}{\partial T} + \frac{\partial p^\alpha}{\partial v^\alpha} \cdot \frac{\partial v^\alpha}{\partial T} \right]^2 - \left[c_v^\alpha - T \frac{\partial p^\alpha}{\partial v^\alpha} \left(\frac{\partial v^\alpha}{\partial T} \right)^2 \right] \left[\frac{\partial p^\alpha}{\partial v^\alpha} - \frac{T}{c_v^\alpha} \left(\frac{\partial p^\alpha}{\partial T} \right)^2 \right].$$

Rearranging the brackets leads to considerable simplification and finally to

$$c_v \frac{\partial p}{\partial v}(v, T) - c_v^\alpha \frac{\partial p^\alpha}{\partial v^\alpha}(v^\alpha, T) \left[1 + \frac{T}{c_v^\alpha} \frac{\partial p^\alpha}{\partial T}(v^\alpha, T) \cdot \frac{\partial v^\alpha}{\partial T}(v, T) \right]^2, \qquad \text{(XII.45)}$$

all the quantities being taken on the line $\mathscr{L}$.

In the case considered by Bethe of a single chemical species, the variance is 1 in the two-phase region, so that p does not depend on v and the first term of (XII.45) vanishes: it is then a simple matter to decide on the sign of the jump in the slope of the isentrope

$$\frac{\partial p}{\partial v}(v, s) - \frac{\partial p^\alpha}{\partial v^\alpha}(v^\alpha, s^\alpha) > 0, \tag{XII.46}$$

and thus to define a situation where condition (B.1) (II.16a) is violated: when the isentropic *compression* leads from region α to region $(\alpha + \beta)$, $\partial^2 p/\partial v^2(v, s)$ is infinitely negative on $\mathscr{L}$.

In the case considered of a fluid consisting of detonation products, the situation is less clear; the expression (XII.45) only allows us to think that a violation of condition (B.1) may not be excluded. Two examples prove the reality of this possibility.

The first, and the one studied first (see Chéret [13]) concerns explosives whose detonation products comprise a gas phase (H_2O, CO, CO_2, N_2, etc.), with or without a solid phase (carbon) depending on the region of $(\mathscr{D})$. For PETN, the sign of $\partial^2 p/\partial v^2(v, s)$ was studied directly by Fickett [25], who calculated the isentropes numerically and observed that *compression* is accompanied by a positive jump of $\partial p/\partial v(v, s)$—thus a violation of (B.1)—on the line where carbon appears, as in the case considered by Bethe with regard to a single chemical species. It must therefore be expected that the (H_+) arc of the PETN does not uniformly satisfy the properties demonstrated in Section II.2 under the (B.1) condition. Confirmaton of this assumption is found in the numerical results obtained with the LA MINEUR code. Figure XII.2(a), taken from [13], demonstrates that:

(a) the curve representing the variation of detonation velocity D as a function of v along (H) shows a kink at its intersection with the line where carbon disappears; and

(b) there is an interval of variation of the density ρ_0 for which two C–J states coexist on (H_+)—one situated in $(\mathscr{D}; c > 0)$, the other in $(\mathscr{D}; c < 0)$—separated by the kink mentioned in (a).

Figure XII.2(b) specifies the variation of the C–J characteristics as a function of ρ_0.

The second example, more recently evoked, (see Ree [48d]), concerns explosives whose detonation products comprise states cold enough and compressed enough for preferential *segregation* of nitrogen in gas phase to occur during isentropic *compression*. Having modified the CHEQ code to take account of such a possibility, Ree specially considered the cases of cyclonite (RDX) and calculated D_* from $\rho_0 = 0.6$ to $\rho_0 = 1.8$ g/cm^3. Figure XII.3, taken from [48d], shows that these values are everywhere in agreement with the experimental values found by a number of authors, which is an argument in favor of the segregation hypothesis. It also shows the existence of two kinks at $\rho_0 = 1.2$ g/cm^3 and $\rho_0 = 1.6$ g/cm^3, and a correlation of these kinks with the intervals where the C–J state splits on (H_+) in the vicinities, respectively, of the line where carbon appears and the nitrogen segregation line, according to the mechanism shown in Figure XII.2(a) for PETN.

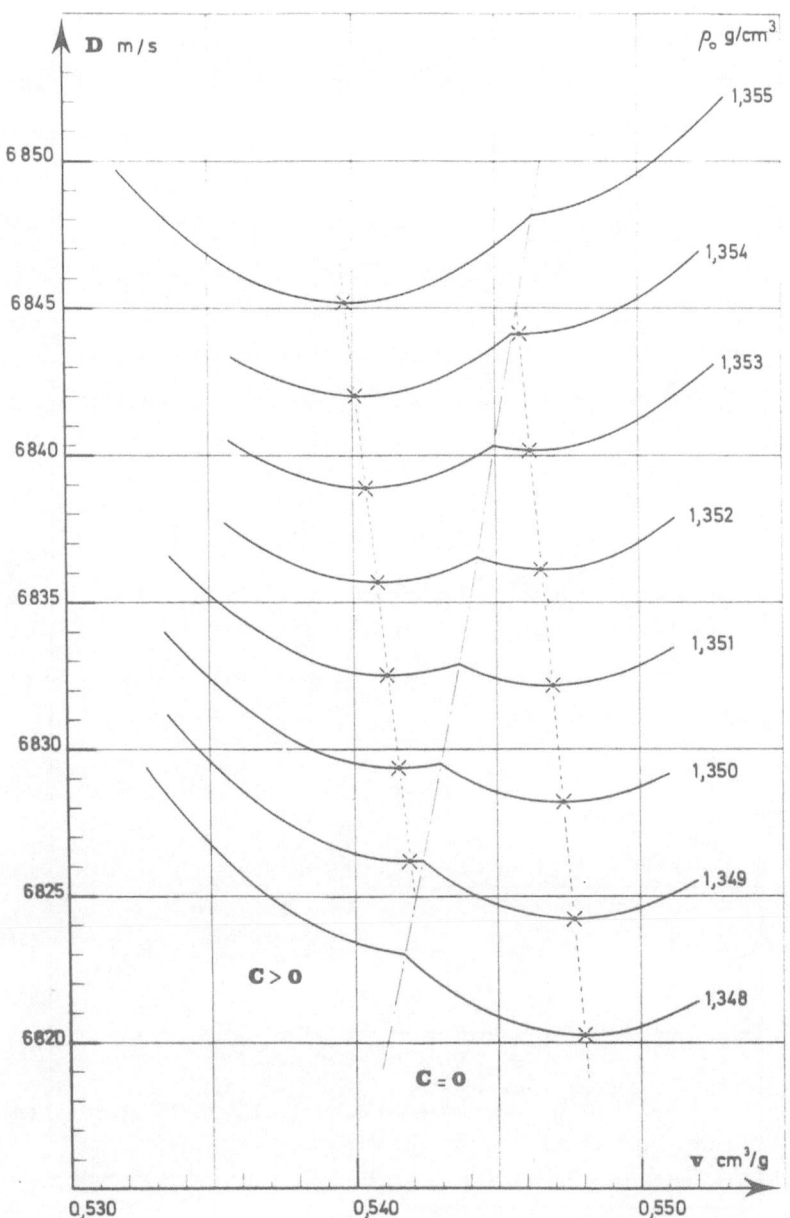

Figure XII.2(a). Penta-erythritol tetranitrate (PETN). Crussard curve in the vicinity of C–J detonations. Variation of detonation velocity with specific volume.

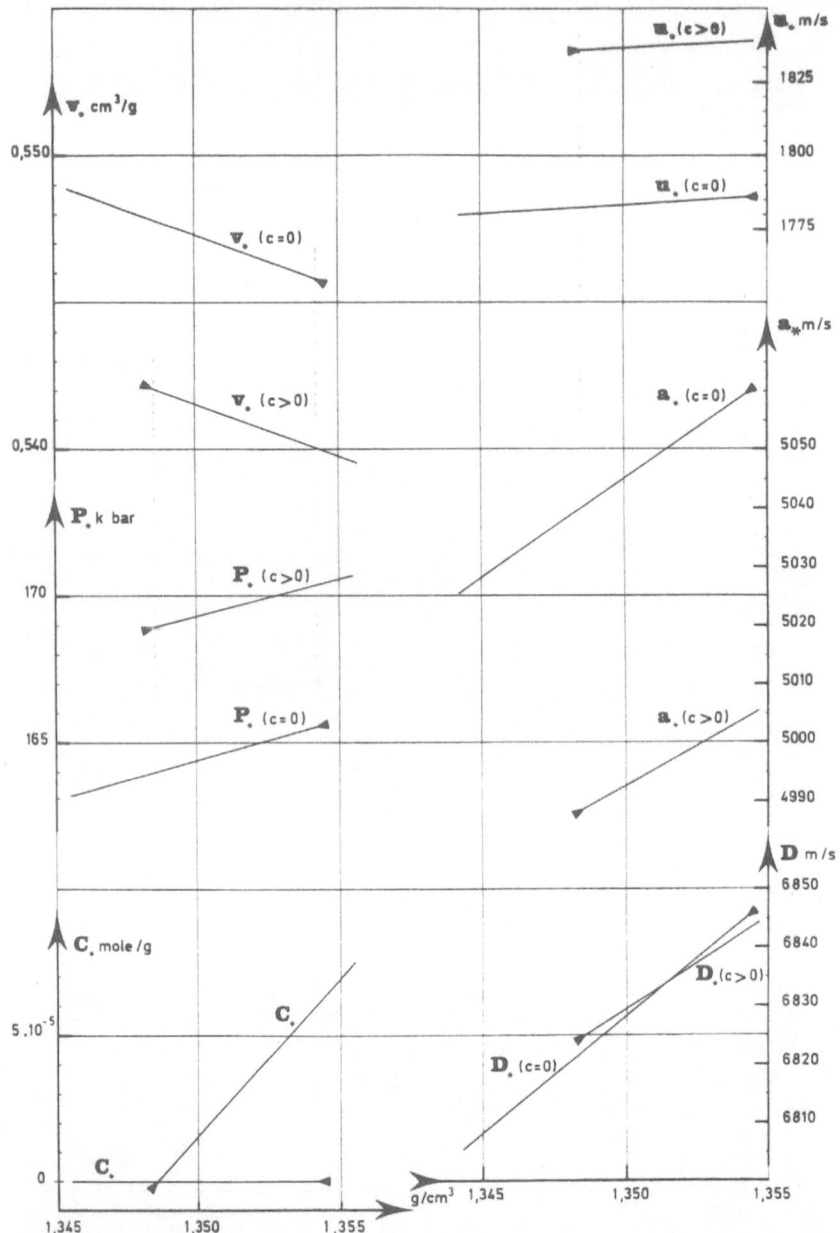

Figure XII.2(b). Variation of the C–J characteristics with the initial density of penta-erythritol tetranitrate (PETN).

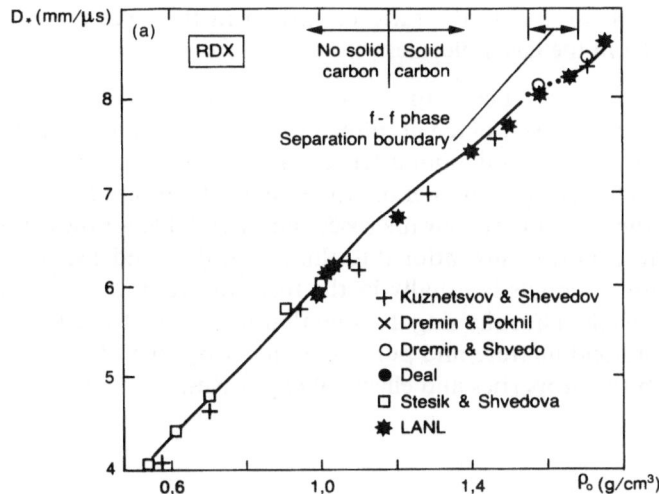

Figure XII.3. C–J detonation velocity D_* for cyclonite (RDX) for different densities ρ_0. Comparison between experimental estimations (different symbols as shown) and values calculated at the absolute minimum velocity (continuous curve where the dashed part corresponds to the uncertainty interval in the position of the segregation line f–f) (after [48d]).

The same remark is true for both examples cited. The violation of the condition (B.1) leads to the coexistence of two C–J states in only a very small interval of variation of ρ_0, the differences in velocity being a few parts per thousand, while the differences in pressures amount to a few per cent. This "indeterminacy" of the C–J state, taking account of the other sources of error fully discussed previously, is thus without serious practical consequence. However, it does pose a difficult theoretical problem:

When two C–J states coexist, which one is actually produced in a quasi C–J detonation?

whose solution might throw light on an aspect of detonations not approached in this work, that of their stability.

Finally, we return to a question already reviewed twice (§1.2 and §2.5), that of the crystalline phase of solid carbon. For this, it is necessary to start by recalling experimental details familiar to flame specialists:

(a) the stage preliminary to the formation of soot is the formation of aromatic hydrocarbons $C_6H_5^-$ or C_6H_6 derived from polyacetylenes $(C_2H_2)_n$;

(b) the principal stage is the polymerization of the aromatic molecules with elimination of hydrogen, resulting in the formation of a graphite structure containing peripheral hydrogen atoms as well as free links for interaction with the ambient gas phase; and

(c) pressure changes only the ratio of carbon in the solid phase to carbon combined in the gas molecules.

Moreover, it is instructive to consider a recent calculation by Ree [48e] who shows that a cluster of ten graphite atoms is more stable than these same ten atoms arranged in a diamond lattice and that this comparative stability only reverses for a larger number of atoms. With these conditions, and taking account of the short time (a few microseconds) available for the carbon atoms to aggregate, it now seems rational to think that the condensation of carbon in detonation products is usually in the form of graphitic soot rather than diamond crystals. Thus the singular line that marks the transition between a graphite/diamond mixture and pure diamond will generally be without consequence on the properties and effects of explosives.

References

[1] ALLAN, J.W.S., LAMBOURN, B.D. An equation of state of detonation products at pressures below 30 kbar. *Proc. 4th Symposium on Detonation*, White Oak/MD (1965), p. 52.

[2] ANDERSEN, H.C., WEEKS, J.D., CHANDLER, D. Relationship between the hard-sphere fluid and fluids with realistic repulsive forces. *Phys. Rev. A*, **4** (1971), p. 1597.

[3] BARKER, J.A., HENDERSON, D.
   a. Perturbation theory and equation of state for fluids I. The square well potential. *J. Chem. Phys.*, **47** (1967), p. 2856.
   b. Perturbation theory and equation of state for fluids. II. A successful theory of liquids. *J. Chem. Phys.*, **47** (1967), p. 4714.
   c. Perturbation theory of fluids at high temperatures. *Phys. Rev. A*, **1** (1970), p. 1266.

[4] BAUER, P.A., VIDAL, P., MANSON, N., HEUZÉ, O. An approach to the determination of gaseous explosive mixtures at elevated initial pressures by means of the inverse method. *Proc. 11th Colloquium on Dynamics of Explosions and Reactive Systems*, Varsovie (1987).

[5] BAUTE, J., CHIRAT, R. Which equation of state for carbon in detonation products? *Proc. 8th Symposium on Detonation*. Albuquerque/NM (1985), p. 287.

[6] BECKER, R.
   a. *Z. Physik*, **4** (1921), p. 393.
   b. *Z. Physik*, **8** (1922), p. 321.
   c. *Z. Technische Physik*, **3** (1922), p. 249.

[7] BERGER, J., FAVIER, J., NAULT, Y. Détermination des caractéristiques de détonation des explosifs solides. *Ann. Phys.* 13th séries (1960), pp. 771–803, 1144–1176.

[8] BRINKLEY, S.R. JR. AND WILSON, E.B.
   a. OSRD Report no. 905 (1942).
   b. OSRD Report no. 1707 (1943).

[9] BRINKLEY, S.R. JR. Calculation of the equilibrium composition of systems of many constituants. *J.Chem. Phys.*, **15** (1947), p. 107.

[10] BROCHET, C., PRESLES, H.-N., CHERET, R. Detonation characteristics of some liquid mixtures of nitromethane and chlorform or bromoform. *Proc. 15th Symposium (International) on Combustion*, Pittsburgh/PA (1974), p. 29.

[11] CHÉRET, R. Représentation quasi polytropique de la surface d'état des produits de détonation d'un explosif condensé. *C. R. Acad. Sci. Paris*, **305**, Series II (1987), p. 1337.

[12] CHERET, R. Contribution à l'étude numérique des produits de détonation d'une substance explosive. Rapport CEA-R-4122 (1971).

[13] CHERET, R. Sur un cas où l'unicité de la détonation de Chapman–Jouguet est en défaut. *C. R. Acad. Sci. Paris*, **274** (1972), p. 1347.

[14] CHERET, R. Le code ARPÈGE; application à l'étude d'un explosif à l'aluminium. *Acta Astronautica* **1** (1974), p. 893.

[15] CHIRAT, R., PITTION-ROSSILLON, G. A new equation of state for detonation products. *J. Chem. Phys.*, **74** (1981), p. 4634.

[16] CHIRAT, R., PITTION-ROSSILLON, G. Detonation properties of condensed explosives calculated with an equation of state based on intermolecular potentials. *Proc. 7th Symposium on Detonation.* Annapolis/MD (1981), p. 703.

[17] CHIRAT, R., BAUTE, J. An extensive application of WCA4 equation of state for explosives. *Proc. 8th Symposium on Detonation.* Albuquerque/NM (1985), p. 94.

[18] COOK, M.A. *et al.* Velocity-diameter curves, velocity transients and reaction rates in PETN, RDX, EDNA and tetryl. *J. Amer. Chem. Soc.*, **79** (1957), p. 32.

[19] COWAN, R.D., FICKETT, W. Calculation of the detonation properties of solid explosives with the Kistiakowsky–Wilson equation of state. *J. Chem. Phys.*, **24** (1956), p. 932.

[20] COWPERTHWAITE, M., ZWISLER, W.H.
   a. Theoretical and mathematical formulations for the TIGER computer program. SRI Publications, no. Z106, January 1973.
   b. The JCZ equation of state for detonation products and their incorporation in the TIGER Code. *Proc. 6th Symposium on Detonation*, San Diego/CA (1976), p. 162.

[21] DAVIS, W.C. Failure of Chapman–Jouguet theory for solid and liquid explosives. *Phys. Fluids*, **8** (1965), p. 2169.

[22] DORN, W.S. Variational principles for chemical equilibrium. *J. Chem. Phys.*, **32** (1960), p. 1490.

[23] EDWARDS, J.C., CHAIKEN, R.F. Detonation calculations with Percus–Yevick equation of state. *Combustion and Flame*, **22** (1974), p. 269.

[24] FICKETT, W., WOOD, W.W. A detonation products equation of state obtained from hydrodynamic data. *Phys. Fluids*, **1** (1958), p. 528.

[25] FICKETT, W. Detonation properties of condensed explosives calculated with an equation of state based on intermolecular potentials. LASL Report LA 2712 (1962).

[26] FINGER, M. *et al.* The effect of elemental composition on the detonation behaviour of explosives. *Proc. 6th Symposium on Detonation*, San Diego/CA (1976), p. 710.

[27] HEUZÉ, O. Contribution au calcul des caractéristiques de détonation de substances explosives gazeuses ou condensées. Thèse de Doctorat, Poitiers (1985).

[28] HIRSCHFELDER, J.O., CURTISS, C.F., BIRD R. BYRON. *Molecular Theory of Gases and Liquids.* Wiley, New York (1954).

[29] HORNIG, H.C. *et al.* Equation of state of detonation products. *Proc. 5th Symposium on Detonation.* Pasadena/CA (1970), p. 503.

[30] JACOBS, S. On the equation of state for detonation products at high density. *Proc. 12th Symposium on Combustion,* Poitiers (1968), p. 501.

[31] JANAF. *Thermochemical Tables.* Dav. Chemical Company, Midland, Michigan.

[32] KISTIAKOWSKY, G.B. AND WILSON, E.B.
    a. Report on the prediction of detonation velocities of solid explosives. O.S.R.D. Report 69 (1941).
    b. The hydrodynamic theory of detonation and shock waves. O.S.R.D. Report 114 (1941).

[33] LEE, E.L., HORNIG. H.C., KURY, J.W. Adiabatic expansion of high explosive detonation products. UCRL Report no. 504122 (1968).

[34] LEE, L.L., LEVESQUE, D. Perturbation theory for mixtures of simple liquids. *Molecular Phys.,* **26** (1973), p. 1351.

[35] LENNARD-JONES, J.E., DEVONSHIRE, A.F. Critical phenomena in gases. *Proc. Roy. Soc. A,* **163** (1937), p. 53; **165** (1938), p. 1; **169** (1938), p. 317; **170** (1939), p. 464.

[36] LEVINE, H.B., SHARPLES, R.E. Operator's manual for RUBY. Report UCRL no. 6815 (1962).

[37] LEVINE, H.B. Final report on the method of univariant descent for solving problems in heterogeneous chemical equilibria. JAYCOR Report, J510-82-008154 (1982).

[38] MADER, C.L.
    a. Detonation performance calculations using the Kistiakowsky–Wilson equation of state. LASL Report LA 2613 (1961).
    b. STRETCH BKW. A code for computing the detonation properties of explosives. LASL Report LADC 5691 (1962).
    c. Detonation properties of condensed explosives computed using the Becker–Kistiakowsky–Wilson equation of state. LASL Report LA-2900 (1963).
    d. FORTRAN BKW. A code for computing the detonation properties of explosives. LASL Report LA 3704 (1967).

[39] MADER, C.L. *Numerical Modeling of Detonations.* Los Alamos Series, University of California (1979).

[40] MANSON, N. Détermination par la méthode inverse des caractéristiques des ondes explosives. Publication no. 366 du Ministère de l'Air, Paris (1960).

[41] MANSOORI, G.B., CANFIELD, J.B. Variational approach to the equilibrium thermodynamic properties of simple liquids. *J. Chem. Phys.,* **51** (1969), p. 4958.

[42] MANSOORI, G.A., CARNAHAM, N.G., STARLING, K.E., LELAND, T.W. Equilibrium properties of the mixture of hard spheres. *J. Chem. Phys.,* **54** (1971), p. 1523.

[43] MEDARD, L. Tables thermochimiques à l'usage des techniciens des substances explosives. *Memorial de l'Artillerie Française,* **28** (1954), p. 415.

[44] MEDARD, L. Les explosifs occasionnels. Librairie Lavoisier, Technique et Documentation, Paris (1987).

[45] MURNAGHAN, F.D. *Finite Deformation of an Elastic Solid.* Wiley, New York (1951).

[46] PERCUS, J.K., YEVICK, G.J. Analysis of classical statistical mechanics by means of collective coordinates. *Phys. Rev.,* **110** (1958), p. 1.

[47] RASAIAH, J.C., STELL, G. Upper bounds of free energies in terms of hard spheres results. *Molecular Phys.,* **18** (1970), p. 249.

[48] REE, F.H. (and VAN THIEL M. for c.).

a. Post detonation behavior of condensed high explosives by modern methods of statistical mechanics. *Proc. 7th Symposium on Detonation*, Annapolis/MD (1968), p. 646.

b. A statistical theory of chemically reacting multiphase mixture: application to the detonation properties of PETN. *J. Chem. Phys.*, **81** (1984), p. 1251.

c. Detonation behavior of LX-14 and PBX 9404: theoretical aspect. *Proc. 8th Symposium on Detonation*, Albuquerque/NM (1985), p. 501.

d. Supercritical fluid phase separation: implications for detonation properties of condensed explosives. *J. Chem. Phys.* **39** (1963), p. 474.

e. Phase changes and chemistry at high pressures and temperature. *Proc. 5th APS Conference on Shock Waves in Condensed Matter*, Monterey/CA (1987).

[49] Ross, M. A high-density fluid perturbation theory band on an inverse 12th-power hard-sphere reference system. *J. Chem. Phys.*, **71** (1979), p. 1567.

[50] Thiele, E. Equation of state for hard spheres. *J. Chem. Phys.*, **39** (1963), p. 474.

[51] Verlet, L., Weiss, J.-J. Equilibrium theory of simple liquids. *Phys. Rev. A*, **5** (1972), p. 939.

[52] Vidart, A. Calcul des caractéristiques de détonation des explosifs condensés. Mémorial des Poudres, XLII (1960), p. 83.

[53] Weeks, J.D., Chandler, D., Andersen, H.C. Role of repulsive forces in determining the equilibrium structure of simple liquids. *J. Chem. Phys.*, **54** (1971), p. 5237.

[54] Wertheim, M.S. Exact solution for the Percus–Yevick integral equation for hard spheres. *Phys. Rev. Lett.*, **10** (1963), p. 321.

[55] White, W.B., Johnson, S.M., Dantzig, C.B. Chemical equilibrium in complex mixtures. *J. Chem. Phys.*, **28** (1958), p. 751.

[56] Wilkins, M.L., Sqier, B., Halperin, B. The equation of state of PBX 9404 and LX04-01. UCRL Report no. 7797 (1964).

Epilogue

The thermodynamic and mechanical aspects of the propagation of detonations (Part One), and the molecular mechanisms of explosive decomposition (Part Two), lie at the two extreme ends of the scale of explosive phenomena. The macroscopic aspects of the generation of detonation (Part Three), and the dynamic characterization of explosives (Part Four), are the two poles of interest for the "practitioner" of explosives. Ordering the work in such a way shows concern for demonstrating that:

- thorough understanding of detonation proceeds via a modeling where the existence of multiple scaling factors is fundamental, and where the continuum and molecular descriptions back up one another; and
- good practice in the use of explosives is not restricted to access to the appropriate numerical and experimental tools, but also requires preliminary familiarization with the theoretical results.

While the Summary given at the beginning was content to give the tenor of the chapters, this Epilogue underlines the spirit in which they were written, and attempts to lay a firm foundation for the answers which this work proposes to certain questions that have been debated for almost an entire century.

Since the publications of Chapman in 1899 and Jouguet in 1917, their names are found associated to denote a condition which neither of them had pretended to demonstrate. In this regard, it is instructive to cite the beginning of §216 (page 315) of *Mécanique des Explosives*. "If the preceding considerations are exact ..." After reading the entire text, there is scarcely any doubt that neither Jouguet, nor *a fortiori* Chapman (for whom the argumentation does not exist) had a clear vision of the condition known today as C–J, and neither of them ascribed to it—in every causal state—the quasi-magical virtue which some successors, less anxious about exactness than about simplicity, have ascribed in the notes and papers published in the 1950s.

However, the principle of a rigorous analysis exists from 1944 in a note by H. Weyl which, in order to describe a "shock layer" utilizes a procedure related to that used by Prandtl at the start of the century to describe a "boundary layer." We had to wait again until it was formalized in the current method of asymptotic expansions, *à propos* of more or less simple cases of aerodynamics, before imagining its extension to the mechanics of reactive fluids, as is performed in Part One of this work.

To model the "detonation layer" as a zone of steep gradients in a reactive dissipative fluid allows us to renew from top to bottom the description of the phenomena. Thus the idea of a uniform and permanent velocity is excluded in favor of a local and instantaneous velocity. Thus appears the notion of *quasi C–J detonation*, defined as a propagation which scarcely deviates in velocity (in time and space) from the C–J value, and moreover deviates all the less as the value of the local curvature is less. Thus too the propagation rules are established over an extremely wide domain of validity since the only requisite conditions are: uniformity and rest of the initial state (*simple* detonation), the lack of any compression induced in the detonation products by the boundary conditions of the flow of these products (*autonomous* detonation).

Owing to these definitions and concepts, the problems of the spherically diverging wave, the spherically converging wave, and the autonomous permanent axisymmetric wave are solved completely. For this last geometry, which is that of a very large number of experiments carried out over the course of a century, two essential results are established. First, it is shown that it is the *autonomous* condition which is fundamental to the existence of an eigenvalue of the propagation velocity for given diameter, explosive and adjacent medium. We also show the concurrent character of two critical events well known to the experimenter: pure and simple extinction on one hand, and bifurcation towards a three-dimensional structure on the other.

The reader familiar with the standard works on detonation will doubtless be astonished that nowhere is the concept of the "reaction zone" invoked. There are two reasons for this absence—which is neither an error nor an omission. First, there is the ambiguity of the expression: does this zone terminate when the initial substance has completely "disappeared," or whenever the chemical reactions "cease" in the very core of the detonation products? Above all there is the incompatibility of the concept with the method itself of matched asymptotic expansions, the foundations of which aim precisely at substituting for an ill-posed problem of *linking up* on imaginary surfaces, a well-posed problem of *matching* between dissipative flow with steep gradients and nondissipative flow with negligible gradients.

On a completely different tone—the explanation and the forecasting of sensitivity to "shock" of explosive molecules—Part Two of this book jostles equally with the received ideas. The search for criteria, which was still at the patchwork stage because it was limited to looking in the core of the properties of a molecule in its fundamental state, is renewed by taking account of the properties of the molecule after "shock."

Experimental agreement, the unification of primary and secondary explosives, and more generally the coherence of the consequences, are further arguments in favor of:

- the importance of electron distribution in the explosive molecule after "shock";
- the existence, for each molecular species, of a privileged bond, whose maximum relative variation of polarity under "shock" (relative to "rest") conditions sensitivity to "shock" and whose potential rupture is at the origin of the molecules in a dissociative state; and
- the role of the molecules which occupy a nondissociative state after "shock," in energy transfer, and therefore also in the emergence of a cooperative process which underlies the propagation phenomenon itself.

Since the publications of Bowden in 1952 and Campbell in 1960, a large number of the constituents for a comprehensive view of the transition from shock to detonation have existed and reached a large audience. However, lacking a theoretical framework into which they could be incorporated, the works of chemists physicists and engineers have, since then, progressed without great coherence as far as ideas and terminology are concerned; but, *a contrario*, the theory of the "detonation layer" and the molecular theory of sensitivity to "shock" of explosives, which are explained in Parts One and Two, justify the attempt at unification throughout Parts Three and Four. With this aim in view, there take place:

- an attempt to clarify the nature and the chronology of the molecular and/or mesoscopic phenomena which cause a continuous stress on the boundary of an explosive structure to result in propagation in the mass;
- attention is paid, in the reactive flow codes, to a consideration of that which depends on physical modeling and that which results from economy of computing;
- care is taken to show the coherence attained—by means of the notions of *built-up, strong, quasi C–J, autonomous, automorphous, guided* detonation—in interpreting the enormous body of experimental results accumulated for the elementary configurations of simple detonation; and
- care is taken to emphasize the incontrovertible character—in every attempt to "predict" the performance of an explosive—of an adjustment versus the measurements of velocity, and thus of an appropriate interpretation of the flows, etc. which leads the reader back to the first part of this work.

The same reader may be wondering why the word *initiation* does not appear anywhere. Neither random nor set purpose account for that absence, for which two strong reasons exist. The first arises from etymology, and the incorrectness of the word. The second arises from physics itself: the new description needs two distinct terms: *ignition* and *induction* so as to distinguish *ignition threshold pressure*, and *critical induction pressure*.

If someone keen on nuclear fusion should happen to see this book, it is very probable that he will be struck by the analogy between the necessity to control the "confinement" of plasma to ensure fusion and that of controlling "expansion" to ensure detonation. It is also very probable that he will take some pride in the difficulty of his problem compared to that of the detonation expert, simply because of the orders of magnitude of temperature and pressure required. But he will probably also find it difficult to suppress a feeling of envy while considering that an explosive substance—as opposed to plasma—keeps a "human face" insofar as, in its intimate reactions to stresses, it remembers all the stages of its manufacture, and not only the properties of the explosive molecule from which it takes its name, and to which it is—often destructively and sometimes dramatically—reduced!

Appendix A

Some calculations in Chapters I, II, III, and XII suppose a knowledge not only of the definition of the essential quantities of traditional thermodynamics, but also the existence of identities among the derivatives of these quantities for various selections of the state variables. In this appendix, we aim to establish, *with reference to a fluid*, a coherent resume of definitions and a guide in the search for identities and in their treatment.

1. State Variables and Associated Thermodynamic Potentials

1.1. Variables v and s

From the very definition of a fluid, its state depends on only two variables: from the very definition of absolute temperature T and pressure p, the following relationship holds between the derivatives of the mass volume v, the specific entropy s and the specific internal energy e

$$de = -p \, dv + T \, ds. \tag{e.1}$$

Relative to the derivatives dv and ds, de is thus a homogeneous linear form whose coefficients involve only p and T, to the exclusion of all derivatives: e is the thermodynamic potential associated with the variables v and s. From (e.1) it follows directly that

$$\left. \begin{array}{l} p = -\dfrac{\partial e}{\partial v}(v, s), \\[2mm] T = +\dfrac{\partial e}{\partial s}(v, s). \end{array} \right\} \tag{e.2}$$

Thus the specific free energy $f = e - Ts$ may be expressed

$$f = e - s\frac{\partial e}{\partial s}(v, s) \equiv -s^2 \frac{\partial (e/s)}{\partial s}(v, s). \tag{e.3}$$

1.2. Variables v and T

By coupling the relationship obtained by differentiating f to $(e.1)$, we obtain

$$df = -p\,dv - s\,dT, \qquad (f.1)$$

which shows that f is the thermodynamic potential associated with the variables v and T. From $(f.1)$ the relationships

$$\left.\begin{array}{l} p = -\dfrac{\partial f}{\partial v}(v,\,T), \\[3mm] s = -\dfrac{\partial f}{\partial T}(v,\,T), \end{array}\right\} \qquad (f.2)$$

follow directly, and also the relationship

$$e = f - T\frac{\partial f}{\partial v}(v,\,T) \equiv -T^2\frac{\partial(f/T)}{\partial T}(v,\,T). \qquad (f.3)$$

Let us now examine the coefficients of linear form $de = \mathscr{L}(dv,\,dT)$; by differentiating $(f.3)$ and taking account of $(f.2)$, there results

$$\left.\begin{array}{l} \dfrac{\partial e}{\partial v}(v,\,T) = +T^2\dfrac{\partial(p/T)}{\partial T}(v,\,T), \\[3mm] \dfrac{\partial e}{\partial T}(v,\,T) = -T\dfrac{\partial^2 f}{\partial T^2}(v,\,T), \end{array}\right\} \qquad (f.4)$$

so that

$$de = T^2\frac{\partial(p/T)}{\partial T}\cdot dv - \frac{\partial^2 f}{\partial T^2}(v,\,T)\cdot dT. \qquad (f.5)$$

Bringing $(e.1)$ and $(f.5)$ together then makes it possible to find an expression for ds

$$ds = \frac{\partial p}{\partial T}(v,\,T)\cdot dv - \frac{\partial^2 f}{\partial T^2}(v,\,T)\cdot dT. \qquad (f.6)$$

1.3. Variables p and s

When the relation obtained by differentiating the mass enthalpy $h = e + pv$ is linked to $(e.1)$, we obtain

$$dh = v\,dp + T\,ds, \qquad (h.1)$$

which shows that h is the thermodynamic potential associated with the variables p and s. The immediate result of this is

$$
\left.\begin{array}{l}
v = \dfrac{\partial h}{\partial p}(p, s), \\[2mm]
T = \dfrac{\partial h}{\partial s}(p, s).
\end{array}\right\} \tag{h.2}
$$

So the specific free enthalpy $g = h - Ts$ is written

$$
g = h - s\frac{\partial h}{\partial p}(p, s) \equiv -s^2 \frac{\partial (h/s)}{\partial s}(p, s). \tag{h.3}
$$

1.4. Variables p and T

By coupling the relation obtained by differentiating g to (e.1), we obtain

$$
dg = v\,dp - s\,dT, \tag{g.1}
$$

which shows that g is the thermodynamic potential associated with the variables p and T. The immediate result of this is

$$
\left.\begin{array}{l}
v = +\dfrac{\partial g}{\partial p}(p, T), \\[2mm]
s = -\dfrac{\partial g}{\partial T}(p, T),
\end{array}\right\} \tag{g.2}
$$

and also

$$
h = g - T\frac{\partial g}{\partial T}(p, T) \equiv -T^2 \frac{\partial (g/T)}{\partial T}(p, T). \tag{g.3}
$$

We now find the coefficients of the linear form $dh = \mathscr{L}(dp, dT)$; by differentiating (g.3) and taking account of (g.2), we obtain

$$
\left.\begin{array}{l}
\dfrac{\partial h}{\partial p}(p, T) = -T^2 \dfrac{\partial (v/T)}{\partial T}(p, T), \\[2mm]
\dfrac{\partial h}{\partial T}(p, T) = -T\dfrac{\partial^2 g}{\partial T^2}(p, T),
\end{array}\right\} \tag{g.4}
$$

so that

$$
dh = -T^2 \frac{\partial}{\partial T}\left(\frac{v}{T}\right)\cdot dp - T\frac{\partial^2 g}{\partial T^2}(p, T)\cdot dT. \tag{g.5}
$$

Bringing (h.1) and (g.5) together also makes it possible to find the form of ds

$$
ds = -\frac{\partial v}{\partial T}(p, T)\cdot dp - T\frac{\partial^2 g}{\partial T^2}(p, T)\cdot dT. \tag{g.6}
$$

2. The Isentropes

2.1. Preliminary Observation

For specific heats with constant volume and pressure which appear in the second line of $(f.4)$ and $(g.4)$, respectively, we introduce the usual notations

$$c_v = \frac{\partial e}{\partial T}(v, T),$$

$$c_p = \frac{\partial h}{\partial T}(p, T).$$

So the expressions $(f.6)$ and $(g.6)$ of ds lead to the double equality

$$ds = \frac{\partial p}{\partial T}(v, T)\, dv + \frac{c_v}{T}\, dT = -\frac{\partial v}{\partial T}(p, T)\, dp + \frac{c_p}{T}\, dT. \qquad (i.1)$$

The right-hand equality leads to the differential relation

$$\frac{c_p - c_v}{T}\, dT = \frac{\partial p}{\partial T}(v, T)\, dv + \frac{\partial v}{\partial T}(p, T)\, dp$$

which, identified with the linear form $dT = \mathscr{L}(dv, dp)$, implies

$$\frac{T}{c_p - c_v}\frac{\partial p}{\partial T}(v, T) = \frac{\partial T}{\partial v}(v, p),$$

$$\frac{T}{c_p - c_v}\frac{\partial v}{\partial T}(p, T) = \frac{\partial T}{\partial p}(v, p),$$

from which it is a simple matter to see that they both lead to

$$c_p - c_v = T \cdot \left[\frac{\partial v}{\partial T}(p, T)\right]\left[\frac{\partial p}{\partial T}(v, T)\right]. \qquad (i.2)$$

2.2. The "Slopes" of the Isentropes

By making $ds = 0$ in $(i.1)$, we obtain two equalities giving the variation of T according to v on the one hand and p on the other, along an isentrope

$$\frac{\partial T}{\partial v}(v, s) = -\frac{T}{c_v}\frac{\partial p}{\partial T}(v, T), \qquad (i.3)$$

$$\frac{\partial T}{\partial p}(p, s) = \frac{T}{c_p}\frac{\partial v}{\partial T}(p, T). \qquad (i.4)$$

Eliminating $T(\partial p/\partial T)$ from $(i.2)$ and $(i.3)$ on the one hand, and $T(\partial v/\partial T)$ from (i.2) and (i.4) on the other, we obtain

$$\frac{\partial T}{\partial v}(v, s) = \left(1 - \frac{c_p}{c_v}\right)\frac{\partial T}{\partial v}(v, p), \tag{i.5}$$

$$\frac{\partial T}{\partial p}(p, s) = \left(1 - \frac{c_v}{c_p}\right)\frac{\partial T}{\partial p}(v, p). \tag{i.6}$$

Finally, connecting (i.5) and (i.6), we find the well-known relation

$$\frac{\partial p}{\partial v}(v, s) = \frac{c_p}{c_v} \cdot \frac{\partial p}{\partial v}(v, T) \tag{i.7}$$

With the fairly widespread notations

$$\gamma = \frac{c_p}{c_v}, \Gamma = -\frac{\partial \operatorname{Log} p}{\partial \operatorname{Log} v}(v, s), \qquad \Gamma_T = -\frac{\partial \operatorname{Log} p}{\partial \operatorname{Log} v}(v, T)$$

the relation (i.7) takes the form

$$\Gamma = \gamma\Gamma_T, \tag{i.8}$$

the significance of which will be recognized by referring to the ideal gas ($pv \approx T \Rightarrow \Gamma_T = 1$).

The significance of the variation in Log p with Log v along an isentrope suggests that we look at that of Log T with Log v and with Log p, in the same conditions of constant entropy.

Consider the first and note first that

$$\frac{\partial p}{\partial T}(v, T) = \frac{\partial p}{\partial e}(v, e) \cdot \frac{\partial e}{\partial T}(v, T) = c_v \cdot \frac{\partial p}{\partial e}(v, e). \tag{i.9}$$

So the identity (i.3) leads immediately to

$$\frac{\partial T}{\partial v}(v, s) = -T\frac{\partial p}{\partial e}(v, e).$$

By introducing the Grüneisen coefficient G by

$$G = v\frac{\partial p}{\partial e}(v, e), \tag{i.10}$$

finally, we obtain

$$\frac{\partial \operatorname{Log} T}{\partial \operatorname{Log} v}(v, s) = -G. \tag{i.11}$$

Consider the second and observe first that

$$\frac{\partial p}{\partial v}(v, e) = \frac{\partial p}{\partial v}(v, s) + \frac{\partial p}{\partial s}(v, s) \cdot \frac{\partial s}{\partial v}(v, e).$$

Taking account of (e.2) to evaluate the two derivatives in the product of the right-hand member, we obtain

$$\frac{\partial p}{\partial v}(v, e) = \frac{\partial p}{\partial v}(v, s) - \frac{p}{T}\frac{\partial T}{\partial v}(v, s). \tag{i.12}$$

Introducing the square a^2 of the speed of sound

$$a^2 = \frac{\partial p}{\partial \rho}(\rho, s) = -v^2\frac{\partial p}{\partial v}(v, s) \tag{i.13}$$

we can write $(i.12)$ in the form

$$\frac{\partial \operatorname{Log} T}{\partial \operatorname{Log} p}(v, s) = 1 + \frac{v^2}{a^2}\frac{\partial p}{\partial v}(v, e). \tag{i.14}$$

3. Other Identities Involving the Gruneisen Coefficient

3.1. Variables v and p

Use $(i.9)$ to show G in $(i.2)$. We obtain

$$\frac{\partial \operatorname{Log} T}{\partial \operatorname{Log} v}(v, p) = \frac{c_v}{c_p - c_v}G. \tag{j.1}$$

This identity is a first aspect of the role which G plays in the expression of the thermodynamic derivatives for the variables v and p. Continuing in this direction, we shall now attempt to express the derivatives of specific internal energy considered as a function of v and p. First, we obtain, simply from the definition of G

$$\frac{\partial e}{\partial p}(v, p) = \frac{v}{G}. \tag{j.2}$$

We can also write

$$\frac{\partial e}{\partial v}(v, p) = \frac{\partial e}{\partial v}(v, T) + \frac{\partial e}{\partial T}(v, T) \cdot \frac{\partial T}{\partial v}(v, p).$$

The first term of the second member is given by the first equation $(f.4)$ whilst the second is expressed on the basis of c_v and the identity $(i.2)$

$$\frac{\partial e}{\partial v}(v, p) = -p + T\frac{\partial p}{\partial T}(v, T) + \frac{c_v}{c_p - c_v}T\frac{\partial p}{\partial T}(v, T).$$

Introducing G through $(i.9)$ and $(i.10)$, we obtain

$$\frac{\partial e}{\partial v}(v, p) = -p + T\frac{c_p c_v}{c_p - c_v}\frac{G}{v}. \tag{j.3}$$

3.2. Velocity of Sound

Consider the square of the velocity of sound a^2 in the form $-v^2(\partial p/\partial v)(v, s)$ and attempt to transform the partial derivative by:

$$\frac{\partial p}{\partial v}(v, s) = \frac{\partial p}{\partial v}(v, T) + \frac{\partial p}{\partial T}(v, T) \cdot \frac{\partial T}{\partial v}(v, s). \qquad (j.4)$$

Taking account of (i.3), we obtain

$$\frac{\partial p}{\partial v}(v, s) = \frac{\partial p}{\partial v}(v, T) - \frac{T}{c_v}\left[\frac{\partial p}{\partial T}(v, T)\right]^2.$$

Now taking account of (i.7) and (i.9) to eliminate the derivatives of p outside the isentrope, we arrive at

$$\frac{c_p - c_v}{c_p} \cdot \frac{\partial p}{\partial v}(v, s) = -Tc_v\frac{G^2}{v^2},$$

which leads to the expressions of a^2

$$a^2 = T\frac{c_p c_v}{c_p - c_v}G^2. \qquad (j.5)$$

Appendix B

Quotations refer to the list of references for Part One

1. Introduction

Since the variables $\bar{\xi}'$, $\bar{\xi}''$, $\bar{t}$ play no part—unless as parameters—in the problem posed in §III.2.1 *in fine*, we can ignore them for the time being and symbolize the derivation in ζ by a point over the dependent variables. With these notations, it is a question of finding a solution in $(\zeta; v, T, m)$ of the system

$$\left.\begin{array}{l} \dfrac{M\mu''}{L}\dot{v} = p - p_0 + M^2(v - v_0), \\[3mm] \dfrac{\lambda}{LM}\dot{T} = e - e_0 + (v - v_0)\left[p_0 + \dfrac{M^2}{2}(v_0 - v)\right], \\[3mm] \dfrac{w}{T}\dot{m} = (1 - m)s, \end{array}\right\} \qquad \text{(B.1)}$$

such that (it is then *regular* according to the terminology of Friedrichs [20]) its limit for ζ tending toward:

$-\infty$ is the singular upstream point $(v_0, T_0, 0)$; and
$+\infty$ is the singular downstream point $(v_1, T_1, 1)$,

where (v_1, T_1) is a point such that $w_0 = Mv_0$ $(= D^{(o)})$ on the detonation arc (H_+) originating from (v_0, p_0).

The problem was analyzed by Friedrichs [20] in the case where $w_1 \neq a_1$ assuming that both the explosive and its detonation products are a polytropic gas, and by using a method which cannot be generalized to the basic case where $w_1 = a_1$. That is why we preferred a more homogeneous treatment inspired by that devised by Weyl [37] for shock waves.

We suggest

$$v - v_i \doteq x, \qquad T - T_i = y, \qquad i = 1 \text{ or } 0.$$

The system (B.1) can be written in the form

$$
\left.
\begin{aligned}
\dot{x} &= A_i x + B_i y + P_i(x, y, m), \\
\dot{y} &= C_i x + D_i y + Q_i(x, y, m), \\
\dot{m} &= (1 - m)\frac{LS}{w},
\end{aligned}
\right\} \tag{B.2}
$$

where the functions $P_i Q_i$ and their initial derivatives in x and y are cancelled out in $(0, 0, 0)$ if $i = 0$, in $(0, 0, 1)$ if $i = 1$. Let K_i be the matrix

$$
\begin{vmatrix} A_i & B_i \\ C_i & D_i \end{vmatrix}.
$$

2. Eigenvalues and Directions of Matrix K_i

(To simplify the notations, the index i will be omitted throughout this subsection.)

Referring to the system (B.1), we find that

$$
\left.
\begin{aligned}
A &= \frac{L}{M\mu''}\left[M^2 + \frac{\partial p}{\partial v}(v, T)\right], & B &= \frac{L}{M\mu''}\frac{\partial p}{\partial T}(v, T), \\
C &= \frac{LM}{\lambda}\left[\frac{\partial e}{\partial v}(v, T) + M^2(v_0 - v) + p_0\right], & D &= \frac{LM}{\lambda}c_v.
\end{aligned}
\right\} \tag{B.3}
$$

Taking account of the conventional thermodynamic identities and $w = Mv$ (see equations (III.22)), we find that the sum $A + D$ and the product $AD - BC$ of the eigenvalues of K are given by

$$
\left.
\begin{aligned}
A + D &= \frac{c_v}{\lambda}\left(1 - \frac{\lambda}{\mu''c_p}\frac{a^2}{w^2} + \frac{\lambda}{\mu''c_v}\right)LM, \\
AD - BC &= \frac{c_v}{\mu''\lambda}\left(1 - \frac{a^2}{w^2}\right)L^2M^2.
\end{aligned}
\right\} \tag{B.4}
$$

Thus the characteristic equation of K is

$$
\kappa^2 - \frac{c_v}{\lambda}\left(1 - \frac{\lambda}{\mu''c_p}\frac{a^2}{w^2} + \frac{\lambda}{\mu''c_v}\right)LM\kappa + \frac{c_v}{\mu''\lambda}\left(1 - \frac{a^2}{w^2}\right)L^2M^2 = 0. \tag{B.5}
$$

The δ discriminant of (B.5) is

$$
\delta = \left(LM\frac{c_v}{\lambda}\right)^2\left[\left(1 - \frac{\lambda}{\mu''c_p}\frac{a^2}{w^2} + \frac{\lambda}{\mu''c_v}\right)^2 - \frac{4\lambda}{\mu''c_v}\left(1 - \frac{a^2}{w^2}\right)\right].
$$

We pose $\mathfrak{N} = a^2/w^2$, $\omega = \mu''c_p/\lambda$ (ω is the longitudinal Prandtl number). Whatever $\mathfrak{N}$ might be, the sign of δ is that of the trinomial

$$
\begin{aligned}
\mathscr{F}(\mathfrak{N}) &= (\omega + \gamma - \mathfrak{N})^2 - 4\gamma\omega(1 - \mathfrak{N}) \\
&= \mathfrak{N}^2 + 2[(\gamma - 1)\omega + \gamma(\omega - 1)]\mathfrak{N} + (\omega - \gamma)^2. \tag{B.6}
\end{aligned}
$$

We find that the reduced discriminant f of $\mathscr{F}$ equals

$$f = 4\omega\gamma(\gamma - 1)(\omega - 1).$$

Taking account of $\gamma > 1$, it is $\omega - 1$ which gives its sign to f, in such a way that:

— if $\omega - 1 < 0$, $\mathscr{F}$ has no real roots, therefore $\mathscr{F}(\mathfrak{N}) > 0$; and
— if $\omega - 1 \geq 0$, $\mathscr{F}$ has two negative roots, therefore $\mathscr{F}(\mathfrak{N}) > 0$ for $\mathfrak{N} \geq 0$.

Consequently, δ is always positive and the characteristic equation of K always has *two distinct real roots*.

Referring to (B.4), we see that the sign of the roots

$$\left.\begin{aligned} n &= \frac{LM}{\mu''}\frac{\omega + \gamma - \mathfrak{N} + \sqrt{\mathscr{F}}}{2\gamma}, \\ r &= \frac{LM}{\mu''}\frac{\omega + \gamma - \mathfrak{N} - \sqrt{\mathscr{F}}}{2\gamma} \end{aligned}\right\} \tag{B.7}$$

of (B.5) depends on that of $\mathfrak{N} - 1$ according to Table B.1.

Table B.1

		AD − BC	A + D	n	r
$\mathfrak{N} < 1$	w > a	+	+	+	+
$\mathfrak{N} = 1$	w = a	0	+	+	0
$\mathfrak{N} > 1$	w < a	−	▨	+	−

We can see that n and r are rational whatever $\mathfrak{N}$ might be if $\omega = 1$. The simplicity of the calculations in this case explains why this hypothesis was developed by Hirschfelder [24], in particular. However it is appropriate not to over-estimate the importance of this particular case.

The eigendirection $(v, \mathscr{T})$ which corresponds to the eigenvalue κ is defined by the first-order system

$$(A - \kappa)v + B\mathscr{T} = 0,$$
$$Cv + (D - \kappa)\mathscr{T} = 0.$$

Taking account of (B.3) and (B.7), we find

$$\mathscr{T}(\kappa) = -T\mathscr{G}(\kappa),$$

$$\mathscr{G}(\kappa) = \frac{2G\omega}{\pm\sqrt{\mathscr{F}} + \mathfrak{N} - \gamma + \omega}, \qquad + \text{ if } \kappa = r, \quad - \text{ if } \kappa = n, \tag{B.8}$$

where G denotes the Grüneisen coefficient $v(\partial p/\partial e)(v, e)$. Returning to the definition (B.6) of $\mathscr{F}$, we obtain

$$\mathscr{G}(n) \cdot \mathscr{G}(r) = \frac{G^2\omega}{\mathfrak{N}(1-\gamma)} < 0,$$

From which

$$G \cdot \mathscr{G}(\kappa) \begin{cases} < 0 & \text{if } \kappa = n, \\ > 0 & \text{if } \kappa = r. \end{cases} \tag{B.9a}$$

Moreover, from (B.6) and (B.8) we deduce

$$\mathfrak{N} = 1 \quad \Rightarrow \quad \frac{\mathscr{F}(r)}{v} = -\frac{TG}{v} = \frac{\partial T}{\partial v}(v, s). \tag{B.9b}$$

3. Properties of the Upstream Singular Point

For $i = 0$ (upstream singular point), the equations (B.4) are written

$$\left.\begin{aligned} \dot{x} &= A_0 x + B_0 y + P_0(x, y, m), \\ \dot{y} &= C_0 x + D_0 y + Q_0(x, y, m), \\ \dot{m} &= h_0(x, y, m), \end{aligned}\right\} \tag{B.10}$$

where the functions P_0 and Q_0 and their first derivatives cancel out at the origin and where h_0 is cancelled for fairly small values of x and y. In (B.10) we perform the transformation

$$\left.\begin{aligned} x &= v_0 X + v_0 Y, \\ y &= \mathscr{F}(n_0)X + \mathscr{F}(r_0)Y, \\ m &= Z. \end{aligned}\right\} \tag{B.11}$$

This transformation leads to the canonical system

$$\left.\begin{aligned} \dot{X} &= n_0 X + F_0(X, Y, Z), \\ \dot{Y} &= r_0 Y + G_0(X, Y, Z), \\ \dot{Z} &= H_0(X, Y, Z), \end{aligned}\right\} \tag{B.12}$$

where the functions F_0, G_0, and their first derivations in X and Y cancel out at the origin, and the function H_0 is cancelled for fairly small values of X and Y.

According to the definition (III.18) of $D^{(\circ)}$ and the properties of (H_+) shown in §II.3.1, we have $D^{(\circ)} > a_0$. So we deduce from Table B.1 that

$$n_0 > r_0 > 0. \tag{B.13}$$

Under these conditions, according to the argument given by Weyl [37], there exists a group with one parameter of trajectories $\mathscr{T}_0$ of (B.12) which reach the

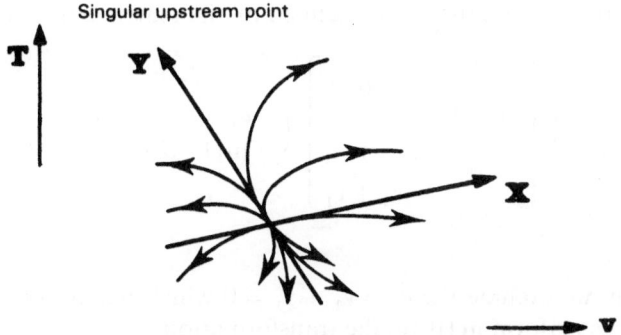

Figure B.1. The arrow indicates the travel direction on a trajectory as ζ increases.

origin for $\zeta \to -\infty$ (see Fig. B.1) in such a way that

$$\left.\begin{array}{l} X = 0(e^{r_0\zeta}), \\ Y = k_0 e^{r_0\zeta} + 0(e^{r_0\zeta}), \\ Z = 0 \text{ for fairly small } \zeta, \end{array}\right\} \tag{B.14}$$

where k_0 is an arbitrary constant. From which, according to (B.11) and (B.8)

$$\left.\begin{array}{l} \dfrac{v - v_0}{v_0} = k_0 e^{r_0\zeta} + 0(e^{r_0\zeta}), \\[2mm] \dfrac{T - T_0}{T_0} = -k_0 \mathscr{G}(r_0) e^{r_0\zeta} + 0(e^{r_0\zeta}), \\[2mm] m = 0 \text{ for fairly small } \zeta. \end{array}\right\} \tag{B.15}$$

4. Properties of the Singular Downstream Point

For $i = 1$, assuming that $m - 1 = z$, (B.2) can be written

$$\left.\begin{array}{l} \dot{x} = A_1 x + B_1 y + P_1 z + f_1(x, y, z), \\ \dot{y} = C_1 x + D_1 y + Q_1 z + g_1(x, y, z), \\ \dot{z} = -z\left(\dfrac{LS_1}{w_1} + h_1(x, y, z)\right), \end{array}\right\} \tag{B.16}$$

where the functions f_1, g_1, and h_1 cancel out at the origin as well as the first derivatives of f_1 and g_1. Let us suppose

$$v_1 = \frac{LS_1}{w_1}, \tag{B.17}$$

and denote by (P_1', Q_1') the vector defined by the matrix equality

$$\begin{pmatrix} v_1 & v_1 \\ \mathscr{T}(n_1) & \mathscr{T}(r_1) \end{pmatrix} \begin{bmatrix} \dfrac{1}{v_1 + n_1} & 0 \\ 0 & \dfrac{1}{v_1 + r_1} \end{bmatrix} \begin{pmatrix} v_1 & v_1 \\ \mathscr{T}(n_1) & \mathscr{T}(r_1) \end{pmatrix}^{-1} \begin{pmatrix} P_1 \\ Q_1 \end{pmatrix} + \begin{pmatrix} v_1 P_1' \\ T_1 Q_1' \end{pmatrix}$$

$$= 0 \tag{B.18}$$

(in so doing, we exclude the case $v_1 + r_1 = 0$ which has no physical significance). Let us perform in (B.16) the transformation

$$\left.\begin{aligned} x &= v_1 X + v_1 Y + P_1' Z, \\ y &= \mathscr{T}(n_1) X + \mathscr{T}(r_1) Y + Q_1' Z, \\ z &= Z. \end{aligned}\right\} \tag{B.19}$$

We are thus led to the canonical system

$$\left.\begin{aligned} \dot{X} &= n_1 X + F_1(X, Y, Z), \\ \dot{Y} &= r_1 Y + G_1(X, Y, Z), \\ \dot{Z} &= -Z[v_1 + H_1(X, Y, Z)], \end{aligned}\right\} \tag{B.20}$$

where the functions F_1, G_1, and H_1 as well as the first derivatives of F_1 and G_1 cancel out at the origin.

In the case where $w_1 \neq a_1$, we can use a reasoning analogous to that made by Weyl [37].

(1) $w_1 > a_1$ (weak detonation)
    According to Table B.1, we have $n_1 > r_1 > 0$ so that only two isolated trajectories of (B.20) reach the origin when $\zeta \to +\infty$ tangentially to OZ.
(2) $w_1 < a_1$ (strong detonation)
    According to Table B.1, we have $n_1 > 0 > r_1$. This time, there exists a group with one parameter of trajectories $\mathscr{T}_1$ which reach the origin for $\zeta \to +\infty$ in such a way that:

$$\left.\begin{aligned} X &= 0(e^{r_1 \zeta}), \\ Y &= k_1' e^{r_1 \zeta} + 0(e^{r_1 \zeta}), &&\text{if } \quad v_1 < -r_1 \\ Z &= k_1 e^{-v_1 \zeta} + 0(e^{-v_1 \zeta}), \end{aligned}\right\} \tag{B.21a}$$

$$\left.\begin{aligned} X &= 0(e^{-v_1 \zeta}), \\ Y &= k_1' e^{r_1 \zeta} + 0(e^{r_1 \zeta}), &&\text{if } \quad v_1 > -r_1 \\ Z &= k_1 e^{-v_1 \zeta} + 0(e^{-v_1 \zeta}). \end{aligned}\right\} \tag{B.21b}$$

The projections of $\mathscr{T}_1$ on the plane (Y, Z) have the outline indicated in Figure

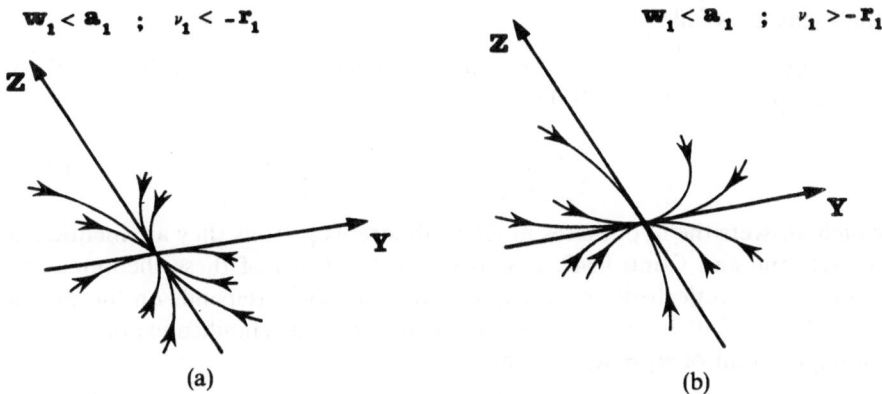

Figure B.2. The arrow indicates the travel direction on a trajectory as ζ increases.

B.2(a) and (b). Returning to the physical variables and taking account of (B.8), the relations (B.21) show that

$$\left.\begin{aligned} \frac{v - v_1}{v_1} &= k_1 P_1' e^{-v_1\zeta} + 0(e^{-v_1\zeta}), \\[6pt] \frac{T - T_1}{T_1} &= k_1 Q_1' e^{-v_1\zeta} + 0(e^{-v_1\zeta}), \qquad \text{if } v_1 < -r_1 \\[6pt] m - 1 &= -v_1 k_1 e^{-v_1\zeta} + 0(e^{-v_1\zeta}). \end{aligned}\right\} \qquad \text{(B.22a)}$$

$$\left.\begin{aligned} \frac{v - v_1}{v_1} &= k_1' e^{r_1\zeta} + 0(e^{r_1\zeta}), \\[6pt] \frac{T - T_1}{T_1} &= -k_1' \mathscr{G}(r_1) e^{r_1\zeta} + 0(e^{r_1\zeta}), \qquad \text{if } v_1 > -r_1 \\[6pt] m - 1 &= -v_1 k_1 e^{-v_1\zeta} + 0(e^{-v_1\zeta}), \end{aligned}\right\} \qquad \text{(B.22b)}$$

We note on (B.22) that in a strong detonation:

— the physical transformation never imposes its rate on the chemical transformation; and
— the chemical transformation dominates the whole phenomenon only if it is sufficiently slow ($v_1 < -r_1$).

In the case where $w_1 = a_1$ (C–J detonation), the system (B.20) is written:

$$\left.\begin{aligned} \dot{X} &= n_1 X + F_1(X, Y, Z), \\[4pt] \dot{Y} &= G_1(X, Y, Z), \\[4pt] \dot{Z} &= -Z[v_1 + H_1(X, Y, Z)]. \end{aligned}\right\} \qquad \text{(B.23)}$$

Since Weyl's argument [37] no longer allows us to draw a conclusion, we shall reason as follows:

(a) Let us consider among the solutions of (B.23) those which are entirely situated in the plane $Z = 0$. They confirm

$$\left.\begin{array}{l} \dot{X} = n_1 X + F_1(X, Y, O), \\ \dot{Y} = G_1(X, Y, O), \end{array}\right\} \tag{B.24}$$

which answers the hypotheses in Keil's theorems [29] as they are mentioned by Sansone and Conti [34]. According to the third of these theorems, the nature of the trajectories of (B.24) nontangent to OX depends on the sign of $n_1/(\partial^2 G_1/\partial Y^2)(0, 0, 0)$. Going back to the physical significance of G_1 and taking account of $w_1 = a_1$, we find

$$\tau_1 \equiv \frac{1}{\dfrac{\partial^2 G_1}{\partial Y^2}(0, 0, 0)} = \left[\frac{\gamma + \omega - 1}{L} \frac{\lambda v}{w c_p} \frac{\rho^3 w^2}{\dfrac{\partial^2 p}{\partial v^2}(v, s, m)} \right]_1. \tag{B.25}$$

So, taking account of $n_1 > 0$ (see Table B.1) and $(\partial^2 p/\partial v^2)(v, s, 1) > 0$ (see hypothesis (II.16a)), Keil's second and third theorems show that only one trajectory $\mathscr{T}_1^0$ of (B.24) reaches the origin for $\zeta \to +\infty$, and this in such a way that

$$\lim_{\zeta \to +\infty} \zeta X = 0, \qquad \lim_{\zeta \to +\infty} \zeta Y = -\tau_1, \tag{B.26}$$

(the outline of the trajectories in the neighborhood of 0 is indicated on Figure B.3).

(b) According to (B.9b) and (B.19), OY is the tangent in 0 at the isentrope originating from 0, with the result that the function $H(X, Y, Z)$ confirms $H_Y(0, 0, 0) = 0$. Then, by integrating (B.23) through successive approxima-

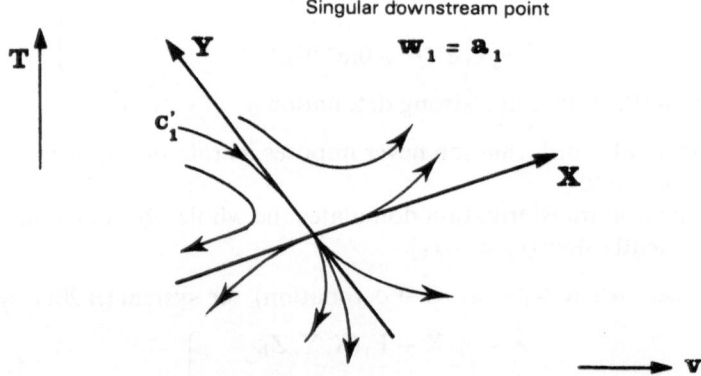

Figure B.3. The arrow indicates the travel directtion on a trajectory as ζ increases.

tions starting from $\mathcal{T}_1^0$, it is demonstrated that there exists, for every $k_1 > 0$, one trajectory $\mathcal{T}_1$ of (B.23) and one only which reaches the origin for $\zeta = +\infty$ and is such that

$$\lim_{\zeta \to +\infty} \zeta X = 0, \qquad \lim_{\zeta \to +\infty} \zeta Y = -\tau_1, \qquad \lim_{\zeta \to +\infty} e^{v_1 \zeta} Z = k_1. \qquad (B.27)$$

With (B.19) and (B.8) we deduce from this that the corresponding $\mathcal{T}_1$ group presents the following asymptotic behavior

$$\left.\begin{array}{l}
\dfrac{v - v_1}{v_1} = -\alpha_1 \dfrac{\ell_1}{L\zeta} + 0\left(\dfrac{1}{\zeta}\right), \\[3mm]
\dfrac{T - T_1}{T_1} = \mathcal{G}(r_1)\alpha_1 \dfrac{\ell_1}{L\zeta} + 0\left(\dfrac{1}{\zeta}\right), \\[3mm]
m - 1 = -v_1 k_1 e^{-v_1 \zeta} + 0(e^{-v_1 \zeta}),
\end{array}\right\} \qquad (B.28)$$

where the length ℓ_1 and the number α_1 were defined by

$$\left.\begin{array}{l}
\ell_1 = (\omega + \gamma - 1)_1 \left(\dfrac{\lambda v}{w c_p}\right)_1, \\[5mm]
\alpha_1 = \left[\dfrac{\rho^3 w^2}{\dfrac{\partial^2 p}{\partial v^2}(v, s, m)}\right]_1
\end{array}\right\} \qquad (B.29)$$

5. Discussion

Let $\mathcal{L}^i$ be the arc swept in the plane $m = 0$ on the curve $s(v, T) = s^i$ by the $\mathcal{T}_0$ trajectories when k_0 varies. Let K be the point where the $\mathcal{T}_1$ trajectory meets the plane $m = 0$. In order that $\mathcal{T}_1$ corresponds to a regular solution, it is necessary and sufficient for K to be on $\mathcal{L}^i$. The preceding paragraphs show that we must distinguish the two cases:

(1) $w_1 > a_1$

$\mathcal{T}_1$ depends on zero parameter; it is the same for K which therefore can be on $\mathcal{L}^i$ only for isolated values of $D^{(o)}$. These depend not only on the thermodynamic state q_0 of the explosive substance but also on its ignition entropy s^i. To each of these, if they exist, there corresponds a regular solution.

(2) $w_1 \leq a_1$

$\mathcal{T}_1$ depends on one parameter; it is the same for K which therefore can be on $\mathcal{L}^i$ for an appropriate value of the parameter. Under these conditions a regular solution can exist for any value of $D^{(o)}$ greater than or equal to D_*.

Friedrichs [20] and Hayes [23], who studied the system (B.1) as being representative of the permanent plane wave, concluded that weak detonations could not be excluded on the basis of a thermodynamic reasoning, but should be excluded in view of the extremely high reaction rates which were necessary. Following them, we shall admit that weak detonations are a spurious solution of the problem posed.

Appendix C

We now consider the system (III.28)

$$
\left.
\begin{aligned}
(a^2 - w^2)\frac{\partial u}{\partial Z} &= 2\frac{ua^2}{R_m} + \cdots, \\[2mm]
(a^2 - w^2)\frac{\partial \operatorname{Log} v}{\partial Z} &= 2\frac{uw}{R_m} + \cdots,
\end{aligned}
\right\}
\tag{C.1}
$$

where the ellipses denoted the quantities which tend towards 0 when $Z \to +0$. We are seeking a solution of (C.1) which tends towards (u_1, v_1) when $Z \to +0$, it being understood that

$$
\begin{aligned}
m &= 1, \\
s &= s_1, \\
w &= a_1.
\end{aligned}
\tag{C.2}
$$

From the equations (C.1) we deduce that $U = u - u_1$ and $V = v - v_1$ must confirm

$$
\left.
\begin{aligned}
&\lim_{Z \to +0} \frac{V}{U} = \frac{v_1}{a_1}, \qquad &\text{(a)} \\[2mm]
&\lim_{Z \to +0}\left[(w^2 - a^2)\frac{\partial U}{\partial Z}\right] = -2\frac{u_1 a_1^2}{R_m}.\text{(b)}
\end{aligned}
\right\}
\tag{C.3}
$$

Taking account of the thermodynamic identity

$$
\frac{\partial a}{\partial v}(v, s) = \frac{a}{v} - \frac{v^2}{2a}\frac{\partial^2 p}{\partial v^2}(v, s),
\tag{C.4}
$$

and of (C.2), we obtain

$$
w^2 - a^2 \underset{Z \to +0}{\sim} v_1^2 \frac{\partial^2 p}{\partial v^2}(v_1, s_1, 1) \cdot V,
\tag{C.5}
$$

or again, according to (C.3a)

$$w^2 - a^2 \underset{z \to +0}{\sim} \frac{v_1^3}{a_1} \frac{\partial^2 p}{\partial v^2}(v_1, s_1, 1) \cdot U. \qquad (C.6)$$

Therefore, (C.3) is equivalent to

$$
\left.
\begin{array}{ll}
\displaystyle \lim_{z \to +0} \frac{V}{U} = \frac{v_1}{a_1}, & \text{(a)} \\[4mm]
\displaystyle \lim_{z \to +0} U \frac{\partial U}{\partial Z} = -2 \frac{u_1(\rho_1 w_1)^3}{R_m \cdot \dfrac{\partial^2 p}{\partial v^2}(s_1, v_1, 1)}. & \text{(b)}
\end{array}
\right\}
\qquad (C.7)
$$

Let us suppose

$$\beta = -\frac{w}{u}\alpha, \qquad (C.8)$$

where α is the quantity defined in Table III.1 of §III.2.2

$$\alpha = \frac{\rho^3 w^2}{\dfrac{\partial^2 p}{\partial v^2}(v, s, m)}.$$

With these notations (C.7b) is written

$$U \underset{z \to +0}{\sim} \pm 2u_1 \sqrt{\beta_1 \frac{Z}{R_m}}. \qquad (C.9)$$

From which

$$
\left.
\begin{array}{l}
\displaystyle \frac{u - u_1}{u_1} = \pm 2 \sqrt{\beta_1 \frac{Z}{R_m}} + 0\left(\sqrt{\frac{Z}{R_m}}\right), \\[5mm]
\displaystyle \frac{v - v_1}{v_1} = \pm 2 \frac{u_1}{a_1} \sqrt{\beta_1 \frac{Z}{R_m}} + 0\left(\sqrt{\frac{Z}{R_m}}\right).
\end{array}
\right\}
\qquad (C.10)
$$

Appendix D

The effective solution of the differential equation (III.69)

$$\frac{\sin \eta}{\sin \hat{\eta}} \left[\frac{d\hat{\theta}}{d\eta} + \frac{\hat{v}}{|\hat{\mathbf{W}}|^2} \cotan \hat{\eta} \frac{d\hat{p}}{d\eta} \right] \frac{d\eta}{d\hat{r}} = \frac{1}{\hat{c}^2} \frac{\hat{S}}{\hat{\rho}|\hat{\mathbf{W}}|} \frac{\widehat{\partial p}}{\partial m} - \frac{\sin \hat{\theta}}{\hat{r}}, \qquad (\text{D.0})$$

demands that the derivatives $d\hat{\theta}/d\eta$ and $d\hat{p}/d\eta$ be clarified using the equations (III.67), then that the form thus obtained be transformed by substituting for the Hugoniot curve (III.67a) the linear relation—within experimental reach—between the relative normal upstream velocity U and the absolute value u of the normal material velocity jump.

1. Transformation of (D.0)

● Differentiating (III.67b) by considering $\hat{p}$ and $\hat{v}$ as linked by (III.67a), we obtain

$$\frac{d\hat{p}}{d\eta} \left(1 + \rho_0^2 |\mathbf{W}_0|^2 \sin^2 \eta \frac{d\hat{v}}{d\hat{p}} \right) = 2\rho_0^2 |\mathbf{W}_0|^2 (v_0 - \hat{v}) \sin \eta \cos \eta,$$

which, by eliminating $\rho_0^2 W_0^2$ using (III.67b), is also written

$$\frac{\frac{d\hat{p}}{d\eta}}{\hat{p} - p_0} = 2 \cotan \eta \left[1 - \frac{\hat{p} - p_0}{\hat{v} - v_0} \Big/ \frac{d\hat{p}}{d\hat{v}} \right]^{-1}. \qquad (\text{D.1})$$

Differentiating (III.67c), we obtain

$$\frac{1}{\sin^2 \hat{\theta}} \frac{d\hat{\theta}}{d\eta} = \frac{\rho_0 |\mathbf{W}_0|^2}{(\hat{p} - p_0)^2} \tan \eta \frac{d\hat{p}}{d\eta} - \frac{1}{\cos^2 \eta} \left(\frac{\rho_0 |\mathbf{W}_0|^2}{\hat{p} - p_0} - 1 \right), \qquad (\text{D.2})$$

with the result that the square brackets [ ] which appear in the first member

409

of (D.0) confirm

$$\frac{[\ ]}{\sin^2\hat{\theta}} = \frac{d\hat{p}}{d\eta}\left[\frac{\rho_0|\mathbf{W}_0|^2}{(\hat{p}-p_0)^2}\tan\eta + \frac{\hat{v}}{|\hat{\mathbf{W}}|^2}\frac{\cotan\hat{\eta}}{\sin^2\hat{\theta}}\right] - \frac{1}{\cos^2\hat{\eta}}\left(\frac{\rho_0|\mathbf{W}_0|^2}{\hat{p}-p_0}-1\right).$$
(D.3)

The expression of the equations (III.67b) and (III.67c) suggests the introduction of

$$\Lambda = \frac{\rho_0|\mathbf{W}_0|^2}{\hat{p}-p_0}-1,$$

which gives, for the said equations, the equivalent forms

$$\left.\begin{array}{ll}\dfrac{1}{\Lambda+1} = \left(1-\dfrac{\hat{v}}{v_0}\right)\sin^2\eta, & \text{(a)}\\[3mm]\Lambda = \cotan\hat{\theta}\cdot\cotan\eta, & \text{(b)}\end{array}\right\}$$
(D.4)

This notation also makes it possible to write (D.3) in the form

$$\frac{[\ ]}{\sin^2\hat{\theta}} = \frac{\dfrac{d\hat{p}}{d\eta}}{\hat{p}-p_0}\left[(\Lambda+1)\tan\eta + \frac{\hat{v}(\hat{p}-p_0)}{|\hat{\mathbf{W}}|^2}\frac{\cotan\hat{\eta}}{\sin^2\hat{\theta}}\right] - \frac{\Lambda}{\cos^2\eta}.$$
(D.5)

Starting from $\hat{\eta} = \eta - \hat{\theta}$, and using (D.4b), we can establish

$$\frac{\cotan\hat{\eta}}{\sin^2\hat{\theta}} = (\Lambda+1)\frac{\Lambda^2\tan^2\eta+1}{\Lambda\tan^2\eta-1}\tan\eta.$$
(D.6)

On the other hand, starting from (III.67d) and using (D.4a) we can establish

$$\frac{\hat{v}(\hat{p}-p_0)}{|\hat{\mathbf{W}}|^2} = \frac{1}{\Lambda\dfrac{v_0}{\hat{v}}-1}.$$
(D.7)

Taken together, the relations (D.1), (D.6), and (D.7) make it possible to transform (D.5) into

$$\frac{[\ ]}{\sin^2\hat{\theta}} = 2(\Lambda+1)\left[1-\frac{\hat{p}-p_0}{\hat{v}-v_0}\bigg/\frac{d\hat{p}}{d\hat{v}}\right]^{-1}\left[1+\frac{1}{\Lambda\dfrac{v_0}{\hat{v}}-1}\frac{\Lambda^2\tan^2\eta+1}{\Lambda\tan^2\eta-1}\right]$$

$$-\frac{\Lambda}{\cos^2\eta}.$$
(D.8)

The next step consists of transforming the second brackets of the righthand member of (D.8) using (D.4a)

$$1+\frac{1}{\Lambda\dfrac{v_0}{\hat{v}}-1}\frac{\Lambda^2\tan^2\eta+1}{\Lambda\tan\eta-1} = \frac{2}{1+\left(\dfrac{v_0}{\hat{v}}\cotan\eta\right)^2}.$$

Thus (D.8) becomes

$$[\ ] = \Lambda \frac{\sin^2 \hat{\theta}}{\cos^2 \eta} \left\{ 4 \frac{\Lambda + 1}{\Lambda} \frac{\cos^2 \eta}{1 + \left(\dfrac{v_0}{\hat{v}} \cotan \eta\right)^2} \left[1 - \frac{\hat{p} - p_0}{\hat{v} - v_0} \bigg/ \frac{d\hat{p}}{d\hat{v}} \right]^{-1} - 1 \right\}$$

The last step consists of establishing, with the help of (D.4a)

$$\frac{\Lambda + 1}{\Lambda} \frac{\cos^2 \eta}{1 + \left(\dfrac{v_0}{\hat{v}} \cotan \eta\right)^2} = \left[1 + \frac{\hat{v}}{v_0} \tan^2 \eta + \frac{v_0}{\hat{v}}\left(1 + \frac{v_0}{\hat{v}} \cotan^2 \eta\right) \right]^{-1}.$$

From which the new expression in the square brackets

$$[\ ] = \Lambda \frac{\sin^2 \hat{\theta}}{\cos^2 \eta} \left\{ 4 \left[1 + \frac{\hat{v}}{v_0} \tan^2 \eta + \frac{v_0}{\hat{v}}\left(1 + \frac{v_0}{\hat{v}} \cotan^2 \eta\right) \right]^{-1} \right.$$
$$\left. \times \left[1 - \frac{\hat{p} - p_0}{\hat{v} - v_0} \bigg/ \frac{d\hat{p}}{d\hat{v}} \right]^{-1} - 1 \right\}. \tag{D.9}$$

• Let us now return to the differential equation (D.0). Taking account of the last form given to [ ], the expression (D.4b) of Λ and the jump relation $\rho_0 |\mathbf{W}_0| \sin \eta = \hat{\rho} |\hat{\mathbf{W}}| \sin \hat{\eta}$, the equation (D.0) is written

$$\frac{\cotan \hat{\theta} \cotan \eta}{\cos^2 \eta} \left(\frac{\sin \eta \sin \hat{\theta}}{\sin \hat{\eta}}\right)^2 \{\ \} \frac{d\eta}{d\hat{r}} + \frac{\sin \eta \sin \hat{\theta}}{\sin \hat{\eta}} \frac{1}{\hat{r}} = \frac{\hat{S}}{\rho_0 |\mathbf{W}_0| \hat{c}^2} \frac{\widehat{\partial p}}{\partial m},$$

where the braces are those in (D.9). It now remains to express the trigonometric functions as factors of $d\eta/d\hat{r}$ and $1/\hat{r}$ by using the relation

$$\cotan \hat{\theta} = \left(\cotan \eta + \frac{\hat{v}}{v_0}\right)\left(1 - \frac{\hat{v}}{v_0}\right)^{-1},$$

obtained by eliminating Λ from (D.4a) and (D.4b); we obtain

$$\frac{\sin \eta \sin \hat{\theta}}{\sin \hat{\eta}} = \left(\frac{v_0}{\hat{v}} - 1\right) \sin \eta \cos \eta,$$

$$\frac{\cotan \hat{\theta} \cotan \eta}{\cos^2 \eta} = \left(\frac{v_0}{\hat{v}} - 1\right)^{-1} \frac{\dfrac{v_0}{\hat{v}} \cos^2 \eta + \sin^2 \eta}{\sin^2 \eta \cos^2 \eta},$$

$$\frac{\cotan \hat{\theta} \cotan \eta}{\cos^2 \eta} \left(\frac{\sin \eta \sin \hat{\theta}}{\sin \hat{\eta}}\right)^2 = \left(\frac{v_0}{\hat{v}} - 1\right)\left(\frac{v_0}{\hat{v}} \cos^2 \eta + \sin^2 \eta\right).$$

Finally, observing that the definition itself of $\hat{c}^2$ leads to

$$\left(\frac{v_0}{\hat{v}} - 1\right)^{-1} \frac{1}{\rho_0 \hat{c}} \frac{\widehat{\partial p}}{\partial m} = -\frac{\partial \text{Log}(\rho - \rho_0)}{\partial m} \qquad (p = \hat{p}, s = \hat{s}, m = 0)$$

for (D.0) we arrive at the expression

$$
\left.\begin{aligned}
&g\frac{d\eta}{d\hat{r}} + \frac{\sin 2\eta}{2\hat{r}} + \frac{\hat{S}}{|W_0|}\frac{\partial \operatorname{Log}(\rho - \rho_0)}{\partial m}(p = \hat{p}, s = \hat{s}, m = 0) = 0, \\
&g = \left(\frac{v_0}{\hat{v}}\cos^2 \eta + \sin^2 \eta\right)\left\{4\left[1 + \frac{\hat{v}}{v_0}\tan^2 \eta + \right.\right. \\
&\left.\left. + \frac{v_0}{\hat{v}}\left(1 + \frac{v_0}{\hat{v}}\cotan^2 \eta\right)\right]^{-1}\left[1 - \frac{\hat{p} - p_0}{\hat{v} - v_0}\Big/\frac{d\hat{p}}{d\hat{v}}\right]^{-1} - 1\right\}.
\end{aligned}\right\} \quad (D.10)
$$

2. Explicit Form of (D.0) Using a Relation $U = A + Bu$

From the jump relations for the shock Σ_0, let us extract those which correspond to conservation of mass and of normal momentum at Σ_0

$$
\left.\begin{aligned}
\hat{p} - p_0 &= -\rho_0 w_0(\hat{w} - w_0), \\
\rho_0 w_0 &= \hat{\rho}\hat{w}.
\end{aligned}\right\}
$$

Let us assume $w_0 = U$ and $w_0 - \hat{w} = u$. We obtain

$$
\hat{p} - p_0 = \rho_0 Uu,
$$

$$
\hat{v} = v_0\left(1 - \frac{u}{U}\right).
$$

Let us suppose as valid, as experiments suggest (see [1], for example), a relation $U = A + Bu$ in the domain concerning the shock Σ_0 in the explosive substance. Then

$$
\frac{d\hat{p}}{d\hat{v}} = -\frac{\rho_0^2}{A}(A + 2Bu)(A + Bu)^2
$$

is established without difficulty, from which comes

$$
1 - \frac{\hat{p} - p_0}{\hat{v} - v_0}\Big/\frac{d\hat{p}}{d\hat{v}} = \frac{2Bu}{A + 2Bu}.
$$

In order to complete the explicit writing of g in (D.10), it is sufficient to observe that u is linked to $|W_0|$ and η by

$$
A + Bu = |W_0| \sin \eta,
$$

with the result that

$$
\left.\begin{aligned}
\left[1 - \frac{\hat{p} - p_0}{\hat{v} - v_0}\Big/\frac{d\hat{p}}{d\hat{v}}\right]^{-1} &= 1 + \frac{1}{2\left(\dfrac{|W_0|}{A}\sin \eta - 1\right)}, \\
B\frac{\hat{v}}{v_0} &= B - 1 + \frac{A}{|W_0| \sin \eta}.
\end{aligned}\right\} \quad (D.11)
$$

Appendix E

Transformation of μ_G^i

The leading idea in the transformation of μ_G^i is to express, as *a function of $\mathcal{F}$*, the divergence from the ideal state for each species present in the gas phase of the detonation products. The essential tool is identity ($g.2$) of Appendix A, rewritten by introducing the volume v' of N' moles at p and T

$$v' = \frac{\partial g'}{\partial p}(p, T, .N^i.)$$

which, on differentiating with respect to N_i, gives

$$\frac{\partial \mu^i}{\partial p}(p, T, .N^i.) = \frac{\partial v'}{\partial N^i}(p, T). \tag{E.1}$$

The first stage is to write μ_G^i in the form

$$\mu_G^i(p, T) = \tilde{\mu}^i(p, T) + \Re T \, \text{Log} \, \frac{N^i}{N'} + \Re T \, \text{Log} \, \frac{f^i(p, T)}{p}, \tag{E.2}$$

where $\tilde{\mu}^i$ is the proper molar chemical potential that (i) would have if it were ideal at conditions p and T of pressure and temperature; by definition $f^i(p, T)$ is the fugacity of the gas (i) in the mixture.

The second stage is to define $\tilde{\mu}^i$ by integration of (E.1) from p^0 to p for the gas (i) supposed ideal and on its own: taking account of the value $N^i \Re T/p$ for v' we find

$$\tilde{\mu}^i(p, T) = \tilde{\mu}^i(p^0, T) + \Re T \, \text{Log} \, \frac{p}{p^0}. \tag{E.3}$$

The third stage is to evaluate $\text{Log}(f^i/p)$. For this (E.2) is differentiated with respect to p, then (E.1) is applied to the mixture of N' moles occupying volume v'. By using (E.3) we find

$$\frac{\partial}{\partial p} \, \text{Log} \, \frac{f^i}{p} = \frac{1}{\Re T} \frac{\partial v'}{\partial N^i}(p, T, .N^i.) - \frac{1}{p}. \tag{E.4}$$

Returning to the equation (XII.5) which defines the function $\mathscr{F}$ by

$$\mathscr{F}(v', T, . N^i .) = pv'/N'\mathfrak{R}T, \tag{E.5}$$

and differentiating at constant T, $N^1, \ldots, N^I$ we obtain

$$\frac{dp}{p} = \left(\frac{\partial \operatorname{Log} \mathscr{F}}{\partial \operatorname{Log} v'} - 1\right) d \operatorname{Log} v',$$

so that (E.4) takes the form

$$d_p \operatorname{Log} \frac{f^i}{p} = \left[\frac{p}{\mathfrak{R}T}\frac{\partial v'}{\partial N^i}(p, T, . N^i .) - 1\right]\left(\frac{\partial \operatorname{Log} \mathscr{F}}{\partial \operatorname{Log} v'} - 1\right) d \operatorname{Log} v'. \tag{E.6}$$

Evaluating $\partial v'/\partial N^i$ by differentiating (E.5) at constant p and T, the first form is

$$\frac{dv'}{v'}\left(1 - \frac{\partial \operatorname{Log} \mathscr{F}}{\partial \operatorname{Log} v'}\right) = dN^i\left(\frac{1}{N'} + \frac{\partial \operatorname{Log} \mathscr{F}}{\partial N^i}\right),$$

so that, after further calculation, (E.6) takes the final form

$$d_p \operatorname{Log} \frac{f^i}{p} = -\left(\mathscr{F} - 1 + N'\frac{\partial \mathscr{F}}{\partial N^i}\right) d \operatorname{Log} v' - d_{v'} \operatorname{Log} \mathscr{F}. \tag{E.7}$$

Now, because of the property of a "real mixture" to tend, at low pressures, to an 'ideal mixture" of "ideal gases," the function $\mathscr{F}$ and the fugacities f^i necessarily prove

$$\left.\begin{aligned}&\lim_{p\to 0} v'(p, T, . N^i .) = \infty, \\[2mm] &\lim_{p\to 0} \mathscr{F} - 1 = \lim_{p\to 0}\frac{\partial \mathscr{F}}{\partial N^i} = \lim_{p\to 0}\left(\frac{f^i}{p} - 1\right) = 0,\end{aligned}\right\} \tag{E.8}$$

from which the thermodynamic identity follows

$$\operatorname{Log} \frac{f^i}{p} = \int_{v'}^{\infty}\left(\mathscr{F} - 1 + N'\frac{\partial F}{\partial N^i}\right) d \operatorname{Log} v' - \operatorname{Log} \mathscr{F}. \tag{E.9}$$

From (E.3), (E.5), and (E.9) the required expression for μ_G^i is

$$\frac{1}{\mathfrak{R}T}\mu_G^i(p, T) = \frac{1}{\mathfrak{R}T}\tilde{\mu}^i(p^0, T) + \operatorname{Log}\frac{N^i\mathfrak{R}T}{p^0v'} + \int_{v'}^{\infty}\left(\mathscr{F} - 1 + N'\frac{\partial \mathscr{F}}{\partial N^i}\right) d \operatorname{Log} v'. \tag{E.10}$$

Transformation of μ^j

The leading idea for the evaluation of μ^j is to express, as a function of φ^j, the divergence from the standard conditions (p^0, T) which the condensed species (j) presents. The essential tool is again identity $(g.2)$ of Appendix A, rewritten by introducing the proper volume $N^j M^j v^j$ of N^j moles at p and T

$$N^j M^j v^j = \frac{\partial g^j}{\partial p}(p, T, N^j),$$

from which it follows, by differentiating with respect to N^j

$$\frac{\partial \mu^j}{\partial p}(p, T) = M^j v^j. \tag{E.11}$$

By integrating (E.11) between p^0 and p we obtain

$$\mu^j(p, T) = \mu^j(p^0, T) + M^j \int_{p^0}^{p} v^j(p, T)\, dp. \tag{E.12}$$

By using the equation of state $p = \varphi^j(v^j, T)$ introduced in (XII.3) it is possible to pass from variable p to variable v^j at constant T and arrive at

$$\mu^j(p, T) = \mu^j(p^0, T) + M^j \int_{v^j(p^0, T)}^{v^j(T)} v^j \frac{\partial \varphi^j}{\partial v^j}(v^j, T) \cdot dv^j. \tag{E.13}$$

Transformation of $e - e_0$

The leading idea is to transform $e_G - e_0$ and $e^j - e_0$ separately by a splitting which introduces both the terms derived from the functions of state $\mathscr{F}$ and φ^j and the quantities relative to the standard state tabulated in specialist works.

Consider for the gas phase an arbitrarily large volume $\tilde{v}'$ which ensures ideal state ($p\tilde{v}' = N'\mathscr{R}T$, i.e., $\mathscr{F}(\tilde{v}', T, \cdot N^i \cdot) = 1$); according to identity (f.4) of Appendix A, the corresponding specific internal energy $\tilde{e}_G$ depends only on the temperature, so that the difference $e_G - e_0$ may be written

$$e_G(v', T, \cdot N^i \cdot) - e_0 = e_G(v', T, \cdot N^i \cdot) - \tilde{e}_G(T, \cdot N^i \cdot)$$
$$+ \tilde{e}_G(T, \cdot N^i \cdot) - \tilde{e}_G(T^0, \cdot N^i \cdot)$$
$$+ \tilde{e}_G(T^0, \cdot N^i \cdot) - e_0. \tag{E.14}$$

To calculate the first difference in (E.14) we use identity (f.4) of Appendix A, rewritten using (XII.5) in the form

$$\left(\sum N^i M^i \right) \frac{\partial e_G}{\partial v'}(v, T) = N' \frac{\mathscr{R}T^2}{v'} \frac{\partial \mathscr{F}}{\partial T}.$$

To calculate the second difference of the second member, the definition of enthalpy is used which, because of the ideal state, leads immediately to

$$\left(\sum N^i M^i \right) \Delta \tilde{e}_G = \sum_i N^i \Delta \tilde{H}^i(T) - N' \mathscr{R} \Delta T,$$

where $\tilde{H}^i(T)$ is the molar enthalpy of species (i) in the ideal gas state. Thus

(E.14) becomes

$$\left(\sum N^i M^i\right)[e_G(v', T, . N^i .) - e_0]$$

$$= N'\Re T^2 \int_{\infty}^{v'} \frac{\partial \mathscr{F}}{\partial T} \frac{dv'}{v'} + \sum_i N^i \Delta_{T^0}^T \tilde{H}^i(T)$$

$$- N'\Re(T - T^0) + \sum_i N^i M^i [\tilde{e}_G(T^0, . N^i .) - e_0]. \qquad (E.15)$$

In a similar manner we consider for the condensed species (j) the state (p^0, T) intermediate between the real state and the standard state, and obtain

$$e^j(p, T) - e_0 = e^j(p, T) - e^j(p^0, T) + e^j(p^0, T) - e^j(p^0, T^0)$$

$$+ e^j(p^0, T^0) - e_0. \qquad (E.16)$$

To calculate the first difference we again make use of identity $(f.4)$ of Appendix A, this time written in the form

$$\frac{\partial e^j}{\partial v}(v^j, T) = T^2 \frac{\partial}{\partial T}\left(\frac{\varphi^j}{T}\right).$$

The second difference is calculated by using the definition of molar enthalpy H^j of the species (j); for transformation at constant pressure p^0, this definition leads immediatey to

$$\Delta e^j = \frac{\Delta H^j}{M^j} - p^0 \Delta v^j.$$

(E.16) thus becomes

$$e^j(p, T) - e_0 = T^2 \int_{v^j(p^0, T)}^{v^j(T)} \frac{\partial}{\partial T}\left(\frac{\varphi^j}{T}\right) dv^j + \frac{1}{M^j} \Delta_{T^0}^T H^j(p^0, T)$$

$$- p^0 [v^j(p^0, T) - v^j(p^0, T^0)] + e^j(p^0, T^0) - e_0. \qquad (E.17)$$

Evaluating (XII.12) by using (E.15) and (E.17) leads to the quantity

$$\left(\sum_i N^i M^i\right)[\tilde{e}_G(T^0, . N^i .) - e_0] + \sum_j N^j M^j [e^j(p^0, T^0) - e_0], \qquad (E.18)$$

which may be interpreted as the difference in internal energy between a unit of mass of the detonation products in the standard state and a unit of mass of the explosive substance in the initial state, provided that pressure p^0 has been chosen sufficiently low for $v'(p^0, T^0, . N^i .)$ itself to be large enough to ensure the ideal state of the gas phase. With this assumption, quantity (E.18) can then be obtained as the sum of three differences:

- that of $e^0 - e_0$ when the explosive substance passes from the initial state to the standard state (p^0, T^0);

- that of $-e_f^0$ when the explosive substance decomposes at (p^0, T^0) into its constituent elements; and
- that of $\sum_\lambda N^\lambda[(H_f^0)^\lambda - v_f^\lambda \Re T^0]$ when the elements recombine to form the detonation products at (p^0, T^0).

The following approximate expression is thus established

$$
\left.\begin{aligned}
e - e_0 \simeq{} & -e_f^0 + \sum_\lambda N^\lambda[(H_f^0)^\lambda - v_f^\lambda \Re T^0] + e^0 - e_0 \\
& + \sum_i N^i \Delta_{T^0}^T \tilde{H}^i(T) + \sum_j N^j \Delta_{T^0}^T H^j(p^0, T) - N'\Re(T - T^0) \\
& + N'\Re T^2 \int_\infty^{v'} \frac{\partial \mathscr{F}}{\partial T} \frac{dv'}{v'} + T^2 \sum_j N^j M^j \int_{v^j(p^0, T)}^{v^j} \frac{\partial}{\partial T}\left(\frac{\varphi^j}{T}\right) dv^j \\
& - p^0 \sum_j [v^j(p^0, T) - v^j(p^0, T^0)].
\end{aligned}\right\} \quad \text{(E.19)}
$$

Signs, Symbols, and Characters

0. Note

Particular care has been taken in the choice of notation, with three aims in mind:

- to find a compromise which would suit mathematicians, engineers, physicists, and chemists;
- to give coherence and unity to all the chapters as a whole, yet without excluding the flexibility which, for each domain, allows the complexity of the notation to be limited to what is strictly required; and
- to put the resources of typography to the service of a visual familiarity of the nature of the quantities and/or the indices.

1. Principal Symbols

1.1. Universal Constants

h	Planck constant
k	Boltzmann constant
$\mathscr{N}$	Avogadro number
$\mathfrak{R} = k\mathscr{N}$	ideal gas constant
$N = 0, 1, 2$	one-dimensional flow constant
c	velocity of light
$\mathscr{C}$	Euler constant

1.2. Scaling Factors, Perturbations, and Order

L	range of thermomechanical dissipation
L'	range of deviation from chemical equilibrium
$\tilde{L}$	characteristic macroscopic dimension

| $\tilde{\tilde{L}}$ | characteristic mesoscopic dimension |
| $\hat{L}$ | characteristic dimension downstream of ignitor shock |

$$\left.\begin{array}{l} \varepsilon = L/\tilde{\tilde{L}} \\ \varepsilon' = L'/\tilde{L} \\ \tilde{\varepsilon} = \tilde{L}/L' \end{array}\right\} \quad \text{perturbation parameters}$$

| $0(\ ), \mathbf{0}(\ )$ | of the order of ( ) |

1.3. Classical Mechanics

$\mathscr{R}$	referential of space
$\mathscr{S}$	material system
t	time
M $(X_i; i = 1, 2, 3)$	"particle" (Lagrange variables)
P $(x_i; i = 1, 2, 3)$	position (Euler variables) of M at instant t
$\vec{u}/\mathbf{u}$ $(u_i; i = 1, 2, 3)$	absolute velocity vector
$\vec{W}/\mathbf{W}$ $(w_i; i = 1, 2, 3)$	velocity vector relative to a wave surface
$\vec{\delta}$ $(\delta_i; i = 1, 2, 3)$	absolute velocity of a wave surface
$\vec{f}$ $(f_i; i = 1, 2, 3)$	volume density of external forces
$\vec{g}$ $(g_i; i = 1, 2, 3)$	mass density of external forces
$\vec{m}$ $(m_i; i = 1, 2, 3)$	volume density of external moments
ℓ	rate of volume heat received from the exterior
t_{ij} $(i, j = 1, 2, 3)$	stress tensor
σ_{ij} $(i, j = 1, 2, 3)$	Cauchy stress tensor
q_j	heat flux vector
$\mathscr{T}$	trajectory
V	speed of a moving body

1.4. Wave Geometry and Mechanics

D	connected domain of R, R^2, or R^3
∂D	frontier of D
$\vec{n}$ $(n_i; i = 1, 2, 3)$	normal external to D
DV	differential element of volume
DS	differential element of surface
Σ	wave surface
$\Sigma(D)$	portion of Σ contained in D
$\vec{N}$ or N	unit vector normal to Σ
$\vec{N}', \vec{N}''$ or N', N''	unit vectors on principal lines of Σ
R', R''	principal radii of curvature on Σ
R_m	average radius of curvature on Σ
(ξ', ξ'', z)	Euler variables linked to Σ
$Z = z/\tilde{L}$	local external variable along N
$\zeta = z/L$	local internal variable along N

$\eta = (\pi/2) - (\vec{N}, \vec{W})$	co-incidence of a wave Σ on a flow $\mathbf{W}$
θ	deflection of $\mathbf{W}$ across Σ
$\mathbf{M}$	normal mass flux across Σ
$\mathfrak{C}^+, \mathfrak{C}^-$	characteristic curves of a one-dimensional expansion
$\mathfrak{C}$	one-dimensional shock or its flow diagram
$\mathcal{L}$	geometric locus
$\varnothing,$	diameter
(h) or (h)	Hugoniot curve
(H)	Crussard curve
(H$_-$)	deflagration arc of (H)
(H$_+$)	detonation arc of (H)
g	magnification
v	scanning speed on a film or screen

1.5. Thermodynamics

ρ	density
p	pressure
T	absolute temperature
$v = 1/\rho$	mass volume
s	specific entropy
e	specific internal energy
h	specific enthalpy
f	specific Helmholtz free energy
g	specific Gibbs free enthalpy
c_v	specific heat at constant volume
c_p	specific heat at constant pressure
σ	specific surface
σ	surface mass
$\gamma = c_p/c_v$	ratio of specific heats
a	velocity of sound in a fluid
$\mathfrak{N} = a/\mathbf{W} \cdot \mathbf{N}$	ratio of a to the normal relative velocity
c^L	longitudinal velocity of sound in a solid
Γ	$[\mathrm{d}\log p/\mathrm{d}\log v](v, s)$
Γ_T	$[\mathrm{d}\log p/\mathrm{d}\log v](v, \mathrm{T})$
$G = v[\partial p/\partial e(v, e)]$	Grüneisen coefficient
λ	thermal conductivity
μ	dynamic shear viscosity
μ''	longitudinal dynamic viscosity
$\omega = \mu'' c_p/\lambda$	longitudinal Prandtl number
$\mathscr{P}_e = \mathrm{u}L\rho c_p/\lambda$	Péclet number
$\mathscr{D}$	surface of state of detonation products
α	compactness

1.6. Chemical Thermodynamics

N^λ	mass density of moles of species (λ)
$\mu^\lambda = \partial g^\lambda / \partial N^\lambda$	chemical potential of species (λ)
v_q^λ	algebraic coefficient of (λ) in the q-th reaction
m	mass fraction of decomposed explosive
$S = dm/(1-m)\,dt$	reactivity
$\mathscr{E}$	constituent(s) of a substance
$\underset{\sim}{N}(N^\lambda; \lambda = 1, \ldots)$	molar mass composition of a substance
$\overset{\approx}{N}$	composition at chemical equilibrium

1.7. Statistical Mechanics

$(\mathbf{x}, \mathbf{u})$	point in phase space
$f^k(\mathbf{x}, \mathbf{u}, t)$	probability density for molecules (k)
n^k	volume density of molecules (k)
m^k	mass of molecule (k)
$M^k = \mathscr{N}m^k$	mass of mole (k)
$\rho^k = n^k m^k$	mass of (k) per unit of volume
$\rho = \Sigma \rho^k$	specific mass (density)
$n = \Sigma n^k$	number of molecules per unit of volume
$X^k = n^k/n$	molar fraction of (k)
$Y^k = \rho^k/\rho$	mass fraction of (k)
ε^k	internal energy of a molecule (k)
M_i^k	diffusive mass flux of (k)
$\underline{\varepsilon}^k$	potential well between molecules (k)
$\underline{r}^k$	equilibrium distance between molecules (k)
$\mathscr{V}^k(r)$	interaction potential between molecules (k)
P^k	volume rate of production in mass of (k)

1.8. Electro-Optics

I	intensity
V	potential difference
R	resistance
L	self-inductance
B	magnetic induction
n	refractive index
r	reflectivity
φ	phase
i	incidence
D	optical density
p	interference order

2. Use of Signs Accompanying Symbols

2.1. Index Used as Subscript

i, j, k	refers to	Cartesian coordinates in real space
m	...	an average
q	...	a chemical reaction
0	...	initial state of an explosive substance
1, 2	...	upstream and downstream states of a shock
c	...	a critical condition
f	...	the reaction of formation of a chemical species
i	...	an isentropic condition
$k, k + 1$	...	a node of a mesh
$k + 1/2$	...	a mesh
m	...	a maximum or a medium adjacent to the explosive
p	...	a derivative at constant pressure
v	...	a derivative at constant volume
A	...	the axis of a flow or the point A of $(\mathscr{D})$
B	...	a flow boundary or the point B of (H)
C	...	a shock $\mathfrak{C}$ or the point C of (H)
D	...	a point D of $(\mathscr{D})$
I	...	the interface between explosive and adjacent medium
(0), (1)	...	terms of a limited expansion
*	...	Chapman–Jouguet state
∞	...	an extrapolation for an infinite diameter

2.2. Index Used as Superscript

a	refers to	the explosive substance
b	...	the detonation products in chemical equilibrium
α, β	...	the phases of the system
i, j, λ, p, q	...	the chemical species of the system
k	...	a partition of the molecules of a system
n	...	a "normal" wave
a	...	the priming
i	...	the ignition conditions
j	...	a sequence of experimental values
k	...	the critical edge velocity
ℓ	...	a free edge
p	...	a plane detonation
s	...	a spherical detonation
$n, n + 1$	...	the stepwise instants of a numerical calculation
D	...	a point of $(\mathscr{D})$
I	...	the induction phase

S	. . .	a sonic flow
0	. . .	a reference state, or initial conditions
^{0}	. . .	the zero-order structures of a detonation
*	. . .	the build-up of a detonation
		or the excitation of an explosive molecule
$\neq$	. . .	the activation of an explosive molecule

2.3. Signs Above a Symbol

$\wedge$	specifies	the downstream state of an ignitor shock
$-$	specifies	$\begin{cases} \text{a reduced quantity in classical mechanics} \\ \text{an average in statistical mechanics} \end{cases}$
$\sim$	$\begin{cases} \text{specifies} \\ \text{refers to} \\ \text{marks} \end{cases}$	a macroscopic scaling in laminar flow / the averages and fluctuations in turbulent flow / ideal state of a gas in thermodynamics
$\approx$	$\begin{cases} \text{specifies} \\ \text{refers to} \end{cases}$	chemical equilibrium in laminar flow / Reynolds tensor in turbulent flow
$\bullet$	marks	a specific derivative

2.4. Signs Below a Symbol

| $-$ | specifies | the intermolecular equilibrium |

3. Use of Characters According to Nature of Quantities

3.1. Scalar Quantities

We use

- italic letters for
  | extensive quantities in mass | $e, v, s, f, g, h, c_v, c_\mathrm{p}, g_i, \ldots$ |
  | elecro-optical quantities | $I, V, R, B, n, i, r, \varphi, L, D, \ldots$ |
- roman letters for
variables in space and time	$\mathrm{x}_i, \mathrm{X}_i, \mathrm{t}, \mathrm{r}, \mathrm{e}, \varphi$
extensive quantities in volume	$\mathrm{f}_i, \mathrm{m}_i, \mathrm{n}^k$
intensive quantities	p, T
kinematic quantities	$\mathrm{u}_i, \mathrm{w}_i, \delta, \mathrm{a}, \mathrm{c}, \mathrm{v}, \mathrm{V}$
molecular quantities	m^k
- upper case for
"elementary" quantities	$\mathrm{D}S, \mathrm{D}V$
"molar" quantities	M^k, H^k, S^k, N^k
surface rates of mass	M_i^k
volume rates of mass	P^k
reactivity	S

3.2. Vector Quantities

We use:

- plain characters with the sign $\rightharpoonup$ above or bold characters for geometric or kinematic quantities or dynamic quantities in real space: $\overline{OM}$, $\vec{T}$, $\vec{n}$; N or $\vec{N}$, u or $\vec{u}$, W or $\vec{W}$; and
- bold characters in all other cases.

3.3. Tensor Quantities

Bold characters are used in all cases.

Index

This index is designed as a complement to the "Contents": it allows the reader to look up the meaning of common terms, as well as the acceptation, of abbreviations and proper names of which the author, by tradition or by convention, has made special use. The arabic numerals refer to pages.